This book is to be returned on
or before the date stamped below

UNIVERSITY OF PLYMOUTH

32

COMPOSITE STRUCTURES

5

Proceedings of the 5th International Conference on Composite Structures, held at Paisley College of Technology, Scotland, from 24th to 26th July 1989, co-sponsored by the Scottish Development Agency, the National Engineering Laboratory, the US Air Force European Office of Aerospace Research and Development, the US Army, Research, Development and Standardisation Group, UK, the Strathclyde Regional Council and Renfrew District Council.

Also published by Elsevier Science Publishers:

COMPOSITE STRUCTURES
(Proceedings of 1st International Conference, 1981)

COMPOSITE STRUCTURES 2
(Proceedings of 2nd International Conference, 1983)

COMPOSITE STRUCTURES 3
(Proceedings of 3rd International Conference, 1985)

COMPOSITE STRUCTURES 4
(Proceedings of 4th International Conference, 1987)

COMPOSITE STRUCTURES

5

Edited by

I. H. MARSHALL

*Department of Mechanical and Production Engineering,
Paisley College of Technology, Scotland*

ELSEVIER APPLIED SCIENCE
LONDON and NEW YORK

ELSEVIER SCIENCE PUBLISHERS LTD
Crown House, Linton Road, Barking, Essex IG11 8JU, England

Sole Distributor in the USA and Canada
ELSEVIER SCIENCE PUBLISHING CO., INC.
655 Avenue of the Americas, New York, NY 10010, USA

WITH 110 TABLES AND 451 ILLUSTRATIONS

British Library Cataloguing in Publication Data

International conference on Composite Structures (5th:
1989 : Paisley College of Technology)
Composite Structures 5.
1. Structures. Composite materials
I. Title   II. Marshall, I. H.
624.1′8

Library of Congress Cataloging-in-Publication Data

International Conference on Composite Structures (5th : 1989 : Paisley
College of Technology)
Composite structures 5 / edited by I. H. Marshall.
p. cm.
"Proceedings of the 5th International Conference on Composite
Structures, held at Paisley College of Technology, Scotland, from the
24th to 26th July 1989, co-sponsored by the Scottish Development
Agency..."—Prelim. p.
Bibliography: p.
Includes index.
1. Composite construction—Congresses. 2. Composite materials—
Congresses. I. Marshall, I. H. (Ian H.) II. Scottish Development
Agency. III. Title. IV. Title: Composite structures five.
TA664.I574 1989
620.1′18—dc20   89-12014

ISBN 1-85166-362-2

Printed in Great Britain by Galliard (Printers) Ltd, Great Yarmouth

Preface

The papers contained herein were presented at the Fifth International Conference on Composite Structures (ICCS/5) held at Paisley College of Technology, Scotland in July 1989. The conference was organised and sponsored by Paisley College of Technology. It was co-sponsored by the Scottish Development Agency, the National Engineering Laboratory, the US Air Force European Office of Aerospace Research and Development, the US Army Research, Development and Standardisation Group—UK, Strathclyde Regional Council and Renfrew District Council. It forms a natural and ongoing progression from the highly successful ICCS/1/2/3 and /4 held at Paisley in 1981, 1983, 1985 and 1987 respectively.

It has often been said that at the end of each decade there is much to be gained by reflecting on the manifold factors that have influenced the path of one's life. Doubtless there is much truth in that ancient adage, since it is only by subjectively considering the past that a glimmer of the future is possible. Now that volume five of the 'Composite Structures' series has been assembled, thereby cataloging the progress of composites as structural materials in the 1980s, it is surely fitting to reflect on the advances of that decade. There is little doubt that composites have come a long way in the last ten years, from a newly established maturity with many exciting applications such as the Space Shuttle Program to a position of unrivalled superiority in a great many applications. Indeed, it is often difficult to comprehend the advances made by these fibre reinforced materials in a space of half a century. In human terms, carbon fibre is only now reaching the age at which life is said to begin!

What of the future? Although personally handicapped by a lack of mystic clairvoyance, it is nevertheless certain that composites will play an increasingly dominant role in virtually every area of engineering. The present and potential properties are increasing in leaps and bounds to an extent that could not have been envisaged a decade ago. All indications are that this rate of progress, far from abating, will continue to increase well into the foreseeable future. Many seemingly impossible engineering challenges have already been successfully surmounted. Doubtless today's dreams will prove to be tomorrow's reality.

It is our earnest hope that the present series of International Conferences will play a small part in disseminating knowledge of the structural applications of composites and thereby assist in widening their role in general engineering.

With over twenty countries represented by authors and delegates, the present conference is truly an international forum for specialists in this expanding area of technology.

To the authors and session chairmen go our sincere thanks for their efforts, without their active participation there would be no conference.

As always, an international conference can only achieve success through the willing and enthusiastic contributions of a number of individuals. In particular, our thanks are due to the following:

The International Advisory Panel

J. Anderson	Paisley College of Technology (UK)
W. M. Banks	University of Strathclyde (UK)
A. M. Brandt	Polish Academy of Sciences (Poland)
A. R. Bunsell	Ecole des Mines de Paris (France)
W. S. Carswell	National Engineering Laboratory (UK)
T. Hayashi	Japan Plastic Inspection Association (Japan)
R. M. Jones	Virginia Polytechnic & State University (USA)
L. N. Phillips, OBE	Consultant, Farnborough (UK)
J. Rhodes	University of Strathclyde (UK)
S. W. Tsai	Air Force Materials Laboratory (USA)
J. A. Wylie	Paisley College of Technology (UK)

The Local Organising Committee

S. K. Harvey
J. Kirk
G. Macaulay
J. S. Paul

The Conference Secretary

Mrs C. A. MacDonald

Grateful thanks are due to other individuals who contributed to the success of the event. A final thanks to Nan, Simon, Louise and Richard for their support during the conference.

I. H. MARSHALL

Contents

Failure Studies

Environmental Effects

Plate Vibrations

Shell Vibrations

Damage Tolerance

Transport Applications

Finite Element Studies

Structural Stability

Platework Structures

Aerospace Applications

Mechanical Fasteners

Pipework Analysis

Fatigue and Creep

Theoretical Studies

Cementitious Structures

Laminate Analysis

Matrix Considerations

Complementary Studies

Plenary Paper

Fibre Composite Repairs to Damaged Structural Components

R. Jones and J. Paul

*Aircraft Structures Division, Aeronautical Research Laboratory,
Defence Science and Technology Organisation, Department of Defence,
Melbourne, Australia*

ABSTRACT

This paper outlines recent developments in bonded repair technology. Repairs to both metallic and composite structures are discussed, and the role of lap joint theory in the design of bonded repairs is outlined. This technology is illustrated by considering the recent repair to F111C aircraft in service with the RAAF.

1 INTRODUCTION

The Aeronautical Research Laboratory (ARL) in Australia has pioneered the use of adhesively bonded boron fibre reinforced plastic (BFRP) patches to repair cracks in aircraft components.[1] The procedure has been successfully used in several applications to RAAF aircraft, including the field repair of fatigue cracks in the lower wing skin of Mirage III aircraft[2] and in the landing wheels of Macchi aircraft.[1] In each case repairs were made by adhesively bonding a BFRP patch to the component with the fibres spanning the crack, the aim being to restrict the opening of the crack under load, thereby reducing the stress intensity factor and thus reducing further crack growth.

The present paper presents the design philosophy used to design repairs to thick sections and illustrates this approach by considering repairs to surface cracks, cracked bolt holes and the recent reinforcement to the wing pivot fitting of F111C aircraft. In the case of composite structures attention is focused on repairs to impact damage.

1

2 BONDED REPAIRS TO CRACKED METALLIC STRUCTURES

In recent years a number of boron fibre reinforced plastic (i.e. BFRP) patches have been successfully used to repair cracks in aircraft components (Table 1).

Initial design approaches, for repairs to metallic wing skins, were based on a two-dimensional finite element approach.[3] However, for thicker structures or surface flaws a three-dimensional approach is required. The adhesive and the boron epoxy laminate must be modelled separately. In order to avoid numerical ill-conditioning which may arise due to the large aspect ratios resulting from this approach, it is necessary to evaluate the stiffness matrix using either a $2 \times 2 \times 2$ or a $2 \times 2 \times 3$ Gaussian quadrature formulation. Here the three Gauss points would be in the direction perpendicular to the plane of the laminate. The $2 \times 2 \times 3$ system is more reliable than the fully reduced integration approach that can give rise to non-physical solutions.

When designing an externally bonded patch the main requirements are:

(1)   A significant reduction in the stress intensity factor after patching, i.e. K_{1p}, must be achieved. Wherever possible it is desirable to reduce K_{1p} to below the fatigue threshold of the material.

(2)   The energy density in the adhesive, W, must be kept below its critical value, W_{crit}^{adh}, such that fatigue damage does not occur in the adhesive, i.e.

$$W|_{r=r_o} \leq W_{crit}^{adh} \tag{1}$$

Here r_o is a material constant.

(3)   The maximum fibre strain must be below critical; for boron epoxy this value is approximately $5000 \, \mu\varepsilon$. To prevent delamination in the laminate the available energy density must be below its critical value, W_{crit}^{boron}, i.e.

$$W|_{r=r_o} \leq W_{crit}^{boron} \tag{2}$$

where, as above, r_o is a material constant.

The relationship between energy density theory and other commonly used theories is given in Ref. 8.

To illustrate the development of this methodology we will consider three specific examples, viz.:

(1)   Repair of a semi-elliptical surface flaw.
(2)   Repair of a cracked fastener hole.
(3)   The proposed reinforcement to the F111C wing pivot fitting.

TABLE 1
Current applications of crack patching

Cracking	Material	Component	Aircraft	Comments
Stress–corrosion	7075-T6	Wing plank[a]	Hercules	Over 300 repairs since 1975
Fatigue	Mg alloy	Landing wheel[a]	Macchi	Life doubled at least
Fatigue	2024-T3	Fin skin	Mirage	In service since 1978
Fatigue	AU4SG	Lower wing	Mirage	Over 500 repairs since 1979
Fatigue	2024-T3	Upper wing skin	Nomad (fatigue test)	Over 105 900 simulated flying hours
Fatigue	2024-T3	Door frame	Nomad (fatigue test)	Over 106 619 simulated flying hours
Stress–corrosion	7075-T6	Console truss	F111	Recent repair
Lightning burn	2024-T3	Fuselage skin	Orion	Recent repair[b]

[a] Now RAAF standard procedures.
[b] CFRP patches and ambient temperature curing epoxy adhesive to simulate rapid repair of battle damage.

3 ILLUSTRATIVE EXAMPLES

3.1 Repair of a Semi-elliptical Surface Flaw

Let us consider the repair of a surface flaw with a surface length of 40·1 mm and a depth of 5·72 mm centrally located in a 2024-T4 aluminium alloy section, 11·1 mm thick, 108·3 mm wide and 304 mm long. The crack is repaired by a ten-ply boron epoxy laminate 180 mm long and 108·3 mm wide and where the boron properties are:

$$E_{11} = 208 \cdot 3 \, \text{GPa} \qquad v_{12} = v_{13} = 0 \cdot 183$$
$$E_{22} = E_{33} = 25 \cdot 4 \, \text{GPa} \qquad v_{23} = 0 \cdot 1667$$
$$G_{12} = G_{13} = G_{23} = 7 \cdot 24 \, \text{GPa}$$

whilst the adhesive has $G = 0 \cdot 7 \, \text{GPa}$ and $E = 1 \cdot 89 \, \text{GPa}$.

It should be noted that crack patching was initially developed for repairs to thin metallic sheets, i.e. wing skins. Indeed it was intuitively thought that this approach would be inapplicable for the repair of thick sections. However, the numerical results presented in Ref. 3 showed a significant reduction in the stress intensity factors for surface flaws up to 6 mm deep in a 12·7 mm thick aluminium alloy. For a uniform applied stress the reduction in the stress intensity factors, K_{1p}/K_{1u}, obtained for the present problem is given in Table 2. Here the subscripts p and u refer to the patched and unpatched values respectively.

It should also be noted that for a surface flaw, or a through crack, under

TABLE 2
Ratio of patched to unpatched stress intensity factors
(K_{1p}/K_{1u})

At surface	0·22
At maximum depth	0·44

uniform stress the value of the stress intensity factor K_{1p} asymptotes to a constant value as the crack length, on the surface, increases.[3]

Reductions in K of the magnitude given in Table 2 should lead to a significant increase in fatigue life. Indeed to confirm this prediction a series of constant amplitude fatigue tests were performed. The stress amplitude was 64·9 MPa with $R = 0·01$. The results of these tests are shown in Table 3. Whilst there is a large scatter in the fatigue lives of the patched specimens even specimen number (2P), which has the lowest life, still exhibits more than five times the mean life of the unpatched specimens. To avoid obtaining unrealistically long lives for the patched specimens a small amount of crack growth was allowed prior to patching. Unfortunately in the case of specimens (2P) and (6P) the crack grew faster, and hence longer, than anticipated before patching with the result that their subsequent fatigue lives were substantially reduced.

3.2 Repair of Cracked Fastener Holes

To investigate the behaviour of cracked fastener holes let us consider a 2024-T4 aluminium alloy 72 mm wide, 11·2 mm thick and with a working length of 200 mm. The specimen contains a centrally located 10 mm diameter hole which in turn contains two diametrically opposed 3 mm

TABLE 3
Fatigue lines of patched and unpatched specimens

Specimen number		Cycles to failure	Specimen number		Cycles to failure
Unpatched	(1U)	21 500	Patched	(1P)	728 000
	(2U)	21 800		(2P)	137 000
	(3U)	30 439		(3P)	728 000
	(4U)	27 980		(4P)	451 000
	(5U)	29 000		(5P)	1 253 000[a]
				(6P)	170 000
Mean life		26 144	Mean life		577 000

[a] Specimen did not fail.

TABLE 4
Stress intensity factors for σ = 68·9 MPa

	Angle position[a] (ϕ)	Stress intensity factor K (MPa $\sqrt{m}$)			
		Basic specimen	Bonded patch only	Bonded sleeve only	Bonded patch and sleeve
Specimen A	0	6·74	3·17	4·81	2·43
	0·25	7·06	4·27	4·84	3·20
	0·5	7·34	4·94	4·75	3·42
	0·75	8·56	6·17	4·93	3·78
	1·0	10·61	7·97	4·60	3·66
Specimen B	0	7·89	3·80	5·78	2·96
	0·25	8·34	5·18	5·93	3·94
	0·5	8·79	6·09	6·02	4·45
	0·75	10·37	7·67	6·53	5·14
	1·0	12·63	9·72	5·77	4·70

[a] ϕ is a non-dimensional angle equal to 0 at the free surface and 1 at the bore of the hole.

radius quadrant cracks on one face of the plate. The repair methods to be investigated include:

(a) Bonding a 1 mm thick steel sleeve into the hole using a 0·126 mm thick acrylic adhesive.
(b) Bonding a ten-ply boron epoxy laminate over the cracked hole.
(c) Using both (a) and (b) as described above.

The resulting stress intensity factors for each repair configuration are shown in Table 4. Here σ is the remote applied stress, K_{1s} and K_{1b} are the values of K_1 at the surface and down the bore of the hole respectively. We see that the adhesively bonded sleeve gives the greatest reduction in K_{1b} whilst the boron patch is the most effective in reducing K_{1s}.

If an external patch is not used the stresses in the adhesive bonding the steel sleeve to the structure are very high and may cause premature failure of the adhesive. However, the boron patch, as well as further lowering the stress intensity factor, reduces the maximum value of the strain energy density in the adhesive by approximately 75%. This should significantly increase the fatigue life of the adhesive. In order to confirm this phenomena four of these specimens were tested at a constant amplitude cyclic tensile stress of 68·9 MPa with a stress ratio $R = 0·01$. The fatigue lives for these

TABLE 5

Experimental fatigue test results for specimen with two quadrant cracks at a 10 mm diameter hole

Specimen condition	Specimen number	Fatigue life (cycles)	Average (cycles)
Unrepaired	A-1	94 240	
	A-2	104 150	113 537
	A-3	137 830	
	A-4	117 930	
Repaired	A-9	634 630[a]	N/A
	A-10	2 347 200[a]	N/A
	A-11	2 253 500[a]	N/A
	A-12	5 078 200[b]	N/A

[a] Specimen failed in grips; no cracking of notched hole observed.
[b] Test stopped; no cracking of notched hole observed.

specimens are given in Table 5. Four additional specimens were repaired as outlined in (c) above and their fatigue lives are also given in Table 5.

Although there is scatter in the comparative fatigue lives the results show a significant increase in the average fatigue life of the repaired holes. This is consistent with earlier experimental work, which compared the fatigue lives of an open hole, a cold expanded hole and a hole with a bonded insert.[6]

From the work outlined above and that contained in the open literature we see that:

(1) For a bonded repair as the crack length increases the stress intensity factor asymptotes to a constant value. Closed form solutions for this asymptotic value are given in Ref. 4.

(2) A bonded insert significantly reduces the stress intensity factor. This is true even after a significant proportion of the adhesive has failed.

(3) The values of the stress intensity factors for the case of a crack on one side of a bonded insert and for the case of a crack, of the same length, on both sides of the hole are very similar (see Refs 5 and 6).

(4) As the crack length increases, a crack in a bonded rivet hole behaves as if the specimen does not contain a hole at all (see Refs 5 and 6). This is particularly important since a handbook solution is available for the 'no hole' case;[7] it appears that this solution could be conveniently used in some instances for estimating the stress intensity factor for a cracked hole containing a bonded insert.

4 BORON/EPOXY REINFORCEMENT FOR THE F111C WING PIVOT FITTING

Following the development of this methodology ARL was tasked with the development of an external reinforcement for the wing pivot fitting (WPF) of F111C aircraft in service with the RAAF. The results of a preliminary design concept, which proposed that a boron/epoxy doubler be bonded to the upper WPF surface over the critical regions, is presented in Ref. 9. This paper summarises the design process and the method by which a structural finite element model was tailored to reproduce strain survey results obtained from an aircraft under cold proof load test (CPLT) conditions.

The structure of the wing is comprised of a 2024-T856 wing skin spliced onto a D6ac steel WPF by a series of flush fasteners. Cracking is occurring inboard from the splice in the runout region of the top surface integral stiffeners. The stiffener in which failures have occurred is designated stiffener No. 2 although stiffeners 4 and 5 are also thought to be critical. The general geometry of the critical region is shown in Fig. 1. A detailed review of the WPF manufacturing processes and a subsequent failure case investigation is presented in Ref. 10.

The proposed repair, as described in Ref. 9, consists of two discrete boron/epoxy doublers bonded to the upper surface of the F111 WPF over the stiffeners known to be subjected to the high strains, as shown in Fig. 2. One large doubler covers stiffeners Nos 4 and 5, and a smaller doubler covers stiffener No. 2. Uni-directional boron/epoxy composite is used to form the doublers and is bonded with an adhesive at elevated temperatures. Each doubler consists of two segments. The lower segment is bonded to the D6ac steel to provide a bridge level with the aluminium skin step. The upper segment is bonded to the aluminium skin and the lower segment. Each segment of the preferred design consists of approximately 60 plies of boron, suitably tailored to achieve single plies at the extremities of the doublers. The lower segment contains a cut-out region adjacent to the aluminium skin step, designed to reduce the stress concentration produced by the discontinuity of the load path. Figure 3 details the ARL proposed reinforcement.

4.1 Finite Element Model

The finite element models used for this analysis are described in detail in Ref. 9. Here both the unreinforced stiffener section and the reinforced section (incorporating the preferred ARL boron/epoxy doubler) models

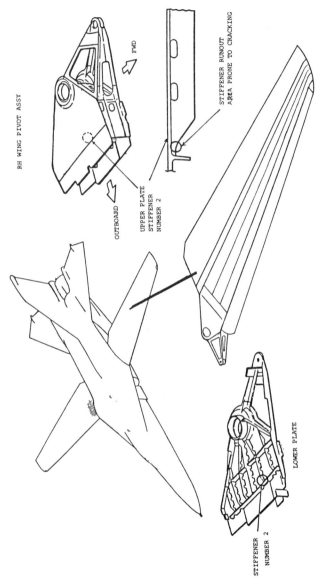

FIG. 1. View of F111 aircraft and wing showing location of critical area.

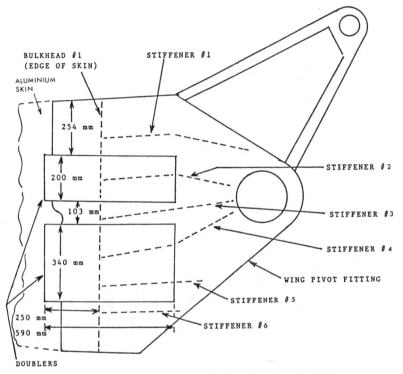

FIG. 2.　Doubler locations on the F111 surface wing pivot fitting.

were analysed. These models were developed to represent the geometries of the USA strain survey wings. Both right- and left-hand wings were modelled, in order to predict the results of that survey.

For the reinforced model, the left-hand wing was deemed to be the design case, as, from all available data at present, this wing recorded the highest strain levels and thus the most severe problem.

In Ref. 9 a three-dimensional model was also analysed and found to produce similar, yet conservative, results to the two-dimensional model. For efficiency, only the two-dimensional case (assuming plane stress) is

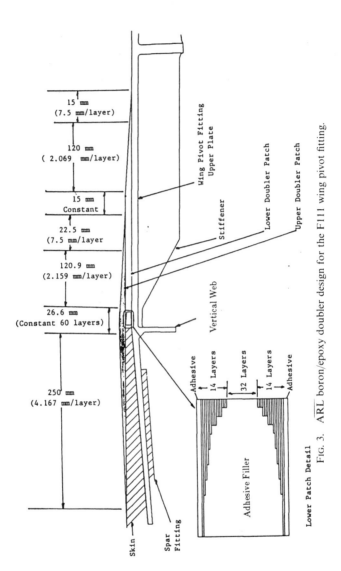

FIG. 3.   ARL boron/epoxy doubler design for the F111 wing pivot fitting.

considered in this paper. A plate width of 165 mm was chosen as a representative case for the proposed doubler widths.

4.2 Material Properties and Failure Criterion

The tensile stress–strain curve for D6ac steel used in this analysis is as given in Ref. 9. The material properties used for the boron/epoxy system and the adhesive are presented in Table 6. Young's modulus and Poisson's ratio for the aluminium were taken to be 71×10^3 MPa and 0.33, respectively.

The doubler design must achieve a stress reduction sufficient to prevent yielding of the steel, and to withstand an applied load of 7·3 and $-2·4$ g. The design allowables for the adhesive shear stress τ, peel stress σ_p and the fibre stress σ_f for the boron/epoxy system were taken to be

$$\tau < 40 \, \text{MPa} \qquad \sigma_p < 40 \, \text{MPa} \, \text{(tensile values)} \qquad \sigma_f < 1000 \, \text{MPa}$$

The interlaminar stresses in the doubler must also be below their critical values. Values for the critical interlaminar stresses are usually derived from the short beam shear test. The manufacturers' value for the boron/resin system used is $\tau_{xy} < 90$ MPa and in tension we use $\sigma_y < 70$ MPa.

The strain energy density criterion is used to estimate critical values. This criterion takes the form

$$W_{av}|_{r=r_o} = (\tfrac{1}{2}\sigma_{ij}\varepsilon_{ij} - W_f)|_{r=r_o} = W_{\text{critical}} \tag{3}$$

where r_o is the characteristic damage size for boron (typically 1 mm), W_{av} is the available energy density and W_f is the energy density in the fibres. The

TABLE 6
Boron/epoxy and adhesive properties compliances $(MPa)^{-1}$

Quantity	Adhesive	Boron
S_{XX}	$0·529\,10 \times 10^{-3}$	$0·480\,50 \times 10^{-5}$
S_{YY}	$0·529\,10 \times 10^{-3}$	$0·393\,08 \times 10^{-4}$
S_{ZZ}	$0·592\,910 \times 10^{-3}$	$0·393\,08 \times 10^{-4}$
S_{XY}	$-0·185\,19 \times 10^{-3}$	$-0·805\,86 \times 10^{-6}$
S_{YZ}	$-0·185\,19 \times 10^{-3}$	$-0·985\,15 \times 10^{-5}$
S_{ZX}	$-0·185\,19 \times 10^{-3}$	$-0·805\,86 \times 10^{-6}$
SH_{XY}	$0·142\,86 \times 10^{-2}$	$0·138\,12 \times 10^{-3}$
SH_{YZ}	$0·142\,86 \times 10^{-2}$	$0·138\,12 \times 10^{-3}$
SH_{ZX}	$0·142\,86 \times 10^{-2}$	$0·138\,12 \times 10^{-3}$

strain energy density criterion is further described in Ref. 8. For the purposes of this design study, the critical interlaminar energy density $W_{critical}$ was taken to be 0·56 MPa in shear and 0·097 MPa in tension.

4.3 Elastic–Plastic Analysis

Several elastic–plastic analyses were conducted and the finite element model was 'tuned' so as to reproduce measured strain data. The value of the load applied to the model to give the recorded strains at 7·3 g was determined to be 238 MPa.

As a further verification of the finite element model, the far field strain was computed and compared to the recorded far field strain in the steel (see Ref. 12). From the elastic–plastic analysis, the far field strain in the steel was computed to be $-3500\,\mu\varepsilon$. This can be compared to -3577 and $-3061\,\mu\varepsilon$ for gauges which lie either side of stiffener No. 2 and which are at an equivalent distance from the vertical web.

4.4 Reinforced Structure

In order to evaluate the effectiveness of the proposed ARL boron/epoxy reinforcement an elastic analysis of the reinforced finite element model was conducted. The effectiveness of the doubler is based on the stress (or strain) reduction achieved in the critical runout region. A reduction of approximately 36% in stress was achieved at a location corresponding to gauge 9L in the stiffener runout. This gauge is located on the lower surface of the runout approximately 16 mm from the vertical web. This reduction compares favourably with the value as measured after application of the small doubler to a test wing at ARL (see Ref. 9).

4.5 Doubler Stresses

The location of the most critical points in the doubler were investigated in Ref. 9. The design of the reinforcement was developed so as to minimise the energy density at these locations. From the elastic finite element analysis of the various runout radius configurations, the peel and shear stresses at the design points were tabulated. The effect of an 'adhesive filler' in the cutout region of the doubler was also investigated. In order to simulate the effect of the adhesive filler, this region in the finite element mesh was partially filled with an adhesive element. This adhesive filler is present in the doubler cutout to provide it with stability. With this configuration it was found that the boron fibre stresses and the energy density were below the critical design values.

TABLE 7
Comparison of experiment and analysis

% Strain reduction at gauge 9L	
Predicted	36·5
Measured	34·0

4.6 Experimental Verification

These doublers have now been installed on an F111C wing at ARL and, as shown in Fig. 4, the measured strain reduction is in good agreement with prediction (see Table 7). Utilising an insulated chamber cooled by liquid nitrogen a uniform temperature distribution of −40°C across the test article WPF was achieved. Subsequently a 100% proof download was applied successfully. However, at approximately 98% proof upload the wing failed at a mechanically fastened repair. Although the shock loading experienced by the wing upon failure served to debond the small doubler, sufficient data were collected to give confidence in the strength of the reinforcement scheme.

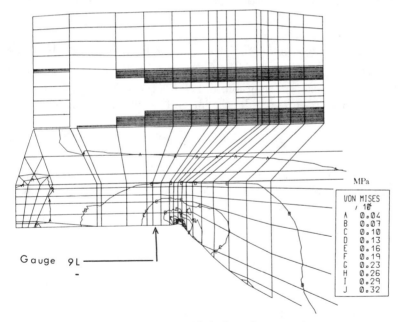

MPa

VON MISES
/ 10⁴

A	0.04
B	0.07
C	0.10
D	0.13
E	0.16
F	0.19
G	0.23
H	0.26
I	0.29
J	0.32

Gauge 9L

FIG. 4. F111 wing with the boron/epoxy repairs.

5 IMPACT DAMAGE

5.1 Residual Strength

For metallic structures the stress intensity factor is known to characterise both crack growth and failure. For composite materials there are two fracture parameters which are widely used to predict the residual strength of delaminated composite laminates, viz.

(i)   the energy release rate approach, and
(ii) the strain energy density approach.

5.1.1 The energy release rate

For Mode 1 self-similar crack growth of a through crack in the absence of body forces, the energy release rate G[15] can be written as

$$G = \lim_{\varepsilon \to 0} \int_{\Gamma_\varepsilon} W n_1 - t_i \frac{\partial u_i}{\partial x_1} \, \mathrm{d}s \tag{4}$$

where W is the energy density, Γ_ε is a vanishing small closed path around the tip with normal n and t_i are the components of the traction vector on the path.

Here W is defined as

$$W = \tfrac{1}{2}\varepsilon_{kl} C_{ijkl} \varepsilon_{ij} - \beta_{ij}\varepsilon_{ij}(T - T_0) - \phi_{ij}\varepsilon_{ij}(M - M_0) + C_1(T, M) \tag{5}$$

where C_{ijkl} is the stiffness tensor, β_{ij} and ϕ_{ij} are related to the coefficients of thermal and moisture expansion, and C_1 is a function of temperature and moisture. Let us now consider the integral J_S, which we will define as

$$J_S = \int_{\Gamma_S} (W n_1 - t_i \, \partial u_i / \partial x_1) \, \mathrm{d}s \tag{6}$$

where Γ_S is the external boundary of the body. There is often a tendency to drop the subscript S and refer to J_S as J. Using Green's theorem, it follows that

$$G = J_S - \int_{v_S - v_\varepsilon} \left\{ \tfrac{1}{2}\varepsilon_{ij} \frac{\partial C_{ijkl}}{\partial x_1} \varepsilon_{kl} - \varepsilon_{ij} \frac{\partial}{\partial x_1} \right.$$
$$\left. \times [\beta_{ij}(T - T_0) + \phi_{ij}(M - M_0)] + \frac{\partial C_1}{\partial x_1} \right\} \mathrm{d}V \tag{7}$$

where the integral of the second term in eqn (7) is over the area between the two contours. Thus J_S will not equal the energy release rate G unless the area integral vanishes. At constant moisture and temperature, J_S may be equal to G. However, in general the area integral will be non-zero and J_S, which is measured experimentally from the movement of the load points,[16] will not equal G.

In service aircraft heating is often localised and the moisture content varies. This will produce a spatial variation in the tensor C_{ijkl} with the result that the area integral will, in general, be non-zero. Consequently, when designing laboratory tests, care should be taken to reproduce the near tip stress, strain, temperature and moisture fields rather than reproducing the 'global behaviour'. This is particularly true if the aim of the test is to establish such quantities as the critical damage size or the maximum permissible load.

For a three-dimensional fracture problem, the integral on the right-hand side of eqn (4) is no longer equal to the energy release rate and is referred to as T^*. The integral of the first term on the right-hand side of eqn (4) is also referred to as W_t.

5.1.2 Strain energy density

In the strain energy density approach, failure is assumed to occur when the available energy density W_{av} at a distance r_0 in front of the delamination in the direction of growth reaches a critical value W_c. The value W_c is dependent both on the value of $dV (= \varepsilon_{11} + \varepsilon_{22} + \varepsilon_{33})$ and dA, the change of area per unit area.

For thermomechanical problems

$$W_{av} = \tfrac{1}{2}\sigma_{ij}\varepsilon_{ij} - \tfrac{1}{2}\sigma_{ij}\alpha_{ij}(T - T_0) - \tfrac{1}{2}\sigma_{ij}\Psi_{ij}(M - M_0) - W_f \qquad (8)$$

where $\beta_{kl} = \alpha_{ij}C_{ijkl}$, $\phi_{kl} = \Psi_{ij}C_{ijkl}$ and W_f is the energy density in the fibre and α_{ij}, Ψ_{ij} are the coefficients of thermal and moisture expansion.

As shown in Refs 17 and 18, the strain energy density approach has certain advantages over the energy release rate approach. For self-similar growth, both approaches give residual strengths in good agreement with experimental results.[17,18] However, for non-self-similar growth, energy release rate concepts must be used with caution.

It should be noted that finite element studies and experimental test results have shown that for delaminations in

(a)   composite laminates,[19]
(b)   at step lap joints[30] or
(c)   at mechanically fastened composite joints[18,21]

a stage is reached after which a significant increase in the size of the damage

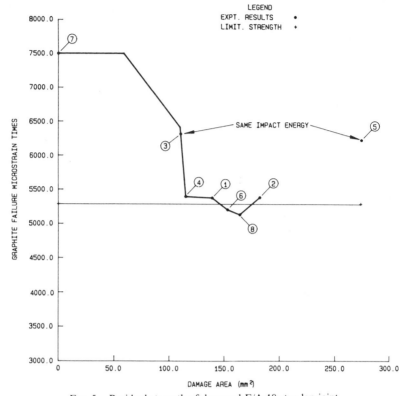

FIG. 5.   Residual strength of damaged F/A-18 step lap joint.

does not significantly reduce the residual strength. This phenomena can be seen in Fig. 5, which shows the change in failure strain with damage area for the step lap joint in the F/A-18.[30] Closed form analytical solutions which reveal the underlying mechanisms for this phenomena are given in Refs 13 and 14.

5.2 Effect of Impact Damage on Fatigue Life

Cyclic fatigue tests play a central role in the assessment of fatigue life and the specification of inspection intervals for metallic components. This has led to similar tests for composites with impact damage. It has been found that the S–N curves generated from experimental data are often over a large range of cycles.[19-23] Tension–tension fatigue loading is considered to be less critical than compression–compression and tension–compression loading.[24-26] This paper will concentrate on the latter two loading cases.

TABLE 8
Summary of experimental data on S–N curves derived from the literature

Reference	Maximum compressive strain range	Maximum compressive stress range	Range of cycles to failure	Comments
Rosenfeld and Gause[31]	0·002 4–0·003 $\Delta\varepsilon = 0·000\,6$		10^3–10^6	Laminated plate specimens. Delamination area 12·9 cm², 16·4 J impact, $R = -\infty$
Demuts et al.[28]	Strain at 0·003		10^4–10^6	Multirib specimens, AS6/2220-3. 136 J impact, $R = 10$
Demuts and Horton[19]	$S = 0·6$–$0·7$		5×10^4–10^6	Three-stiffener panel, AS6/2220-3. 136 J impact, $R = 10$
Walter et al.[29]	$S = 0·28$–$0·30$		10^4–10^6	Laminated plate specimens, T300/5208. 6·2 J impact, $R = -1·0$
Stellbrink and Aoki[27]		200–250 MPa $\Delta\sigma = 50$ MPa	0–10^5	Laminated plate specimens, T300/CODE69. Impacted with ball of mass 255 g and radius of 5 mm at velocity of 6·15 m/s. $R = -1·0$
Potter[20]		Stress level fixed at 206 MPa	622–8 242	Tapered-thickness specimens, Fibredux XAS/9. 14·5 J impact, $R = 10$
Ryder et al.[21]		Stress level fixed at 241 MPa	2×10^4–10^6	24-ply damaged hole specimens. $R = -1·0$
		Stress level fixed at 152 MPa	2×10^4–5×10^5	32-ply damaged hole specimens. $R = -1·0$

For compression–compression and tension–compression loading, the maximum residual compressive load as a fraction of the static failure load (S) decreases from 1·0 to typically 0·6 for N in the range 1–10^6, depending on the initial damage size.

The greatest rate of degradation appears to be during the early part of cyclic loading, for N up to about 100 cycles. As the number of cycles, and hence the damage size, is increased, very little further residual strength degradation is observed. In fact, it has been observed that $S = 0·6$ may be taken as the 'fatigue threshold' value, below which the component may be assumed to have 'infinite life'. Various researchers report threshold values of between 0·6 and 0·8.[33,35,41-44] S–N curves are invariably drawn from a best fit of limited experimental data, often with large scatter, and their shapes vary from straight slopes to some form of curve.

The differences in fatigue lives resulting from changes in maximum compressive stress or strain from different researchers are summarised in Table 8. From the values shown in the table it can be seen that an uncertainty in the magnitude of the applied maximum compressive stress, for example 10%, can lead to a 100-fold difference in fatigue life. Of particular interest is the considerable scatter in the data recorded by Ryder et al.[21] and Potter,[20] which show an order of magnitude of difference in the fatigue lives of specimens at a nominally constant stress level.

For service aircraft, the in-flight loads on a component will not be accurately known. There is often a significant variation between aircraft, with the result that the stress levels will always be subjected to a degree of uncertainty. We have seen that a small change in stress can result in a very large change in fatigue life. This implies that laboratory testing cannot be expected to provide accurate estimates of the fatigue lives of impact damage laminates and therefore cannot be expected to provide accurate information on inspection intervals.

If this is true, then the rationale for conducting such fatigue tests becomes questionable. Indeed, there are two natural consequences of this phenomenon.

(1) During the design of a component the stresses should be kept below the threshold value, thus eliminating the possibility of a fatigue failure due to BVID.

(2) If an aircraft is already in service and the design procedure had not included step (1), then there is little point to fatigue testing of impact damaged laminates. In this case the emphasis for research and development should be moved towards

(a) static strength testing to determine critical damage size and loads, allowing for representative service conditions, and

(b) developing rational reject or repair criteria for the component which includes a methodology for determining inspection intervals.

Indeed, as a result of this work, it is clear that further research is required in order to understand the physical reasons why the pronounced fatigue threshold exists. Such an explanation should involve the ratio of the relative length scales of the damage size to the laminate thickness and the far-field strain energy density relative to the critical strain energy density of the undamaged laminate.

6 BONDED REPAIR TO IMPACT DAMAGE

6.1 Numerical Analysis of Bonded Repairs

In order to understand the mechanisms governing failure and the increase in residual strength of a repaired structure a detailed three-

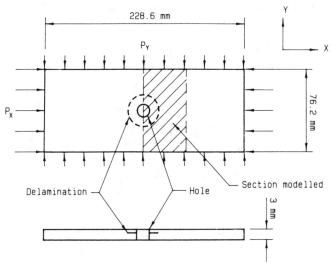

FIG. 6. Damaged hole structure.

dimensional finite element analysis was undertaken. The problem chosen for this comparative analysis is similar to that used in Ref. 32 as a benchmark problem to compare the predictive capabilities of T^* and energy density. However, with the purchase of the F/A-18 by the Royal Australian Air Force (RAAF) attention has been focused on the use of AS4/3501-6 rather than the T300/5208 graphite/epoxy which was considered in Ref. 32. For this reason the present investigation will use the mechanical properties corresponding to AS4/3501-6.

The problem considered in this paper is an impact damaged laminate with a fastener hole loaded under compression. The dimensions of the model are the same as those used in the experimental work of Ref. 21. The structure modelled in this analysis can be seen in Fig. 6 and the finite element mesh for the unrepaired case can be seen in detail in Fig. 7. The specimen analysed was a $(0/45/0_2/-45/0_2/45/0_2/-45/0)_s$ AS4/3501-6 graphite/epoxy laminate and contained a centrally located hole of 9·5 mm diameter, surrounded by delamination damage due to impact and poor drilling.

The elements used were mostly 20-noded isoparametric elements with directionally reduced integration and $2 \times 2 \times 3$ Gaussian quadrature points, with the three points being taken through the ply thickness. The crack tip elements along the circular delamination were 15-noded isoparametric wedge elements. The finite element model contained 2763

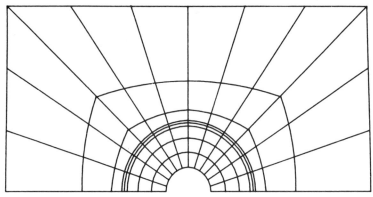

Fig. 7. Plan view of the mesh.

nodes, 528 elements and had 7644 degrees of freedom. Restraints were applied along the unloaded edges so as to achieve various levels of restraint representative of a real structure.

As in Ref. 32 the initial damage around the fastener hole was modelled as a circular delamination of radius 13·75 mm between the second and third plies (i.e. between the 45° and 0° plies). This is an approximate simulation of the initial damage found in Refs 21 and 33.

The material properties used are those of AS4/3501-6 and are

$$E_{11} = 128\,200 \text{ (MPa)} \qquad G_{12} = G_{13} = G_{23} = 5857 \text{ (MPa)}$$
$$E_{22} = E_{33} = 13\,800 \text{ (MPa)} \qquad v_{12} = 0·3$$

As mentioned in Ref. 32 it is important that, in the finite element model, the faces of the delamination are prevented from overlapping. Otherwise non-physical solutions may be obtained. To eliminate this effect the present work uses non-linear gap elements between nodes on opposite sides of the delamination. A compressive load of 150 kN was applied to the ends of the model in the x-direction (see Fig. 6), and to investigate biaxial effects a tensile load of either 17 or 0 kN was applied in the y-direction.

To investigate the effect of repairing the specimen a 12-ply $(0/+45/0_2/-45/0)_s$ laminate was used as an externally bonded patch covering the damaged area. The patch contained a hole to allow for the possible insertion of a fastener. To evaluate fastener/structure interaction effects the conditions applied to the hole were chosen to represent

(1) an open hole,
(2) an interference fit fastener, and
(3) a bonded insert.

6.1.1 Results and discussion

For each configuration the value of T^* and W_t was computed at each node around the crack tip, as was the local minimum of W_{av}, which in each case corresponded to the local maximum of dV. A summary of the maximum values of T^*, W_{av}, W_t and the corresponding value of dV_{max} are given in Table 9. The angles at which the respective maximum occurs, which are referred to as Θ_T and Θ_W, are also shown.

It was observed that the two plies above the delamination move out of plane, i.e. crack opening and/or closure, and that for all of the cases analysed this out-of-plane movement was non-symmetric.

The results of this analysis suggest that:

(1) Load biaxiality has a marked effect on the fracture parameters and hence on the failure of structural components. The fracture parameter T^* is less affected by load biaxiality than are W_{av} and W_t.

(2) For all cases considered the location of the maximum values of both parameters were similar.

(3) When the patch material has the same stiffness as the parent laminate the reductions in T^*, W_{av} and W_t can be estimated (see Table 9) by multiplying the values corresponding to the unrepaired structure with the square of the reduction in the net sectional stress. This infers that the residual strength of a repaired structural component can be estimated by the following simple formula:

$$\frac{\text{Residual strength (repaired)}}{\text{Residual strength (unrepaired)}} = \left[\frac{W \text{ (unrepaired)}}{W \text{ (repaired)}} \right]^{1/2} \quad (9)$$

where W is the energy density in the laminate in the region of the patch. Whilst this formula can account for multiaxial loading in the case of uniaxial loading, when the patch has the same effective moduli as the parent laminate, it reduces to

$$\frac{\text{Residual strength (repaired)}}{\text{Residual strength (unrepaired)}} = \frac{\sigma \text{ (unrepaired)}}{\sigma \text{ (repaired)}} \quad (10)$$

The energy density in both the repaired and the unrepaired structure can be computed from a knowledge of strain gauge results before and after repair.

(4) Prohibiting local bending at the hole had an insignificant effect on the fracture parameters. However, prohibiting in-plane movement and local bending reduced T^* whilst not reducing W_{av} and increasing W_t. This reduction in T^* would correspond to an

TABLE 9
Summary of T^*, W_{av}, dV_{max} and W_t

Case (MPa mm)	T^*	Θ_T	Predicted T^*	dV_{max}	W_{av} (MPa)	Θ_w	Predicted W_{av}	W_t (MPa mm)	Predicted W_t
1UN	0·0907	30°	—	−0·001 26	0·517 6	30°	—	0·050 3	—
2UN	0·0901	30°	—	−0·001 25	0·517 5	30°	—	0·050 8	—
3UN	0·0687	45°	—	−0·001 18	0·519 4	30°	—	0·064 8	—
1UR	0·0471	45°	0·040 3	−0·001 21	0·267 2	30°	0·2300	0·027 2	0·0224
2UR	0·0470	45°	0·040 0	−0·001 19	0·267 3	30°	0·2300	0·027 4	0·0226
3UR	0·0376	45°	0·030 5	−0·001 13	0·269 3	30°	0·2308	0·028 3	0·0288
1BN	0·1390	30°	—	0·004 48	1·075 3	30°	—	0·1160	—
2BN	0·1390	30°	—	0·004 47	1·073 2	30°	—	0·1150	—
3BN	0·0992	30°	—	0·003 58	1·072 8	30°	—	0·1720	—
1BR	0·0669	30°	0·061 8	0·002 90	0·496 8	30°	0·4779	0·056 1	0·0516
2BR	0·0669	30°	0·061 8	0·001 76	0·496 8	30°	0·4779	0·055 1	0·0511
3BR	0·0514	30°	0·044 1	0·002 36	0·502 5	30°	0·4768	0·079 0	0·0764

Case description:
1, open hole; 2, interference fit fastener; 3, bonded insert.
U, uniaxial load; B, biaxial load; N, no repair; R, repaired.

increase in the failure load of between 11 and 16%. Unlike T^*, W_{av} was essentially unaffected by this restraint. As a result, experimental results are required to evaluate this effect.

(5) Energy density theory predicted that, for most cases, damage growth would be predominantly in-plane.

6.2 Specimen Fabrication

To confirm this prediction a series of laboratory tests were undertaken. The graphite/epoxy material used throughout these tests was AS4/3501-6 with a ply configuration of $[(+45_2/-45_2/0_4)_3/90]_s$. Before specimens were cut from these panels, the panels were C-scanned to determine the void content of the material.

Each specimen was impacted with a $\frac{1}{2}$ in diameter ballbearing with a mass of 1 kg and from a height of 1·3 m. The specimens were placed between two thick steel plates, with holes drilled to the size of damage required. The absorbed energy was calculated by differencing the initial kinetic energy and the rebound kinetic energy. A laser located on the bottom right-hand side of the rig was triggered when the impactor cut the beam. The initial and rebound pulses were recorded on a digital oscilloscope (Nicolet 2090, Model 207).

Each specimen contained a 6 mm diameter centrally located hole, simulating a fastener hole, as was previously analysed. The hole was drilled using a diamond-tipped core drill and was cooled by water, restrained by a plasticine dam. All specimens were subjected to a C-scan of the impacted area and the damage size was approximated from the C-scan results.

6.3 Repair Fabrication

In order to validate the simple design rule previously postulated it was required that the effective stiffness and ply configuration of the patch be representative of the parent material. The material used for the patch was AS4/3501-6 and was 16 plies thick, with a ply configuration of $[0_2/\pm45/\pm45/0_2]_s$. The length of the patches and distance to edge of the grips were 190 and 60 mm respectively. The patches were bonded to the parent laminate using the cold setting acrylic adhesive Flexon 241. This adhesive was chosen for its shear strength, ease of application and because environmental effects were not an issue in this test series.

6.4 Test Methodology

Four strain gauges were bonded to each of the unrepaired specimens whilst each repaired specimen had two strain gauges located on the patched

TABLE 10
Results of static compression tests

Specimen number	Impactor diameter (mm)	Absorbed energy (J)	Damaged area (mm²)	Unrepaired (U) Repaired (R)	Failure load (kN)	Failure strain (με)	Predicted strain (με)
1	19·8	7·59	453	U	−190·9	−4 503	—
2	19·8	7·55	453	U	−213·6	−4 680	—
3	19·8	7·96	453	U	−213·1	−4 826	—
4	19·8	5·60	479	R	−238·1	−4 993[a]	−6 164
5	19·8	8·21	428	R	−238·1	−4 993[a]	−6 164
6	19·8	7·84	448	R	−289·3	−6 699	−6 164
7	19·8	7·88	458	R	−289·3	−6 699	−6 164
8	30·0	7·46	733	U	−168·0	−4 097	—
9	30·0	6·93	718	U	−191·5	−4 375	—
10	30·0	7·88	665	U	−173·2	−4 305	—
11	30·0	7·54	761	U	−195·6	−4 350	—
12	30·0	7·74	761	U	−194·7	−4 274	—
13	30·0	7·67	761	U	−178·0	−4 025	—
14	30·0	7·66	800	R	−233·2	−5 293	−5 594
15	30·0	7·35	800	R	−233·2	−5 293	−5 594
16	30·0	7·45	739	R	−227·0	−5 554	−5 594
17	30·0	7·64	704	R	−227·0	−5 554	−5 594
18	39·7	6·25	1 252	U	−178·5	−4 061	—
19	39·7	5·76	1 252	U	−204·0	−4 445	—
20	39·7	5·87	1 252	U	−180·9	−4 090	—
21	39·7	7·83	1 385	R	−237·6	−5 337	−5 542
22	39·7	7·58	1 212	R	−237·6	−5 337	−5 542
23	39·7	7·70	1 290	R	−222·4	−5 099	−5 542
24	39·7	7·59	1 120	R	−222·4	−5 099	−5 542

[a] Specimens exhibited extensive bending prior to failure (anti-buckling rig was distorted).

side. In each case the gauges were 110 mm from the edge of the hole. This meant that both the repaired and unrepaired specimens had the gauges located in the same position relative to the hole.

All specimens, except two which exhibited extensive bending, produced load versus strain curves which were essentially linear to failure. (The patched specimens 4 and 5 had vastly different absorbed energies and the patches were not exactly in line, thereby inducing extensive bending near and up to failure.) To alleviate the problem of global buckling the specimens were tested back to back and the strain at failure of the first specimen was taken to be the failure strain for both specimens. The failure strains for each specimen can be found in Table 10. From this table we see that the failure strains follow the asymptotic nature outlined in Section 4. For a given damage size the different forms of internal damage, due to impact, were reflected in a slight variation in the failure strains. In all cases the patches and adhesive bond failed after the failure of the parent laminate.

The failure load of the repaired specimen can be predicted (see Table 10) using eqn (10) and requires only a knowledge of the unrepaired residual strength and the change in net sectional stresses due to patching. The change in net sectional stress can be readily calculated from the change in net sectional area. This result is believed to substantiate the trends predicted in the previous section and significantly simplifies the repair design philosophy.

As outlined in Section 4.2 a stress reduction of 10% can produce a 100-fold change in life. This reduction in stress can easily be obtained using an external patch. The use of an external patch method avoids the requirement for complete removal of the damaged area, resulting in speedier and simpler repairs for in-service structural components.

7 LAP JOINT THEORY

As mentioned in Ref. 34 bonded repairs are often designed using lap joint theory. Indeed, this often provides a good first estimate of the critical design parameters. However, it should be noted that lap joint theory is an approximation and does not yield exact values. To illustrate this, Ref. 35 considered the surface flaw problem discussed in Section 2.1. The specimen consisted of two 11·2 mm thick aluminium alloy plates bonded back to back with an aluminium honeycomb core. Each plate has a centrally located surface crack with a surface length of 40 mm and a depth of 5·7 mm. The cracks were covered with a bonded unidirectional composite patch consisting of 10 plies of boron/epoxy.

The testing machine employed was an Instron 500 kN electrohydraulic fatigue testing machine. The loads were applied sinusoidally at a frequency of 10 Hz. After mounting the specimen in the machine the surface of the patch was cleaned and sprayed with a matt black paint. The applied cyclic loads were ± 170 kN with a mean of zero.

In Ref. 35 changes in stress were calculated by measuring changes in the infrared emission. This was achieved using a system marketed under the trade name SPATE 8000, which has the potential for a spatial and temperature resolution of $0.25\,\text{mm}^2$ and $0.001\,\text{K}$ respectively. Under adiabatic conditions it is possible to use SPATE output, in conjunction with

R. *Jones and J. Paul*

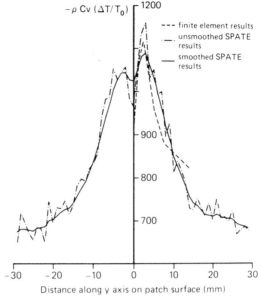

FIG. 8. Calculated and predicted SPATE values.

eqn (11), to obtain stress information regarding the problem under consideration:

$$\frac{\Delta T}{T_0} = \frac{1}{\rho C_v}\left[\frac{\partial C_{ijkl}}{\partial T}\varepsilon_{kl} - \left(\beta_{kl} + \frac{\Delta T \partial \beta_{kl}}{\partial T}\right) - \left(\phi_{ij}\frac{\partial M}{\partial T} + \frac{\Delta M \partial \phi_{ij}}{\partial T}\right)\right]\Delta \varepsilon_{ij} \quad (11)$$

where C_{ijkl} is the elasticity tensor, M is the moisture content, $\phi_{ij} = \Psi_{kl}C_{ijkl}$, $\beta_{ij} = \alpha_{kl}C_{ijkl}$, Ψ_{kl} are the coefficients of moisture expansion and α_{kl} are the coefficients of thermal expansion.

The results of the numerical analysis and the experimentally obtained contour plots, obtained in Ref. 35, show that the maximum temperature on the surface of the boron patch, and hence the maximum stresses, do not occur along the centreline of the specimen but at a distance of approximately 2 mm on either side of the crack. Furthermore, as can be seen from Fig. 8, there was good agreement between experiment and analysis for both the maximum value obtained and for the shape of the temperature variation for a line plot over the crack along $x = 0$.

The numerical analysis revealed that the stresses on the surface of the patch were predominantly in the fibre direction and as a result the observed temperature field may be used to obtain an estimate of the stresses in the fibre direction. This observation should apply to other problems involving the bonded repair of damage using unidirectional patches. This raises the

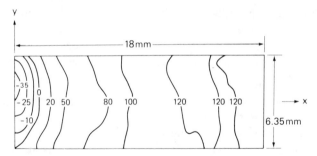

FIG. 9. Contour plot non-dimensional SPATE values for the gap region of aluminium.

possibility of using the temperature field to determine the stress distribution for composite repairs on operational aircraft.

It should be noted that, in lap joint theory, for a symmetrical double lap joint the maximum fibre stresses occur directly over the joint. For the present problem the occurrence of the maximum stresses on either side of the crack reveals that the central region is not behaving as a lap joint.

Let us now examine the stress distribution in the adherents of a lap joint. In Ref. 35 a symmetric double overlap joint was used to investigate the stress distribution in the vicinity of the gap. The inner adherand was a 2024-T4 aluminium alloy, whilst the outer adherand was an 8-ply unidirectional boron/epoxy laminate.

The numerical and the experimental results, given in Ref. 35, show that the temperature in the vicinity of the gap in the aluminium adherand was negative, indicating that this region was in compression (see Fig. 9). The shape and size of this compressive zone was essentially the same in both analyses. The magnitude of the temperature in the vicinity of the gap was similar to that obtained in the aluminium remote from the gap. Since the direct stress σ_{xx} is zero at the gap this implies that the σ_{yy} stress in the adherand is of a similar magnitude to the applied stress.

Most theories for the stress analysis of bonded joints assume the σ_{yy} stress in the adherands to be zero even if peel stresses in the adhesive have been accounted for. As can be seen from the present investigation theories which do not allow for the σ_{yy} stresses in both the adherands and the adhesive may give erroneous results in the vicinity of the gap.

8 DIRECTIONS FOR FUTURE WORK

It is highly desirable that procedures are available so that the possible occurrences of delamination-type defects are allowed for in the design and certification of composite aircraft structures. The development of approaches for through-life support, to provide a rationale for setting inspection intervals, for highly stressed regions and a repair/reject criteria is also required. We have seen that a variety of approaches are currently being evaluated, viz.

(1) path-independent integrals, which in certain instances reduce to the energy release rate;
(2) the concept of a critical energy density; and
(3) the use of effective stiffness measurements using a simple extensometer.

For laboratory tests the effective stiffness approach is a valuable tool for assessing damage severity. However, the use of an extensometer to measure changes in stiffness is highly directional and for service aircraft, which may have a damaged component in a complex multi-axial stress state, the direction in which to monitor the changes in stiffness will not generally be known. A method for measuring changes in stiffness which is non-directional would be particularly useful. One possible method was proposed in Ref. 35 and is based on the coupling between thermal and mechanical energy.

This coupling has enabled the development of a unique non-contact method for investigating the stress distribution over the surface of a body. However, until recently the formulation[36] did not fully account for the interaction between the available mechanical energy and the thermal energy in the case of a composite material (see Ref. 39). In particular, the effects of moisture and mean load had previously been neglected. This revised theory predicts that when a body is subjected to a single frequency cyclic loading, the temperature response should not only be at the loading frequency, as inferred by Kelvin's law, but there should also be a component at twice that frequency. Subsequent experiments were performed to detect this predicted second harmonic response[40] and the results conclusively confirmed such a non-linear response, thereby strongly supporting the validity of the revised theory.

This theory has recently been used to illustrate how temperature measurements can be used to determine the surface stresses for a problem associated with bonded repair technology and to characterise impact damage in graphite/epoxy laminates.[35]

In Ref. 35 it was shown that impact damage caused a significant change in the surface temperature profile. It seems reasonable to expect that the greater the damage severity, the greater this change will be. This raises the possibility of obtaining an estimate of damage severity by determining a parameter which characterises the magnitude of the change in the surface temperature profile, from the undamaged to the damaged state. Reference 35 proposed the parameter

$$D_t^2 = \frac{1}{T_r^2} \left\{ \iint_S [T(x, y) - T_r]^2 \, dx \, dy - \frac{A_S}{A_R} \iint_R [T(x, y) - T_r]^2 \, dx \, dy \right\} \quad (12)$$

where $T(x, y)$ is a function representing the surface temperature field in the region S. Region S is the integration area (A_S) which includes and extends beyond the damaged area to where temperature values have become essentially constant and region R is an integration area (A_R) remote from the damaged region which has an average value of approximately T_r (the variance of temperature values in this region is assumed to be due to random experimental error).

Equation (12) was formulated to yield a measure of damage that is invariant to (i) the average value of the far field temperatures, (ii) the magnitude of the random experimental error in the temperature values (i.e. variance), and (iii) how far the calculation region S extends past the damaged region (see Table 11 and Ref. 35 for further details).

TABLE 11
Damage severity parameter D_t for impact surface of impact damaged specimen

Remote mean temperature (μ)	Remote variance of temperatures (σ_r^2)	Damage severity parameter D_t (mm)	
		Calculation area $A_S = 2\,006\,\text{mm}^2$	Calculation area $A_S = 851\,\text{mm}^2$
Remote region 1 $0 < x < 10,\ 0 < y < 32$			
99·7	198	14·6	14·0
100·4[a]	1 150[a]	14·5[a]	14·0[a]
Remote region 2 $0 < x < 50,\ 0 < y < 8$			
104·8	294	13·7	13·2
105·6[a]	1 150[a]	14·3[a]	13·5[a]
Remote region 3 $40 < x < 50,\ 0 < y < 32$			
100·3	255	14·1	14·7
100·5[a]	1 127[a]	13·8[a]	14·1[a]

[a] Values obtained after adding Gaussian noise to all data points.

Equation (12) can be numerically evaluated from the experimentally determined discrete surface temperature values over a rectangular region S as

$$D_t^2 = \frac{1}{T_r^2} \left[\sum_{i=1}^{N_x} \sum_{j=1}^{N_y} (T_{ij} - T_r)^2 \, \Delta x \, \Delta y - N_x N_y \sigma_r^2 \, \Delta x \, \Delta y \right] \tag{13}$$

where

$$\sigma_r^2 = \frac{1}{N_r} \sum_{k=1}^{N_r} [T_k - T_r]^2 \tag{14}$$

T_{ij} are temperature values at points in the region S, T_k are temperature values at points in the remote region, N_x is the number of values taken in the x direction sampled at distance Δy, σ_r^2 is an estimate of the variance in readings due to random experimental noise (this estimate is obtained using far field values) and N_r is the number of readings used for the calculation of σ_r^2.

As can be seen from eqns (11) and (12) D_t involves the product of the compliance tensor and the strain tensor, and as such represents a measure of the effective compliance of the region over which the scan is performed. This is particularly interesting since, as mentioned above, the change in the effective compliance has been found to correlate with failure (see Ref. 41), and in the case of an edge delamination can be related to the energy release rate.[42] Indeed, D_t behaves in a fashion similar to the effective compliance measured in Ref. 41 in that there appears to be a rapid increase in the rate of change of D_t as the specimen approaches the onset of failure. In this respect D_t may be thought of as an alternative to the use of an extensometer for measuring changes in compliance. It has the advantage of being non-directional and relatively independent of the size of the region considered.

Subject to further experimental validation, the authors believe that this parameter may prove to be very useful in the analysis of problems associated with damaged composite laminates. It is also thought that, when attempting to use laboratory coupon tests to represent components on in-service aircraft, it is essential that the coupon test reflect the interaction of load, geometry, material and damage seen in the aircraft. If this is to be achieved it is clear that, for the critical load case, the surface temperature profile of the coupon and the structure should coincide. In this case the damage parameter D_t obtained in each case would also coincide. It is clear that for a given geometry, restraint condition, structure and load, changes in the damage state will be reflected in changes in the surface temperature field. In this case the term 'damage' may include cracking, hole elongation, impact damage, residual stresses, environmental degradation, etc. As a result, the present methodology may form the basis of a non-destructive technique for comparing the results of coupon and full-scale test results.

In this context it should be noted that recent work[43] has revealed that, as in Ref. 35, a knowledge of the local temperature field and the relevant boundary conditions is sufficient to determine the individual stress components.

To illustrate how a knowledge of the thermal emission profile can be used to determine the individual stress components let us first consider the case when the material is isotropic. In this case if the test frequency is such that the structure is behaving in an adiabatic fashion then the thermal emission can be related to the bulk stress via eqn (11). The finite element method can now be used to determine the stress components.

In order to obtain the appropriate stiffness matrix we follow the approach outlined by Reddy[44] and Tay[45] and formulate the penalty functional

$$\Pi_p = \Pi + \frac{1}{2\kappa} \int_v [\sigma - \mathbf{C}\mathbf{u}^\mathrm{T}]^2 \, dV \qquad (15)$$

where, as outlined in Ref. 45, the penalty number is taken as $1/2\kappa$, where κ is the bulk modulus, σ is the bulk stress,

$$\mathbf{C}^\mathrm{T} = \kappa \left[\frac{\partial}{\partial x}, \frac{\partial}{\partial y}, \frac{\partial}{\partial z} \right] \qquad (16)$$

$$\mathbf{u}^\mathrm{T} = (u, v, w) \qquad (17)$$

and Π is the usual form for the strain energy

$$\Pi = 1/2 \int C_{ijkl} \varepsilon_{ij} \varepsilon_{kl} \, dV \qquad (18)$$

Now consider an arbitrary L-noded element within which the following interpolations are assumed:

$$\mathbf{u} = N\mathbf{u}_n \qquad (19)$$

$$\sigma = \mathbf{M}^\mathrm{T}\sigma_n \qquad (20)$$

where

$$\mathbf{u}_n^\mathrm{T} = [u_1, v_1, w_1, \ldots, u_i, v_i, w_i, \ldots, u_L, v_L, w_L] \qquad (21)$$

$$\sigma_n = [\sigma_1, \sigma, \ldots, \sigma_i, \ldots, \sigma_L] \qquad (22)$$

Here the subscript i denotes the value at the ith node. If we now define a vector $\boldsymbol{\delta}$, which contains the nodal degrees of freedom u, v, w and σ, as

$$\boldsymbol{\delta}^{\mathrm{T}} = [\mathbf{u}_n^{\mathrm{T}}, \boldsymbol{\sigma}_n^{\mathrm{T}}] \tag{23}$$

then the associated stiffness matrix for the element, which we will define as K_e, is given by

$$K_e = \begin{vmatrix} K & 0 \\ 0 & 0 \end{vmatrix}^{+\kappa^{-1}} \begin{vmatrix} k_u & -k_{\sigma u}^{\mathrm{T}} \\ -k_{\sigma u} & k_\sigma \end{vmatrix} \tag{24}$$

where K is the usual form for the stiffness matrix. When the nodal degrees of freedom are u, v, w and σ we obtain

$$K_u = \int N^{\mathrm{T}} \mathbf{C}^{\mathrm{T}} \mathbf{C} N \, \mathrm{d}V \qquad K_\sigma = \int \mathbf{M} \mathbf{M}^{\mathrm{T}} \, \mathrm{d}V \qquad K_{\sigma u} = \int \mathbf{M} \mathbf{C}^{\mathrm{T}} N \, \mathrm{d}V \tag{25}$$

8.1 Illustrative Example

Let us first consider the problem of an edge crack in a rectangular specimen 20 mm long, 10 mm wide and 2 mm thick. The crack is centrally located and is 1·666 mm long. The structure is loaded with a constant displacement of 0·01 mm. The problem is assumed to behave elastically and

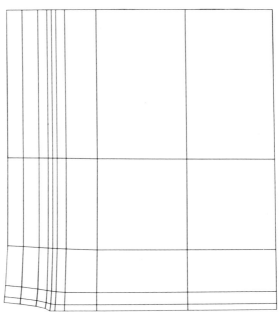

Fig. 10.   Displaced shape: direct solution.

Young's modulus and Poisson's ratio of the material are 1·89 GPa and 0·35 respectively. The finite element model consisted of approximately 50 three-dimensional elements described above. Each element had 20 nodal points and used a quadratic form for the displacement field and the bulk stress.

This approach gave a value of K_1 of 0·142 MPa $\sqrt{\text{m}}$ at the free surface and 0·145 MPa $\sqrt{\text{m}}$ at the midsurface, which compares very well with the analytical solution of 0·145 MPa $\sqrt{\text{m}}$.

As a second problem the values of the bulk stress on the external surfaces, as previously calculated, were used as input data together with the surface strains at three equally spaced points on each external surface. For this problem neither the loads or the displacements were specified. The formulation described above was able to determine the individual stress components, the energy density field, the displacement field and the total load applied to the structure. The accuracy of the solution can be judged from Figs 10 and 11, which show the displaced shape as calculated in the direct approach and as calculated from a knowledge of bulk stress, and from Figs 12 and 13, which show contours of the energy density field around the crack tip.

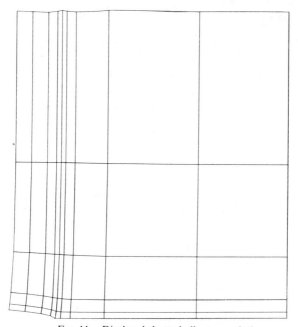

FIG. 11.   Displaced shape: bulk stress solution.

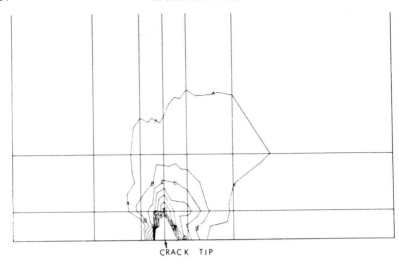

CRACK TIP

FIG. 12.   Energy density field: direct solution.

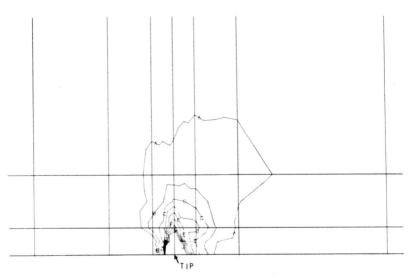

TIP

FIG. 13.   Energy density field: bulk stress solution. $A = 0.19$ MPa, $B = 0.28$ MPa, $C = 0.37$ MPa, $D = 0.45$ MPa.

Slight errors, in the energy and displacement fields, occur in the region of the loaded edge. However, the energy field around the crack is particularly accurate. The stress intensity factors, as calculated from a knowledge of the bulk stress alone, differ from those previously calculated by less than 1%.

It thus appears that, as first postulated in Refs 35 and 43, a knowledge of the thermal emission profile and a number of strain gauge values may enable the individual stress components to be evaluated. This approach, when coupled with existing C-scan techniques, raises the possibility of the structural characterisation of damage, and the non-destructive evaluation of critical damage sizes and loads.

9 CONCLUSION

This paper has presented a unified approach to the repair of metallic and composite structures. Attention has been focused on the use of externally bonded repairs. This methodology has been illustrated by considering the recent repair to F111C aircraft in service with the RAAF.

REFERENCES

1. BAKER, A. A., A summary of work on applications of advanced fibre composites at the Aeronaurical Research Laboratories, Australia. *Composites,* **9** (1978) 11–16.
2. BAKER, A. A., CALLINAN, R. J., DAVIS, M. J., JONES, R. and WILLIAMS, J. G., Repair of Mirage III aircraft using BFRP crack patching technique. *Theoret. Appl. Fract. Mech.,* **2**(1) (1984) 1–16.
3. JONES, R. and CALLINAN, R. J., Bonded repairs to surface flaws. *Theoret. Appl. Fract. Mech.,* **2**(1) (1984) 17–126.
4. JONES, R., Bonded repair of damage. *J. Aeronaut. Soc. India,* **3**(3) (1984) 193–201.
5. HELLER, M., JONES, R. and WILLIAMS, J. F., Analysis of bonded inserts for the repair of fastener holes. *Engng Fract. Mech.,* **24**(4) (1986) 523–32.
6. MANN, J. R., PELL, R. A., JONES, R. and HELLER, M., Reductions in fatigue life of rivet holes by adhesive bonding. *Theoret. Appl. Fract. Mech.,* **3** (1985) 113–24.
7. ROOKE, D. P. and CARTWRIGHT, D. J., *A Compendium of Stress Intensity Factors.* Her Majesty's Stationery Office, London, 1970.
8. MOLENT, L., PAUL, J. and JONES, R., Criteria for matrix dominated failure. Aeronautical Research Laboratory, Melbourne, Australia, Aircraft Structures Report 432, 1988.
9. MOLENT, L. and JONES, R., Design of a boron epoxy reinforcement for the F111C wing pivot fitting. Aeronautical Research Laboratory, Melbourne, Australia, Aircraft Structures Report 426, 1987.

10. ANON., Final engineering report: A8-112 wing pivot fitting failure investigation (manufacturing processes), FZM-12-5130A. General Dynamics, Fort Worth, USA, 1982.

11. BUNTON, W. D., Concept and conduct of proof test of F-111 production aircraft. Paper presented to Royal Aeronautical Society, London, 1971.

12. SUSANS, G. R., PATCHING, C. A. and BECKETT, R. C., Visit by an Australian team to the US to discuss the failure of A8-112 during CPLT. Department of Defence, AF1511/912/100 Pt 3, Canberra, Australia.

13. WILLIAMS, J. F., STOUFFER, D. C., ILLIC, S. and JONES, R., An analysis of delamination behaviour. *Comp. Struct.*, **5** (1986) 203–16.

14. EVANS, A. G. and HUTCHINSON, J. W., On the mechanics of delamination and spalling in compressed films. *Int. J. Solids Struct.*, **20** (1984) 455–66.

15. ATLURI, S. N., *Computational Methods in the Mechanics of Fracture*. North-Holland, Amsterdam, 1986.

16. ATLURI, S. N., NAKAGUKI, M., NISHIOKA, T. and KUANG, Z. B., Crack tip parameters and temperature rise in dynamic crack propagation. *Engng Fract. Mech.*, **23**(1) (1986) 167–82.

17. JONES, R., TAY, T. E. and WILLIAMS, J. F., Thermo-mechanical behaviour of composites. *Proc. US Army Workshop on Composite Material Response: Constitutive Relations and Damage Mechanisms*, Glasgow, 1987, ed. G. C. Sih, G. F. Smith, I. H. Marshall and J. J. Wu. Elsevier Applied Science, London, 1988, pp. 49–61.

18. JANARDHANA, M. N., BROWN, K. C. and JONES, R., Designing for tolerance to impact damage at fastener holes in graphite/epoxy laminates under compression. *Theoret. Appl. Fract. Mech.*, **5**(1) (1986) 51–6.

19. DEMUTS, E. and HORTON, R. E., Damage tolerant composite design development. In *Composite Structures—Vol. 3*, ed. I. H. Marshall. Elsevier Applied Science, London, 1985.

20. POTTER, R. T., The interaction of impact damage and tapered-thickness sections in CFRP. *Comp. Struct.*, **3** (1985) 319–39.

21. RYDER, R. J., LAURAITIS, K. N. and PETTIT, D. E., Advanced residual strength degradation rate modeling for advanced composite structures. AFWAL-TR-79-3095, Vol. II, Tasks II and III, 1981.

22. DOREY, G., Structural life predictions for fibre reinforced composite materials. ONR Workshop on Damage Tolerance of Fibre Reinforced Composite Materials, Glasgow, Scotland, 1985.

23. O'BRIEN, T. K., Interlaminar fracture of composites. *J. Aeronaut. Soc. India*, **37** (1985) 61–9.

24. BLACK, N. F. and STINCHCOMB, W. W., Compression fatigue damage in thick, notched graphite–epoxy laminates. *Long-term Behaviour of Composites*, ASTM STP 813, 1983, pp. 95–115.

25. ROSENFELD, M. S. and HUANG, S. L., Fatigue characteristics of graphite–epoxy laminates under compression loading. *J. Aircraft*, **15** (1978) 264–8.

26. RATWANI, M. M. and KAN, H. P., Compression fatigue analysis of fibre composites. NADC-78047-69, 1979.

27. STELLBRINK, K. K. and AOKI, R. M., Effect of defect on the behaviour of composites. *Progress in Science and Engineering of Composites*, ICCM—IV, Tokyo, 1982.

28. DEMUTS, E., WHITEHEAD, R. S. and DEO, R. B., Assessment of damage tolerance in composites. *Comp. Struct.* **4** (1985) 45–58.
29. WALTER, R. W., JOHNSON, R. W., JUNE, R. R. and MCCARTY, J. E., Designing for integrity in long-term composite aircraft structures. *Fatigue of Filamentary Composite Materials*, ASTM STP 636, 1977, pp. 228–47.
30. VAN BLARICUM, T. J., BATES, P. and JONES, R., An experimental investigation into the effect of impact damage on the compressive strength of step lap joints. *Engng Fract. Mech.* (in press).
31. ROSENFELD, M. S. and GAUSE, L. W., Compression fatigue behaviour of graphite epoxy in the presence of stress raisers. *Fatigue of Fibrous Composite Materials*, ASTM STP 7231, 1981, pp. 174–96.
32. TAY, T. E., WILLIAMS, J. F. and JONES, R., Application of the T^x integral and S criteria in finite element analysis of impact damaged fastener holes in graphite/epoxy laminates under compression. *Comp. Struct.*, **7**(4) (1987) 233–53.
33. PAUL, J., Damage assessment and repair of thick composite structures. MEngSc Thesis, University of Melbourne, Australia, December 1988.
34. BAKER, A. A. and JONES, R., *Bonded Repair of Aircraft Structures*. Martinus Nijhoff Publishers, Dordrecht, The Netherlands, 1988.
35. HELLER, M., DUNN, S., WILLIAMS, J. F. and JONES, R., Thermomechanical analysis of metallic and composite specimens. *J. Comp. Struct.* (in press).
36. THOMSON, W., On the thermoelastic and thermomagnetic properties of matter. *Q. J. Maths*, **1** (1855) 55–77.
37. MACHIN, A. S., SPARROW, J. G. and STIMSON, M. G., Mean stress dependence of thermoelastic constant. *Strain*, **23** (1987) 27–30.
38. WONG, A. K., JONES, R. and SPARROW, J. G., Thermoelastic constant or thermoelastic parameter. *J. Phys. Chem. Solids*, **48** (1988) 749–53.
39. JONES, R., TAY, T. E. and WILLIAMS, J. F., Thermomechanical behaviour of composites. *Constitutive Response: Constitutive Relations and Damage Mechanisms*, ed. G. C. Sih, G. F. Smith, I. H. Marshall and J. J. Wu. Elsevier Applied Science, London, 1988, pp. 49–61.
40. WONG, A. K., SPARROW, J. G. and DUNN, S. A., On the revised theory of the thermoelastic effect. *J. Phys. Chem. Solids*, **49**(1) (1988) 395–400.
41. SAUNDERS, D. S. and VAN BLARICUM, T., Effect of load duration on the fatigue behaviour of graphite/epoxy laminates containing delaminations. *Composites*, **19**(3) (1988) 217–28.
42. O'BRIEN, T. K., Characterization of delamination onset and growth in a composite laminate. *Damage in Composite Materials*, ASTM STP 775, 2, 1982, pp. 140–67.
43. RYALL, T. G. and WONG, A. K., Determining stress components from thermoelastic data: a theoretical study. *J. Mech. Mater.* (in press).
44. REDDY, J. N., Simple finite elements with relaxed continuity for non-linear analysis of plates. *Proc. Third Int. Conf. in Australia on Finite Elements*, University of New South Wales, Australia, July 1979.
45. TAY, T. E., Characterisation and analysis of fracture in layered composite materials. PhD Thesis, Melbourne University, Department of Mechanical Engineering, August 1988.

Plenary Paper

Composite Intensive Automobiles—An Industry Scenario*

P. Beardmore, C. F. Johnson and G. G. Strosberg

Ford Motor Company, Dearborn, Michigan, USA

ABSTRACT

The application of structural composites to automotive structures will require: (1) clear evidence of the functional capability of the composite structures, including long-term effects; (2) the development of high speed, demanding manufacturing and assembly processes with associated quality control; and (3) evidence of economic incentives which will be sensitively dependent on the manufacturing processes.

This report summarizes the current status of these issues and considers possible future developments. In particular, the economics will not become clear until the manufacturing developments and associated experience are in hand. Economic indicators suggest that structural composites would initially only be used for low-volume vehicles, and thus the most likely scenario would be limited usage of these materials. The corresponding effect on the industry would, therefore, be minor.

Clearly, the large-scale adoption of composite construction for automobile structures would have a major impact on fabrication, assembly and supply network. The infrastructure of today would completely change; for example, the multitude of metal forming presses would be replaced by a much smaller number of molding units, multiple welding machines would be replaced by a limited number of adhesive bonding fixtures, the assembly sequence would be modified to reflect the tremendous reduction in parts, etc. This complete revamping of the stamping and body construction facilities would clearly entail a revolution (albeit at an evolutionary pace) in the industry. However,

* Funded in part by the Office of Technology Assessment, US Congress.

39

there would probably not be a significant impact on the size of the overall labor force or the skill level required.

Extensive use of structural composites by the automotive industry would necessitate the development of completely new supply industries geared to providing inexpensive structural composites. Such an industry does not exist at present and would need to evolve to satisfy the demands. It is anticipated that such developments would take place by two mechanisms. First, current automotive suppliers, particularly those with plastics expertise, would unquestionably expand and/or diversify into structural composites to maintain and possibly increase their current level of business. Second, the currently fragmented structural composites industry, together with raw material suppliers, would generate new integrated companies with the required supply capability. In addition, completely new industries would have to be developed; for example, a comprehensive network of composite repair facilities, a recycling industry based on new technologies, etc.

The study includes discussion of the above issues with probable scenarios for the initial introduction of structural composites to the automotive industry. Economics, customer perception and functional improvements will dictate the eventual extent of usage, but the effect of various volumes of composite intensive vehicles on the automotive and supply industry is considered. It is clear that a significant fraction of the annual car production would have to convert to large-scale composite usage before a major impact is felt by the steel industry. Likewise, similar volume sensitivity would apply to related supply industries such as tool and die manufacturers and chemical manufacturers.

1 INTRODUCTION

The use of plastic-based composite materials in automotive applications has gradually evolved over the past two decades. With new materials and processing techniques being continuously developed within the plastics and automotive industries, there is potential for a more rapid expansion of these types of applications in the future. Structural composite applications, both for components and for major modular assemblies, appear to be a potential major growth area that could have a significant impact on the automotive industry and associated supply industries if the required developments result in cost-effective manufacturing processes.

Extensive research and development efforts, currently under way, are aimed at realizing the following *potential* benefits of composite structures to the automotive industry:

—Weight reduction, which may be translated into improved fuel economy and performance.
—Improved overall product quality and consistency in manufacturing.
—Part consolidation resulting in lower product and manufacturing costs.
—Improved function (reduced noise, vibration and harshness for improved ride performance).
—Product differentiation with acceptable or reduced cost.
—Lower investment costs for plants, facilities and tooling—depends on cost/volume relationships.
—Corrosion resistance.
—Lower cost of ownership.

However, there are areas where major uncertainties exist which will require extensive research and development prior to resolution. For example:

—High speed, high quality manufacturing processes with acceptable economics.
—Satisfaction of all functional requirements, particularly crash integrity and long-term durability.
—Repairability.
—Recyclability.
—Customer acceptance.

The intent of this report is to project scenarios showing the potential impact of structural composites as related to the automotive industry, the supplier base and customers. These scenarios will depict the likely effects generated by implementation of the types of materials and process techniques that could find acceptance in the manufacturing industries during the late 1990s provided development and cost issues are favorably resolved.

In order to project the potential of structural composites for automotive applications, it is necessary to provide a reasonably extensive summary of the current state of these materials from an automotive perspective. Consequently, the first part of the report provides this information and, in particular, the specific types of composites showing the most promise for structural applications (Section 3) and the viable and developable fabrication processes most likely to be used (Section 5) will be summarized.

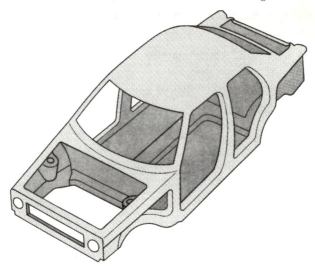

FIG. 1.   Schematic of steel body shell structure.

The particular properties of most relevance to automotive applications will be presented, together with a scenario for the necessary extension of knowledge required before high confidence can be placed in the structural application of the materials. The projections outlined here are intended to form a basis for assessing the potential impact of structural composites implementation on the domestic economy.

This report will illustrate the potential of composites by concentrating on a highly integrated composite body shell (Fig. 1). Basically, this body shell is the major load-bearing structure of the automobile and does not include the closure panels (hood, decklid and doors). This basic structure has been chosen as representative of the type of assembly which might be produced in moderate volumes as composite materials begin to penetrate the automobile industry.

2 BACKGROUND

The economic constraints in a mass-production industry such as the automotive business are quite different than those in the aerospace or even the specialty vehicle business. This is particularly true in the potential application of high-performance composite materials, which to date have primarily been developed and applied in the aerospace industry in a cost-intensive manner, both from a basic materials cost and from a fabrication viewpoint. Virtually all uses of plastics and composites in high-volume vehicles are restricted to decorative or semi-structural applications. Sheet molding compound (SMC) materials are the highest performance composites in general automotive use for bodies today, and the most widely used SMC materials contain approximately 25 wt% chopped glass fibers and cannot really be classified as a high-performance composite. Typically, SMC materials are used for grille opening panels on many car lines and closure panels (hoods, decklids, doors) on a few select models. A characteristic molding time for SMC is of the order of two to three minutes, which is on the borderline of viability for automotive production rates.

The next major step for composites in the automotive business is extension of usage into truly structural applications such as the primary body structure (Fig. 1) and to chassis/suspension systems. These are the areas which have to sustain the major road load inputs and crash loads. In addition, these major structures must deliver an acceptable level of vehicle dynamics such that the passengers enjoy a comfortable ride. These functional requirements must be totally satisfied for any new material to find extensive application in body structures and it is no small challenge to structural composites to effectively meet these criteria. Of course, these criteria must also be satisfied in a cost-effective manner. Appropriate composite fabrication procedures must be applied or developed which satisfy high production rates but still maintain the critical control of fiber placement and distribution.

Composite body structures have been used in a variety of specialty vehicles for the past three decades, Lotus cars being a particularly well known example. The composite material used in these specialty vehicles is invariably glass fibers, typically in a polyester resin. A variety of production methods have been used but perhaps the only common thread is that all the processes are slow, primarily because of the very low production rate (typically to a maximum of 5000 per annum). Thus there has been no incentive to accelerate the development of these processes for volumes at this level. The other common factor among these vehicles is the general use of some type of steel backbone or chassis which absorbs most of the road loads and crash impact energy. Thus, while the FRP body can be considered somewhat structural, the major structural loads are not imposed on the FRP materials.

High fiber reinforced plastic (FRP) content vehicles in existence today were designed specifically for FRP materials, as opposed to being patterned after a steel vehicle. Consequently, there is no direct comparison available

FIG. 2.   Photograph of the Ford GrFRP vehicle.

between an FRP vehicle and an identical steel vehicle to relate baseline characteristics. Perhaps the best comparison is the prototype Graphite LTD built by Ford to directly compare a production steel vehicle with a high tech FRP vehicle.[1] Although graphite fibers were used and the vehicle was fabricated by hand lay-up procedures, several interesting vehicle features were evaluated.

The graphite fiber composite (GrFRP) car is shown in Fig. 2, and an exploded schematic showing the composite parts is shown in Fig. 3. The

TABLE 1
Major weight savings in GrFRP

Component	Weight in pounds[a]		
	Steel	GrFRP	Reduction
Body-in-white	423·0	160·0	263·0
Front end	95·0	30·0	65·0
Frame	283·0	206·0	77·0
Wheel(s)	91·7	49·0	42·7
Hood	49·0	17·2	31·8
Decklid	42·8	14·3	28·5
Doors (4)	141·0	55·5	85·5
Bumpers (2)	123·0	44·0	79·0
Driveshaft	21·1	14·9	6·2
Total vehicle	3 750	2 504	1 246

[a] 1 lb = 0·454 kg.

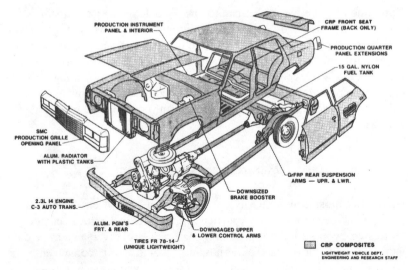

FIG. 3. Exploded schematic view of the Ford GrFRP vehicle with the composite components shown.

weight savings for the various structures are given in Table 1. While these weight savings (of the order of 55–65%) might be considered optimal because of the use of graphite fibers, other cost-effective fibers can achieve a major portion of these weight savings (see Section 4.3). Although the GrFRP vehicle weighed 2504 lb compared to a similar steel production vehicle of 3750 lb, vehicle evaluation tests indicated no perceptible difference between the vehicles. Ride quality and vehicle dynamics were judged at least equal to top quality production steel LTD cars. Thus, on a direct comparison basis, a vehicle with an entire FRP structure was proven at least equivalent to a steel vehicle from a vehicle dynamics viewpoint at a weight level only 67% of the steel vehicle.

The GrFRP car clearly showed that high-cost fibers (graphite) and high-cost fabrication techniques (hand lay-up) can yield a perfectly acceptable vehicle based on handling, performance and vehicle dynamics criteria. However, crash and durability performance were not demonstrated and these issues will need serious development work to achieve better-than-steel results. An even bigger challenge is to translate that performance into realistic economics by the use of cost-effective fibers, resins and fabrication procedures.

High composite content specialty cars utilize steel as the major load-carrying structure and are not generally priced on a competitive basis.

Consequently, these vehicles cannot be used to generate guidelines for extensive composite usage in high-volume applications. The governing design guidelines for composites need to be further developed to ascertain, for example, how to design utilizing low-cost composite materials, and to ascertain allowables for stiffness in situations where major integration of parts in composites eliminates a myriad of joints. The following sections will summarize the extant knowledge in composites potential for automotive structures.

3 COMPOSITE MATERIALS

By far the most comprehensive property data have been developed on aerospace-type composites, in particular graphite fiber reinforced epoxies fabricated by hand lay-up of prepreg materials. Relatively extensive data banks are available on these materials and it would be very convenient to be able to build off this data base for less esoteric-type applications such as automobile structures. Graphite fibers are the favored choice in aerospace because of the superb combination of stiffness, strength and fatigue resistance exhibited by these fibers. Unfortunately, for the cost-conscious mass-production industries, these properties are only attained at significant expense. Typical graphite fibers cost in the range of $25 per lb. There are intensive research efforts devoted to reducing these costs by utilizing a pitch-based precursor but the most optimistic predictions are in the $5–10 per lb range, which still keeps these fibers in the realm of very restricted potential for consumer-oriented industries.

The fiber with the greatest potential for automobile structural applications is E-glass fiber (currently ~ $0·80 per lb), based on the optimal combination of cost and performance. Likewise, the resin systems likely to dominate at least in the near term are polyester and vinyl ester resins based primarily on a cost/processability tradeoff versus epoxy. Higher performance resins will only find specialized applications (in much the same way as graphite fibers) even though the ultimate properties may be somewhat superior.

The form of the glass fiber used will be very application-specific and both chopped and continuous glass fibers should find extensive use. It might be expected that most structural applications involving significant load inputs will utilize a combination of both chopped and continuous glass fibers with the particular proportions of each depending on the component or structure. Since all the fabrication processes anticipated to play a significant role in automotive production are capable of handling mixtures of continuous and chopped glass (see Section 5), this requirement should not present major restrictions. One potential development which is likely if glass fiber composites come to occupy a significant portion of the structural content of an automobile is the tailoring of glass fibers and corresponding specialty resin development. The size of the industry (each pound of composite per vehicle translates to approximately 10 million pounds per year in North America) dictates that it would be economically feasible to have fiber and resin production tailored exclusively for the automotive market. The advantage of such an approach is that these developments will lead to incremental improvements in specific composite materials which in turn should lead to increased applications and increased cost effectiveness.

While the thermoset matrix composites discussed above will probably constitute the bulk of the structural applications, thermoplastic-based composites formed by a compression molding process may well have a significant, but lesser, role to play. Most of the compression molded thermoplastics in commercial use today tend to concentrate on poly-propylene or nylon as the base resin. The reason is simply the economic fact that these materials tend to be the most inexpensive of the engineering thermoplastics and are easily processed. Both these materials are somewhat deficient in heat resistance and/or environmental sensitivity relative to vehicle requirements for high-performance structures. Other thermoplastic matrices for glass fiber reinforced composites are under development, and materials such as polyethylene terephthalate (PET) hold significant promise for the future. The extent to which thermoplastic composites will be used in future structures will be directly dependent on material developments and the associated economics.

4 PERFORMANCE CRITERIA

From a structural viewpoint, there are two major categories of material response which are critical to applying composites to automobiles. These are fatigue (durability) and energy absorption. In addition, there is another critical vehicle requirement, namely ride quality, which is usually defined in terms of noise, vibration and ride harshness, and generally perceived as directly related to vehicle stiffness. Obviously, material characteristics play a significant role in this category of vehicle response. These three areas will be summarized separately.

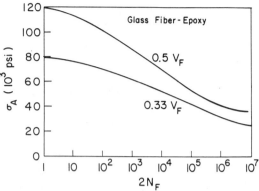

FIG. 4.   Typical fatigue curves for unidirectional GlFRP as a function of volume fraction of glass fibers (V_F).

4.1 Fatigue

The specific fatigue resistance of glass fiber reinforced composites is a sensitive function of the precise constitution of the material but there are preliminary research indications of the sensitivity to cyclic stresses. For unidirectional glass FRP materials, the fatigue behavior can be characterized as illustrated in Fig. 4. The most important characteristic in Fig. 4 is the fairly well defined fatigue limit exhibited by these materials and as a guiding principle this limit can be estimated as approximately 35–40% of the ultimate strength.[2] A chopped glass composite, by contrast, would have a fatigue limit closer to 25% of the ultimate strength[3,4] and would exhibit much greater scatter in properties (Fig. 5). It is also important to realize that the different failure modes in composites (in comparison to

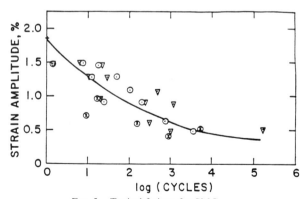

FIG. 5.   Typical fatigue for SMC.

metals) can result in different design criteria for these materials depending on the functionality involved. For example, a decrease in stiffness can occur under cyclic stressing long before physical cracking and strength deterioration occur. If stiffness is a critical part of the component function, the loss in stiffness under the cyclic road loads could result in loss of the stiffness-controlled function with no accompanying danger of any loss in mechanical function. This phenomenon does not occur in steel.

As a guiding principle, it follows from the above that, wherever possible, automotive structures should be designed such that continuous fibers take the primary stresses, and chopped fibers should be present to generate some degree of orthotropic behavior. It is critical to minimize the stress levels, particularly fatigue stresses, which have to be borne by the chopped fibers.

There is emerging evidence from both fundamental research data and field experience with composite components that glass fiber reinforced composites can be designed to withstand the rigorous fatigue loads experienced under vehicle operating conditions. The success of composite leaf springs and SMC componentry attests to the capability of composites to withstand service environments.

While the data for all combinations of composite materials are not yet available, a sufficient data base is available such that conservative estimates can be generated and lead to reliable designs. It cannot be overemphasized, however, that the mechanical properties of composites (much more than isotropic materials) are very sensitive to the fabrication process. It is imperative that properties be related to the relevant manufacturing technique to prevent misuse of baseline data.

4.2 Energy Absorption

The stress–strain curves of all high-performance composites are essentially linear in nature (Fig. 6). This is in contrast to most metals which exhibit a high degree of plasticity (Fig. 7) and is much more akin to the behavior of so-called brittle materials such as ceramics. The point to be drawn from comparison of the stress–strain behavior is that materials which are essentially elastic to failure (composites, ceramics) might be concluded to have no capacity for energy absorption since no plastic deformation energy is available to satisfy such requirements. However, ceramics have been used for decades as armored protection against high-velocity projectiles. Such energy absorption is achieved by spreading the localized impact energy into a high-volume cone of fractured ceramic material, as shown schematically in Fig. 8. A large amount of fracture surface area is created by fragmentation of the solid ceramic and the impact

P. Beardmore, C. F. Johnson and G. G. Strosberg

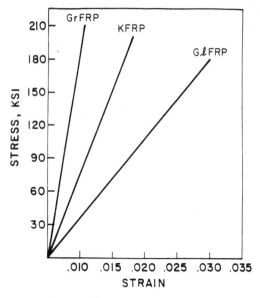

FIG. 6.   Tensile stress–strain curves for graphite fiber composite (GrFRP), Kevlar fiber composite (KFRP) and glass fiber composite (GlFRP).

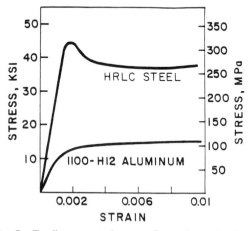

FIG. 7.   Tensile stress–strain curves for steel and aluminum.

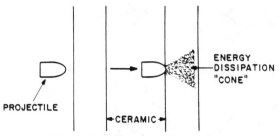

FIG. 8. Schematic illustration of energy dissipation mechanism in ceramics.

energy is converted into surface energy resulting in successful protection and very efficient energy dissipation. Elastic (so-called 'brittle') materials can be very effective energy absorbers but the mechanism is fracture surface energy rather than plastic deformation.

The analogy given above leads directly to the conclusion that, in a similar manner, high-performance composites may well be able to absorb energy by a controlled disintegration (fracture) process. Evidence is emerging from laboratory test data on the axial collapse of composite tubes that efficient energy absorption needed for vehicle structures can be achieved in these materials.

A comparison between the energy absorption mechanisms in metals and composites is shown in Fig. 9. Glass fiber and graphite fiber reinforced composites behave as shown in Fig. 9(b). By contrast, composites utilizing fibers consisting of highly oriented long-chain polymers (e.g. Kevlar) collapse in a metal-like fashion utilizing plastic deformation as the energy absorbing mechanism. The fragmentation/fracture mechanism typical of glass fiber composites can be very weight effective and Table 2 illustrates this point by giving typical relative values for different materials. It is particularly significant that while high-performance, highly oriented composites generate the maximum energy absorption, commercial-type composites yield specific energy numbers considerably superior to metals.

TABLE 2
Energy absorption (typical properties)

Material	Relative energy absorption (per unit weight)
High-performance composites	100
Commercial composites	60–75
Mild steel	40

FIG. 9. Photographs illustrating the different collapse mechanisms in (a) metal and (b) composite tubes for energy absorption. ·

Thornton and co-workers[5-7] have accumulated extensive data on energy absorption in composites.

Virtually all the energy absorption data generated to date have been developed for axial collapse of relatively simple structures, usually tubes. The ability to generate the same effective fracture mechanisms in complex structures is still unresolved. In addition, it is well known from observations on metal vehicles that bending collapse normally plays a significant part in the collapse of the vehicle structure and it is consequently of considerable importance to evaluate energy mechanisms in bending failure. Just as in metals, little data are available on energy absorption characteristics in bending. There is no reason to believe that the energy absorption values of metals relative to composites in bending should change significantly from the ratios in axial collapse except that bending failure (fracture) in composites may tend to occur on a more localized basis than plastic bending in metals. If indeed this does occur, then the ratio could change in the favor of metals.

Other crash issues involve the capacity to absorb multiple and angular impacts, and the long-term effects of environment on energy absorption capability. In general, practical data from a realistic vehicle viewpoint are not yet in hand. An additional significant factor is the consumer acceptance of these materials as perceived in relation to safety. A negative perception would be a serious issue on something as sensitive as safety, and better-than-steel crash behavior will be required before wide-scale implementation of composites can occur. Conversely, a positive perception would be a valuable marketing feature and provide an additional impetus for composite applications.

4.3 Stiffness and Damping

Glass fiber reinforced composites are inherently less stiff than steel. Some typical values for various types of composite are listed in Table 3. There are two offsetting factors to compensate for these material limitations. First, an increase in wall thickness can be used to partially offset the lesser material stiffness. Also local areas can be thickened as required to optimize properties. Since the composite has a density approximately one-third that of steel, a significant increase in thickness can be achieved while maintaining an appreciable weight reduction. The second, and perhaps the major, compensating factor is the additional stiffness attained in composites by virtue of part integration. This integration leads directly to the elimination of joints which results in significant increases in effective stiffness. It is becoming increasingly evident that this synergism is such that structures of acceptable stiffness and considerably reduced weight are feasible in glass fiber reinforced composites. As a rule of thumb, a glass FRP structure with significant part integration relative to the steel structure being replaced can be designed for a nominal stiffness level of 50–60% of the steel structure. Such a design procedure should lead to adequate stiffness and to typical weight reductions between 20 and 50%.

The stiffness requirement for vehicles is normally dictated by vehicle dynamics characteristics. The historical axiom in the vehicle engineer's

TABLE 3

Typical stiffness of composites

Material	Modulus (10^6 psi)
Unidirectional GrFRP	20
Unidirectional GlFRP	6
Unidirectional Kevlar	11
XMC	4·5
SMC-R50	2·3
SMC-25	1·3

design principles is the stiffer the better. However, there are some intangible factors which enter the overall picture, in particular the damping factor. Obviously, it is an oversimplification to assume stiffness alone dictates vehicle dynamics, although it unquestionably dominates certain categories. Damping effects can play an equally significant role in many categories of body dynamics, and the fact that the damping of composite materials considerably exceeds that of metals is relevant to the overall scenario. Most people involved with composite component/structure prototype development express the opinion that some aspect of body dynamics (usually noise or vibration) is improved but little quantitative data are available to document the degree of improvement. If this translates to a similar customer perception of improved dynamics, composites will receive added impetus for structural usage. There seems, however, to be little doubt that this increased damping capacity can only be a positive asset. It remains to be seen whether or not the additional damping can be of sufficient benefit as to become a cost-effective asset in that it allows beneficial iterative modifications to the vehicle.

5 POTENTIAL MANUFACTURING TECHNIQUES

The successful application of structural composites to automotive structures is more dependent on the ability to use rapid, economic fabrication processes than on any other single factor. The fabrication processes must also be capable of close control of composite properties to achieve lightweight, efficient structures. Currently, the only commercial process which comes close to satisfying these requirements is compression molding of sheet molding compounds (SMC) or some variant on the process. There are, however, several developing processes which hold distinct promise for the future in that these techniques have the potential to

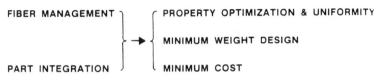

DESIGN / FABRICATION / PERFORMANCE / COST

FIBER MANAGEMENT ⎱ ⎧ PROPERTY OPTIMIZATION & UNIFORMITY
              ⎰ → ⎨ MINIMUM WEIGHT DESIGN
PART INTEGRATION ⎱ ⎩ MINIMUM COST

Fig. 10.   Overriding requirements of fiber control and part integration for optimizing composite economics.

AUTOMOTIVE COMPOSITE PROCESS DEVELOPMENT

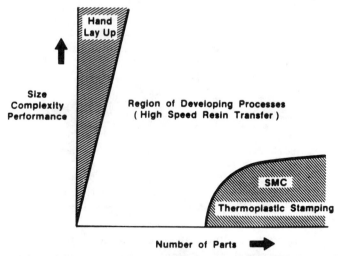

FIG. 11. Schematic illustration of the relationship between performance and fabrication for composites.

combine high rates of production, precise fiber control and high degrees of part integration. The overall philosophy behind composites fabrication for automobile body structure is summarized in Fig. 10. The requirements for precise fiber control, rapid production rates and high complexity demand that automotive processes be in the 'region of developing processes', shown schematically in Fig. 11. The evolving processes are thermoplastic compression molding, high speed resin transfer molding and filament winding. Each of these processes will be summarized separately with comments relative to the merits and potential limitations.

5.1 Compression Molding

A schematic of the sheet molding compound (SMC) process, depicting both the fabrication of the SMC material and the subsequent compression molding into a component, is shown in Fig. 12. This technology has been widely used in the automobile industry for the fabrication of grille opening panels on many car lines, and for some exterior panels on selected vehicles. Tailgates (Fig. 13) and hoods (Fig. 14) are examples on cars and light trucks, while the entire cab on some heavy trucks (Fig. 15) is constructed in this manner. The process consists of placing sheets of SMC (1–2-in chopped glass fibers in chemically thickened thermoset resin) into a heated mold

SHEET MOLDING

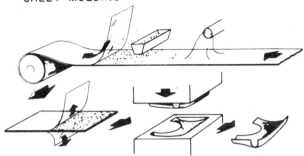

FIG. 12.   Schematic illustration of SMC material preparation and component fabrication.

(typically at 300°F (149°C)) and closing the mold under pressures of 1000 psi for about 2–3 min to cure the material. Approximately 80% of the mold surface is covered by the SMC charge and the material flows to fill the remaining mold cavity as the mold closes.

The above description of the SMC process delineates material primarily used for semi-structural applications rather than high load-bearing segments of the structure which must satisfy severe durability and energy absorption requirements. To sustain the more stringent structural demands, it is normally necessary to incorporate appreciable amounts of continuous fibers in predesignated locations and orientations. The same basic SMC operation can be utilized to incorporate such material modifications either by formulating the material to include the continuous fibers along with the chopped fibers or by using separate charge patterns of two different types of material. The complexity of shape and degree of flow possible are governed by the amount and location of the continuous glass material. Careful charge pattern development is necessary for components of complex geometry. A typical example of a prototype rear floor pan fabricated by this technique[8] is shown in Fig. 16.

The limitations for the usefulness of compression molding of SMC-type materials in truly structural applications have yet to be established. Provided continuous fiber is strategically incorporated, these materials promise to be capable of providing high structural integrity and may well prove to be the pioneering fabrication procedure in high load-bearing applications. The advanced state of commercialization of this process relative to other evolving techniques will also provide a lead time for compression molding to branch into higher performance parts.

Although compression molding of SMC-type materials is an economically viable, high production rate process in current usage, there are some

Fig. 13.   Typical SMC production Aerostar tailgate.

FIG. 14.   Typical SMC Aerostar hood.

FIG. 15. Typical SMC truck cab.

limitations inherent in the process which, in the longer term, will restrict applications and tend to favor the developing processes. For example, the degree of flow required to optimize the mechanical properties results in a variation in mechanical properties due to imprecise control of fiber location and orientation. Typically, variations in mechanical properties of a factor of two throughout the component are not unusual based on an initial charge pattern coverage of approximately 70%. Such uncertainty in properties introduces reliability issues and conservative design allowables yielding a heavier-than-necessary component or structure. Currently, extensive research efforts are under way to develop SMC-type materials which will allow 100% charge pattern coverage and which attain high, uniform mechanical properties with minimal flow—these materials can also be molded at lower pressures on smaller capacity presses. Materials developments such as these may well make the newer breed of 'SMCs' much more applicable to highly loaded structures than has ever been envisaged up to the present.

Fig. 16.   Prototype Escort composite rear floor pan.

Another potential limitation of compression molding is the degree of part integration attainable. The basic strategy in composite applications is to integrate as many individual (steel) pieces as possible to minimize fabrication and assembly costs (which offsets increased material costs) and to minimize joints (which increases 'effective' stiffness). Compression molding requires fairly high molding pressures (about 1000 psi) and thus limits potential structures in areal size and complexity (particularly in 3-D geometries requiring foam cores). Consequently, while compression molding is likely to play a key role in the development of composites in structural automotive applications in the next decade, ultimately the process is unlikely to provide optimum structural efficiency and weight. It should be pointed out that this statement will only prove true if the alternate, more optimal processes undergo the developments required because currently compression molding is the *only* commercial structural composite process capable of satisfying the economic constraints of a mass-production industry.

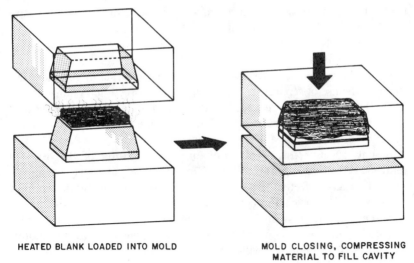

HEATED BLANK LOADED INTO MOLD   MOLD CLOSING, COMPRESSING
                                   MATERIAL TO FILL CAVITY

FIG. 17.   Schematic illustration of thermoplastic compression molding.

5.2 Thermoplastic Compression Molding

The process of thermoplastic compression molding (stamping) is attractive to the automotive industry because of the rapid cycle time and the potential utilization of some existing stamping equipment. Thermoplastic compression molding at its current level of development achieves cycle times of 1 min for large components. Figure 17 presents a schematic of the process. Typically, a sheet of premanufactured thermoplastic and reinforcement is preheated above the melting point of the matrix material and then rapidly transferred to the mold. The mold is quickly closed until the point where the material is contacted and then the closing rate is slowed. The material is formed to shape and flows to fill the mold cavity much the same as thermoset compression molded SMC. The material is cooled in the mold for a short period of time to allow the part to harden, and then the mold opens and the component is removed.

Thermoplastic compression molding is currently used in automobiles to form low-cost semi-structural components such as bumper backup beams, seats and load floors. Commercially available materials range from wood-filled polypropylene and short glass-filled polypropylene with relatively low physical properties to continuous random glass-reinforced materials based on polypropylene or PET which offer somewhat higher physical properties. Other materials, based on highly oriented reinforcements and such resins as polyetheretherketone (PEEK) and polyphenylene sulphide

(PPS), are in use in the aerospace industry. These materials are expensive and limited in their conformability to complex shapes.

Higher levels of strength and stiffness must be developed in low-cost materials before they can be used in structural automotive applications. Attempts have been made to improve the properties of stampable materials through the use of separate preimpregnated, unidirectional reinforcement tapes. These materials are added to the heated material charge at critical locations to locally improve the strength and stiffness. Using these materials adds to the cost of the material and increases cycle times slightly. Although effective for simple configurations, location of the oriented reinforcement and reproducibility of location are problems in complex parts. To be cost effective, these types of reinforcements will ultimately have to be part of the premanufactured sheet or robotically applied. Current research is in progress in the area of thermoplastic sheet materials with oriented reinforcement in critical areas. For application to automotive structures these materials will have to retain the geometric flexibility in molding (i.e. ability to form complex shapes with the reinforcement in the correct location) exhibited by today's commercial materials.

. The question of part integration is a major issue in the expanded use of this process. The high pressures (1000–3000 psi) required limit the size of components which can be manufactured on conventional presses. Thermoplastic compression molding is also limited in capability to incorporate complex 3-D cores required for optimum part integration. If very large integrated structures are required from an overall economic viewpoint, thermoplastic compression molding will be restricted to smaller components where geometry is relatively simple and/or part integration is limited due to physical part constraints, such as door, hood and decklid inner panels. If very large scale integration proves too expensive, then thermoplastic compression molding will exhibit increased penetration. Ongoing long range research in the area of low pressure systems and incorporation of foam cores in moldings could significantly alter this outlook in the longer term.

5.3 High-speed Resin Transfer Molding

Fabrication processes allowing precise fiber control with rapid processability would overcome many of the deficiencies outlined above. The use of some kind of preform of oriented glass fibers preplaced in the mold cavity followed by the introduction of a resin with no resultant fiber movement would satisfy the requirements for optimum performance and high reliability. The basic concepts required for this process are practiced fairly widely today in the boat-building and specialty car business—however, with few exceptions, the glass preform is hand constructed and the resin injection and cure times are of the order of tens of minutes or greater. Also dimensional consistency, which is necessary for assembling high quality products, has not been studied for this process. Major time contractions and automation of all phases of the process are necessary to generate automotive production rates. However, the basic ingredients of precise fiber control and highly integrated complex part geometries (including, for example, box sections) are an integral part of this process and offer potential large cost benefits.

There are two basic elements to the high-speed resin transfer molding (HSRTM) process which must be developed. The assemblage of the glass preform must be developed such that it can be placed in the mold as a single piece. In addition, the introduction of the resin into the mold must be rapid and the cure cycle must be equally fast to generate a mold closed/mold open cycle time of only a few minutes. A schematic of the process is shown in Fig. 18.

There are two processes currently in use which may have the potential to offer rapid resin injection and cure times. Resin transfer molding, currently in wide usage at slow rates, could be accelerated dramatically by the use of

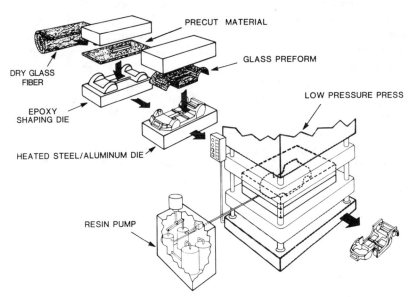

FIG. 18.   Schematic illustration of high-speed resin transfer molding (HSRTM).

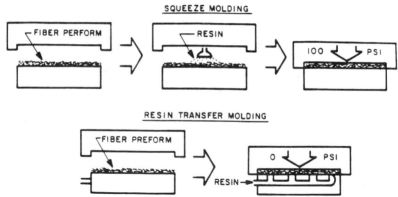

FIG. 19.   Schematic illustration of squeeze molding and RTM.

low viscosity resins, multi-port injection sites, computer-controlled feedback injection controls and sophisticated heated steel tools. There do not appear to be any significant technological barriers to these kinds of developments, but it will require a strong financial commitment to prove out such a system. A schematic of the process is shown in Fig. 19, which also illustrates a variant on the process usually termed squeeze molding.

The second technique which promises rapid injection and cure cycles is reaction injection molding (RIM). Once the dry glass preform is in the mold, the resin can be introduced by any appropriate procedure and reaction injection would be ideal provided the resultant resin has adequate mechanical properties. The inherent low viscosity of RIM resins would be ideal for rapid introduction into the mold. Full 3-D geometries including box sections are attainable by preform molding. In addition, only low pressure presses are necessary. The high degree of part integration maximizes 'effective' stiffness and minimizes assembly. In principle, major portions of vehicles could be molded in one piece—for example, Lotus car body structures consist of two major pieces (albeit molded very slowly) with one circumferential bond. If similar size complex pieces could be molded in minutes, a viable volume production technique would result.

5.4 Filament Winding

Filament winding is a composite fabrication process which for some geometric shapes can bridge the gap between slow labor-intensive aerospace fabrication techniques and the rapid automated automotive fabrication processes. The basic process utilizes a continuous fiber reinforcement to form a shape by winding over some predetermined path.

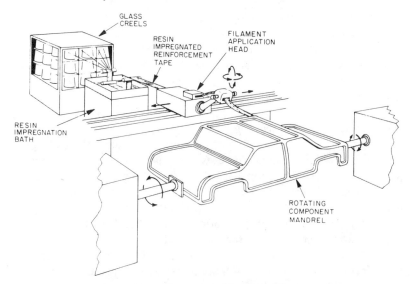

Fig. 20.   Schematic illustration of filament winding.

Figure 20 provides a process schematic. The complexity and accuracy of the winding path are highly controllable with microcomputer-controlled winding machines. Thin hollow shapes having high fiber/resin ratios are possible, making the process well suited to light-weight, high-performance components. Glass fiber, aramid, or carbon fiber can be utilized as the winding material. Filament winding can use either thermosetting or thermoplastic resin systems. The uniform fiber alignment afforded by the process provides high reliability and repeatability in filament-wound components. Some simple shapes, such as leaf springs, can be fabricated by this process and are currently in production on a limited basis.

Thermoset filament winding

The majority of filament winding done today uses thermosetting resin systems. In the thermoset filament winding process the resin and reinforcement fibers are combined (referred to as wetting out of the reinforcement) immediately prior to the winding of the fibers on to the part. The wet out and winding process require precise control of several variables. Reinforcement tension, resin properties and the fiber/resin ratio all relate directly to the final physical properties of the part. As the winding speed and part complexity increase, these variables become increasingly difficult to control. Winding of complex parts at automotive production rates will require major process developments. The likely potential of thermoset filament winding in automotive applications is in the fabrication of simple shapes such as leaf springs where the cost penalty involved due to slow production speeds can be offset by a need for high reliability and maximum use of material properties.

Thermoplastic filament winding

Filament winding of thermoplastic materials represents the leading edge of this technology, an area where its potential benefit to the automotive industry is largest. Thermoplastic filament winding uses a reinforcement preimpregnated with a thermoplastic resin rather than a reinforcement which is impregnated with a thermosetting resin at the time of winding. The preimpregnated filament or, more often, tape is wound into the appropriate shape in a manner similar to the thermosetting filament winding process, with the exception of a local heating and compaction step. The reinforcement tape is heated with hot air or laser energy as it first touches the mandrel. The heat melts the thermoplastic matrix on both the tape and locally on the substrate, allowing a slight pressure applied by a following roller to consolidate the material in the heated area.

Since the reinforcement filament is already wet out in a prepregging process, the substrate is solidified over its entirety with the exception of a small zone of molten material in the area of consolidation. The limitations with respect to filament tension, filament wet out and speed at which a wet filament can be pulled through a pay out eye are no longer a major concern with thermoplastic filament winding. Thick sections can be rapidly wound, and nongeodesic and concave sections can be formed; but the applicability to geometries as complex as body structures is not established.

Currently, problem areas exist in the translation of physical properties due to the limited amount of research done to date on the process. There are problems with the control of the heating and consolidation of the thermoplastic materials that yield less than predicted values for tensile strength and interlaminar shear strength. Although the thermoplastic materials will likely expand the structural component applications in which filament winding can economically compete when compared to thermoset filament winding, the increased speed and shape capabilities will not entirely offset the limited degree of part integration possible. The inability to integrate box sections with large flat panels in a one-piece structure having complex geometry will tend to limit the penetration of filament winding in body structures.

6 IMPACT ON PRODUCTION METHODS

6.1 Manufacturing Approach

The manufacturing approach for producing structural composite cars will be considerably different than the methods employed for building conventional vehicles with steel bodies. Currently, many domestic automotive assembly plants for steel vehicles are used for very little basic manufacturing. Instead high-volume sheet steel stamping plants, geographically located to service several assembly plants, produce body components and small assemblies, and ship them to the assembly plants. The plants assemble the sheet metal components into an auto body structure as shown in Table 4 and Figs 21(a) and (b).

The body is a complex structure and the design is influenced by many demanding factors. At the present time, it appears that two systems can be considered as possible processes to build the structural panels for composite auto bodies, namely:

—Compression-molded thin-walled panels are first bonded together to form structural panels and then the structural panels are assembled to form auto bodies (see Fig. 22).

—HSRTM in which preforms of fiber reinforcement are combined with foam cores, placed in a mold, and resin injected to form large 3-D structural panels. These panels are then assembled into auto bodies (see Fig. 23).

TABLE 4

Typical body construction system (steel panels)

Build front structure—assemble aprons, radiator support, torque boxes, etc., to dash panel
Build front floor pan assembly
Assemble front and rear floor pan into underbody assembly
Complete spot welding of underbody assembly
Transfer underbody to skid
Build left-hand and right-hand bodyside assemblies
Move underbody, bodyside assemblies, cowl top, windshield header, rear header, etc., into body buck line and tack weld parts
Complete spot welding of body
Braze and fusion weld sheet metal where required
Assemble, tack weld and respot roof panel to body
Assemble front fenders to body
Assemble closure panels (doors, hood, decklid) to body
Metal finish exterior surface where required

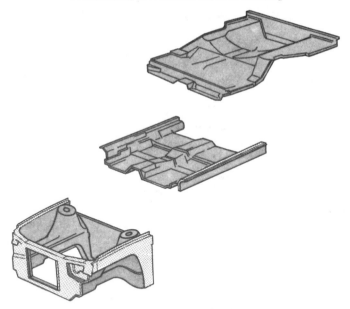

FIG. 21(a).   Typical assembly of steel underbody.

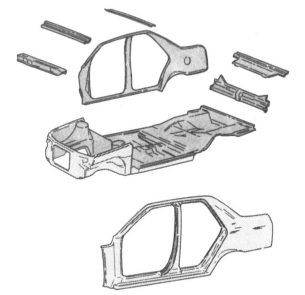

FIG. 21(b).   Typical assembly of underbody, bodyside assemblies, etc., into the completed steel body shell.

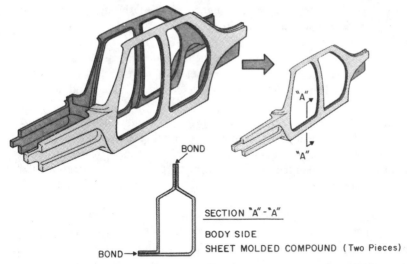

FIG. 22.   Typical assembly of composite bodyside panel by adhesive bonding of inner and
outer SMC molded panels.

It is important to note that the filament winding process is not viewed as
a viable construction method largely due to the restricted complexity which
can be generated by this technique. Therefore it cannot now be considered
as a competitive process for high-volume body structure fabrication.
The following is a brief outline of the two processes.

Compression molded structural bodies
Compression molded structural panels might be produced as shown in
Table 5 and Fig. 22. Panel assemblies (side panels, floor pans, roof panels,

TABLE 5
Compression molded structural panels

Fabricate and prepare SMC charge
Load charge into die and mold panel
Remove components from molding machine (inner and outer panels), trim and
   drill parts as required
Attach reinforcements, latches, etc., to inner and outer panels with adhesive/
   rivets
Apply mixed, two-component adhesive to outer
Assemble inner to outer, clamp and cure
Remove excess adhesive
Remove body panel from fixture and transport to body construction line

---

etc.) are produced from inner and outer components. First, SMC sheet must be manufactured and blanks of proper size and weight arranged in a large predetermined pattern on the die surface. The panels are then molded in high tonnage presses. After molding, the parts are processed in a number of secondary operations, bonded together and transported to the body assembly line. The process sequence is described in Table 5 and illustrated in Fig. 22.

Body construction commences with the floor panel being placed in an assembly fixture. Side panel assemblies and mating components are bonded in place. Attachment points for the exterior body panels are drilled and the body structure is painted prior to transporting to the trim operations. (The body construction sequence is described later; see Table 9 and Fig. 24.)

High-speed resin transfer molded structural bodies

The HSRTM processing consists of making a dry glass fiber preform, combining the preform with foam cores and finally resin injection in a mold to infiltrate the preform. First, dry glass preform reinforcements must be fabricated. The preform may be composed of primarily randomly oriented glass fibers with added directional glass fibers or woven glass cloth for local reinforcement of high stress areas. To be economically viable, preform fabrication must be accomplished by a highly automated technique. The process sequence is described in Table 6.

Secondly, foam core reinforcements must be fabricated to obtain 3-D inserts, such as those used in rockers and pillars. Local metal reinforcements and fasteners can be prepositioned in the mold at this point during foam fabrication. The chemicals are subsequently introduced into the mold and the resulting reaction yields an accurately shaped foam part. The process sequence is described in Table 7.

In the third stage, the body structural panel is molded. A gel coat may be sprayed into both lower and upper mold cavities to obtain a smoother surface on the finished part. Preforms, foam cores and any necessary

TABLE 6
Preform fabrication sequence

Apply random glass fibers over mandrel
Apply directional fibers (or woven cloth) for local reinforcement
Stabilize preform
Remove preform and transfer to trim station
Trim excess fibers
Transfer to HSRTM body panel molding line

TABLE 7
Foam core fabrication sequence

Clean mold and apply part release
Install reinforcements and basic fasteners
Close mold
Mix resins and inject into mold
Chemicals react to form part
Open mold
Unload part and place on trim fixture
Trim and drill excess material
Transport to HSRTM body panel line

additional local reinforcements and fasteners are placed in the respective lower and upper mold cavities. In high production these operations must be carried out robotically. The mold is closed and a vacuum may be applied. The resin is injected, infiltrates the preform and cures to form the body structural panel. The molding is then removed from the die and trimmed. This process sequence is described in Table 8 and illustrated in Fig. 23.

The final phase of body construction is the assembly of the individual panels. The underbody panel is placed on the body construction line.

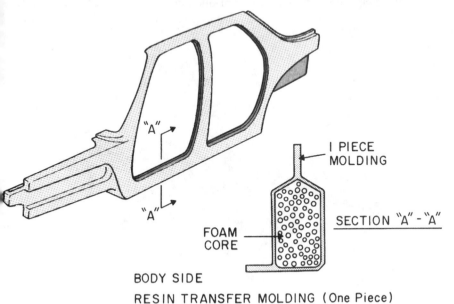

FIG. 23.   Typical one-piece composite bodyside panel with foam core molded by HSRTM.

TABLE 8
HSRTM molded structural panels

Clean mold and apply part release
Spray gel coat into mold (optional)
Insert lower preforms and local woven fiber reinforcements
Insert specialized reinforcements and fasteners
Insert foam cores
Insert upper preforms and local woven fiber reinforcements
Close mold
Apply vacuum to mold (optional)
Inject resin and allow chemicals to react
Open mold
Remove assembly and place in trim fixture
Trim and drill body panel
Transport to body assembly line

Adhesive is applied to the side panels by robots in the appropriate joint locations, the side panels are mated to the underbody, clamped in place and fasteners added. Similarly, the remainder of the body structure is located and bonded to form a complete auto body. After curing, the body is washed and dried, prior to transfer to trim operations. This process sequence is described in Table 9 and illustrated in Fig. 24. (Note that the body assembly

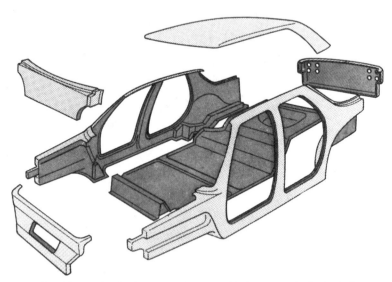

FIG. 24.   Typical body construction assembly for composite body shell.

TABLE 9
Body construction assembly sequence

Place underbody in body build line
Apply adhesive to bond lines of side panels and cowl
Mate side panels to underbody, clamp and insert fasteners
Apply adhesive to cowl top assembly, lower back panel, and
  mate
Apply adhesive to roof panel and mate to body side panels
Allow adhesive to cure (using heated fixtures)
Transfer body to final trim operations

NB: After final trim, the painted exterior body panels are
          assembled to the auto body

sequence (Table 9) is essentially common to both panel molding procedures, namely compression molding and HSRTM. This is only one illustration of the assembly of a number of moldings to form the body—the number of moldings could vary from 2 to 10 depending on the specific design and manufacturing details.)

Figure 24 is a schematic of a multi-piece composite body. For comparison, a two-piece HSRTM body construction is illustrated in Fig. 25 to indicate the various levels of part integration which might be achieved using composites. In both these scenarios, the body shell would consist of

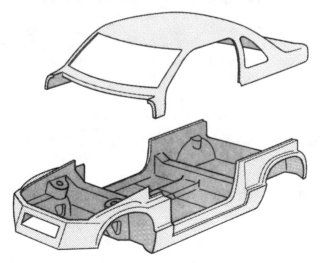

FIG. 25. Typical body construction assembly using two major composite moldings.

composite structure with no metal parts except for molded-in steel reinforcements.

6.2 Assembly Operation Impact

A review of the manufacturing techniques previously described indicates that in both scenarios there would be a considerable change in the assembly operations for structural composite vehicles versus conventional steel cars.

The number of component parts in a typical body structure (body-in-white, less closure panels, i.e. hoods, doors and decklids) varies with vehicle design and material, as shown below in Table 10.

The dramatic reduction in the number of parts to be assembled in a composite vehicle would result in a corresponding reduction in the number of subassembly operations and subassembly equipment, as well as a reduction in the required floorspace. For example, robots used to produce composite assemblies can lay down an adhesive bead considerably faster than robots can spot weld a comparable distance on mating components. Therefore it is anticipated that there would be a considerable reduction in the number of robots and complex welding fixtures required.

6.3 Labor Impact

The design and engineering skills which would be required to apply structural composites would be somewhat different than the skills utilized for current metal applications. A broader materials training curriculum would be required because the chemical, physical and mechanical properties of composites differ significantly from metals. These programs are currently being developed at some universities but significant expansion would be necessary to ensure this trend on a broader scale.

The overall labor content for producing a composite body will likewise

TABLE 10
Effect of composites on body complexity

Vehicle design and material	Typical number of major parts in body structure	Typical number of assembly robots
A. Conventional welded steel body structure	250–350	300
B. Molded SMC body structure	10–30	100
C. HSRTM composite/body structure	2–10	50
Estimated part reduction (A less B)	240–320	200–250
Estimated part reduction (A less C)	248–340	250–280

be reduced as numerous operations would be eliminated. However, it is important to note that body assembly is not a labor-intensive segment of total assembly. Other assembly operations which are more labor intensive (e.g. trim) would not be significantly affected. Thus the overall labor decrease due to composites might be relatively small. The level of skills involved in composite assembly line operations (e.g. bonding operations) is not expected to be any more demanding than the skill level currently required for spot welding conventional body assemblies. In either case, good product design practice dictates that product assembly skill requirements be matched with the skill levels of the available workers to obtain a consistently high quality level.

7 EFFECT ON SUPPORT INDUSTRIES

7.1 Material Suppliers, Molders and Fabricators

Composite vehicle body structures of the future, whether built with compression molded or HSRTM parts, are expected to be designed with components integrated into modular subassemblies. Therefore these units will be of considerable size and are not conducive to long-range shipping. Manufacturing on site in a dedicated plant for molding body construction and assembly operations, just prior to trim and final assembly of the vehicle, becomes a very practical consideration.

Composite automobile production, in relatively high volumes, will require additional qualified supplier capacity. The extent of these requirements will be dependent upon the economic attractiveness and incentives for developing in-house capacity by the automotive manufacturers. Fortunately, these demands would be evolutionary, since the automobile industry would undoubtedly commence with low volume (10 000–60 000) composite body production units. When (and if) higher volume production of composite bodies is planned, additional supplier capacity can be put in place as a result of supplier/manufacturer cooperation throughout the normal lead time (4–6 years) for the planning, design and release of a new vehicle to manufacturing.

Although compression molding is the most mature of the evolving composite production techniques, a major conversion to SMC composites for body structures would require a substantial increase in resin and reinforcement output, mold building capacity, molding machine construction, component molding and subassembly facilities, adhesives and quality control tools, etc.

If the high-speed resin transfer molding processing concept is utilized, resin and reinforcement suppliers would need to substantially expand their output (as in the case of compression molding) and perhaps develop new products to meet the unique demands of the process. Tooling is similar in construction to that used in injection molding and, therefore, manufacturers of this type of tooling would likely expand to fill the need. If inexpensive electroformed molds which consist of 0·25-in (6·35-mm) electroplated nickel facing on a filled epoxy backing (used today for low-pressure or vacuum-assisted molding) become feasible, this phase of the industry would have to be developed and expanded. Since molding pressures for this process are low, high tonnage hydraulic presses would not be needed. However, companies specializing in automation and resin handling equipment would play an increased role.

If composite structures were suddenly implemented, there would be an expected shortfall of qualified molders and fabricators regardless of the process chosen. It is more likely, however, that implementation would be evolutionary and the supply base will be addressed during the composite vehicle planning and design stage. To enlarge the supply base, the auto industry is currently working with, and encouraging, qualified vendors to expand and/or diversify, as required, to support product plans. Additionally, with growth in composite demand, new suppliers would be expected to become qualified. Current suppliers may also form joint ventures and/or make acquisitions to expand capabilities during the phase-in period of composite structures. Along with the need to expand materials supply and facilities is the significant need to develop and retrain qualified personnel to provide support for both supplier and automotive industry operations.

The current molding capacity for SMC, which is devoted to the automotive industry, is of the order of 500–750 × 10^6 lb annually. Figure 26 projects the additional volume of structural composites which would be needed as a function of production volume of high composite content cars. The data are based on a typical vehicle weight of 2700 lb containing 1600 lb steel, in which 1000 lb of steel (out of the 1600 lb) are replaced by 650 lb of structural composites. Note the relatively massive incremental amount of structural composites for even 10^6 vehicles per year—approximately 300 000 tons (650 × 10^6 lb), that is roughly a doubling of the current SMC capacity. Each additional 10^6 vehicles would require the same increase, obviously creating (ultimately) an enormous new industry.

The implementation of composites in automobiles would clearly have an impact on the steel industry. As with the plastics industry, this would be volume dependent. In the initial stages, with volumes in the range of

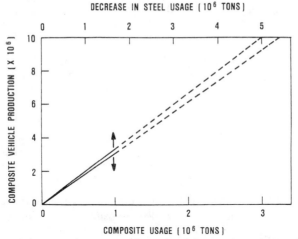

FIG. 26. Increase in composites usage and associated decrease in steel usage for various production volumes of composite-intensive vehicles. Assumes 1000 lb of steel is replaced by 650 lb of composites in a typical vehicle weighing 2700 lb which contains 1600 lb steel.

10 000–60 000 vehicles per annum, there would be a minimal effect. A small loss in steel tonnage, some excess press capacity and some additional stamping die building capacity would result.

If a vehicle replacement is designed for composites instead of steel and, for example, the volume is of the order of 500 000 units, the impact would become substantial. The loss of steel tonnage from the steel mills would cause additional problems for this already beleaguered segment of industry. Both captive and/or supplier stamping plants would have idle capacity. Stamping die builders would lose orders, unless they also build molds for plastics. However, these potential problems for the steel industry would be evolutionary and would take several years to occur after the successful introduction of vehicles using this technology.

Reference to Fig. 26 will put some perspective on the potential impact. The decrease in steel usage as a function of increasing production volume of high composite content cars would only become significant at intermediate volume levels. The data reproduced in Table 11 show that at volumes up to 500 000 composite vehicles per year automotive use of steel would drop approximately 250 000 tons per year (or 6·25%). Major steel production decreases would only result from a major change in composite vehicle volume (e.g. 2×10^6 or more).

It is also recognized that the steel industry and its suppliers are aggressively seeking methods to reduce costs and provide a wider range of

TABLE 11

*Automotive steel usage
(based on annual volume of 10^7 vehicles)*

Production volume of composite-intensive vehicles	Steel usage (10^6 tons)
0	8
50 000	7·975
500 000	7·75
5×10^6	5·5
10^7	3·0

steels. This healthy economic competition between steel and composites will have an effect on the timing of composite car introductions and the volumes to be produced. World competition in the steel industry is leading to the availability of a wider range of high strength steels with improved quality resulting in an increase in productivity for the end user. The balance between improved economic factors for steel usage and the rate of improvement of composites processing will dictate the usage, rate of growth and timing of the introduction of composite vehicles.

7.2 Repair Service Requirements

Another support industry of importance is the composite repair service required to repair vehicle damage caused by in-service accidents, etc. During the design process, automobile engineers will design components and body parts to simplify field repairs. For repair of major damage, large replacement components or modules will have to be supplied. At the present time, the comparative expense of repairing composite structures relative to steel is unclear. Replacement parts or sections will tend to be more expensive. However, low energy impacts are expected to result in less damage to the vehicle. Thus the overall repair costs across a broad spectrum of vehicles and damage levels are not anticipated to be any higher than current levels.

Exterior SMC body panel repair procedures are already in existence at automotive dealerships and independent repair shops specializing in 'fiberglass' repairs. With additional composite usage, it is expected that the number of these repair service facilities will increase. To repair the composite structure, dealers and independent repair centers will need new repair procedures and repair materials. The development of the appropriate repair procedures will be contained well within the time frame of composite vehicle introduction. There will be new business opportunities to establish additional independent shops.

As with any new vehicle design, composite repair procedures must be fully developed and standardized, along with additional training of repair personnel. This training will require skill levels equivalent to those required for steel repair.

7.3 Composite Vehicle/Component Recycling

The recycling industry is another support industry which will undergo significant change in the longer term if composite vehicles become a significant portion of the volume of scrapped vehicles. Current steel vehicle recycling techniques (shredders and magnetic separators) will not be applicable, and cost-effective recycling methods will need to be developed exclusively for composites. Low volumes of scrap composite vehicles will have minimal effect, but the problems will increase as the volume reaches a significant level. This level is estimated to be in the range of 20% or more of the total vehicles to be salvaged.

Plastics must be segregated into types prior to recycling. Currently, only clean, unreinforced thermoplastic materials can readily be reclaimed, but the techniques are somewhat inadequate and tend to be very expensive. Fortunately, the calorific value of many thermoplastics approaches that of fuel oil. Some waste incinerating plants now operate with various types of plastics as fuel. A developing use for plastic waste involves pulverizing plastics and utilizing their calorific value as a partial substitute for fuel oil in cement kilns. Some plastics in automobiles can be recycled by melt recovery, pyrolysis and hydrolysis but economic methods are not yet in place.

Thermoset plastic components, such as SMC panels and other composite body components, currently cannot easily be reclaimed because of their relatively infusible state and low resin content. Grinding, followed by reuse in less demanding applications such as roadfill or building materials, is possible, but is not currently economical. Consequently, these kinds of parts are used as landfill or are incinerated. Again, at low volumes these recycling procedures are acceptable, but would not be viable at high scrap rates. Without the advent of some unforeseen recycling procedures, it appears likely that incineration will have to be the major process for the future.

The viability of incineration will either have to be improved by more efficient techniques for more favorable economics or some penalty will likely have to be absorbed by the product.

8 POTENTIAL EFFECTS OF GOVERNMENTAL REGULATIONS

Any major potential changes in industrial practice, particularly in large-scale industries such as the automotive industry, must take into consideration not only current regulations but also future regulations which (1) may already be under consideration, (2) may be on the horizon irrespective of contemplated industrial changes, and (3) may be initiated because of the potential changes in industrial practice. An example of (1) is the increased safety (crash resistance) requirements already under active consideration. Potential increases in CAFE standards for the 1990s is a typical example of (2). The third category is the one which requires a crystal ball and obviously is the most difficult. However, all three types of regulation must be put into the scenario of widespread composites usage because of the implications in terms of limitations and cost resulting from such legislation. While the following discussion of such potential regulations may not be complete, it will serve to illustrate the importance of such considerations in introducing major materials changes to the automotive industry. Implicit throughout this discussion is the underlying principle that all current government and company requirements would have to be met or exceeded by the composite structures.

8.1 Health Safeguards

The introduction of fiber reinforced plastics in significant volumes into the automotive workplace may raise health and environment concerns. Any fibers in very fine form have the potential to create lung and skin problems, and while glass fibers may be among the more inert types of fiber, there must still be adequate precautions taken in handling these fibers. Glass fibers are, of course, widely used currently in various industries (molding industries, boat building, home construction, etc.) and extension to the automotive industry would probably not require development of safeguards other than those utilized within those industries. However, the widespread nature of the automotive business would undoubtedly raise awareness of potential health risks and may well precipitate more stringent requirements for the workplace. While such additional precautions may not be a technological problem, the extent of the regulations could have a significant effect on the economic viability of material usage.

Similarly, the same implications may be drawn regarding the resin matrix materials of these plastic-based composites. While it is not clear which specific resin materials will be dominant for composites usage, there is widespread concern regarding all chemicals in the workplace and in the environment. The are already strong regulations concerning chemical usage and handling, but again the magnitude of the automotive industry is likely to propel such requirements to the forefront of interest and may result in additional legislation. This could result in limitation of the types of resins used and implementation of additional safeguards. The impact is less likely to prevent implementation of composites technology and more likely to affect the economic viability and timeliness of the introduction of this technology.

8.2 Recyclability

The current recycling of scrap automobiles is a major industry. Sophisticated techniques have been developed for separation of the various materials, and cost-effective recycling procedures are an integral part of the total automotive scene. Obviously, since steel constitutes 60% by weight of a current automobile, the recycling of steel is the major portion of the recycling industry. Recycling of composites is a radically different proposition and usage of these materials will necessitate development of new industrial recycling processes *if* large volumes are generated. (If structural composites are only applied in low-volume specialty vehicles, current recycling techniques of landfill and incineration will probably be adequate.)

The potential for large numbers of composite-intensive vehicles to be scrapped will undoubtedly raise the issue of disposal to the national spotlight. One concern would be that scrapped cars would become environmental eyesores because of the lack of economic incentives to recycle. Consequently, there is a distinct probability of the government forestalling a future recycling problem by insisting the issue be resolved before allowing widespread usage of these materials. The result of such legislation would be the commitment of finances and resources at an early stage of development with no method of payback except as part of the overall composite development cost.

8.3 Crash Regulations

There are government regulations in existence, and others proposed, that set or would set impact survival criteria in frontal impacts, side crash tests and vehicle-to-vehicle impacts. Various scenarios have been proposed for making these standards more stringent (for example, raising frontal impact criteria from 30 to 35 mph and possibly higher). Basic experience to date in crash energy management has primarily been with steel vehicles and, consequently, the specific wording of the regulations is based on the characteristics of these vehicles. Vehicles consisting largely of composite materials absorb energy by significantly different mechanisms, and the details of impact would be very different from those of steel vehicles even though the objective of occupant protection would be the same. Thus new rulemaking in this area could well contain provisions which would preclude the use of composites because of the lack of information. For example, if a requirement were promulgated that stated no fracture of a major body structure shall occur during a certain impact, composites would be excluded because, unlike steels, *internal* fracture of the composites is a critical part of energy absorption. Thus detailed wording of crash regulations based on steel experience could inadvertently jeopardize the potential use of composite structural materials.

8.4 Fuel Economy Standards

Just as some government regulations might produce deterrents to composites usage, others might promote development and usage. If CAFE requirements were drastically increased, there are a limited number of options for increasing fuel efficiency, namely downsizing, increased powertrain efficiency and weight reduction. In the weight reduction arena, aluminum alloys and structural composites represent perhaps the major options for significant weight reduction. Thus legislation for a marked increase in fuel economy might tend to promote the development and use of structural composites, provided functional requirements, manufacturing feasibility and overall economic factors are feasible.

8.5 Anti-trust Legislation

Technologies can be divided into those which are competitive and those which are pre-competitive, either because they are so mature that everyone has the required knowledge or so premature, uncertain and encompassing that no single company can justify the massive investments required to accelerate the pace of the technology. The risks are too large even though the opportunities may be significant. This is especially true when the emerging technology is such that it could not be well protected once developed and applied. Advances in structural composites fall into this pre-competitive category.

In order to spread the risk and accelerate technical developments, Japanese automotive companies, with encouragement, support and coordination by the Japanese Government, work jointly on a variety of emerging technologies while remaining strongly competitive with each other. European automobile manufacturers also do joint generic research, again with governmental blessing and support. Passage of the National Cooperative Research Act in October 1984 might encourage a variety of US companies to pursue joint pre-competitive research. Governmental encouragement for joint research by US manufacturers in structural composites might enhance developments of these materials, including generic developments which could spill over into other industries.

9 TECHNOLOGY DEVELOPMENT AREAS

Although years of development efforts have advanced structural composites to become a significant material for use in the aviation and aerospace industries, the cost-effective use of these materials in the automobile industry requires considerable additional developmental work in the following areas.

9.1 Improved Compression Molding

Improvements in SMC technology are required in the areas of reduced cycle times, reproducibility of physical properties and material handleability. A major step in SMC technology improvement would be a significant cycle time reduction. Current objectives are to cut the conventional time from 2–3 min to 1 min or less. Development of materials requiring less flow to achieve optimum physical properties would allow more reproducible moldings to be made. Other potential technology improvements for SMC include a reduction in the 'aging' time for material prior to molding, an internal mold release in the material, an improved cutting operation to prepare a loading charge for the molding machine, automated loading and unloading of these machines, and improved dimensional control of final parts.

9.2 Improved HSRTM

There are two critical segments of this process which require major developments to achieve viability. Cycle time reduction is one important area and new, faster curing resins currently under development should make significant contributions in this area. The development of automated preform technology (foam core development and subassembly) is crucial to achieving cost effectiveness. This is perhaps the area requiring major innovation and invention, and has yet to be perceived by the existing fiber and fiber manipulation industry to be a major area for development.

9.3 Improved Fiber Technology

There are three major areas in fiber technology which must be optimized to promote usage in high production industries, namely improved translation of physical properties, improved fiber handling and placement techniques, and improved high volume production techniques providing fibers at lower cost.

Superior physical properties would result from improved sizings (fiber coatings) to reduce fiber damage during processing and provide improved mechanical properties in the finished components. Fiber-handling equipment allowing high-speed, precise fiber placement with minimal effect on fiber properties is a vital requirement for the development of stable, 3-D preforms. Cost minimization is obviously a critical factor in utilizing glass fibers but also should be considered from the viewpoint of generating other fibers (e.g. carbon fibers) at costs amenable to mass-production usage.

9.4 Joining

Two key areas dominate the category of joining technology. Adhesives and mechanical fasteners must be tailored for composite construction, to generate the necessary combination of production rate and mechanical reliability. The second major requirement involves the generation of design criteria and design methodologies for adhesive joints. Both these areas have not been systemically developed for a mass-production industry and far greater attention must be paid to joining materials and to methods to alleviate any problems in this element of the overall composites technology.

9.5 General Technology Requirements

Design methodologies for use with structural composites to cover all aspects of vehicle requirements must be generated to a degree comparable to current steel knowledge. The ability to tailor composites for specific requirements must be integrated into such design guides, and makes the task more complex than the equivalent guidelines for isotropic materials (metals). Manufacturing knowledge and experience, which provide constraints for the design process, must be fully documented to optimize product quality, reliability and cost effectiveness. The degree of component integration must obviously be a key factor in determining manufacturing rates, and this interdependence of design and manufacturing will only evolve over a protracted time period. This build-up of experience will be the key factor in resolving overall economic factors for composite vehicle production.

9.6 Standards

The complexity of composite materials relative to metals will require the development of standard testing procedures and material specifications. The rapid proliferation of materials in the composite arena will not permit final establishment of these generic standards until composite technology matures. It is likely that specific corporate standards will be used in the interim prior to professional society actions.

10 SUMMARY

The extension of composites use to automotive structures will require an expanded knowledge of the design parameters for these materials together with major innovations in fabrication technologies. There is abundant laboratory evidence and some limited vehicle evidence which strongly indicates that glass fiber reinforced composites are capable of meeting the functional requirements of the most highly loaded automotive structures. There are, however, sufficient uncertainties (e.g. long-term environmental effects, complex crash behavior) that will force applications to be developed slowly until adequate confidence is generated. Nevertheless, it seems inevitable that the functional questions will be answered; it only remains to be seen how soon. Perhaps the more imperative requirement is the cost-effective fabrication advancement that appears to be a prerequisite to widespread use of composites in automotive structures. High-volume, *less stringent performance* componentry can be manufactured by variations of compression molding techniques. It is the high-volume, *high-performance* manufacturing technology which needs development and the HSRTM process appears to be a 'sleeping fabrication giant' with the potential of developing into just such a process. All the elements for resolving the rapid, high-performance issue are scattered around the somewhat fragmented composites industry. It will require the appropriate combination of fiber manufacturers, resin technologists, fabrication specialists and industrial end-users to generate the necessary developments.

The advent of composite-intensive vehicles will be evolutionary. The most likely scenario would be pilot programs of large composite substructures as initial developments to realistically evaluate these materials in field experience. This would be followed by low production (20 000–60 000) volumes of a composite-intensive vehicle(s) which would achieve the extensive manufacturing experience that is vital to the determination of realistic fabrication guidelines and true economics. The data derived from such introduction would determine the potential for high-volume production. Currently, the industry infrastructure is not in place for composite-intensive vehicles. In addition, the supply base could presently respond to only low-volume production of composite vehicles. Neither of these situations is a major restriction in the sense that the time element would allow the appropriate changes to occur over a protracted period. Rather the initial decisions to make the necessary changes and (substantial) financial commitments will have to be based on *significant* evidence that composite vehicles are viable economically and offer customer and product benefits.

Irrespective of the scenario for the eventual introduction of composite vehicles, there are probably some general conclusions that can be drawn relative to the impact of these materials. Composite applications are unlikely to have a large effect on the size of the labor force because the major changes are not in labor-intensive areas of vehicle build and assembly. Likewise, the necessary skill levels for both the fabrication of the composite parts and assembly of the body should not be significantly changed. Engineering know-how would be very different, but similar skill levels today would be needed. Perhaps the largest effect would be on the supply industries, which would need to implement production of structural composites. This would involve both a change in technology for many current suppliers, together with the development of a new supply base. The steel industry would experience a corresponding decrease in output, but the decrease would only be of major proportion if composite vehicle volumes became a significant part of the total vehicle output. The repair and recycling industries would similarly undergo a major change to accommodate the radical change in the materials of the vehicles.

REFERENCES

1. BEARDMORE, P., HARWOOD, J. J. and HORTON, E. J., Int. Conf. Composite Materials, Paris, August 1980.
2. DHARAN, C. K. H., *J. Mats Sci.*, **10** (1975 1665–70.
3. SMITH, T. R. and OWEN, M. J., *Modern Plastics* (April 1969) 124–9.
4. POTEM, A. and HASHIN, Z., *AIAA J.*, **14** (1976) 868–72.
5. THORNTON, P. H., *J. Comp. Mat.*, **13** (1979) 274.
6. THORNTON, P. H. and EDWARDS, P. J., *J. Comp. Mat.*, **16** (1982) 521–45.

7. THORNTON, P. H., HARWOOD, J. J. and BEARDMORE, P., *Int. J. Comp. Struct.*, (1985).
8. JOHNSON, C. F. and CHAVKA, N. G., An Escort rear floor pan. Paper presented at 40th Society of the Plastics Industry Annual Tech. Conf., Atlanta, Georgia, January 1985.

1

FRP Reinforcement of Externally Pressurised Torispherical Shells

A. MUC

Institute of Mechanics and Machine Design,
Technical University of Cracow,
31–155 Cracow, ul. Warszawska 24, Poland

G. D. GALLETLY and D. N. MORETON

Department of Mechanical Engineering, University of Liverpool,
PO Box 147, Liverpool L69 3BX, UK

ABSTRACT

The aim of the paper is to analyse the effectiveness of FRP reinforcement of steel torispherical shells. The failure modes of unreinforced and reinforced torispheres are studied, taking into account the various geometrical and mechanical properties of the shells and reinforced layers. Both local and full reinforcement are considered as well as perfect and imperfect shells. The influence of fibre orientation on the limit pressure is also investigated in order to find the optimal stacking sequence. The theoretical analysis is verified by various experimental tests. However, the tests were limited to just one geometry of steel torisphere, i.e. $t_s/R = 0.01$, $R/D = 0.6$ and $r/D = 0.24$.

1 INTRODUCTION

Steel shell structures in service may incur local dents due to various causes, e.g. tugs may collide with legs of offshore platforms, domes of flotation tanks may suffer local impact damage, or may be manufactured outside the specified tolerances. Thus such steel structures will not be able to withstand the loads which they were supposed to. The simplest method to counter the weakening effects of foreseen damage zones is to design local or global reinforcing layers or a system of meridional and circumferential stiffeners. Calladine[1] presented the effects of a reinforcement of spherical shells with the use of layers made of the identical material as dome itself. Tillman[2] and Amiro *et al.*[3] examined theoretically and experimentally buckling problems for spherical shells reinforced by the set of meridional and circumferential stiffeners.

Another approach is to apply as reinforcing layers fibre reinforced plastics (FRP), which offer a significant amount of material savings (over that for isotropic materials) and additionally a great simplicity in tailoring and bonding them to steel shells. In this area the considerations were mainly directed to the building of plastic interaction surfaces in the space of membrane stresses and moments based on the Tresca–Guest hypothesis (see Capurso and Gandolfi,[4] Nemirovsky[5] and Vasiliev[6]). The investigations dealt with steel shells of revolution reinforced by glass fibre. The numerical analysis of failure modes for steel torispheres reinforced with the use of FRP layers carried out by Muc[7] showed evidently that the maximal carrying capacity was the dominant failure mode.

The aim of this paper is to study the influence of different properties of composite reinforcement and of its length in the meridional direction on the value of limit loads of the hybrid (steel/FRP) shells and then on the verification of the obtained numerical results with the experiment. It should be emphasised that the main attention will be focused on the analysis of clamped perfect and imperfect torispherical shells with the following geometry: $R/D = 0.6$, $r/D = 0.24$, $R/t = 100$ (i.e. identical to that used in the experiment) and subjected to external pressure. A broader review of the problems discussed here can be found in Refs 7 and 8.

2 FAILURE MODES OF HYBRID (STEEL/FRP) TORISPHERICAL SHELLS

In order to distinguish various failure modes which occur for hybrid structures, at first let us discuss briefly the case of a simply supported, reinforced in the middle, beam with rectangular cross-section loaded uniformly by a pressure p (Ref. 7) (Fig. 1(a)). It is assumed that the steel part of the beam may yield whereas the composite one is made of a perfectly elastic material. For such a structure the limit carrying capacity will be reached at the junction of thicker and thinner segments, i.e. for $x = l_c$. Comparing the value of bending moment at the point $x = l_c$ with the limit

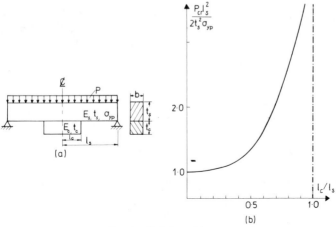

FIG. 1.   Reinforced beam.

carrying capacity in pure bending for steel beams ($bt_s^2\sigma_{yp}$), one can obtain the limit value of pressure:

$$p_{cr} = 2t_s^2\sigma_{yp}/\{l_s^2[1-(l_c/l_s)^2]\} \qquad (1)$$

The distribution of p_{cr} versus l_c/l_s (Fig. 1(b)) shows that at the beginning p_{cr} is insensitive to the changes of l_c/l_s, then the limit pressure increases reaching infinity for $l_c = l_s$. Such a character of limit loads versus the parameter describing geometry of a reinforcement is typical for ribbed plates and shells in the plastic range (see, for example, Save[9]). It is obvious that the critical pressure p_{cr} cannot achieve the infinite value for $l_c = l_s$ due to other failure mechanisms which can precede limit states. It concerns mainly typical failure modes for composite materials such as debonding of layers, their delamination or cracking.

Let us consider now the effects of local or global composite reinforcement for perfect and imperfect torispheres on the types of failure modes. For both perfect and imperfect steel torispherical shells three dangerous regions have been reported (see, for example, Ref. 10):

(1) the apex region (axisymmetric collapse);
(2) the knuckle region (axisymmetric collapse or bifurcation);
(3) the clamped edge (axisymmetric collapse or bifurcation).

Their positions depend on the shell geometry and the length of imperfection (see Fig. 2). The imperfection considered here is of the increased-radius polar type (see Fig. 3(a)).

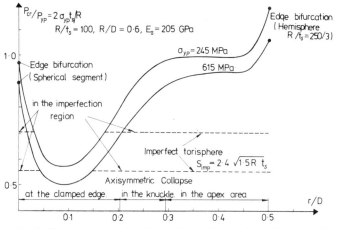

FIG. 2. Collapse pressures of perfect and imperfect steel torispherical shells.

As was discussed previously for hybrid (steel/FRP) structures, one can expect the appearance of two new failure modes:

(4) the maximal carrying capacity (plastic hinge) at the junction of thinner (steel) and thicker (steel/FRP) shell segment (axisymmetric collapse);

(5) debonding of steel and composite layers at the junction of two materials that is the most dangerous region in the hybrid structure.

Since the numerical analysis was carried out with the help of the BOSOR5 program,[11] enabling calculation of buckling loads only (under the assumptions of yielding of steel layer and of elasticity of FRP layers), the last failure pressure could not be found directly and thus the real value of limit loads corresponding to the debonding could be overestimated if the debonding preceded buckling. In order to determine critical pressures for the fifth type of failure mode, we propose to apply the Hoffmann criterion:[12]

$$\left(\frac{1}{X_t} - \frac{1}{X_c}\right)\sigma_1 + \left(\frac{1}{Y_t} - \frac{1}{Y_c}\right)\sigma_2 + \frac{\sigma_1^2 - \sigma_1\sigma_2}{X_t X_c} + \frac{\sigma_2^2}{Y_t Y_c} = 1 \qquad (2)$$

where σ_1, σ_2 are stresses in the meridional and circumferential directions, respectively, and X_t, Y_t (X_c, Y_c) denote allowable tensile and compressive stresses, respectively, in the meridional and circumferential directions. The stresses σ_1, σ_2 are evaluated along the shell meridian at the

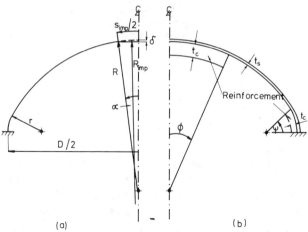

FIG. 3.   (a) Geometry of steel torispherical shell with an increased-radius polar imperfection.
(b) Geometry of reinforced torispheres.

junction of the steel and composite layers for the values of X_t, X_c, Y_t, Y_c
corresponding to the epoxy resin. If the left-hand side of eqn (2) is less than
one for the value of buckling pressure calculated with the use of the
BOSOR5 program, it means that the first four types of failure modes are
possible. In the opposite case debonding precedes buckling and the
debonding pressure is determined from the equality (2).

3   FAILURE PRESSURES AND MODES FOR LOCALLY OR FULLY REINFORCED STEEL TORISPHERES

There is a variety of geometrical and mechanical parameters (see Fig. 2)
which affect the values of buckling pressures and the types ((1)–(3)) of failure
modes, even for pure steel perfect or imperfect torispheres. Reinforcement
of torispherical shells with the use of FRP makes the discussed problem
more difficult due to the existence of some new parameters (see Fig. 3(b))
such as semi-angles of reinforcement (ϕ, ψ), total thickness of composite
layers t_c and their Young's modulus E_c. The effects of fibre orientations are
completely neglected in this chapter due to the lack of a numerical program
which takes into account coupling terms in the stiffness matrix for elastic–
plastic shells. The majority of the numerical examples studied below deal
with the materials for which mechanical properties are summarised in
Table 1.

TABLE 1
Mechanical properties of the materials considered in the paper

Material	E_1 (GPa)	E_2 (GPa)	G_{12} (GPa)	v_{12}	v_{21}	σ_{yp} (MPa)
Mild steel	207	207	81·5	0·27	0·27	245·6
Woven roving HS CFRP	65	65	3·85	0·045	0·045	—
Unidirectional HM CFRP	182	12	8·51	0·021	0·32	—

3.1 Effects of ϕ and ψ

For steel torispherical shells which, when unreinforced, fail in the knuckle region the reinforcement starting from the shell pole does not increase the limit pressure p_{cr} in comparison with the unreinforced torispheres. On the contrary, it even reduces p_{cr} insignificantly (see Fig. 4) due to the concentration of stresses at the point of abrupt change of the shell thickness. The failure mode is of the fourth type up to $\phi = 46°$. The steel torispheres with the increased-radius polar imperfections fail in the shell pole (the type (1) failure mode) and the failure mode does not change with the extension of ϕ to $\phi = 10°$ ($s_{imp} = 2·4\sqrt{1·5Rt_s}$) and $\phi = 18°$ ($s_{imp} = 4\sqrt{1·5Rt_s}$). However, collapse pressures increase, almost reaching the value for $\phi = 10°$ (and 18° respectively) equal to that for perfect steel unreinforced shells and for the above values of ϕ the failure mode switches to the fourth type. For $\phi > 46°$ (the end of the spherical segment) the failure occurs again in the knuckle region (type (2)) both for perfect and imperfect shells and the critical pressure increases rapidly, reaching a value 2·51 times that for unreinforced shells ($\phi = 90°$) independently on the values of s_{imp}. For $\phi = 90°$ the failure mode is bifurcation buckling at the clamped edge (type (3)); however, this buckling load is preceded by debonding (type (5)) which starts at $\phi = 84°$.

When the reinforcement starts from the clamped edge ($\psi = 0°$) for perfect shells, the critical pressure increases more quickly with the growth of ψ than in the previous case because the most dangerous (knuckle) region is more quickly covered by an additional layer (see Fig. 4). The failure mode is of type (4); only in the surrounding of $\psi = 90°$ does it switch to bifurcation buckling (type (3)).

For imperfect torispherical shells the local reinforcement starting from the clamped edge ($\psi = 0°$) has no influence on the change of the critical pressure and the failure modes; they are identical to those for unreinforced shells (Fig. 4). When the reinforcement length ψ reaches the imperfect region (i.e. $\psi > 90°(\alpha)$, see Fig. 3), the critical pressure rapidly increases,

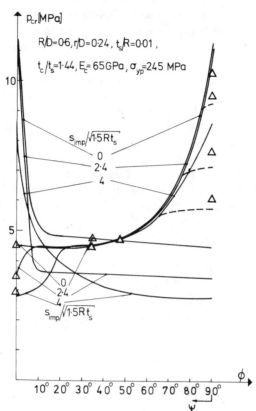

FIG. 4. Influence of a reinforcement length on failure pressures for perfect and imperfect shells. (———, Axisymmetric collapse—1, 2, 3 or 4; – – –, debonding—5; △, experimental results.)

which is simultaneously associated with the change of the failure mode to type (4).

It is evident that the composite reinforcement starting from the shell pole is more effective from the viewpoint of the values of critical pressures (both for perfect and imperfect torispheres) than that which is introduced from the shell clamped edge. Let us notice also that the character of variations of the critical pressure p_{cr} versus the reinforcement length ϕ (or ψ) (Fig. 4) resembles entirely that presented for reinforced beams in Fig. 1(b). The type of failure mode varies with the change in the $\phi(\psi)$ parameters. The broader discussion of the above problems for other geometrical parameters is presented in Ref. 7.

3.2 Effects of t_c/t_s and E_c for Locally Reinforced Shells

The values of both t_c/t_s and E_c were varied in order to analyse the influence of these factors on the values of limit pressures p_{cr} of locally reinforced imperfect steel torispherical shells. It should be emphasised that the considerations are restricted to the analysis of woven roving composite materials having $E_{c1} = E_{c2} = E_c$ (see Table 1). In the numerical calculations E_c varies whereas Kirchhoff's modulus G_{12} and Poisson's ratio are kept constant. The analysis carried out for beams shows the independence of p_{cr} on the mechanical properties of the reinforcement (eqn (1)). The detailed numerical investigations for imperfect torispheres with $R/D = 0.6$ and $r/D = 0.24$ confirm the above conclusions. For each $\phi < 90°$ the limit values of t_c/t_s and E_c/E_s exist, for which the limit pressures are insensitive to the variations of these variables (see Fig. 5). It looks very strange from the point of view of classical buckling problems, where p_{cr} is always a function of Young's modulus; however, it can be easily explained on the basis of limit

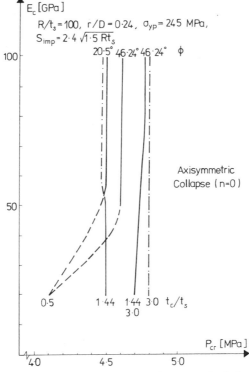

FIG. 5.   Variations of limit pressures with t_c/t_s and E_c. (——, In the knuckle region; - - -, at the apex; —·—, at the junction.)

states—this is the main reason that we prefer to call p_{cr} here the limit pressures (which are more general) than buckling ones. For locally reinforced shells the abrupt change of shell thickness results in the possibility of the appearance of plastic hinges in this region, which increases with the growth of the t_c/t_s ratio. The results plotted in Fig. 5 show that the plastic hinges arise more quickly for smaller values of ϕ when the ratios t_c/t_s are identical. For $t_c/t_s > 5$ and $E_c/E_s > 0.25$ the values of limit pressures as well as the positions of the plastic hinges are independent of these variables; they are dependent only on the mechanical and geometrical properties of the steel shells and on the length of the composite reinforcement. In this case the values of the limit pressures p_{cr} increase with the growth of ϕ (or l_c for beams).

3.3 Effects of r/D

Figure 6 is a plot of the limit pressures p_{cr} divided by the collapse pressures for unreinforced steel shells p_{un} versus ϕ. As may be seen, the

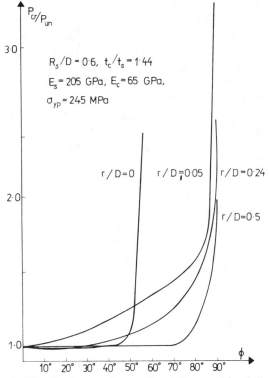

FIG. 6. The effectiveness of composite reinforcement for shells.

effectiveness of the reinforcement is a function of r/D and in this way of failure modes (see Fig. 2). The biggest and quickest effects of local composite reinforcement occur for torispheres which, when unreinforced, collapse at the clamped edge. In this case the collapse pressure for steel shells is the smallest one (see Fig. 2). Then the effects of reinforcement decrease, reaching their minimum for hemispherical shells ($r/D = 0.5$). It is worth noting that the strength-to-weight ratio is rather good in the whole range of r/D. The increase of weight for entirely reinforced shells varies from 20 to 28% (it increases with the growth of r/D). Comparing this with the results obtained by Amiro *et al.*,[3] one can easily notice the advantage of applications of fibre reinforced plastics as reinforcing materials.

4 OPTIMAL FIBRE ORIENTATION FOR THE HYBRID TORISPHERES

One of the aims of the study is to see how fibre orientation affects the value of critical pressures of torispherical shells made of a steel layer and several composite layers. Since, as far as the authors are concerned, any numerical buckling program which is available takes into account plastic deformations and coupling terms in the stiffness matrix, the current analysis is limited to the elastic range only with the use of the BOSOR4 program.[11] The investigations are carried out under the assumption of the shell axisymmetry, i.e. fibres are oriented with respect to an arbitrary shell meridian. The position of the fibre in the *i*th layer is denoted by the angle θ_i, where number 1 is the innermost layer and number N corresponds to the layer closest to steel. The composite reinforcement is made of boron/epoxy having the following properties:

$$E_{c1} = 207\,\text{GPa} \qquad E_{c2} = 20.7\,\text{GPa} \qquad G_{c12} = 7\,\text{GPa} \qquad v_{c12} = 0.3$$

Some optimal stacking sequences of laminated structures can be easily found under the assumptions applied in the BOSOR4 program. In this case critical pressures p_{cr} are always functions of the mechanical properties which, written in the invariant form, depend on $\cos 2\theta_i$ and $\cos 4\theta_i$—the detailed discussion of this problem is presented in Ref. 13. Thus one may write

$$p_{cr} = p_{cr}[f_1(\cos 2\theta_i), f_2(\cos 4\theta_i)] \tag{3}$$

By differentiating the above formula with respect to θ_i, one obtains

$$dp_{cr}/d\theta_i = -2\sin 2\theta_i \times \text{(other derivatives)} \tag{4}$$

Thus the optimal stacking sequences with respect to θ_i occur for $\theta_i = 0°$ or

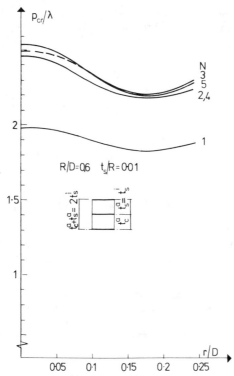

FIG. 7. Optimal buckling pressures for $R/D = 0.6$. (——, Bifurcation buckling; ---, axisymmetric collapse.)

90°, or for roots of the expression in the brackets. The maximum critical pressure p_{cr} corresponds to one of the above values of the angle θ_i.

It is clear that any optimisation of the laminated shells considered herein should take into account two variables, i.e. the thickness of a particular laminate and the fibre orientation θ_i, in a global system of coordinates. However, from the practical point of view, the above-mentioned variables differ in the field of admissible solutions; the first quantity has a discrete spectrum of solutions for composites (the thicknesses of composite fabrics come only in certain sizes), whereas the second has a continuous spectrum in the interval $0°, 90°$ due to the simplifications introduced in the BOSOR4 program. Of course, the thickness of the steel layer can be changed in an arbitrary way. Additionally, it is assumed that the actual total thicknesses (i.e. $t^a = t_s^a + t_c^a$) of the reinforced torispheres are always equal to $2t_s^i$.

The results of some numerical calculations are presented in Fig. 7 for torispherical shells having $R/D = 0.6$, $t_c^a/t_s^a = 1$, $R/t_s^i = 100$, various values of r/D and with the optimal fibre configurations. It is seen that the variations of lamina thicknesses (i.e. N), for $N > 2$, have only a small influence on the value of the buckling pressures. The optimal fibre configuration for the shells considered herein is the combination of the fibres oriented in the $0°$ (the innermost surface) and in the $90°$ (near the steel layer) directions with the exception of one-layered composite reinforcement when the fibres are oriented along the shell meridian ($\theta = 0°$). Thus in this case the optimal stacking sequence is identical to that obtained directly from eqn (4). More numerical examples dealing with the problems discussed are presented in Ref. 14.

5 EXPERIMENTAL RESULTS

The theoretical analysis was confirmed by various experimental tests. The steel torispheres with the geometry $t_s/R = 0.01$, $R/D = 0.6$, $r/D = 0.24$ ($t_s = 1.35$ mm, $D = 225$ mm) were machined using a numerically-controlled lathe

TABLE 2
Comparison of experimental and theoretical failure pressures for unreinforced and reinforced torispherical shells subjected to external pressure

Dome number	ϕ	Failure pressures (MPa)			Experimental failure mode
		$p_{cr} = p_{thu}$	p_{thl}	p_{exp}	
Perfect shells					
A	$0°$	4·50 (0)	—	4·50	Single unsymmetric buckle
C	$90°$	11·28 (4)	9·24	9·55	Debonding
Shells with increased-radius polar imperfections, $s_{imp} = 2.4\sqrt{1.5Rt_s}$					
G	$0°$	3·45 (0)	—	3·46	Inward axisymmetric dimple
I	$20° 30'$	4·46 (0)	—	4·15	Single unsymmetric buckle
L	$35° 40'$	4·48 (0)	—	4·75	Single unsymmetric buckle
H	$90°$	11·20 (4)	7·15	7·68	Debonding
O[a]	$90°$	13·72 (2)	9·87	10·31	Inward axisymmetric dimple
Shells with increased-radius polar imperfections, $s_{imp} = 4\sqrt{1.5Rt_s}$					
A1	$0°$	2·79 (0)	—	2·90	Inward axisymmetric dimple
B1	$35° 40'$	4·47 (0)	—	4·47	Single unsymmetric buckle
E1	$46° 24'$	4·74 (0)	—	4·76	Single unsymmetric buckle
C1	$90°$	8·59 (4)	5·74	6·06	Debonding

[a] This dome was reinforced by unidirectional HM CFRP.
Note: The numbers in parentheses show the predicted number of circumferential waves at buckling.

on both the inside and the outside surfaces. Twenty-one domes were produced and tested. Four of these were perfect and the rest were imperfect, with the increased-radius imperfections located at the shell apex. The imperfect models were made with the length of the dent, s_{imp}, equal to $2.4\sqrt{1.5Rt_s}$ (thirteen shells) and $4\sqrt{1.5Rt_s}$ (four shells). The composite reinforcement was applied on the inside surface of the shells using the vacuum bag technique. Three different types of pre-impregnated composite materials were used as the reinforcement, viz. (i) woven roving CFRP (thirteen shells), (ii) unidirectional high modulus CFRP (two shells) and (iii) woven roving GFRP (one shell). In all the experiments the average thickness of the composite reinforcement was equal to 1·95 mm ($t_c/t_s =$ 1·44), whereas the number of composite layers, their orientations and the meridional extent of the reinforcement, ϕ, varied. The most representative experimental results are presented in Table 2, the rest of them in Ref. 8. The reinforcement made of woven roving CFRP consisted of six layers oriented at 0°/9°/18°/27°/36°/45° (the innermost layer) whereas the optimal

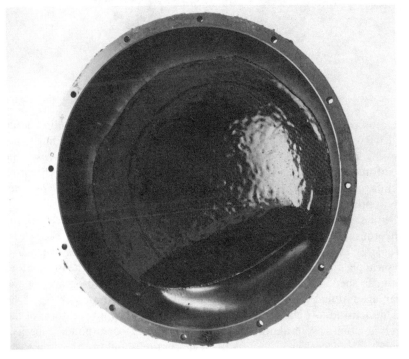

Fig. 8. A torispherical shell after failure (model E1).

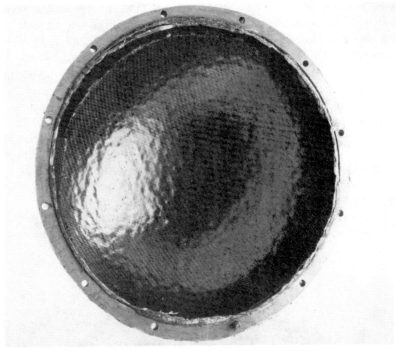

FIG. 9.   A torispherical shell after failure (model C).

reinforcement (with respect to limit pressures) made of unidirectional CFRP was prepared of fifteen layers oriented at $90_7^\circ/0_8^\circ$ (i.e. according to the results of the previous section). Additionally, in the latter case, the reinforcement was cut off in the form of curved triangles (each layer of the reinforcement consisted of twelve triangles along circumferential) in order to approximate fibre orientation to the assumed axisymmetry.

For steel and locally reinforced torispheres, the agreement between the theoretical results and experimental tests was very good (see Fig. 4 and Table 2). It was mainly caused by the type of failure modes which had typical plastic character—a plastic hinge at the junction of the thicker and thinner shell segments that is evidently seen in the photograph of the damaged dome (Fig. 8). For the locally reinforced domes with the increased-radius polar imperfections at the apex, the limit pressures of the hybrid domes were almost the same as those of corresponding perfect unreinforced steel shells for $\phi > 10°$, which entirely confirms the previously presented numerical results. However, the discrepancy between theory and

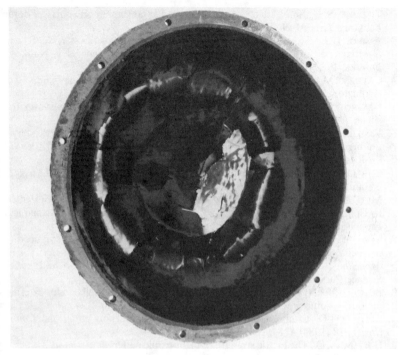

FIG. 10.   A torispherical shell after failure (model O).

experiment increased for shells reinforced over the whole surface area due to the appearance of the debonding (see Fig. 9). In this case the lower bound of the failure pressure (p_{thl}), calculated with the use of eqn (2), gives the lower estimation of the experimentally obtained values (see Table 2). The experimental results were found to lie in the interval $p_{thl} < p_{exp} < p_{thu} = p_{cr}$. For hybrid domes ($\phi = 90°$) the experimental failure pressure was 2·1 times higher than for the unreinforced shells. The factor 2·1 increased to 3·0 for optimally-oriented composite reinforcement. However, in the latter case, the experimental failure mode is completely different than for domes A, H or C1 (see Table 2). The shell O failed almost axisymmetrically, in the same manner as unreinforced steel imperfect torispheres G and A1, i.e. at the pole (see Fig. 10).

REFERENCES

1. CALLADINE, C. R., *Theory of Structures*. Cambridge University Press, Cambridge, 1983.

2. TILLMAN, S. C., Some effects of rib-reinforced arrangement on spherical dome buckling. *Exp. Mech.*, **18** (1978) 396–400.
3. AMIRO, J. J., GRATCHEV, O. A., ZARUTSKI, V. A., PALTCHEVSKY, A. S. and SANNIKOV, YU. A., *Stability of Ribbed Shells of Revolution.* Kiev, Naukova Dumka, 1983 (in Russian).
4. CAPURSO, M. and GANDOLFI, A., Sull collasso rigide-plastico dei gusci nervati di rivoluzione. *Costruzioni Metalichi*, **19** (1967) 418–24.
5. NEMIROVSKY, YU. V., Limit equilibrium of multi-layer reinforced axisymmetric shells. *Izv. AN SSR MTT*, **6** (1969) 80–9 (in Russian).
6. VASILIEV, Y. V., Elastic–plastic deformations of pressure vessels reinforced by a unidirectional system of glass fibres. *Mekhanika Polymerov*, **6** (1969) 1069–74 (in Russian).
7. MUC, A., Reinforcement of externally-pressurised steel domes. *Proc. I.Mech.E.* (submitted).
8. MUC, A., GALLETLY, G. D. and MORETON, D. N., On the buckling of dented torispherical domes reinforced by composites. Department of Mechanical Engineering, University of Liverpool, Report A/138/88, 1988.
9. SAVE, M. A., Limit analysis and design: an up-to-date subject of engineering plasticity. In *Proceedings of the International Symposium Plasticity Today, Modelling Methods and Applications*, ed. A. Sawczuk and G. Bianchi. Elsevier Science Publishers, London, 1985, pp. 767–85.
10. BŁACHUT, J. and GALLETLY, G. D., Clamped torispherical shells under external pressure—some new results. *J. Struct. Analysis*, **23** (1988) 9–24.
11. BUSHNELL, D., Computerized analysis of shells—governing equations. *Comp. Struct.*, **18** (1984) 471–536.
12. HOFFMANN, O., The brittle strength of orthotropic materials. *J. Comp. Mater.*, **1** (1967) 200–14.
13. MUC, A., Optimal fibre orientation for simply-supported, angle-ply plates under biaxial compression. *Comp. Struct.*, **9** (1988) 161–72.
14. GALLETLY, G. D. and MUC, A., Buckling of fibre reinforced plastic/steel torispherical shells under external pressure. *Proc. I.Mech.E.* (in press).

2

The Mechanical Behavior Peculiarities of Angle-Ply Reinforced Shells Made out of Glass and Organic Fiber Reinforced Plastics

V. B. ANTOKHONOV

East-Siberia Technological Institute, Ulan-Ude, USSR

V. N. BULMANIS and A. S. STRUCHKOV

Institute of Physical and Technical Problems of the North, Yakutsk, USSR

ABSTRACT

We have investigated axial tension of winding glass and organic fiber plastics at normal and low temperatures. Deformation of the specimens up to the point of strength loss has been analyzed. Experiments have been carried out at temperatures of $+23°C$ and $-60°C$. The results of the experiments compared to each other have shown an increase of the elasticity modulus. They have also shown an increase in stress. All this corresponds to an increase in the cracking moment and strength of the glass plastics of 22, 147 and 112%, respectively. Organic fiber reinforced plastics have been shown to increase their modulus by 20% and to decrease their strength almost by 40%.

The influence of the extent and width of the reinforced belt interlacing on properties of the specimens have been investigated. Now it is clear that the glass plastics strength depends on given structural parameters. Their influence on the failure modulus and the beginning of failure has not been proved. An approach has been worked out to describe the mechanical behavior of the composites. It will include two models. The first one will represent the layer's shear non-linearity and instantaneous effects before cracking. The second one will show the belt width and interlacing extent as calculated parameters. The calculations have proved the results of the experiments.

Though a great many experimental data on glass plastics have been obtained, scientific literature has not yet offered comprehensive diagrams of the deformation of winding composites under tension. The well-known diagrams of $\sigma \sim \varepsilon$ have been obtained only for the flat and ring-shaped specimens (the edge effect) or by circular tension of the tubes. All the other diagrams are not perfect.[1,2]

Investigations have revealed great deformation anisotropy and low strength under transverse shear and tension of unidirectional organic fiber reinforced plastics. Transverse shear and tension result in a split of the filaments and never result in failure of the adhesive and surface. Most studies consider the problem of the behavior of elastic filament reinforced plastics. It should be noted that we have not obtained enough experimental data on deformation and strength properties of the winding glass and organic fiber reinforced plastics at a low temperature. One can get an idea of the advantages and disadvantages of the composites only by studying their mechanical behavior before failure.[3,4]

The specimens of the winding composites have been obtained by tube section, the tubes being made by the method of 'wet' belt winding. The belts are made out of the glass and organic filaments soaked with the epoxy adhesive, the following structural parameters being varied: the winding angle, $\pm \phi$; the coefficient of reinforcing, ψ; the width of the wound belt (the number of filaments in the belt), b; and 'the extent' of interlacing (the number of interlacings per unit of the material volume), κ.

Figure 1 shows a typical scheme of glass plastics deformation at a temperature of $+23°C$. The diagram of the composite tension at the winding angle of $\pm 30°$ (curve 1) shows that the strength reserve before failure is about 100% with respect to the point of cracking. The specimens fail in a brittle way, no shaftjournal being formed. Under the stress which corresponds to the field BC in the tension diagram one can observe small cracks along the boundaries of the reinforced belts. Break of the fibers shows complete failure of the specimens.

The second type of failure and deformation, $\phi = \pm 45°$ (curve 2), differs greatly from the previous one. The strength reserve before the loss of strength is $\sim 25\%$, with respect to the point of cracking. As a rule shaftjournals are formed on the failed specimens (the field of global failure of the adhesive).

The composites with a reinforcing angle of $\pm 60°$ fail in a brittle way in all cases, i.e. no shaft is formed. In fact the specimens fail in a mono-lithic manner up to the point of failure. This is shown in the diagram of tension (curve 3), in which the point of cracking is at the point of

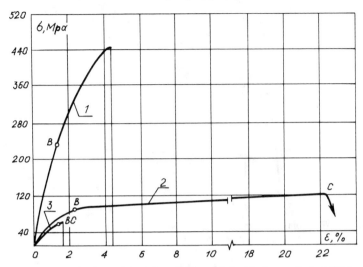

FIG. 1. Diagrams of deformation of glass plastics with the winding angle: 1, ± 30°; 2, ± 45°; 3, ± 60°.

complete failure. The failed specimens have large cracks which are parallel to the direction of reinforcing. The cracks spread along the perimeter at full thickness of the specimen wall.

Let us analyze the diagrams of the tension of specimens made out of glass and organic fiber reinforced plastics at the angle of reinforcing of ±45°. The complete diagram of the glass plastics can be divided into fields of non-linear elastic deformation and growing failure of the composite. However, the organic fiber reinforced plastics can hardly be divided in the same way (only the first field is clearly seen).

If we compare the results of the experiments carried out at normal and low temperatures ($-60°C$) we will observe an increase of the elasticity modulus and strength degree which corresponds to the point of cracking and strength of the glass plastics with $\phi = \pm 45°$, the temperature falling by 22, 147 and 112%, respectively. For organic fiber reinforced plastics with $\phi = \pm 45°$ the elastic modulus increases by 20%, the strength decreasing almost by 40%. The unidirectional organic fiber reinforced plastics also decrease in strength (under transversal loading), $\phi = 90°$.

Diagrams comparing deformation of glass and organic fiber reinforced plastics with $\phi = \pm 45°$ at low and normal temperatures are shown in Figs 2 and 3. Non-linear behavior of the glass plastics before cracking does not change at a low temperature, though it weakens greatly (Fig. 2). This may be

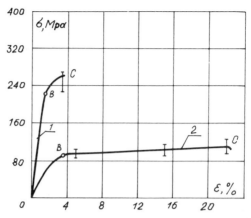

FIG. 2.   Diagrams of the deformation of glass plastics with the winding angle $\pm 45°$ at low (1) and normal (2) temperatures.

accounted for by a decrease of non-linearity of the shear component of the composite elasticity tensor, which greatly depends on mechanical behavior of the adhesive, its thermomechanical dependence being well known. In the field BC intensive failure of the material takes place both at normal and low temperatures. It is accompanied by an increase of the strain level up to the point of complete failure, just as happens when the temperature is normal. The great value of strength may be accounted for by an increase of the adhesive bond at the boundary 'fiber matrix'.

Some peculiarities of mechanical behavior of the organic fiber reinforced plastics are shown in Fig. 3. A decrease of strength takes place at the same time as a decrease of temperature occurs. This may be accounted for in the following way: the fiber strength during the shear and transverse break is a weak point of organic fiber reinforced plastics. It may result in a decrease of the strength properties. (It has been proved by the results of the experiments made under transverse tension.)

Such parameters as the width and extent of interlacing of the belts have not been investigated yet. Their dependence on mechanical properties has been studied by testing the winding composites at low temperature. No influence of the parameters on the elasticity modulus and strain level has been fixed, the strain level corresponding to the point of cracking. At a winding angle of $\phi = \pm 45°$ the glass plastics strength depends on the width and extent of the belt interlacings. This is shown in Fig. 4. The first cracks are formed and spread at the places where the transverse reinforced belts are formed. This is typical of the glass plastics which we have investigated.

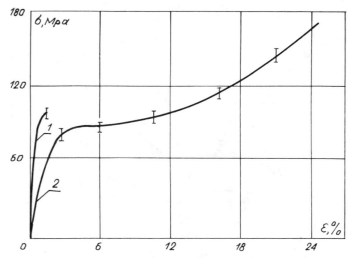

FIG. 3. Diagrams of the deformation of organic fiber reinforced plastics with the reinforcing angle ±45° at low (1) and normal (2) temperatures.

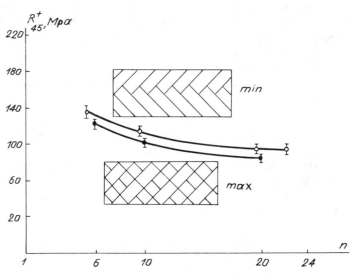

FIG. 4. Dependence of strength of glass plastics with the reinforcing of ±45° on the width and 'extent' of the wound belt's interlacing.

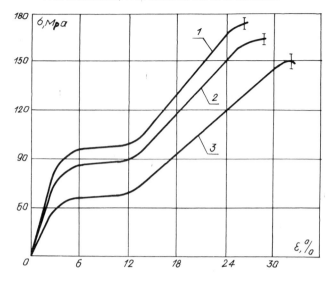

FIG. 5.   Curves of deformation of the organic fiber reinforced plastics with the reinforcing of
±45° with different 'extents' of the belt's interlacing: 1, min; 2, mid; 3, max.

Figure 5 shows mechanical behavior of the organic fiber reinforced plastics which depends on the extent of interlacing. The curves in the non-linear field of deformation have turned out to be identical.

An increase of the reinforcing coefficient results in an increase of the tension level, which corresponds to the point of cracking, the winding angle being ±30°. It causes a decrease at angles of ±45° and ±60° (Fig. 6). The experiments on strength have given the same results. It proves our supposition that final failure of the composites at a reinforcing angle of 45° or more is of the interlayer type. It is caused by an increase of the reinforcing coefficient, which results in a decrease of thickness of the adhesive layer between the composite layers.

When the winding angles are small an increase of strength is accompanied by an increase of the reinforcing coefficient. It is accounted for by the fact that up to the point of failure it is the reinforced layers that have the greater part of the loading.

Let us consider the approach which makes it possible to describe the process of deformation of the angle-ply reinforced shell made out of glass plastics under tension in the axial direction at all stages of loading up to the loss of strength. The approach was worked out on the basis of Bolotin's method. Later on it was developed by Partsevsky. He investigated the

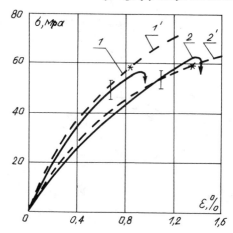

FIG. 6. Experimental (——) and calculated (– – –) deformation curves for glass plastics with the reinforcing of $\pm 60°$: 1, $\psi = 0.70$; 2, $\psi = 0.62$ ($*$, the calculated value).

instantaneous effects which resulted from a mutual turn of the rigid filaments.[1,5]

To describe the curve of the glass plastics deformation before and after cracking two models have been suggested. The first one differs greatly from that suggested by Partsevsky (before cracking). It reflects the idea that rigidity of the reinforced elements is much greater than that of the adhesive. The thickness of the reinforced and adhesive layers of the transverse shear is assumed to be constant. The composites are characterized by a regular reinforcing structure. They are made out of the belt winding; the latter consists of filaments with the final shear rigidity. In addition, non-linearity of the material during shear in the layer plane has been modeled by introducing a non-linear connection between the components $\langle\!\langle \tau_{12} \rangle\!\rangle \sim \langle\!\langle \gamma_{12} \rangle\!\rangle$ of the unidirectional layer into the calculations. In the numerical form the non-linear dependence was substituted by the piece-linear function which was made more precise at every stage of loading.

We have used the well-known characteristics of the EDT-10 adhesive and those of the unidirectional glass plastics. The tensor-polynomial criterion of Malmeister, which is a kind of singular criterion of Khashin, has also been employed. Figure 6 shows calculation of the glass plastics with a winding angle of $\phi = \pm 60°$. The calculations give us a good idea of the processes and effects which can be observed in the case where the glass plastics are deformed up to the 'point' of cracking. The model of the

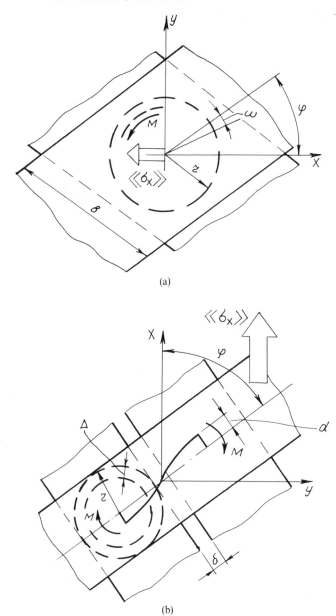

(a)

(b)

FIG. 7.   Fragments of the belt's interlacing: (a) model of mutual turn of the angle-ply belts; (b) model of the bending.

deformation of the glass plastics before cracking has been worked out. It has made it possible to extend the method to the deformation region which is characterized by growing failure of the composite. We have used the fact that the reinforced layers acquire a specific structure which consists of the reinforced belts disconnected along the edges. The shell element under the conditions of heterogeneous values x and y and strain deformation has been investigated. The element consists of the angle-ply belts which are h in thickness and b in width. They are connected by the adhesive layer which is $(c - h)$ in thickness.

The general relative material deformation is assumed to consist of three components:

$$\langle\!\langle \varepsilon_x \rangle\!\rangle = \langle\!\langle \varepsilon_1 \rangle\!\rangle + \kappa(\langle\!\langle \varepsilon_2 \rangle\!\rangle + \langle\!\langle \varepsilon_3 \rangle\!\rangle) \tag{1}$$

where $\langle\!\langle \varepsilon_1 \rangle\!\rangle$ is the fragment of deformation caused by the middle tangential stress $\langle\!\langle \sigma_x \rangle\!\rangle$, $\langle\!\langle \varepsilon_2 \rangle\!\rangle$ is the deformation caused by mutual turn of the transverse belts to the angle ω, $\langle\!\langle \varepsilon_3 \rangle\!\rangle$ is the deformation caused by curvature and shear of the belt in its plane (Fig. 7(b)), and κ is the coefficient of the belt interlacing extent.

The conditions under which the layer fails will be sought in the inequality

$$\tau_{max} > R_{m\#} \qquad \tau_{max} = r M_k / J_p \tag{2}$$

where $R_{m\#}$ is the strength of the matrix during the shear, M_k is the twisting moment, and J_p is the polar moment of inertia.

The numerical form of the method has been found by the 'step-by-step' method. The strain increments were given and the relative deformation increments were to be sought, with respect to (1). We assumed the layer

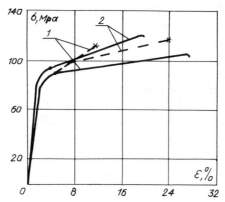

FIG. 8. Experimental (——) and calculated (– – –) curves for glass plastics with the reinforcing of $\pm 45°$: 1, $b = 8$ mm (24 filaments); 2, $b = 2$ mm (6 filaments).

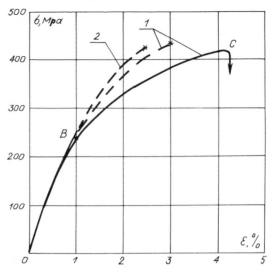

FIG. 9. Experimental (——) and calculated (– – –) curves for glass plastics with the reinforcing of $\pm 30°$: 1, $b = 2$ mm (8 filaments); 2, $b = 3$ mm.

failure to be a discrete process, therefore we interrupted the calculation when the layer failed (2). We decreased the radius r by the value $d - h$. We repeated the procedure once and again till the layer disappeared and the belt broke. By comparing the curves we tested deformation of the glass plastics after cracking. The model proved to be correct. Figure 8 shows the shell deformation which is reinforced at an angle of $\pm 45°$. The maximum deformation and shell strength are inversely proportional to the belt width. The analysis of the calculations shows that general deformation mainly depends on the deformation which is caused by the fragment tension, belt curvature and shear in the plane. The belt's mutual turn depends on the adhesive shear modulus. In fact its values are very small. The material failure is dependent on the method of the layer torsion. Figure 9 illustrates the method for the reinforcing material at $\pm 30°$. Change of the inclined angle of the region BC has turned out to be small. Failure of the composite results in breaking of the belts, strain in the reinforcing direction being larger than the strength limit of the unidirectional belt.

REFERENCES

1. BOLOTIN, V. V. and NOVICHKOV, J. N., *The Mechanics of Laminated Structures* (in Russian). Mashinostrojenije, Moscow, 1980.

2. ROTEM, A. and HASHIN, Z., Failure modes of the angle-ply laminates. *J. Comp. Mater.*, **9** (1975) 191–206.
3. UDRIS, A. O., UPITIS, S. T. and TETERS, G. A., Investigation of deformation and destruction of the glass plastics with spiral reinforcing ±45° under simple and complex loading. *Mekh. Kompos. Mater.* (1984) 805–13.
4. BULMANIS, V. N., GUSEV, J. N., STRUCHKOV, A. S. and ANTOKHONOV, V. B., Experimental investigation of some peculiarities of deformation and destruction of the angle-ply reinforced winding composites. *Mekh. Kompos. Mater.* (1985) 1020–4.
5. PARTSEVSKY, V. V., Instantaneous effects in the flat problem of the angle-ply reinforced composite. *Mekh. Kompos. Mater.* (1979) 45–50.

3

Failure Modes of Laminated Axisymmetric Shells of Revolution Subjected to External Pressure

A. MUC

Institute of Mechanics and Machine Design,
Technical University of Cracow, Cracow, ul. Warszawska 24, Poland

ABSTRACT

The paper deals with the investigations of failure modes of laminated shells of revolution subjected to external pressure. These are analysed for two typical shell forms such as fully clamped hemispheres and torispheres made of unidirectional or woven roving glass/epoxy and carbon/epoxy. The buckling pressures of composite shells are compared with the failure pressures obtained from the various failure criteria in the stress space. The effects of fibre orientation and initial geometric imperfections are also discussed. The imperfections considered are in the form of increased-radius polar imperfections and Legendre polynomial imperfections. For CFRP perfect torispherical shells the numerical results are verified by experimental tests.

1 INTRODUCTION

It is well known that a significant amount of material savings can be achieved by the use of fibre reinforced plastics (FRP). Although overshadowed by the advance of and the interest in aerospace technology, some attempts have been made towards the application of FRP in submersible structures. In such structures, it is particularly important to develop lightweight highly reliable pressure hulls from structural materials possessing a high strength-to-density ratio. As it is pointed out by Hom,[1] this is the main factor in the selection and evaluation of materials for submersible structures.

However, it is clear that both theoretical and experimental investigations involving composites, particularly in application to laminated axisymmetric shells such as hemispheres or torispheres, still far lag behind those for isotropic shells. This is particularly evident from the review by Almroth[2] dealing with buckling studies of composite structures. The majority of the works discussed there is devoted to buckling problems of cylindrical or conical shells, and only two works[3,4] refer to doubly curved shells such as hemispheres, spherical sections, barrels, etc. Now, there are available also several general purpose computer codes for solving composite shell problems (see Refs 5–9, for example). In this area, some analytical or semi-analytical solutions of buckling problems are also proposed (especially for spherical shells, Refs 10–13), however they differ significantly in the basic assumptions such as, for example, the applied shell theory, the formulation of buckling problems, etc.

As far as the author is concerned, the experimental analysis of composite shells of revolution is even more scarce than theoretical investigations in this area. Bert[14] conducted the experiments on the failure of bathyscaphes made of glass fibre. Ivanov[15–17] studied buckling problems of ellipsoidal GFRP dished ends under external pressure, whereas torispherical (perfect and imperfect) composite domes were analysed by Galletly and Muc.[18]

This particular study is concerned with the effects of laminate configurations and imperfection sensitivity on failure modes and loads of clamped hemispherical and torispherical shells under external pressure. The investigations are carried out under the assumptions of shell axisymmetry, linear elasticity of composites and with the use of three parametrical shell theory (the Love–Kirchhoff hypothesis). Some new experimental results for torispherical shells made of CFRP are also reported here.

2 FAILURE MODES OF COMPOSITE SHELLS

The majority of the existing computer codes (see, for example, Refs 5–9) takes into consideration just one failure mode—buckling which additionally is sometimes limited to axisymmetric collapse only (i.e. $n = 0$, where n denotes the wavenumber in buckling).

In the present paper the buckling analysis (including bifurcation failure modes) is carried out with the use of the BOSOR4 program,[6] which is not very general (its disadvantages are discussed in Ref. 18), however its numerical results show rather good agreement with experiment (see Ref. 18). However, the analysis of laminated, anisotropic shells of revolution

TABLE 1
Typical ply mechanical properties

No.	Materials	E_1	E_2 (GPa)	G_{12}	v_{12}	X_t	X_c (MPa)	Y_t	Y_c	S
1[a]	CFRP	203	11·2	8·4	0·32	3 500	1 540	56	150	98
2[a]	GFRP	38·6	8·27	4·14	0·26	1 062	610	31	118	72
3	CFRP	65	65	3·85	0·045	468	364	468	369	61
4[b]	GFRP	15·1	15·1	3·1	0·16	353	304	353	304	35
5[b]	CFRP	68	68	4·25	0·10	590	562	590	562	88

[a] Results presented by Tsai.[20]
[b] Materials tested by the author.

requires to take into account some additional effects as for example fracture or delamination.

Two failure strength quadratic criteria are primarily employed in this study. Both are in-plane fracture predictions based on the stress interaction criteria[19] as proposed by Tsai and Wu:

$$\frac{\sigma_1^2}{X_t X_c} - \frac{\sigma_1 \sigma_2}{\sqrt{X_t X_c Y_t Y_c}} + \frac{\sigma_2^2}{Y_t Y_c} + \left(\frac{1}{X_t} - \frac{1}{X_c}\right)\sigma_1 + \left(\frac{1}{Y_t} - \frac{1}{Y_c}\right)\sigma_2 + \frac{\tau^2}{S^2} = 1 \quad (1)$$

and Hoffmann

$$\frac{\sigma_2^2 - \sigma_1 \sigma_2}{X_t X_c} - \frac{\sigma_2^2}{Y_t Y_c} + \left(\frac{1}{X_t} - \frac{1}{X_c}\right)\sigma_1 + \left(\frac{1}{Y_t} - \frac{1}{Y_c}\right)\sigma_2 + \frac{\tau^2}{S^2} = 1 \quad (2)$$

In the above relations σ_1, σ_2, τ are the applied direct and shear stresses and $X_{t,c}, Y_{t,c}, S$ are the corresponding strengths. In this investigation, the stresses appearing in eqns (1) and (2) are determined from the simplest membrane shell theory. Of course, this assumption can involve some errors, particularly in the clamped regions or in the knuckle regions where bending effects can be dominant. The validity of such an approach will be verified later on the basis of experimental results.

The mechanical properties of materials used in our analysis are presented in Table 1.

3 OPTIMAL FIBRE ORIENTATION OF COMPOSITE HEMISPHERES AND TORISPHERES

In order to undertake a parameter study for clamped hemispherical and torispherical laminated shells of revolution, the BOSOR4 program was

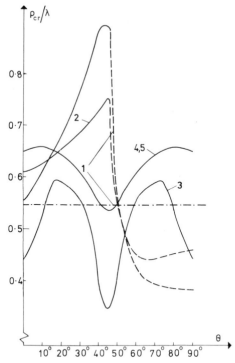

FIG. 1. Variation of failure pressures versus fibre orientation θ for hemispherical shells made of various FRP: 1, $E_1/E_2 = 18 \cdot 125$, $E_1/G_{12} = 24 \cdot 167$; 2, $E_1/E_2 = 4 \cdot 667$, $E_1/G_{12} = 9 \cdot 32$; 3, $E_1/E_2 = 1$, $E_1/G_{12} = 16 \cdot 88$; 4, $E_1/E_2 = 1$, $E_1/G_{12} = 4 \cdot 87$; 5, $E_1/E_2 = 1$, $E_1/G_{12} = 16$. (——, Bifurcation buckling; - - -, axisymmetric collapse; - · -, fracture.)

used to calculate the variations of buckling loads with fibre orientations θ. Figures 1 and 2 are the plots of the results for hemispherical and torispherical shells, respectively. In each case results are referenced to angle-ply laminate $\pm \theta$.

The theoretical values of the elastic buckling pressures p_{cr} are always related to the elastic buckling pressures of a perfect spherical shell

$$\lambda = \frac{2t^2}{R^2} \sqrt{\frac{E_1 E_2}{3(1 - \nu_{21}\nu_{12})}} \tag{3}$$

It is immediately evident that by varying the fibre orientations one can more than double the buckling strength. For unidirectional composites ($E_1 > E_2$) the maximal buckling pressures always occur at $\theta = 45°$ (independently of the material used, i.e. glass or carbon), whereas for woven

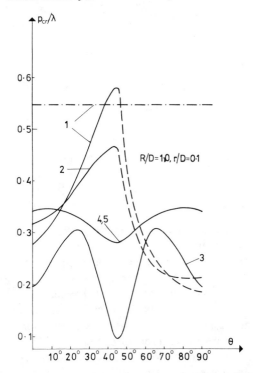

FIG. 2. Effects of fibre orientation θ on failure pressures for FRP torispherical shells: 1, $E_1/E_2 = 18\cdot125, E_1/G_{12} = 24\cdot167$; 2, $E_1/E_2 = 4\cdot667, E_1/G_{12} = 9\cdot32$; 3, $E_1/E_2 = 1, E_1/G_{12} = 16\cdot88$; 4, $E_1/E_2 = 1, E_1/G_{12} = 4\cdot87$; 5, $E_1/E_2 = 1, E_1/G_{12} = 16$. (———, Bifurcation buckling; ———, axisymmetric collapse; —·—, fracture.)

rovings $(E_1 = E_2)$ the positions of the maxima vary (depending on the E_1/G_{12} ratio). In the latter case the decrease of E_1/G_{12} moves the positions of $\theta = \theta(p_{max})$ towards $\theta = 0°$ with a simultaneous decrease of the difference between the values of maximal and minimal buckling pressures. This can be simply explained by the analysis of the elements of the stiffness matrix for laminated shells. As may be seen in Figs 1 and 2 for all plotted curves, some local maxima and minima of buckling pressures occur at $\theta = 0°$ or $45°$ or $90°$. First of all, let us discuss the problem when the buckling pressures will be equal to each other at the positions of the above local maxima and minima.

Thus, let us assume that

$$A_{11}(\theta = 0°) = A_{11}(\theta = 45°) \tag{4}$$

which leads to the equality

$$U_2 = -2U_3(4) \tag{5}$$

and additionally let

$$A_{22}(\theta = 0°) = A_{22}(\theta = 45°) \tag{6}$$

which is equivalent to

$$U_2 = 2U_3 \tag{7}$$

Both equations are satisfied if $U_2 = U_3 = 0$ which results in:

$$E_1 = E_2 \quad \text{and} \quad G_{12} = \frac{E_1}{2(1 + v_{12})} \tag{8}$$

and it corresponds to isotropic materials. Now, it is obvious that woven roving CFRP, having a bigger E_1/G_{12} ratio than GFRP causes the differences in the maximal and minimal buckling pressures (curves 3, 4 and 5 in Figs 1 and 2) to be bigger than for shells made of glass fibres. Hence, one can conclude that for composite shells having identical E_1/E_2 ratios increase in buckling loads can be achieved by increase of the E_1/G_{12} ratio (see Figs 1 and 2). However, the positions of the global maxima on the curves p_{cr} versus θ are different for various E_1/G_{12}. When the E_1/G_{12} ratio is kept constant, increase of buckling loads can be obtained by increase of the E_1/E_2 ratio (see curves 1 and 2) in Figs 1 and 2. The types of buckling failure mode (bifurcation, $n \neq 0$ or axisymmetric collapse, $n = 0$) depend on the fibre orientation and the mechanical properties of the shells considered herein. For woven roving FRP shells there is one failure mode only (bifurcation buckling) for all values of the angle θ, whereas for shells made of unidirectional FRP the axisymmetric collapse appears in the interval $(45°, 90°)$. Let us notice also that for the laminate considered herein the positions of the global maxima of the buckling pressures and the types of buckling failure modes are insensitive to the variations of the shell form— compare Fig. 1 with Fig. 2.

On the other hand, for CFRP shells the failure pressures obtained from the strength criteria ((1), (2)) are much lower than the maximal buckling pressures (see Figs 1 and 2). This is caused by the very low ultimate strength of the analysed material 1 (see Table 1). In this case even using the simplest maximal stress criterion one can find that

$$\sigma_1 = \sigma_2 = \frac{p_f R}{2t} < Y_c \tag{9}$$

Tsai–Wu's and Hoffmann's criteria (eqns (1) and (2)) give insignificantly lower values of the failure pressures p_f (see Figs 1 and 2). For the other composite materials (2–5) discussed herein, the failure pressures calculated from eqns (1) and (2) are much greater than the maximal buckling loads and they are even out of the scale in Figs 1 and 2. For instance, for woven roving FRP having $X_t = X_c$, $Y_t = Y_c$ and $X_t = Y_t$, one can find from eqns (1) and (2) that

$$\sigma_2 = \frac{p_f R}{2t} = X_t \tag{10}$$

For the real values of the material properties taken from Table 1 (materials (3–5)) the fracture pressure p_f is insignificantly lower than that obtained from eqn (10).

The previous analysis was concerned with just one composite layer ($N = 1$). Now, let us consider the composite shell made of N layers in order to identify the relations between number of layers and optimal design for buckling pressures of laminated shells of revolution. It is assumed that each layer has the same thickness, i.e. t/N. Figures 3 and 4 are the plots of the maximal (optimal) dimensionless buckling pressures p_{cr} versus the r/D ratio for torispherical shells. They include also the cases of the spherical cap ($r/D = 0$) and the hemispherical shell ($r/D = 0.5$). However, in the latter case, the R/t ratio is different to that for torispheres and can be determined from the following relation

$$r/t = \frac{(r/D)(R/t)}{(R/D)} \tag{11}$$

As may be calculated, this ratio is much bigger (twice) than for torispheres and it results in a rapid increase of buckling loads in the surrounding of $r/D = 0.5$.

It is seen in Figs 3 and 4 that the optimal fibre orientation varies with the increasing number N of layers. In addition, for the greater value of N ($N > 2$) the solutions are not unique, i.e. it is possible to obtain almost the same buckling pressures as for the orientations shown in Figs 3 and 4. For spherical caps and hemispherical shells the buckling pressure increases with the increasing number of layers N, reaching its maximum for the quasi-isotropic shells. Roughly speaking in the latter case such a configuration can be obtained when the number of layers increases infinitely, the thickness of each layer decreases infinitely and the ply orientation of each layer is expressed by the relation $\theta_i = \pi/2N$. Thus, the shell becomes quasi-isotropic in the mid-plane and quasi-homogeneous across the thickness.

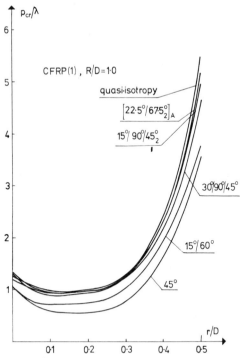

FIG. 3.   Buckling pressures for optimally oriented unidirectional CFRP torispherical shells.

For quasi-isotropic shells the engineering constants are given, for example, by Tsai.[20]

For instance, the increase in buckling pressure (with respect to $N = 1$) is equal to 56·5% (17·7%) for quasi-isotropic hemispheres made of CFRP and GFRP, respectively. Composite shells do not always achieve their maximal buckling pressures for the quasi-isotropic state. In the cases considered herein, the maximum occurs for shells composed of four layers oriented at $30°/90°/45°_2$. However, the CFRP torisphere having $r/D = 0·1$ gives an increase in buckling pressure (with respect to the one-layered torisphere) equal to 66·7% for shells made of four optimally oriented layers and 56·7% for quasi-isotropic torispheres. Thus, the optimal orientation for torispheres depends clearly on the r/D ratio. However, for torispheres the difference in buckling pressures between the optimally oriented shells and quasi-isotropic ones does not exceed 10% in the worst case. On the other hand, quasi-isotropic torispheres can be much more easily produced than others, especially from the shell axisymmetry point of view. In

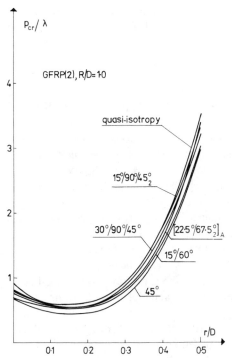

FIG. 4. Buckling pressures for optimally oriented unidirectional GFRP torispherical shells.

addition, these specific optimal configurations have the advantage of ease of analysis since we know much more about isotropic homogeneous materials. Let us notice also that the optimal ply orientations are insensitive to the ply mechanical properties and the r/D ratio for the identical number of layers N (see also Figs 1 and 2). These are also independent of the t/R ratio. The buckling pressures are proportional to $(t/R)^2$ (see the discussion of this problem in Ref. 18) under the Love–Kirchhoff assumption. All optimally oriented shells presented in Figs 3 and 4 fail by bifurcation buckling.

However, from the point of view of the strength of CFRP, the maximal fracture pressure p_f is lower than those presented in Fig. 3 (see the results in Figs 1 and 2 for material (1)). On the other hand, the ultimate strengths for the quasi-isotropic and quasi-homogeneous composite shells should be much higher than those for unidirectional materials. It is worth noting also that the buckling pressures p_{cr} are proportional to $(t/R)^2$, whereas p_f obtained from the strength criteria are a linear function of the t/R ratio.

Thus, for all composite shells there are always some limit values of the t/R ratio where p_f is equal to p_{cr}. For the values of t/R higher than that limit there is only one failure mode—a fracture of composite due to its ultimate strength.

4 THE EFFECTS OF INITIAL IMPERFECTIONS

The initial geometrical imperfections can cause a substantial reduction in buckling strength. The reduction in the buckling loads depends on the magnitudes of the imperfections, their shapes and positions on the shell surface and their extent. In addition, for laminated shells, there are other factors which affect the shell resistance to buckling such as ply orientation and stacking sequence. In the present study two types of axisymmetric imperfections are considered. The first one has a form of an increased-radius polar imperfection (Fig. 5(a)) in order to investigate the effects of localised imperfections, whereas the second one takes the form of Legendre polynomials (extending from the pole to the shell equator—Fig. 5(b)) to describe the random character of imperfections encountered in practice. The main purpose is to determine the lower-bound curve of buckling pressures. With regard to the latter, for Legendre polynomial imperfections having the prescribed magnitude of the imperfection at the pole δ_0, it is necessary to alter l (the degree of the Legendre polynomial $P_l (\cos \phi)$). l should be an odd number in order to satisfy the assumed boundary conditions (i.e. the clamped edges). For the increased-radius imperfections the minimisation procedure (finally leading to the lower-bound envelope of

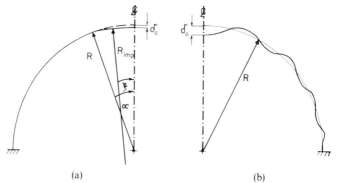

(a)   (b)

FIG. 5.   Geometry of initial imperfections: (a) an increased-radius polar imperfection; (b) a Legendre polynomial imperfection.

buckling pressures) is carried out in the following way: vary the semi-angle ξ and then for the prescribed value of δ_0 calculate buckling pressures. On the basis of the curves p_{cr} versus δ_0 parametrised by the values of ξ it is possible to draw the lower-bound envelope curve. In each case, the radius of imperfection R_{imp} is determined from the simple geometrical relation (see Fig. 5(a))

$$\delta_0 = R_{imp}(1 - \cos \xi) - R[1 - \sqrt{1 - (R_{imp} \sin \xi/R)^2}] \qquad (12)$$

The broader description of the lower-bound envelopes discussed above can be found in Ref. 21. The numerical analysis is limited to just one laminate, i.e. quasi-isotropic. As is shown in the previous section (Figs 3 and 4) such a fibre orientation gives the maximal buckling pressures for composite hemispheres and almost maximal for composite torispheres. Moreover, from the practical point of view, such a ply orientation is the simplest one among others which can offer the exact axisymmetry for doubly curved laminated shells of revolution. Of course, it is always possible to find a ply orientation which can increase buckling pressures for imperfect shells (see for example Tennyson and Hansen[22]) in comparison with the proposed quasi-isotropic. However, it can be done for a particular type of imperfection, and practically all shell structures contain random imperfection distributions.

Figure 6 presents the variations of dimensionless buckling pressures

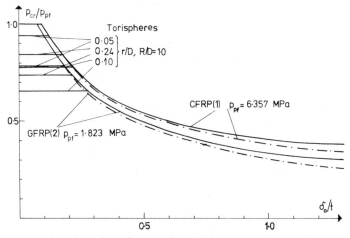

FIG. 6. Lower-bound envelopes for imperfect FRP hemispherical and torispherical quasi-isotropic shells. (——, Increased-radius imperfections; – ·–, Legendre polynomial imperfections.)

p_{cr}/p_{pf} versus the dimensionless magnitude of the imperfections δ_0/t. The calculations are carried out for two unidirectional FRP-glass and carbon (materials (1) and (2) in Table 1). For small values of the δ_0/t ratio, the failure mode corresponds to bifurcation buckling (imperfection insensitivity), and for larger values of δ_0/t, the failure mode is axisymmetric collapse. Similar failure modes were obtained by Galletly and Muc[18] and by Galletly *et al.*[21] It may be seen that imperfections in the form of axisymmetric Legendre polynomials cause a greater reduction in the buckling pressures than the increased-radius polar imperfections. However, the difference between buckling pressures is not very great. It is worth noting that for composite torispheres the regions of the imperfection insensitivity are greater than for hemispherical shells; at the beginning they increase to $r/D \simeq 0.1$ (compare with Figs 3 and 4) and then again decrease with increase of the r/D ratio. In those regions, similarly as for composite hemispheres, the failure mode is also bifurcation buckling.

The mechanical properties of the shells considered (i.e. the E_1/E_2 and E_1/G_{12} ratios) affect the values of the buckling pressures. The failure pressures are lower for shells made of GFRP than for those of CFRP domes. This resembles entirely the variations of the maximal buckling pressures with the E_1/E_2 and E_1/G_{12} ratios (see Figs 1 and 2); they increase with increase of the mentioned ratios, but it needs further study.

5 EXPERIMENTS

A series of buckling tests were undertaken to provide some comparisons with the analytical predictions described in the previous sections. The tested torispherical shells were made of carbon–epoxy (woven roving 98141 fabricated by Interglass with the epoxy resin—Epidian 53 and hardener Z-1—material (5) in Table 1). The manufactured shells had diameter $D = 272$ mm and thicknesses equal to 1·61 mm (three models) and 3·22 mm (three models). One laminate configuration $0°/\pm45°/0°$ only was analysed. Of course, the above notation has a conventional meaning only due to the impossibility of the fabrication of axisymmetric torispheres made of one piece of woven roving. Thus, even the direction $0°$ denotes the direction of an arbitrary chosen shell meridian, and the rest of the layers are oriented with respect to the first one. Failure load summaries for the analysed CFRP torispheres having $R/D = 1.0$ and $r/D = 0.1$ are given in Table 2.

The first purpose of the study was to analyse the effects of the thickness-to-spherical segment radius ratio on the values of buckling pressures. In

TABLE 2

Comparison of theoretical and experimental results for CFRP torispherical shells having R/D = 1·0 and r/D = 0·1

No.	No. of layers	R/t	Failure pressures (MPa)				
			Experimental	Buckling		Strength ((1), (2))	
			P_{exp}	P_{cr}	P_{cr}/P_{exp}	P_f	P_f/P_{exp}
1	4	168·9	1·31	1·365	1·04	6·328	4·83
2	4	168·9	1·28	1·365	1·07	6·328	4·94
3	4	168·9	1·29	1·365	1·06	6·328	4·905
4	8	84·47	5·11	5·46	1·07	12·66	2·48
5	8	84·47	5·03	5·46	1·09	12·66	2·52
6	8	84·47	5·21	5·46	1·05	12·66	2·43

order to examine this problem the shells were made of 4 and 8 layers (the thickness of layers having the same directions was doubled in the second case). The identical problem was considered in Ref. 18, however in that case the compared shells had different orientations which caused automatically the lack of consistency in the considerations. The present experimental study shows evidently that the buckling pressures are proportional to the $(t/R)^2$ ratios.

The repeatability of the failure pressures and modes seems reasonable. As may be seen from Table 2 and Fig. 7 the torispheres analysed should fail

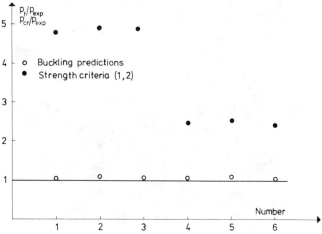

FIG. 7. Comparison of experimental results with the buckling and stress failure criteria predictions.

(a)

(b)

FIG. 8.   A torispherical CFRP shell after failure: (a) model (1); (b) model (4).

by bifurcation buckling rather than by fracture due to the strength criteria—see eqns (1) and (2). However, the photographs presented (Figs 8(a) and 8(b)) of the post-buckled state for all shells show the brittle fractures of shells. For the thinner shells (made of four layers), the fractures were rather small and were in the circumferential direction (Fig. 8(a)). For the thicker composite shells, the cracks appeared also in the circumferential direction but were much bigger than previously (Fig. 8(b)). Since all shells tested failed in the spherical cap region, the previous assumption about taking into account membrane stresses in the strength criteria (eqns (1) and (2)) seems to be reasonable.

The appearance of circumferential cracks in the spherical region may be due to the following reasons: a very low limit strain and strength (especially in compression) for CFRP in comparison with, for example, GFRP, and additionally very high bending stresses can arise in the moment of the loss of stability in buckling wave regions. In this situation for CFRP structures it is very difficult to distinguish the types of the shell load-carrying capacity due to buckling or fracture. The experimental failure loads obtained are lower than the BOSOR4 predictions—see Fig. 7 and Table 2. The difference is not very great, it does not exceed 10%. This can be caused mainly by the shell theory used here and the assumption of shell axisymmetry. Now, it is well known (see, for example, Muc[13]) that using the Love–Kirchhoff hypothesis one can obtain too high buckling pressures and this fact is much more important for shells having high E_1/G_{13} and t/R ratios, so that for the composite materials analysed here, the discrepancy between the experimental results and the theoretical (BOSOR4) predictions is much better than that presented in Ref. 18, probably due to the better method of fabrication. It is much easier to manufacture doubly-curved composite shells having bigger geometrical dimensions and in addition without imperfections—the results presented in Ref. 18 concerned mainly the imperfect torispheres.

With regard to the strength criteria ((1), (2)) for composite shells and to the buckling pressures, it is visible from the fifth and the seventh columns in Table 2 that the buckling pressures p_{cr} increase proportionally with $(t/R)^2$, and p_f with t/R—see also Fig. 7. Hence one can conclude that for the t/R ratios analysed herein the failure mode is bifurcation buckling.

REFERENCES

1. HOM, K., Fibre-reinforced plastics for hydrospace applications. In *Mechanics of Composite Materials*, 1967, pp. 456–66.

2. ALMROTH, B. O., Design of composite material structures for buckling—an evaluation of the state-of-art. *Technical Report AWFAL-TR-81-3102*, 1981.

3. MCELMAN, J. A., Vibration and buckling analysis of composite plates and shells. *J. Comp. Mater.*, **5** (1971) 529–32.

4. OYLER, J. F. and DYM, C. L., The dynamics and stability of composite shells. In *Proc. 13th Midwestern Conf. Developments in Mechanics*, Vol. 7, 1973, pp. 475–85.

5. COHEN, G. A., FASOR—a second generation shell of revolution code. *Comp. Struct.*, **10** (1978) 301–9.

6. BUSHNELL, D., Computerized analysis of shells—governing equations. *Comp. Struct.*, **18** (1984) 471–536.

7. ABI-SHAHEEN, S. A., Buckling of composite shells of revolution. PhD Thesis, Queen Mary College, University of London, 1979.

8. RIKARDS, R. B. and TETERS, G. A., *Stability of Composite Shells*. Ryga, Znatne, 1974 (in Russian).

9. RADHAMOHAN, S. K. and SHIRODE, A. D., Buckling of orthotropic torispherical pressure vessels. *Comp. Struct.*, **5** (1975) 155–9.

10. CHAO, C. C., TUNG, T. P. and CHERN, Y. C., Buckling of thick orthotropic shells. *Comp. Struct.*, **9** (1988) 113–39.

11. BABICH, YU. and GUZ, A. N., Stability of rods, plates and shells made of composite materials (three-dimensional formulation): survey. *Sov. Appl. Mech.*, **10** (1984) 835–49.

12. PETROVSKY, A. V., On the stability of a spherical shell made of layered composite material. *Mekh. Polymerov*, (1976) 459–64 (in Russian).

13. MUC, A., Transverse shear effects in stability problems of laminated shallow shells. *Comp. Struct.*, **12** (1989).

14. BERT, C. W., Analysis of radial-filament-reinforced spherical shells under deep submergence condition. pp. 529–34.

15. IVANOV, O. N., ORLOVSKIJ, L. S. and MYSYK, D. A., Stability of three-layered glass shells of revolution. *Prikladnaya Mekh.*, **20** (1984) 107–9 (in Russian).

16. IVANOV, O. N. and PEREVOSCHIKOVA, V. M., Experimental study of stability of composite shells of revolution subjected to external pressure. *Mekh. Kompozitnych Mat.*, **15** (1985) 1120–1 (in Russian).

17. PEREVOSCHIKOVA, V. M. and IVANOV, O. N., Stability analysis of stiffened composite shells of revolution under external pressure. *VINITI*, (No. 7) (1984) 651 (in Russian).

18. GALLETLY, G. D. and MUC, A., Buckling of externally-pressurised composite torispherical shells, *Proc. I. Mech. E.* (submitted).

19. FUDZI, T. and DZAKO, M., *Fracture Mechanics of Composite Materials*. Mir, Moscow, 1982 (in Russian).

20. TSAI, S. W., *Composite Design*. Think Composites, Dayton, 1987.

21. GALLETLY, G. D., BLACHUT, J. and KRUZELECKI, J., Plastic buckling of imperfect hemispherical shells subjected to external pressure. *Proc. I. Mech. E.*, **201** (C3) (1987) 153–70.

22. TENNYSON, R. C. and HANSEN, J. S., Optimum design for buckling of laminated cylinders. In *Collapse: the Buckling of Structures in Theory and Practice*. ed. J. M. T. Thompson and G. W. Hunt. Cambridge University Press, Cambridge 1983, pp. 409–29.

4

Exact and Approximate Linear and Nonlinear Initial Failure Analysis of Laminated Mindlin Plates in Flexure

G. J. TURVEY and M. Y. OSMAN

*Department of Engineering, University of Lancaster,
Bailrigg, Lancaster LA1 4YR, UK*

ABSTRACT

Exact Navier and approximate finite-difference dynamic relaxation (DR) solutions of the small deflection Mindlin plate equations are combined with six failure criteria to carry out initial flexural failure analyses of nine laminated rectangular simply supported plates previously analysed by Reddy and Pandey.[15] Although the initial failure results derived from the present exact and approximate analyses agree well, in many instances (particularly for the thinner plates) they show poor agreement with corresponding results (derived from finite element analysis) presented in Ref. 15. The results obtained by the authors suggest that all six failure criteria lead to roughly similar initial failure pressures (which are generally lower than given in Ref. 15) and, therefore, they do not support Reddy and Pandey's conclusion that the Tsai–Hill criterion is conservative for failure in flexure.

An approximate large deflection initial flexural failure analysis, also based on DR, is described. The same nine laminated plates with simply supported and clamped edges are reanalysed to highlight the effect of significant membrane action on the initial failure response. Again it is demonstrated that the initial failure pressures are independent of the failure criterion selected. Moreover, comparison of corresponding small and large deflection failure data indicates that small deflection theory significantly underestimates the initial failure pressure and overestimates the maximum deflection, especially when the laminates are thin.

NOTATION

a	Length of plate
A_{ij} $(i,j=1,2,6)$	Extensional stiffnesses
A_{ij} $(i,j=4,5)$	Transverse shear stiffnesses
$\bar{A}_{ij}$ $(=A_{ij}E_T^{-1}h_o^{-1})$	Dimensionless extensional and transverse shear stiffnesses
b	Plate width
B_{ij} $(i,j=1,2,6)$	Coupling stiffnesses
$\bar{B}_{ij}$ $(=B_{ij}E_T^{-1}h_o^{-2})$	Dimensionless coupling stiffnesses
D_{ij} $(i,j=1,2,6)$	Flexural stiffnesses
$\bar{D}_{ij}$ $(=D_{ij}E_T^{-1}h_o^{-3})$	Dimensionless flexural stiffnesses
e_x^o, e_y^o, e_{xy}^o	Plate mid-plane strains
e_{xz}^o, e_{yz}^o	Transverse shear strains
E_L, E_T	Longitudinal and transverse moduli of a unidirectional lamina
F_i $(i=1\text{–}6)$	Coefficients of linear stress terms in the tensor polynomial failure criterion
F_{ij} $(i,j=1\text{–}6)$	Coefficients of quadratic stress terms in the tensor polynomial failure criterion
G_{LT}, G_{LZ}, G_{TZ}	Shear moduli of a unidirectional lamina
h_{k-1}, h_k	Distances of upper and lower surfaces of the kth lamina from the plate mid-plane
h_o	Plate thickness
k	Lamina number
$k_i k_j$ $(i,j=4,5)$	Shear correction factors
$k_u, k_v, k_w, k_{\phi x}, k_{\phi y}$	Damping factors
k_x^o, k_y^o, k_{xy}^o	Plate mid-plane curvatures
M_x, M_y, M_{xy}	Stress couples
N_x, N_y, N_{xy}	Stress resultants
q, q_o	Transverse pressure
$\bar{q}$ $(=qa^4E_T^{-1}h_o^{-4})$	Dimensionless transverse pressure
Q_{ij} $(i,j=1\text{–}6)$	Reduced stiffnesses
Q_x, Q_y	Transverse shear stress resultants
R, S, T	In-plane and transverse shear strengths of a unidirectional lamina
s	Reference number of failure surface within a lamina
u, v	In-plane displacements
w	Deflection

$\bar{w} \ (= wh_o^{-1})$	Dimensionless deflection
x, y, z	Cartesian coordinates
$\bar{x} \ (= xa^{-1}), \ \bar{y} \ (= yb^{-1})$	Dimensionless coordinates
$X_{t,c}, Y_{t,c}$	Longitudinal and transverse tensile/compressive strengths of a unidirectional lamina
δt	Time increment
λ	Scaling factor
ν_{LT}, ν_{TL}	Poisson ratios of a unidirectional lamina
$\rho_u, \rho_v, \rho_w, \rho_{\phi_x}, \rho_{\phi_y}$	Densities
$\sigma_i \ (i = 1\text{--}6)$	Stress components with respect to material principal axes
σ_L, σ_T	Direct stresses parallel and normal to the fibre direction
$\sigma_{LT}, \sigma_{LZ}, \sigma_{TZ}$	In-plane and transverse shear stresses with respect to material principal axes
ϕ_x, ϕ_y	Rotations of the 'original' normal to the plate mid-plane
$(\ )^{\cdot}, (\ )'$	Partial derivatives with respect to x and y respectively
$(\ )_{,t}$	Total derivative with respect to time

Subscripts/Superscripts

c	Value of the variable at the plate centre
f	Value of variable associated with the onset of flexural failure
l	Lower limit of variable
max	Maximum value of variable
s	Value of the variable associated with a small value of the transverse pressure
u	Upper limit of variable

1 INTRODUCTION

Over the past two decades a considerable literature has accumulated on the linear and nonlinear elastic analysis of laminated fibre-reinforced plates of rectangular planform. Useful summaries of the more important analytical techniques and results of this body of research literature may be found in Refs 1 and 2. Somewhat surprisingly, however, the literature dealing with the initial or ultimate failure analysis of laminated plates in flexure is very small indeed, even though the results of such analyses are potentially of equal, if not greater, importance than those of elastic analysis, especially from the design standpoint.

Some 8 years ago the senior author embarked on the first of a series of initial flexural failure analyses of laminated plates and strips. These analyses are based on exact (usually series) solutions of the elastic small deflection laminated thin plate equations combined with the Tsai–Hill failure criterion. Because of the use of exact solutions, the results of the analyses are restricted, with a few exceptions, to simply supported edge conditions. A reasonably extensive set of initial failure results for uniformly loaded CFRP (carbon fibre reinforced plastic) and GFRP (glass fibre reinforced plastic) plates and strips with symmetric and antisymmetric lay-ups have been computed, the details of which are presented in Refs 3–9. A similar approach was used recently by Adali *et al.*[10,11] to explore the effects of moisture and temperature-dependent lamina elastic stiffnesses on the initial failure response of uniformly loaded specially orthotropic rectangular and cross-ply laminated elliptic plates. A further recent development, undertaken by the senior author, was to explore the effect of shear deformation on the initial flexural failure response of symmetrically laminated cross-ply plates.[12] This exact analysis is based on Reddy's higher-order small deflection laminated plate theory.[13]

Although the initial flexural failure results reported in Refs 3–12 are important because they are exact and, therefore, constitute 'benchmarks' for the checking of alternative approximate analyses, they are, nevertheless, limited—in general being valid only for simply supported thin laminated plates operating in the small deflection regime.

In order to overcome these limitations, resort may be made to approximate initial flexural failure analysis. This implies that an approximate rather than an exact solution of the governing plate equations is used. Approximate elastic plate solutions may be derived using various discrete solution techniques, e.g. finite differences, finite elements, etc. The senior author used finite differences in conjunction with the DR algorithm to compute large deflection initial flexural failure data for laminated plate strips subjected to uniform pressure loading.[14] More recently, Reddy and Pandey[15] combined finite element solutions of the small deflection laminated Mindlin plate equations with six failure criteria to generate initial failure data for uniformly loaded rectangular simply supported plates with nine lay-ups taken from the $0°/\pm45°/90°$ family. The extensive initial flexural failure results computed by Reddy and Pandey highlighted the poor performance of the Tsai–Hill failure criterion in comparison to the other five criteria. The Tsai–Hill criterion led to failure pressure estimates which were far too conservative and, moreover, predicted failure to occur at locations which do not agree with the predictions of the other criteria. Reddy and Pandey's results would, therefore, appear to suggest that the Tsai–Hill criterion is unsuitable for the initial flexural failure analysis of laminated rectangular plates.

In the light of their own experience with both exact and approximate initial flexural failure analyses of laminated plates, the authors were doubtful about the implications of Reddy and Pandey's conclusions concerning the Tsai–Hill failure criterion.

Therefore they decided to develop their own initial flexural failure analysis based on laminated Mindlin plate theory in order to be able to check Reddy and Pandey's initial failure results for uniformly loaded simply supported rectangular laminated plates. An approximate small deflection analysis, based on a finite-difference DR solution of the plate equations, was developed first. This was followed by an exact Navier-type analysis valid only for simply supported cross-ply and antisymmetric angle-ply laminates. And, finally, an approximate large deflection failure analysis, also based on DR, was developed. A reasonably comprehensive description of the main constituents of these analyses, i.e. the elastic solutions of the plate equations and the procedures used to 'scale' it to ensure satisfaction of the failure criterion at one or more points in the plate, is presented in each case.

The exact small deflection failure analysis is used primarily to verify the approximate small deflection DR analysis. The results of these two analyses are shown to agree very well indeed. With the accuracy of the latter analysis established, it is then used to compute failure pressures for each of the laminates considered by Reddy and Pandey. The results comparison indicates that the DR failure pressures are generally lower (sometimes substantially lower) than Reddy and Pandey's values. Moreover, and in contradiction to Reddy and Pandey's conclusions, it is found from the present study that the Tsai–Hill criterion predicts failure pressures and locations which are in broad agreement with other failure criteria.

The large deflection DR failure analysis has also been used to compute initial failure pressures and maximum deflections for the same range of plate problems. The small and large deflection results comparison presented for both simply supported and clamped plates subjected to uniform pressure loading shows that small deflection analysis provides a conservative estimate of the failure pressure and an unconservative estimate of the maximum deflection.

Finally, as a postscript, a number of suggestions are offered as a possible partial explanation for some of Reddy and Pandey's failure pressures being too large.

2 PLATE GEOMETRY, LAMINATION SCHEMES AND MATERIAL PROPERTIES

As this paper is concerned with the recomputation of initial failure data for uniformly loaded simply supported rectangular laminated plates originally presented by Reddy and Pandey,[15] it is appropriate, first of all, to reiterate the relevant details of each plate before proceeding to describe various aspects of the present failure analysis and results.

The planform geometry of each plate and the principal geometric and material coordinate axes are shown in Fig. 1. Each plate is assumed to be laminated from the same unidirectional CFRP pre-preg. The elastic and strength properties of the pre-preg are listed in Table 1. A total of nine lay-ups, designated A–I, have been selected for analysis. They all belong to the $0°/\pm45°/90°$ family of laminates and include both symmetric and antisymmetric lay-ups. Details of the lay-up of each laminate together with its corresponding span to thickness ratio are given in Table 2. In order to clarify further the lamina numbering sequence for each lay-up, the lamina numbers and corresponding fibre-angles for laminates A and B are illustrated in Fig. 2.

It is very obvious from the span to thickness ratios quoted in Table 2 that the majority of the plates are very thin indeed. In fact, even the thickest, 16-layer plates, would be regarded as sufficiently thin for shear deformation

TABLE 1
Material properties of unidirectional CFRP pre-preg

(a) *Elastic modular ratios*

E_L/E_T	G_{LT}/E_T	G_{LZ}/E_T	G_{TZ}/E_T	v_{LT}
12·31	0·526	0·526	0·314	0·24

(b) *Strength ratios*

X_t/Y_t	X_c/Y_t	Y_c/Y_t	R/Y_t	S/Y_t	T/Y_t
34·6	38·7	1·000	1·543	1·984	1·984

$E_T/Y_T = 245·7$ (approximately one quarter of the value for mild steel).
Ply thickness = 0·127 mm.

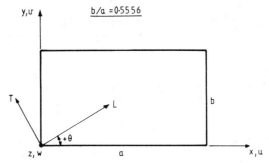

FIG. 1.   Plate geometry and plate and material coordinate systems.

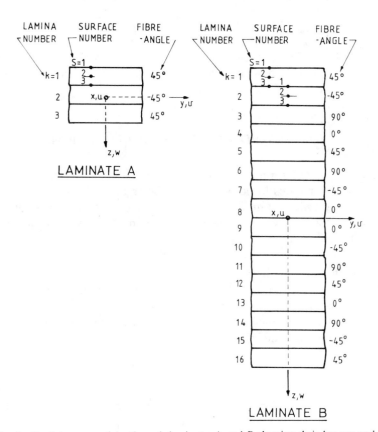

FIG. 2.   Partial cross-sections through laminates A and B showing their lay-ups and the lamina and failure surface numbering systems.

TABLE 2
Plate lay-up and slenderness details

Plate	b/h_o	Lay-up	Type
A	333·3	$(45°/-45°/45°)_t$	Symm., 45° angle-ply
B	62·5	$(45°/-45°/90°/0°/45°/90°/-45°/0°)_s$	Symm.
C	62·5	$(45°/-45°/0°/90°/45°/0°/-45°/90°)_s$	Symm.
D	62·5	$(45°/0°/-45°/0°/-45°/90°/0°/45°)_s$	Symm.
E	62·5	$(45°/0°/-45°/0°/-45°/0°/45°/0°)_s$	Symm.
F	250·0	$(0°/90°/90°/0°)_t$	Symm., cross-ply
G	500·0	$(90°/0°)_t$	Antisymm., cross-ply
H	250·0	$(90°/0°/90°/0°)_t$	Antisymm., cross-ply
I	250·0	$(-45°/45°/-45°/45°)_t$	Antisymm., 45° angle-ply

$(\ )_s$ = symmetric lay-up; $(\ )_t$ = total lay-up.

effects to be negligible. Thus, strictly speaking, there is no need to use the more complicated Mindlin plate theory in an initial flexural failure analysis of these plates.

3 LAMINATED MINDLIN PLATE EQUATIONS

In their 'separated' form the equations which describe the large deflection response of laminated Mindlin plates may be organised into the following groups: equilibrium, strain/curvature, constitutive and boundary condition equations. Each group of equations is presented below in order and without derivation.

3.1 Equilibrium Equations

Mindlin large deflection plate theory attributes five degrees of freedom—three translations and two rotations—to each point in the mid-plane of the plate. Consequently, there are a total of five equations which define the deformed state of equilibrium. They are as follows:

$$N_x^{\cdot} + N_{xy}' = 0$$
$$N_{xy}^{\cdot} + N_y' = 0$$
$$Q_x^{\cdot} + Q_y' + N_x w^{\cdot\cdot} + 2N_{xy}w^{\cdot\prime} + N_y w'' + q = 0 \qquad (1)$$
$$M_x^{\cdot} + M_{xy}' - Q_x = 0$$
$$M_{xy}^{\cdot} + M_y' - Q_y = 0$$

3.2 Strain/Curvature Equations

The relationships between strains and curvatures of the plate mid-plane and derivatives of the mid-plane displacements are:

$$
\begin{aligned}
e^o_x &= u^{\cdot} + \tfrac{1}{2}(w^{\cdot})^2 & k^o_x &= \phi^{\cdot}_x & e^o_{xz} &= \phi_x + w^{\cdot} \\
e^o_y &= v' + \tfrac{1}{2}(w')^2 & k^o_y &= \phi'_y & e^o_{yz} &= \phi_y + w' \\
e^o_{xy} &= u' + v^{\cdot} + \underline{w^{\cdot}w'} & k^o_{xy} &= \phi^{\cdot}_y + \phi'_x &&
\end{aligned}
\qquad (2)
$$

3.3 Constitutive Equations

The general forms of the constitutive equations for arbitrarily laminated shear deformable rectangular plates are:

$$
\begin{bmatrix} N_x \\ N_y \\ N_{xy} \end{bmatrix} =
\begin{bmatrix} A_{11} & A_{12} & A_{16} \\ A_{12} & A_{22} & A_{26} \\ A_{16} & A_{26} & A_{66} \end{bmatrix}
\begin{bmatrix} e^o_x \\ e^o_y \\ e^o_{xy} \end{bmatrix} +
\begin{bmatrix} B_{11} & B_{12} & B_{16} \\ B_{12} & B_{22} & B_{26} \\ B_{16} & B_{26} & B_{66} \end{bmatrix}
\begin{bmatrix} k^o_x \\ k^o_y \\ k^o_{xy} \end{bmatrix}
\qquad (3a)
$$

$$
\begin{bmatrix} M_x \\ M_y \\ M_{xy} \end{bmatrix} =
\begin{bmatrix} B_{11} & B_{12} & B_{16} \\ B_{12} & B_{22} & B_{26} \\ B_{16} & B_{26} & B_{66} \end{bmatrix}
\begin{bmatrix} e^o_x \\ e^o_y \\ e^o_{xy} \end{bmatrix} +
\begin{bmatrix} D_{11} & D_{12} & D_{16} \\ D_{12} & D_{22} & D_{26} \\ D_{16} & D_{26} & D_{66} \end{bmatrix}
\begin{bmatrix} k^o_x \\ k^o_y \\ k^o_{xy} \end{bmatrix}
\qquad (3b)
$$

$$
\begin{bmatrix} Q_x \\ Q_y \end{bmatrix} =
\begin{bmatrix} A_{44} & A_{45} \\ A_{45} & A_{55} \end{bmatrix}
\begin{bmatrix} e^o_{xz} \\ e^o_{yz} \end{bmatrix}
\qquad (3c)
$$

in which

$$
\left.
\begin{aligned}
A_{ij} &= \sum_{k=1}^{n} Q_{ij}(h_k - h_{k-1}) \\
B_{ij} &= \frac{1}{2} \sum_{k=1}^{n} Q_{ij}(h_k^2 - h_{k-1}^2) \\
D_{ij} &= \frac{1}{3} \sum_{k=1}^{n} Q_{ij}(h_k^3 - h_{k-1}^3)
\end{aligned}
\right\}
\qquad \text{for } i,j = 1,2,6
$$

and

$$
A_{ij} = \sum_{k=1}^{n} k_i k_j Q_{ij}(h_k - h_{k-1}) \qquad \text{for } i,j = 4,5
$$

TABLE 3
Plate stiffnesses

(a) Dimensionless extensional and transverse shear stiffnesses

Plate	$\bar{A}_{11}$	$\bar{A}_{12}$	$\bar{A}_{16}$	$\bar{A}_{22}$	$\bar{A}_{26}$	$\bar{A}_{66}$	$\bar{A}_{44}$	$\bar{A}_{45}$	$\bar{A}_{55}$
A	3·987	2·935	0·946	3·987	0·946	3·220	0·350	0·029	0·350
B	5·334	1·588	0·000	5·334	0·000	1·873	0·350	0·000	0·350
C	5·334	1·588	0·000	5·334	0·000	1·873	0·350	0·000	0·350
D	6·753	1·588	0·000	3·915	0·000	1·873	0·328	0·000	0·372
E	8·173	1·588	0·000	2·496	0·000	1·873	0·306	0·000	0·394
F	6·681	0·241	0·000	6·681	0·000	0·526	0·350	0·000	0·350
G	6·681	0·241	0·000	6·681	0·000	0·526	0·350	0·000	0·350
H	6·681	0·241	0·000	6·681	0·000	0·526	0·350	0·000	0·350
I	3·987	2·935	0·000	3·987	0·000	3·220	0·350	0·000	0·350

(b) Dimensionless coupling stiffnesses

Plate	$\bar{B}_{11}$	$\bar{B}_{12}$	$\bar{B}_{16}$	$\bar{B}_{22}$	$\bar{B}_{26}$	$\bar{B}_{66}$
A	0·000	0·000	0·000	0·000	0·000	0·000
B	0·000	0·000	0·000	0·000	0·000	0·000
C	0·000	0·000	0·000	0·000	0·000	0·000
D	0·000	0·000	0·000	0·000	0·000	0·000
E	0·000	0·000	0·000	0·000	0·000	0·000
F	0·000	0·000	0·000	0·000	0·000	0·000
G	1·419	0·000	0·000	−1·419	0·000	0·000
H	0·710	0·000	0·000	−0·710	0·000	0·000
I	0·000	0·000	0·355	0·000	0·355	0·000

(c) Dimensionless flexural stiffnesses

Plate	$\bar{D}_{11}$	$\bar{D}_{12}$	$\bar{D}_{16}$	$\bar{D}_{22}$	$\bar{D}_{26}$	$\bar{D}_{66}$
A	0·332	0·245	0·219	0·332	0·219	0·268
B	0·363	0·169	0·033	0·452	0·033	0·193
C	0·452	0·169	0·033	0·363	0·033	0·193
D	0·589	0·151	0·019	0·263	0·019	0·175
E	0·616	0·153	0·022	0·231	0·022	0·177
F	0·912	0·020	0·000	0·202	0·000	0·044
G	0·557	0·020	0·000	0·557	0·000	0·044
H	0·557	0·020	0·000	0·557	0·000	0·044
I	0·332	0·245	0·000	0·332	0·000	0·268

The product of the shear correction factors, $k_i k_j$, has been assumed equal to 5/6 in evaluating the transverse shear stiffnesses of laminates A–I.

Due to symmetry and/or other features of the lay-ups, none of the laminates A–I is governed by the general form of the constitutive relationships represented by eqns (3). In every case several of the A_{ij}, B_{ij}, etc., terms are zero. For example, the laminates A–F have symmetric lay-ups and, therefore, their B_{ij} stiffnesses are all zero. Details of the normalised values of the A_{ij}, B_{ij}, etc., stiffnesses are given in Table 3 for each of the laminates.

3.4 Boundary Conditions

The principal edge support conditions considered here are those of simple support. However, within the context of laminated Mindlin plate

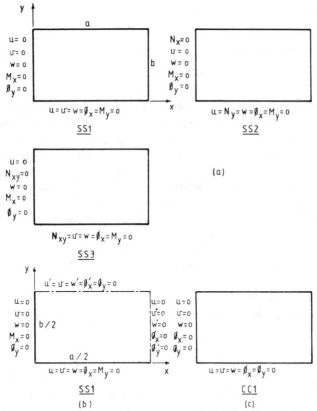

FIG. 3. Boundary conditions for rectangular plates with all edges either simply supported or clamped. (a) SS1, SS2 and SS3 simply supported edge conditions. (b) SS1 edge and centreline conditions for a quarter-plate analysis. (c) CC1 clamped edge conditions.

theory there are six basic types of simply supported edge condition. Three of these are considered here and, for ease of reference, they are designated SS1, SS2 and SS3. Details of the edge constraints for each case are shown in Fig. 3(a). Some of the laminates A–I have cross-ply lay-ups. This feature is exploited to reduce the computational task, so that in these instances only a quarter-plate analysis is carried out. Details of the edge and centreline constraints for such an analysis of an SS1 simply supported plate are shown in Fig. 3(b).

Several initial failure analyses of clamped edge plates have been undertaken as part of the overall study. Six possible types of clamped edge may be defined, but only that shown in Fig. 3(c) and designated CC1 has been used.

4 APPROXIMATE AND EXACT ANALYSIS OF PLATE EQUATIONS

Both approximate and exact numerical analyses of the governing plate equations were developed as part of the initial flexural failure analysis of laminated rectangular plates. Two approximate analyses, each based on a finite-difference version of the dynamic relaxation (DR) method, were developed—one for the solution of the large deflection equations, i.e. eqns (1)–(3), and the other for the solution of the small deflection equations, i.e. eqns (1)–(3) but with the underlined terms absent. The exact analysis is essentially a Navier (double Fourier series) solution of the small deflection equations and is valid for simply supported rectangular plates with cross-ply or antisymmetric angle-ply lay-ups. Further details of both the approximate and exact analyses are presented below.

4.1 Approximate DR Analysis

The DR method, originally developed for static structural analysis by Day[16] and Otter et al.,[17] may be understood in terms of its physical analogy, viz. that of a vibrating system in which the oscillations are attenuated rapidly to yield the static configuration by applying near critical damping. The device of adding damping and inertia terms to produce quasi-dynamic versions of eqns (1) allows the original boundary value problem to be transformed into initial value format for solution using a time-stepping iterative procedure. The transformations are set out below for the large deflection case; the modifications for small deflection analysis are simply to omit the underlined terms in eqns (1) and (2).

Adding damping and inertia terms to the right-hand sides (RHSs) of eqns (1), they become

$$N_x^{\cdot} + N_{xy}' = \rho_u u_{,tt} + k_u u_{,t}$$
$$N_{xy}^{\cdot} + N_y' = \rho_v v_{,tt} + k_v v_{,t}$$
$$Q_x^{\cdot} + Q_y' + N_x w^{\cdot\cdot} + 2N_{xy} w^{\cdot\prime} + N_y w'' + q = \rho_w w_{,tt} + k_w w_{,t} \tag{4}$$
$$M_x^{\cdot} + M_{xy}' - Q_x = \rho_{\phi_x} \phi_{x,tt} + k_{\phi_x} \phi_{x,t}$$
$$M_{xy}^{\cdot} + M_y' - Q_y = \rho_{\phi_y} \phi_{y,tt} + k_{\phi_y} \phi_{y,t}$$

In order to effect the transformation to initial value format, the velocity and acceleration terms on the RHSs of eqns (4) may be approximated as follows:

$$\alpha_{,t} = \tfrac{1}{2}(\alpha_{,t}^a + \alpha_{,t}^b) \quad \text{and} \quad \alpha_{,tt} = \delta t^{-1}(\alpha_{,t}^a - \alpha_{,t}^b) \tag{5}$$

in which $\alpha \equiv (u, v, w, \phi_x, \phi_y)$ and the superscripts a and b refer respectively to the values of the quantities after and before the time step, δt.

Thus now substituting eqns (5) into eqns (4) and rearranging the following expressions are obtained for the new velocities in terms of the old velocities and the 'out of balance forces':

$$u_{,t}^a = \{u_{,t}^b(1 - k_u^*) + \delta t \rho_u^{-1}(N_x^{\cdot} + N_{xy}')\}(1 + k_u^*)^{-1}$$
$$v_{,t}^a = \{v_{,t}^b(1 - k_v^*) + \delta t \rho_v^{-1}(N_{xy}^{\cdot} + N_y')\}(1 + k_v^*)^{-1}$$
$$w_{,t}^a = \{w_{,t}^b(1 - k_w^*) + \delta t \rho_w^{-1}(Q_x^{\cdot} + Q_y' + N_x w^{\cdot\cdot} + 2N_{xy} w^{\cdot\prime} + N_y w'' + q)\}$$
$$\qquad \times (1 + k_w^*)^{-1} \tag{6}$$
$$\phi_{x,t}^a = \{\phi_{x,t}^b(1 - k_{\phi_x}^*) + \delta t \rho_{\phi_x}^{-1}(M_x^{\cdot} + M_{xy}' - Q_x)\}(1 + k_{\phi_x}^*)^{-1}$$
$$\phi_{y,t} = \{\phi_{y,t}^b(1 - k_{\phi_y}^*) + \delta t \rho_{\phi_y}^{-1}(M_{xy}^{\cdot} + M_y' - Q_y)\}(1 + k_{\phi_y}^*)^{-1}$$

in which $k_\alpha^* = \tfrac{1}{2}k_\alpha \delta t \rho_\alpha^{-1}$ and the subscript α is defined as under eqn (5).

Clearly, in order to obtain displacements from eqns (6), it is necessary to integrate the velocities and the following simple set of equations are used to ensure displacement–time compatibility:

$$\alpha^a = \alpha^b + \delta t \alpha_{,t}^a \tag{7}$$

and again α is defined as under eqns (5).

Equations (6), (7), (2) and (3), together with the appropriate boundary conditions (see Fig. 3), constitute the complete set of equations required for the DR iterative procedure, which is as follows:

Step 1   Initialise all variables, apply the load q, and prescribe k_α, ρ_α and δt.
Step 2   Compute the velocities (eqns (6)).
Step 3   Compute the displacements (eqns (7)).
Step 4   Apply displacement boundary conditions, i.e. the relevant conditions of Fig. 3.
Step 5   Compute the strains, curvatures and shear strains (eqns (2)).
Step 6   Compute the stress resultants, stress couples and shear stress resultants (eqns (3)).
Step 7   Apply 'stress' boundary conditions, i.e. the relevant conditions of Fig. 3.
Step 8   Check for solution convergence (usually that the velocities are less than 10^{-6} everywhere). If the convergence criterion is satisfied, then print out the static solution, otherwise return to Step 2 and repeat the sequence of computations.

Because the equations of the DR procedure, as given above, represent the quasi-dynamic response of the plate continuum, they have to be rendered spatially discrete prior to their numerical solution on a digital computer. Central finite differences are used for this purpose, though other discretisation schemes such as finite elements may be used. The finite-difference versions of eqns (6), etc., are not included here for the sake of brevity. Likewise, no details are given of the techniques used to simplify the choice of the parameters k_α, ρ_α and δt, which govern the stability and solution convergence of the DR iterative procedure, as they may be found elsewhere.[18,19]

4.2 Exact Navier Analysis

Reddy[20] has developed exact Navier-type solutions for simply supported rectangular laminated Mindlin plates. These solutions are used in the exact initial failure analysis of laminates F–I. They are only valid for cross-ply lay-ups with SS2 edge conditions and for antisymmetric lay-ups with SS3 conditions. A brief summary of these solutions is presented below.

4.2.1 Cross-ply lay-ups

The small deflection equilibrium equations may be obtained in 'displacement' format by substituting the linear version (underlined terms omitted) of eqns (2) and the cross-ply version ($A_{16} = A_{26} = A_{45} = B_{16} = B_{26} = D_{16} = D_{26} = 0$) of eqns (3) into eqns (1). The solution of the five partial differential equations in terms of the five degrees of displacement freedom is then postulated as

$$u = \sum_{m=1}^{\infty} \sum_{n=1}^{\infty} U_{mn} \cos \alpha x \sin \beta y \qquad \phi_x = \sum_{m=1}^{\infty} \sum_{n=1}^{\infty} \phi_{x_{mn}} \cos \alpha x \sin \beta y$$

$$v = \sum_{m=1}^{\infty} \sum_{n=1}^{\infty} V_{mn} \sin \alpha x \cos \beta y \qquad \phi_y = \sum_{m=1}^{\infty} \sum_{n=1}^{\infty} \phi_{y_{mn}} \sin \alpha x \cos \beta y \qquad (8)$$

$$w = \sum_{m=1}^{\infty} \sum_{n=1}^{\infty} W_{mn} \sin \alpha x \sin \beta y$$

in which $\alpha = m\pi a^{-1}$ and $\beta = n\pi b^{-1}$, and eqns (8) satisfy the SS2-type simply supported edge conditions illustrated in Fig. 3(a).

The transverse load acting on the plate is assumed as follows:

$$q = \sum_{m=1}^{\infty} \sum_{n=1}^{\infty} Q_{mn} \sin \alpha x \sin \beta y \qquad (9)$$

in which $Q_{mn} = 16\pi^{-2} q_0 m^{-1} n^{-1}$ and $m, n = 1, 3, 5, \ldots$ for the case of uniform pressure loading—the only type of load distribution considered here.

Substituting the appropriate derivatives of eqns (8) and (9) into the 'displacement' equilibrium equations allows the following set of algebraic equations to be set up in terms of the unknown displacement amplitudes:

$$\begin{bmatrix} a_{11} & a_{12} & a_{13} & a_{14} & a_{15} \\ a_{12} & a_{22} & a_{23} & a_{24} & a_{25} \\ a_{13} & a_{23} & a_{33} & a_{34} & a_{35} \\ a_{14} & a_{24} & a_{34} & a_{44} & a_{45} \\ a_{15} & a_{25} & a_{35} & a_{45} & a_{55} \end{bmatrix} \begin{bmatrix} U_{mn} \\ V_{mn} \\ W_{mn} \\ \phi_{x_{mn}} \\ \phi_{y_{mn}} \end{bmatrix} = \begin{bmatrix} 0 \\ 0 \\ -Q_{mn} \\ 0 \\ 0 \end{bmatrix} \qquad (10)$$

in which

$$\begin{aligned}
a_{11} &= A_{11}\alpha^2 + A_{66}\beta^2 & a_{25} &= B_{22}\beta^2 + B_{66}\alpha^2 \\
a_{12} &= (A_{12} + A_{66})\alpha\beta & a_{33} &= A_{55}\alpha^2 + A_{44}\beta^2 \\
a_{13} &= 0 & a_{34} &= A_{55}\alpha \\
a_{14} &= B_{11}\alpha^2 + B_{66}\beta^2 & a_{35} &= A_{44}\beta \\
a_{15} &= (B_{12} + B_{66})\alpha\beta & a_{44} &= D_{11}\alpha^2 + D_{66}\beta^2 + A_{55} \\
a_{22} &= A_{22}\beta^2 + A_{66}\alpha^2 & a_{45} &= (D_{12} + D_{66})\alpha\beta \\
a_{23} &= 0 & a_{55} &= D_{66}\alpha^2 + D_{22}\beta^2 + A_{44} \\
a_{24} &= a_{15}
\end{aligned} \qquad (10a)$$

Simple inversion of eqn (10) allows the undetermined coefficients of the displacement functions, i.e. U_{mn}, etc., to be evaluated in terms of the known loading coefficients, Q_{mn}. Other quantities of interest such as stress resultants are then readily computed from eqns (2) and (3).

4.2.2 Antisymmetric angle-ply lay-ups

Again the solution of the 'displacement' equilibrium equations is postulated as eqns (8) with the in-plane displacements redefined as

$$u = \sum_{m=1}^{\infty} \sum_{n=1}^{\infty} U_{mn} \sin \alpha x \cos \beta y \qquad v = \sum_{m=1}^{\infty} \sum_{n=1}^{\infty} V_{mn} \cos \alpha x \sin \beta y \quad (8a)$$

These modified displacement functions satisfy the SS3-type edge conditions shown in Fig. 3(a). Following the substitution procedure described in Section 4.2.1 an equation similar to eqn (10) may be set up. The a_{ij} coefficients in this equation are defined as in eqn (10a) with the following modifications:

$$a_{14} = 2B_{16}\alpha\beta \qquad a_{15} = B_{16}\alpha^2 + B_{26}\beta^2 \qquad a_{25} = 2B_{26}\alpha\beta \quad (10b)$$

Again simple inversion of eqn (10) allows the undetermined coefficients to be defined and hence the complete solution obtained.

5 FAILURE CRITERIA

In their initial flexural failure analyses of laminates A–I, Reddy and Pandey[15] used the following failure criteria: (a) maximum stress, (b) maximum strain, (c) Tsai–Hill, (d) Hoffman and (e) Tsai–Wu. For computational convenience each criterion was defined in truncated tensor polynomial format, i.e. in the form

$$F_i\sigma_i + F_{ij}\sigma_i\sigma_j + \cdots = 1 \qquad (i, j = 1\text{–}6 \quad \text{and} \quad \text{summation convention applies}) \quad (11)$$

The coefficients F_i and F_{ij} in eqn (11) are functions of the unidirectional lamina strengths and are presented below for each of the criteria (a)–(e).

(a) Maximum stress criterion

$$F_1 = X_t^{-1} - X_c^{-1} \qquad F_{11} = (X_t X_c)^{-1}$$
$$F_2 = Y_t^{-1} - Y_c^{-1} \qquad F_{12} = -\tfrac{1}{2}(X_t^{-1} - X_c^{-1})(Y_t^{-1} - Y_c^{-1}) \quad (11a)$$
$$F_{22} = (Y_t Y_c)^{-1}$$

(b) *Maximum strain criterion*

$$F_1 = (X_t^{-1} - X_c^{-1}) - v_{TL}(Y_t^{-1} - Y_c^{-1})$$

$$F_2 = -v_{TL}E_L E_T^{-1}(X_t^{-1} - X_c^{-1}) + (Y_t^{-1} - Y_c^{-1})$$

$$F_{11} = (X_t X_c)^{-1} + v_{TL}(X_t^{-1} - X_c^{-1})(Y_t^{-1} - Y_c^{-1}) + v_{TL}^2(Y_t Y_c)^{-1}$$

$$\begin{aligned} F_{12} = &-v_{TL}E_L E_T^{-1}(X_t X_c)^{-1} - \tfrac{1}{2}(1 + v_{TL}^2 E_L E_T^{-1}) \\ &\times (X_t^{-1} - X_c^{-1})(Y_t^{-1} - Y_c^{-1}) - v_{TL}(Y_t Y_c)^{-1} \end{aligned} \tag{11b}$$

$$\begin{aligned} F_{22} = &v_{TL}^2 E_L^2 E_T^{-2}(X_t X_c)^{-1} + v_{TL}E_L E_T^{-1}(X_t^{-1} - X_c^{-1}) \\ &\times (Y_t^{-1} - Y_c^{-1}) + (Y_t Y_c)^{-1} \end{aligned}$$

(c) *Tsai–Hill criterion*

$$F_{11} = X^{-2} \qquad F_{12} = -\tfrac{1}{2}X^{-2} \qquad F_{22} = Y^{-2} \tag{11c}$$

NB: In eqn (11c) X is set equal to X_t when σ_1 is tensile and equal to X_c when it is compressive. Likewise when σ_2 is tensile $Y = Y_t$ and when it is compressive $Y = Y_c$.

(d) *Hoffman's criterion*

$$\begin{array}{ll} F_1 = X_t^{-1} - X_c^{-1} & F_{11} = (X_t X_c)^{-1} \\ F_2 = Y_t^{-1} - Y_c^{-1} & F_{12} = -\tfrac{1}{2}\{(X_t X_c)^{-1} + (Y_t Y_c)^{-1}\} \\ & F_{22} = Y_t^{-1} Y_c^{-1} \end{array} \tag{11d}$$

(e) *Tsai–Wu*

$$\begin{array}{ll} F_1 = X_t^{-1} - X_c^{-1} & F_{11} = (X_t X_c)^{-1} \\ F_2 = Y_t^{-1} - Y_c^{-1} & F_{12} = -\tfrac{1}{2}(X_t X_c Y_t Y_c)^{-1} \end{array} \tag{11e}$$

In the preceding five failure criteria the F_{44}, F_{55} and F_{66} coefficients are common and are given by

$$F_{44} = R^{-2} \qquad F_{55} = S^{-2} \qquad F_{66} = T^{-2} \tag{11f}$$

All of the remaining F_i and F_{ij} coefficients in each criterion are identically zero.

In addition to the five truncated tensor polynomial failure criteria defined above, Reddy and Pandey[15] also made use of the maximum stress (independent) failure criterion. This is defined in terms of the following limits on the stresses in the principal material directions:

$$\sigma_L \leq X \qquad \sigma_T \leq Y \qquad \sigma_{LZ} \leq R \qquad \sigma_{TZ} \leq S \qquad \sigma_{LT} \leq T \tag{12}$$

Again the strengths X and Y are interpreted as for criterion (c) above.

6 COMPUTATION OF INITIAL FLEXURAL FAILURE DATA

Initial flexural failure data, i.e. pressures and associated plate centre deflections, are computed by determining the particular elastic solution of the plate equations which just satisfies the failure criterion in question at one point in the plate (simultaneous satisfaction at several points is, of course, possible in some circumstances). The means by which the particular elastic solution is determined depends on whether large or small deflection plate theory is used. In the latter case a simple scaling technique may be used, whereas in the former an iterative procedure must be employed. Details of these procedures now follow.

6.1 Scaling Procedure for Small Deflection Initial Failure Analysis

This procedure is used only with the criteria (a)–(e) (the procedure used with the maximum stress (independent) criterion is obvious and does not merit further explanation) and is set out below.

A small transverse pressure, say q_s, is applied to the plate and the stress field, σ_{i_s} ($i = 1, \ldots, 6$), produced within the plate is determined. If these stresses are substituted into the truncated tensor polynomial failure criterion, eqn (11), then

$$F_i \sigma_{i_s} + F_{ij} \sigma_{i_s} \sigma_{j_s} = F_s < 1 \tag{13}$$

Now as the plate response is linear up to the initial failure pressure q_f then $q_f = \lambda q_s$, where λ is a scaling factor and $\lambda > 1$. The stresses produced in the plate by the pressure, q_f, are given by $\sigma_{i_f} = \lambda \sigma_{i_s}$ and correspond to $F_s = 1$. Hence substituting these stresses into eqn (13) gives

$$\lambda F_i \sigma_{i_s} + \lambda^2 F_{ij} \sigma_{i_s} \sigma_{j_s} = 1 \tag{14}$$

Now multiplying eqn (13) by λ^2 and subtracting eqn (14) from the result gives

$$\lambda^2 (F_i \sigma_{i_s} - F_s) - \lambda F_i \sigma_{i_s} + 1 = 0 \tag{15}$$

Clearly eqn (15) is a quadratic in λ^2 and the roots are

$$\lambda = \tfrac{1}{2} A^{-1} \{ B \pm (B^2 - 4A)^{1/2} \} \tag{16}$$

in which $A = B - F_s$ and $B = F_i \sigma_{i_s}$.

The smaller value of λ derived from eqn (15) is used to scale q_s to give q_f— the initial failure pressure. Deflections, etc., at failure are determined by scaling the values determined for $q = q_s$ by λ.

6.2 Iterative Procedure for Large Deflection Initial Failure Analysis

The procedure used to determine the large deflection initial failure pressure is iterative in nature and simple in concept. It is based on the repeated solution of the large deflection plate equations for progressively improving estimates of the failure pressure. For each pressure the stresses are determined at each node of the finite-difference mesh and at each of the surfaces, $s = 1–3$, in each lamina. These stresses are then substituted into the relevant failure criterion to evaluate the failure function, F, for each of these locations. Once the point associated with F_{max} such that $0.99(= F^l) \leq F_{max} \leq 1.01(= F^u)$ has been isolated, then the current pressure is assumed to be equal to the initial failure pressure for the plate and the iterative procedure is terminated.

The efficiency of the iterative procedure is very dependent on the accuracy of the estimated failure pressure used to start the iterative procedure and the strategy for adjusting it after each iteration. Generally one or more trial analyses for arbitrary values of the applied pressure are sufficient to define a reasonable estimate of the initial failure pressure to start the iterative procedure. The strategy used to adjust, and hence improve, the estimated failure pressure during the iterative procedure is as follows.

Assuming that the estimated failure pressure for the first iteration is shown to be an overestimate, i.e. substitution of the stresses into the failure function results in $F > F^u$ at several points in the plate, then for the second iteration the pressure is reduced by 10%. The same pressure decrement is used for subsequent iterations provided $F > F^u$. If, however, an iteration results in $F < F^l$, then for the next iteration the pressure is incremented by half of the value of the current decrement, i.e. 5%. This procedure of successively reducing the pressure increment/decrement by half as the failure function bounds are successively overshot/undershot is continued until the failure function falls within the specified bounds and, hence, the initial failure pressure is determined.

Experience has shown that providing a reasonable estimate of the initial failure pressure is used for the first iteration, then typically between six and ten iterations are sufficient to give an initial failure pressure accurate to about 1%.

7 NUMERICAL RESULTS AND DISCUSSION

It is convenient to arrange and discuss separately the various sets of results produced by the exact and approximate versions of the initial failure analysis programs. First, some results are presented which serve to verify the correctness of the elastic solutions of the small deflection Mindlin plate equations. This is followed by results which confirm that the failure criteria have been correctly incorporated into the exact and approximate small deflection failure programs. The third set of results provide a comparison of the present failure data with that recently reported by Reddy and Pandey.[15] The fourth and final set of results presented is a comparison of the small and large deflection initial failure data for laminates A–I.

7.1 Verification of Elastic Analysis

In Ref. 15 Reddy and Pandey demonstrated the accuracy of their finite element analysis of the small deflection Mindlin plate equations by comparing the finite element results for the centre deflections of uniformly loaded cross- and angle-ply plates with their exact values. Similar comparisons have been made for the present exact and approximate DR Mindlin laminated plate analyses. The present results are compared with the exact and approximate finite element results obtained by Reddy and Pandey in Table 4. As expected the two sets of exact results are almost identical. Moreover, the two sets of approximate results show good agreement and the difference between the exact and approximate centre deflections is less than 1%. Thus the elastic analysis components of the

TABLE 4

Small deflection exact and approximate centre deflection results comparisons for simply supported rectangular laminated plates
$(b/a = 0.5556, k_i k_j = 5/6)$

	$\bar{q}$	Centre deflection ($\bar{w}_c$)			Exact[b]	BC
		DR	FE	Exact[a]		
$(0°/90°)_t$	42 057·7	188·0	188·6	188·30	188·4	SS2
$(0°/90°/0°/90°)_t$	2 628·6	6·75	6·75	6·74	6·70	SS2
$(90°/0°/0°/90°)_t$	2 628·6	11·48	11·50	11·49	11·45	SS2
$(45°/-45°/45°/-45°)_t$	2 628·6	5·45	5·44	5·44	5·43	SS3
$(75°/-75°/75°/-75°)_t$	2 628·6	12·57	12·58	12·58	12·58	SS3
$(45°/-45°)_t$	42 057·7	160·15	160·10	160·54	160·06	SS3
$(75°/-75°)_t$	42 057·7	260·00	260·40	260·27	260·39	SS3

[a] Present exact series solution ($m = n = 19$).
[b] Exact series solution of Ref. 15 ($m = n = ?$).
For the cross-ply lay-ups a 5×5 uniform non-interlacing mesh over one quarter of the plate was used, and for the remaining lay-ups a 10×10 uniform non-interlacing mesh over the complete plate was employed.

exact and approximate small deflection failure analyses appear to be functioning correctly.

Extensive verification of the approximate DR solution of the large deflection Mindlin laminated plate equations has also been undertaken. Results from these verification studies are, however, not reported here because they are in the process of being published elsewhere.[19,21]

7.2 Verification of Small Deflection Initial Failure Analysis

In order to verify that the various failure criteria had been implemented correctly into the programs and, moreover, that the small deflection failure analyses were correct, both approximate and initial failure data were generated for uniformly loaded simply supported rectangular plates with lay-ups corresponding to laminates F–I. Both quarter and whole plate approximate initial failure analyses were carried out for each laminate and failure criterion. The results are compared with corresponding exact initial failure results in Tables 5(a)–5(d). For the plates with cross-ply lay-ups (see Tables 5(a)–5(c)) identical results were obtained for both the quarter and whole plate approximate analyses. Moreover, the approximate initial failure pressures and associated plate centre deflections are marginally greater than the corresponding exact values. Furthermore, identical failure locations are predicted by both the approximate and exact analyses. Only in the case of the antisymmetrically laminated plate (see Table 5(d)) is there any difference between the quarter and whole plate approximate initial failure results. The full-plate failure results are more accurate because the constraints imposed along the plate centrelines (see Fig. 3(c)) in the quarter plate analysis do not accurately model the real situation for an antisymmetric angle-ply lay-up. It is also clear from Table 5(d) that the full-plate approximate initial failure pressures computed in accordance with the maximum strain and Tsai–Wu failure criteria are lower than the corresponding exact values. This is because failure initiates at the plate corners rather than at the centre, as predicted by the other criteria. Extrapolation procedures are used in the finite-difference analysis at the corners of the plate and, therefore, the errors in these regions are likely to be greater than elsewhere in the plate.

7.3 Comparison of DR and Finite Element Initial Failure Results for Laminates A–I

DR small deflection initial failure results for uniformly loaded rectangular plates with SS1 conditions (see Fig. 3(a)) on all edges and with lay-ups corresponding to laminates A–I have been computed. These are

TABLE 5
Comparison of DR and exact small deflection initial failure results

(a) Laminate F $(0°/90°/90°/0°)_t$; SS2 support conditions

Failure criterion	$\bar{q}_f$	$\bar{w}_c^f$	$\bar{x}$	$\bar{y}$	k	s
Maximum stress	11 595·4[a]	50·64	0·5	0·5	4	3
	11 595·4[b]	50·64	0·5	0·5	4	3
	11 295·7[c]	49·43	0·5	0·5	4	3
Maximum strain	12 466·8	54·45	0·5	0·5	4	3
	12 466·8	54·45	0·5	0·5	4	3
	12 141·0	53·13	0·5	0·5	4	3
Tsai–Hill	11 670·1	50·97	0·5	0·5	4	3
	11 670·3	50·97	0·5	0·5	4	3
	11 367·8	49·75	0·5	0·5	4	3
Hoffman	11 610·7	50·71	0·5	0·5	4	3
	11 610·7	50·71	0·5	0·5	4	3
	11 310·6	49·50	0·5	0·5	4	3
Tsai–Wu	12 198·6	53·27	0·5	0·5	4	3
	12 198·6	53·27	0·5	0·5	4	3
	11 881·2	51·99	0·5	0·5	4	3
Maximum stress (independent)	11 715·2	51·16	0·5	0·5	4	3
	11 715·2	51·16	0·5	0·5	4	3
	11 411·3	49·94	0·5	0·5	4	3

(b) Laminate G $(90°/0°)_t$; SS2 support conditions

Failure criterion	$\bar{q}_f$	$\bar{w}_c^f$	$\bar{x}$	$\bar{y}$	k	s
Maximum stress	33 123·3[a]	148·2	0·5	0·5	2	3
	33 123·3[b]	148·2	0·5	0·5	2	3
	32 243·7[c]	144·6	0·5	0·5	2	3
Maximum strain	34 006·5	152·2	0·5	0·5	2	3
	34 006·5	152·2	0·5	0·5	2	3
	33 074·1	148·3	0·5	0·5	2	3
Tsai–Hill	33 205·6	148·6	0·5	0·5	2	3
	33 205·7	148·6	0·5	0·5	2	3
	32 320·7	145·0	0·5	0·5	2	3
Hoffman	33 139·4	148·3	0·5	0·5	2	3
	33 139·4	148·3	0·5	0·5	2	3
	32 258·9	144·7	0·5	0·5	2	3
Tsai–Wu	33 728·0	150·9	0·5	0·5	2	3
	33 728·0	150·9	0·5	0·5	2	3
	32 812·3	147·2	0·5	0·5	2	3
Maximum stress (independent)	33 211·2	148·6	0·5	0·5	2	3
	33 211·2	148·6	0·5	0·5	2	3
	32 325·2	145·0	0·5	0·5	2	3

TABLE 5—*contd.*

(c) *Laminate H (90°/0°/90°/0°)$_t$: SS2 support conditions*

Failure criterion	$\bar{q}_f$	$\bar{w}_c^f$	$\bar{x}$	$\bar{y}$	k	s
Maximum stress	16 714·4[a]	42·90	0·5	0·5	4	3
	16 714·4[b]	42·90	0·5	0·5	4	3
	16 281·6[c]	41·79	0·5	0·5	4	3
Maximum strain	17 410·4	44·69	0·5	0·5	4	3
	17 410·4	44·69	0·5	0·5	4	3
	16 935·5	43·47	0·5	0·5	4	3
Tsai–Hill	16 777·8	43·06	0·5	0·5	4	3
	16 777·8	43·06	0·5	0·5	4	3
	16 340·9	41·95	0·5	0·5	4	3
Hoffman	16 727·0	42·93	0·5	0·5	4	3
	16 727·0	42·93	0·5	0·5	4	3
	16 293·4	41·82	0·5	0·5	4	3
Tsai–Wu	17 193·5	44·13	0·5	0·5	4	3
	17 193·5	44·13	0·5	0·5	4	3
	16 731·8	42·95	0·5	0·5	4	3
Maximum stress	16 792·3	43·10	0·5	0·5	4	3
(independent)	16 792·3	43·10	0·5	0·5	4	3
	16 353·6	41·98	0·5	0·5	4	3

(d) *Laminate I (−45°/45°/−45°/45°)$_t$: SS3 support conditions*

Failure criterion	$\bar{q}_f$	$\bar{w}_c^f$	$\bar{x}$	$\bar{y}$	k	s
Maximum stress	27 664·8[a]	56·43	0·5	0·5	1	1
	28 817·4[b]	59·71	0·5	0·5	4	3
	28 327·2[c]	58·56	0·5	0·5	4	3
Maximum strain	25 974·2	52·99	0·0	0·0	1	1
	25 970·3	53·81	0·0, 1·0	1·0, 0·0	1, 4	1, 3
	27 968·3	57·82	0·0, 1·0	1·0, 0·0	1, 4	1, 3
Tsai–Hill	27 459·5	56·02	0·5	0·5	1	1
	29 138·3	60·38	0·5	0·5	4	3
	28 642·0	59·21	0·5	0·5	4	3
Hoffman	27 760·3	56·63	0·5	0·5	1	1
	28 887·0	59·86	0·5	0·5	4	3
	28 395·4	58·70	0·5	0·5	4	3
Tsai–Wu	27 208·2	55·50	0·0	0·0	1	1
	27 343·6	56·66	0·0, 1·0	1·0, 0·0	1, 4	1, 3
	29 450·7	60·89	0·0, 1·0	1·0, 0·0	1, 4	1, 3
Maximum stress	28 781·5	58·71	0·5	0·5	1	1
(independent)	30 353·1	62·89	0·5	0·5	1, 4	1, 3
	29 853·1	61·72	0·5	0·5	1, 4	1, 3

[a] DR quarter-plate results (5 × 5 uniform non-interlacing mesh).
[b] DR full-plate results (10 × 10 uniform non-interlacing mesh).
[c] Exact results ($m = n = 19$).

compared with Reddy and Pandey's finite element results[15] in Tables 6(a)–6(i). Examination of these results reveals that the DR failure pressures and deflections are consistent, i.e. the values do not vary significantly with the particular failure criterion used, whereas the finite element results show that the failure pressures computed with the Tsai–Hill criterion are significantly lower than those computed with the other criteria which, moreover, are generally consistent. Furthermore, except in the case of the Tsai–Hill criterion, the DR failure pressures are consistently lower than the finite element values.

The DR and finite element results for the four thickest symmetrically laminated plates, i.e. laminates B–E, are in reasonable agreement as far as the magnitude of the initial failure pressure is concerned. However, for laminates B and C, and in some instances for laminates D and E, the DR analysis predicts failure to initiate at the plate corners whereas the finite element analysis predicts it to arise at the centre of the plate. The agreement between the DR and finite element initial failure pressures for the four-layer symmetric cross-ply laminated plate, i.e. laminate F, is rather less good, though the predicted failure locations agree. The finite element failure pressures are, in most instances, about 50% greater than the DR values. For the three-layer symmetrically laminated angle-ply plate (laminate A) there is no agreement between corresponding results. The finite element analysis indicates failure at the centre at a pressure about three times greater than the DR value which is associated with failure at the corners.

Turning now to the two- and four-layer antisymmetrically laminated plates, i.e. laminates G–I, it is clear that much higher (up to three times in some cases) failure pressures are predicted by the finite element analysis though, in general, both analyses predict that failure initiates at the plate centre.

7.4 Comparison of DR Large and Small Deflection Initial Failure Results

A series of uniformly loaded rectangular plates with either SS1 or CC1 conditions (see Fig. 3) applied along all edges and with lay-ups corresponding to laminates A–I have been analysed using the DR small and large deflection failure analysis programs in order to shed light on the significance of membrane action on the initial failure response. The results of these computations are presented in Tables 7(a)–7(i).

Examination of the results for the thicker (16-layer) symmetrically laminated plates, i.e. laminates B–E, reveals that for both SS1 and CC1 edge

TABLE 6
Comparison of DR and finite element small deflection initial failure results

(a) *Laminate A $(45°/-45°/45°)_t$; SS1 support conditions*

Failure criterion	$\bar{q}_f$	$\bar{w}_c^f$	$\bar{x}$	$\bar{y}$	k	s
Maximum stress	$12\,218\cdot7^a$	29·48	0·0, 1·0	0·0, 1·0	1	1
	$36\,221\cdot5^b$		?	?	1	1
	$24\,657\cdot2^c$		$0\cdot5^d$	$0\cdot5^d$	1	1
Maximum strain	$11\,866\cdot9$	28·63	0·0, 1·0	0·0, 1·0	1	1
	$39\,145\cdot9$		0·0	0·0	1	1
	$26\,451\cdot7$		0·0	0·0	1	1
Tsai–Hill	$12\,238\cdot0$	29·52	0·0, 1·0	0·0, 1·0	1	1
	$14\,555\cdot1$		0·5	0·5	3	3
	$9\,836\cdot3$		0·5	0·5	3	3
Hoffman	$12\,211\cdot9$	29·46	0·0, 1·0	0·0, 1·0	1	1
	$35\,955\cdot7$		0·5	0·5	1	1
	$24\,457\cdot9$		0·5	0·5	1	1
Tsai–Wu	$11\,978\cdot0$	28·90	0·0, 1·0	0·0, 1·0	1	1
	$38\,680\cdot6$		0·5	0·5	1	1
	$26\,318\cdot8$		0·5	0·5	1	1
Maximum stress (independent)	$12\,257\cdot0$	29·57	0·0, 1·0	0·0, 1·0	1, 3	1, 3
	$38\,281\cdot9$		0·5	0·5	1	1
	$25\,853\cdot5$		0·5	0·5	1	1

(b) *Laminate B $(45°/-45°/90°/0°/45°/90°/-45°/0°)_s$; SS1 support conditions*

Failure criterion	$\bar{q}_f$	$\bar{w}_c^f$	$\bar{x}$	$\bar{y}$	k	s
Maximum stress	$2\,026\cdot0^a$	3·725 4	0·0, 1·0	0·0, 1·0	1	1
	$2\,092\cdot7^b$		$0\cdot5^d$	$0\cdot5^d$	1	1
	$1\,937\cdot8^c$		0·5	0·5	1	1
Maximum strain	$1\,678\cdot2$	3·085 7	0·0, 1·0	0·0, 1·0	1	1
	$2\,441\cdot5$		0·0	0·0	1	1
	$2\,156\cdot7$		0·0	0·0	1	1
Tsai–Hill	$2\,039\cdot6$	3·750 2	0·0, 1·0	0·0, 1·0	1	1
	$780\cdot0$		0·5	0·5	16	3
	$717\cdot3$		0·5	0·5	16	3
Hoffman	$2\,018\cdot2$	3·711 0	0·0, 1·0	0·0, 1·0	1	1
	$2\,074\cdot3$		0·5	0·5	1	1
	$1\,920\cdot5$		0·5	0·5	1	1
Tsai–Wu	$1\,787\cdot5$	3·286 8	0·0, 1·0	0·0, 1·0	1	1
	$2\,257\cdot3$		0·5	0·5	1	1
	$2\,093\cdot0$		0·5	0·5	1	1
Maximum stress (independent)	$2\,164\cdot1$	3·979 3	0·0, 1·0	0·0, 1·0	1, 16	1, 3
	$2\,292\cdot2$		0·5	0·5	1	1
	$2\,118\cdot1$		0·5	0·5	1	1

(continued)

TABLE 6—*contd.*

(c) *Laminate C* $(45°/-45°/0°/90°/45°/0°/-45°/90°)_s$; *SS1 support conditions*

Failure criterion	$\bar{q}_f$	$\bar{w}_c^f$	$\bar{x}$	$\bar{y}$	k	s
Maximum stress	1 867·4[a]	3·8257	0·0, 1·0	0·0, 1·0	1	1
	1 853·7[b]		0·5[d]	0·5[d]	1	1
	1 714·2[c]		0·5	0·5	1	1
Maximum strain	1 546·7	3·1688	0·0, 1·0	0·0, 1·0	1	1
	2 215·0		0·0	0·0	1	1
	1 967·4		0·0	0·0	1	1
Tsai–Hill	1 879·8	3·8512	0·0, 1·0	0·0, 1·0	1	1
	690·5		0·5	0·5	16	3
	634·1		0·5	0·5	16	3
Hoffman	1 860·1	3·8109	0·0, 1·0	0·0, 1·0	1	1
	1 837·3		0·5	0·5	1	1
	1 698·9		0·5	0·5	1	1
Tsai–Wu	1 647·5	3·3752	0·0, 1·0	0·0, 1·0	1	1
	1 999·9		0·5	0·5	1	1
	1 852·0		0·5	0·5	1	1
Maximum stress (independent)	1 994·6	4·0864	0·0, 1·0	0·0, 1·0	1, 16	1, 3
	2 025·1		0·5	0·5	1	1
	1 870·4		0·5	0·5	1	1

(d) *Laminate D* $(45°/0°/-45°/0°/-45°/90°/0°/45°)_s$; *SS1 support conditions*

Failure criterion	$\bar{q}_f$	$\bar{w}_c^f$	$\bar{x}$	$\bar{y}$	k	s
Maximum stress	1 500·9[a]	3·6752	0·5	0·5	15	3
	1 543·5[b]		0·5[d]	0·5[d]	2	1
	1 453·0[c]		0·5	0·5	2	1
Maximum strain	1 450·9	3·5527	0·0, 1·0	0·0, 1·0	1	1
	1 676·8		0·5	0·5	2	1
	1 580·5		0·5	0·5	2	1
Tsai–Hill	1 510·2	3·6977	0·5	0·5	15	3
	591·9		0·5	0·5	16	3
	553·3		0·5	0·5	16	3
Hoffman	1 502·8	3·6798	0·5	0·5	15	3
	1 537·1		0·5	0·5	2	1
	1 446·8		0·5	0·5	2	1
Tsai–Wu	1 553·2	3·8030	0·0, 1·0	0·0, 1·0	1	1
	1 596·1		0·5	0·5	2	1
	1 503·6		0·5	0·5	2	1
Maximum stress (independent)	1 515·3	3·7103	0·5	0·5	2, 15	1, 3
	1 558·7		0·5	0·5	2	1
	1 467·5		0·5	0·5	2	1

TABLE 6—*contd.*

(e) *Laminate E (45°/0°/−45°/0°/−45°/0°/45°/0°)$_s$; SS1 support conditions*

Failure criterion	$\bar{q}_f$	$\bar{w}_c^f$	$\bar{x}$	$\bar{y}$	k	s
Maximum stress	1 442·1[a]	3·684 3	0·5	0·5	15	3
	1 466·8[b]		0·5[d]	0·5[d]	2	1
	1 372·3[c]		0·5	0·5	2	1
Maximum strain	1 352·3	3·455 0	0·0, 1·0	0·0, 1·0	1	1
	1 595·1		0·5	0·5	2	1
	1 494·6		0·5	0·5	2	1
Tsai–Hill	1 451·1	3·707 2	0·5	0·5	15	3
	561·3		0·5	0·5	16	3
	521·5		0·5	0·5	16	3
Hoffman	1 443·9	3·689 0	0·5	0·5	15	3
	1 460·5		0·5	0·5	2	1
	1 366·5		0·5	0·5	2	1
Tsai–Wu	1 443·5	3·688 0	0·0, 1·0	0·0, 1·0	1	1
	1 517·6		0·5	0·5	2	1
	1 421·1		0·5	0·5	2	1
Maximum stress	1 456·2	3·720 3	0·5	0·5	2, 15	1, 3
(independent)	1 481·6		0·5	0·5	2	1
	1 386·5		0·5	0·5	2	1

(f) *Laminate F (0°/90°/90°/0°)$_t$; SS1 support conditions*

Failure criterion	$\bar{q}_f$	$\bar{w}_c^f$	$\bar{x}$	$\bar{y}$	k	s
Maximum stress	11 595·4[e]	50·64	0·5	0·5	4	3
	15 834·7[b]		0·5[d]	0·5[d]	1	1
	15 288·0[c]		0·5	0·5	1	1
Maximum strain	12 466·8	54·44	0·5	0·5	4	3
	17 222·6		0·5	0·5	1	1
	16 591·8		0·5	0·5	1	1
Tsai–Hill	11 670·1	50·97	0·5	0·5	4	3
	8 642·9		0·5	0·5	4	3
	8 348·5		0·5	0·5	4	3
Hoffman	11 610·7	50·71	0·5	0·5	4	3
	15 750·6		0·5	0·5	1	1
	15 203·9		0·5	0·5	1	1
Tsai–Wu	12 198·6	53·27	0·5	0·5	4	3
	16 381·5		0·5	0·5	1	1
	15 792·7		0·5	0·5	1	1
Maximum stress	11 715·2	51·16	0·5	0·5	1, 4	1, 3
(independent)	16 003·0		0·5	0·5	1	1
	15 414·1		0·5	0·5	1	1

(*continued*)

TABLE 6—*contd.*

(g) *Laminate G (90°/0°)$_t$; SS1 support conditions*

Failure criterion	$\bar{q}_f$	$\bar{w}_c^f$	$\bar{x}$	$\bar{y}$	k	s
Maximum stress	57 633·9[e]	154·5	0·5	0·5	2	3
	171 258·9[b]		?	?	1	1
	161 838·0[c]		?	?	1	1
Maximum strain	60 374·1	161·9	0·5	0·5	2	3
	87 143·5		0·5[d]	0·5[d]	2	3
	87 480·0		0·5	0·5	2	3
Tsai–Hill	57 880·7	155·2	0·5	0·5	2	3
	87 816·5		0·5	0·5	2	3
	86 470·6		0·5	0·5	2	3
Hoffman	57 683·1	154·6	0·5	0·5	2	3
	171 258·9		?	?	1	1
	161 165·1		?	?	1	1
Tsai–Wu	59 522·7	159·6	0·5	0·5	2	3
	174 623·5		?	?	1	1
	165 202·6		?	?	1	1
Maximum stress	57 953·6	155·4	0·5	0·5	2	3
(independent)	173 277·7		?	?	1	1
	162 510·9		?	?	1	1

(h) *Laminate H (90°/0°/90°/0°)$_t$; SS1 support conditions*

Failure criterion	$\bar{q}_f$	$\bar{w}_c^f$	$\bar{x}$	$\bar{y}$	k	s
Maximum stress	20 306·4[e]	46·83	0·5	0·5	4	3
	48 892·1[b]		?	?	1	1
	47 798·6[c]		?	?	1	1
Maximum strain	21 494·3	49·57	0·5	0·5	4	3
	34 592·5		0·5	0·5	4	3
	33 541·0		0·5	0·5	4	3
Tsai–Hill	20 411·3	47·07	0·5	0·5	4	3
	16 465·6		0·5	0·5	4	3
	16 129·1		?	?	4	3
Hoffman	20 327·5	46·88	0·5	0·5	4	3
	48 261·2		?	?	1	1
	47 125·6		?	?	1	1
Tsai–Wu	21 126·8	48·72	0·5	0·5	4	3
	55 200·7		?	?	1	1
	54 149·3		?	?	1	1
Maximum stress	20 454·9	47·17	0·5	0·5	4	3
(independent)	55 453·1		0·5	0·5	4	3
	53 896·9		?	?	4	3

TABLE 6—*contd.*

(i) *Laminate I (−45°/45°/−45°/45°)$_t$; SS1 support conditions*

Failure criterion	$\bar{q}_f$	$\bar{w}_c^f$	$\bar{x}$	$\bar{y}$	k	s
Maximum stress	31 304·8[a]	60·72	0·5	0·5	4	3
	38 072·7[b]		0·5[d]	0·5[d]	1	1
	36 542·9[c]		0·5	0·5	1	1
Maximum strain	31 619·2	61·33	0·1, 0·9	1·0, 0·0	4	3
	43 692·7		0·5	0·5	1	1
	41 792·2		0·5	0·5	1	1
Tsai–Hill	31 738·0	61·56	0·5	0·5	4	3
	16 389·4		0·5	0·5	4	3
	15 918·8		0·5	0·5	4	3
Hoffman	31 391·7	60·89	0·5	0·5	4	3
	37 804·6		0·5	0·5	1	1
	36 306·8		0·5	0·5	1	1
Tsai–Wu	33 293·0	64·58	0·1, 0·9	1·0, 0·0	4	3
	40 278·1		0·5	0·5	1	1
	38 595·8		0·5	0·5	1	1
Maximum stress (independent)	33 339·3	64·67	0·5	0·5	1, 4	1, 3
	40 546·2		0·5	0·5	1	1
	38 829·8		0·5	0·5	1	1

[a] DR full-plate results (10 × 10 uniform non-interlacing mesh).
[b] Four-node finite element quarter-plate results from Ref. 15 (4 × 4 uniform mesh).
[c] Four-node finite element quarter-plate results from Ref. 15 (8 × 8 uniform mesh).
[d] Coordinates of finite element node adjacent to the Gauss point at which the initial failure stresses are evaluated.
[e] DR quarter-plate results (5 × 5 uniform non-interlacing mesh).
? Nodal coordinates nearest to initial failure location ambiguous.

conditions the use of large deflection theory leads to increased initial failure pressures and reduced deflections. Thus for SS1 conditions the large deflection initial failure pressures are between 3·5 and 4 times their corresponding small deflection values and for CC1 conditions they vary between 1·4 and 1·6 times. However, plate centre deflections at failure are overestimated using small deflection theory by a factor of 2 for SS1 conditions and a factor between 1·05 and 1·10 for CC1 conditions.

For all of the thinner plates, i.e. laminates A and F–I, irrespective of whether their lay-ups are symmetric or antisymmetric, the general pattern, as outlined in the previous paragraph, is observed but the factors by which small deflection theory underestimates the failure pressure and overestimates the deflection at failure are much larger. However, it should be appreciated that the results in Tables 7(a) and 7(f)–7(i) have been included largely for the sake of completeness. It is quite obvious that the failure

TABLE 7

Comparison of DR small and large deflection initial flexural failure results[a] for rectangular plates (b/a = 0·5556) with SS1 and CC1 conditions applied along all edges

(a) Laminate A $(45°/-45°/45°)_t$

Failure criterion	$\bar{q}_f$	$\bar{w}_c^r$	$\bar{x}$	$\bar{y}$	k	s	Support condition
Maximum stress	12 218·7[b]	29·48	0·0	0·0	1	1	SS1
	2 825 400·0[c]	14·50	0·8	0·8	3	3	
Maximum strain	11 866·9	28·63	0·0	0·0	1	1	
	3 310 405·0	15·29	0·0	0·1	3	3	
Tsai–Hill	12 238·0	29·52	0·0	0·0	1	1	
	2 881 908·0	14·60	0·8	0·8	3	3	
Hoffman	12 211·9	29·46	0·0	0·0	1	1	
	2 825 400·0	14·5	0·8	0·8	3	3	
Tsai–Wu	11 978·0	28·90	0·0	0·0	1	1	
	3 245 495·8	15·19	1·0	0·9	3	3	
Maximum stress (independent)	12 257·0	29·57	0·0	0·0	3	3	
	3 058 303·5	14·89	0·2	0·2	3	3	
Maximum stress	56 479·5	38·96	0·4	0·0	1	1	CC1
	1 910 000·0	11·68	0·6	0·0	1	1	
Maximum strain	66 416·7	45·80	0·4	0·0	1	1	
	2 259 800·0	12·36	0·6	0·0	1	1	
Tsai–Hill	57 237·7	39·49	0·4	0·0	1	1	
	1 948 200·0	11·76	0·4	1·0	1	1	
Hoffman	56 643·6	39·08	0·4	0·0	1	1	
	1 910 000·0	11·68	0·6	0·0	1	1	
Tsai–Wu	63 622·9	43·87	0·4	0·0	1	1	
	2 150 970·0	12·16	0·6	0·0	1	1	
Maximum stress (independent)	64 223·0	44·29	0·4	0·0	1	1	
	2 894 921·0	13·45	0·4	1·0	1	1	

(b) Laminate B $(45°/-45°/90°/0°/45°/90°/-45°/0°)_s$

Failure criterion	$\bar{q}_f$	$\bar{w}_c^r$	$\bar{x}$	$\bar{y}$	k	s	Support condition
Maximum stress	2 026·0[b]	3·7254	0·0	0·0	1	1	SS1
	9 405·0[c]	1·9415	0·0	0·0	1	1	
Maximum strain	1 678·2	3·0857	0·0	0·0	1	1	
	6 407·0	1·6957	0·0	0·0	1	1	
Tsai–Hill	2 039·6	3·7502	0·0	0·0	1	1	
	9 405·0	1·9415	0·0	0·0	1	1	
Hoffman	2 018·2	3·7110	0·0	0·0	1	1	
	9 216·9	1·9278	1·0	1·0	1	1	
Tsai–Wu	1 787·5	3·2868	0·0	0·0	1	1	
	7 232·7	1·7703	0·0	0·0	1	1	
Maximum stress (independent)	2 164·1	3·9793	0·0	0·0	1	1	
	10 591·6	2·0235	0·0	0·0	1	1	
Maximum stress	2 728·4	1·3453	0·4	0·0	1	1	CC1
	4 660·3	1·1950	0·5	1·0	1	1	
Maximum strain	3 093·7	1·5254	0·4	0·0	13	3	
	4 896·1	1·2239	0·5	1·0	4	1	
Tsai–Hill	2 765·8	1·3637	0·4	0·0	1	1	
	4 753·5	1·2065	0·5	1·0	1	1	
Hoffman	2 736·4	1·3492	0·4	0·0	1	1	
	4 660·3	1·1950	0·5	1·0	1	1	
Tsai–Wu	3 079·2	1·5183	0·4	0·0	1	1	
	4 896·1	1·2239	0·5	1·0	4	3	
Maximum stress (independent)	3 091·0	1·5241	0·4	0·0	13	3	
	4 945·5	1·2298	0·5	0·0	4	1	

TABLE 7—*contd.*

(c) *Laminate C* $(45°/-45°/0°/90°/45°/0°/45°/90°)_s$

Failure criterion	$\bar{q}_f$	$\bar{w}_c^f$	$\bar{x}$	$\bar{y}$	k	s	Support condition
Maximum stress	1 867·4[b]	3·825 7	0·0	0·0	1	1	SS1
	9 000·0[c]	1·921 8	0·0	0·0	1	1	
Maximum strain	1 546·7	3·168 8	0·0	0·0	1	1	
	6 256·2	1·692 9	1·0	1·0	1	1	
Tsai–Hill	1 879·8	3·851 2	0·0	0·0	1	1	
	9 180·0	1·935 0	0·0	0·0	1	1	
Hoffman	1 860·1	3·810 9	0·0	0·0	1	1	
	9 000·0	1·921 8	0·0	0·0	1	1	
Tsai–Wu	1 647·5	3·375 2	0·0	0·0	1	1	
	7 062·5	1·766 4	1·0	1·0	1	1	
Maximum stress (independent)	1 994·6	4·086 4	0·0	0·0	1	1	
	9 000·0	1·921 8	0·0	0·0	1	1	
Maximum stress	2 154·9	1·236 0	0·4	0·0	3	1	CC1
	3 528·4	1·086 0	0·5	0·0	3	1	
Maximum strain	2 165·8	1·242 3	0·4	0·0	14	3	
	3 563·0	1·091 2	0·5	1·0	3	1	
Tsai–Hill	2 155·9	1·236 6	0·4	0·0	14	3	
	3 528·4	1·086 0	0·5	0·0	3	1	
Hoffman	2 155·1	1·236 1	0·4	0·0	3	1	
	3 528·4	1·086 0	0·5	0·0	3	1	
Tsai–Wu	2 162·0	1·240 0	0·4	0·0	3	1	
	3 528·4	1·086 0	0·5	0·0	3	1	
Maximum strain (independent)	2 159·7	1·238 8	0·4	0·0	14	3	
	3 528·1	1·086 0	0·5	0·0	3	1	

(d) *Laminate D* $(45°/0°/-45°/0°/-45°/90°/0°/45°)_s$

Failure criterion	$\bar{q}_f$	$\bar{w}_c^f$	$\bar{x}$	$\bar{y}$	k	s	Support condition
Maximum stress	1 500·9[b]	3·675 2	0·5	0·5	15	3	SS1
	6 370·6[c]	1·864 5	0·5	0·4	15	3	
Maximum strain	1 450·9	3·552 7	0·0	0·0	1	1	
	6 370·6	1·864 5	1·0	1·0	1	1	
Tsai–Hill	1 510·2	3·697 7	0·5	0·5	15	3	
	6 433·0	1·870 9	0·5	0·6	15	3	
Hoffman	1 502·8	3·679 8	0·5	0·5	15	3	
	6 370·6	1·864 5	0·5	0·4	15	3	
Tsai–Wu	1 553·2	3·803 0	0·0	0·0	1	1	
	6 760·5	1·903 5	0·5	0·7	15	3	
Maximum stress (independent)	1 515·3	3·710 3	0·5	0·5	2	1	
	6 370·6	1·864 5	0·5	0·4	15	3	
Maximum stress	1 455·0	1·034 5	0·4	0·0	2	1	CC1
	2 224·8	0·970 9	0·5	1·0	2	1	
Maximum strain	1 462·4	1·039 8	0·4	0·0	2	1	
	2 224·8	0·970 9	0·5	1·0	2	1	
Tsai–Hill	1 455·6	1·035 0	0·4	0·0	15	3	
	2 224·8	0·970 9	0·5	1·0	2	1	
Hoffman	1 455·1	1·034 6	0·4	0·0	2	1	
	2 224·8	0·970 9	0·5	1·0	2	1	
Tsai–Wu	1 459·7	1·037 9	0·4	0·0	2	1	
	2 224·8	0·970 9	0·5	1·0	2	1	
Maximum stress (independent)	1 456·9	1·035 9	0·4	0·0	2	1	
	2 224·8	0·970 9	0·5	1·0	2	1	

(continued)

TABLE 7—contd.

(e) Laminate E $(45°/0°/-45°/0°/-45°/0°/45°/0°)_s$

Failure criterion	$\bar{q}_f$	$\bar{w}_c^f$	$\bar{x}$	$\bar{y}$	k	s	Support condition
Maximum stress	1 442·1[b]	3·684 3	0·5	0·5	15	3	SS1
	4 752·0[c]	1·868 1	0·5	0·3	15	3	
Maximum strain	1 352·3	3·455 0	0·0	0·0	1	1	
	5 042·9	1·908 4	0·5	0·2	15	3	
Tsai–Hill	1 451·1	3·707 2	0·5	0·5	15	3	
	4 798·6	1·874 7	0·5	0·7	15	3	
Hoffman	1 443·9	3·689 0	0·5	0·5	13	3	
	4 752·0	1·868 1	0·5	0·3	15	3	
Tsai–Wu	1 443·5	3·688 0	0·0	0·0	1	1	
	4 992·4	1·901 5	0·5	0·8	15	3	
Maximum stress (independent)	1 456·2	3·720 3	0·5	0·5	2	1	
	4 752·0	1·868 1	0·5	0·3	15	3	
Maximum stress	1 328·0	1·014 3	0·4	0·0	2	1	CC1
	1 835·2	0·959 4	0·5	0·0	2	1	
Maximum strain	1 334·8	1·019 5	0·4	0·0	2	1	
	1 872·7	0·970 6	0·5	1·0	2	1	
Tsai–Hill	1 328·6	1·014 8	0·4	0·0	15	3	
	1 835·2	0·959 4	0·5	0·0	2	1	
Hoffman	1 328·2	1·014 4	0·4	0·0	2	1	
	1 835·2	0·959 4	0·5	0·0	2	1	
Tsai–Wu	1 332·4	1·017 6	0·4	0·0	2	1	
	1 853·6	0·964 9	0·5	0·0	2	1	
Maximum stress (independent)	1 329·7	1·015 6	0·4	0·0	2	1	
	1 872·7	0·970 6	0·5	1·0	2	1	

(f) Laminate F $(0°/90°/90°/0°)_t$

Failure criterion	$\bar{q}_f$	$\bar{w}_c^f$	$\bar{x}$	$\bar{y}$	k	s	Support condition
Maximum stress	11 595·4[b]	50·64	0·5	0·5	4	3	SS1
	950 000·0[c]	8·654 7	0·4	0·4	4	3	
Maximum strain	12 466·8	54·44	0·5	0·5	4	3	
	1 113 076·4	9·123 4	0·5	0·2	4	3	
Tsai–Hill	11 670·1	50·97	0·5	0·5	4	3	
	959 310·0	8·682 8	0·6	0·6	4	3	
Hoffman	11 610·7	50·71	0·5	0·5	4	3	
	950 000·0	8·654 7	0·4	0·4	4	3	
Tsai–Wu	12 198·6	53·27	0·5	0·5	4	3	
	1 069 854·3	9·003 9	0·4	0·6	4	3	
Maximum stress (independent)	11 715·2	51·16	0·5	0·5	1	1	
	969 000·0	8·711 9	0·6	0·4	4	3	
Maximum stress	15 829·5	15·54	0·5	0·0	1	1	CC1
	575 000·0	6·755 2	0·5	0·0	1	1	
Maximum strain	15 910·2	15·62	0·5	0·0	1	1	
	575 000·0	6·755 2	0·5	0·0	1	1	
Tsai–Hill	15 836·7	15·55	0·5	0·0	4	3	
	575 000·0	6·755 2	0·5	0·0	1	1	
Hoffman	15 831·0	15·55	0·5	0·0	1	1	
	575 000·0	6·755 2	0·5	0·0	1	1	
Tsai–Wu	15 881·6	15·60	0·5	0·0	1	1	
	575 000·0	6·755 2	0·5	0·0	1	1	
Maximum stress (independent)	15 842·1	15·56	0·5	0·0	1	1	
	622 398·4	6·938 9	0·5	1·0	1	1	

TABLE 7—contd.

(g) Laminate G $(90°/0°)_t$

Failure criterion	$\bar{q}_f$	$\bar{w}_c^f$	$\bar{x}$	$\bar{y}$	k	s	Support condition
Maximum stress	57 633·9[b]	154·5	0·5	0·5	2	3	SS1
	1 100 000·0[c]	9·503 0	0·5	1·0	2	3	
Maximum strain	60 374·1	161·9	0·5	0·5	2	3	
	1 110 780·0	9·533 2	0·5	0·0	2	3	
Tsai–Hill	57 880·7	155·2	0·5	0·5	2	3	
	1 100 000·0	9·503 0	0·5	1·0	2	3	
Hoffman	57 683·1	154·6	0·5	0·5	2	3	
	1 100 000·0	9·503 0	0·5	1·0	2	3	
Tsai–Wu	59 522·7	159·6	0·5	0·0	2	3	
	1 110 780·0	9·533 2	0·5	0·2	2	3	
Maximum stress (independent)	57 953·6	155·4	0·5	0·5	2	3	
	1 100 000 0	9·503 0	0·5	1·0	2	3	
Maximum stress	46 084·6	45·20	0·5	0·0	2	3	CC1
	9 911 606·0	17·54	0·5	0·4	2	3	
Maximum strain	46 282·4	45·39	0·5	0·0	2	3	
	10 728 640·0	18·02	0·5	0·0	2	1	
Tsai–Hill	46 068·5	45·18	0·5	0·0	2	3	
	10 011 723·0	17·60	0·5	0·4	2	3	
Hoffman	46 088·8	45·20	0·5	0·0	2	3	
	9 911 606·0	17·54	0·5	0·4	2	3	
Tsai–Wu	46 237·8	45·35	0·5	0·0	2	3	
	10 728 640·0	18·02	0·5	0·4	2	3	
Maximum stress (independent)	46 077·8	45·19	0·5	0·0	2	3	
	10 011 723·0	17·60	0·5	0·4	2	3	

(h) Laminate H $(90°/0°/90°/0°)_t$

Failure criterion	$\bar{q}_f$	$\bar{w}_c^f$	$\bar{x}$	$\bar{y}$	k	s	Support condition
Maximum stress	20 306·4[b]	46·83	0·5	0·5	4	3	SS1
	530 000·0[c]	7·255 7	0·5	0·0	4	3	
Maximum strain	21 494·3	49·57	0·5	0·5	4	3	
	530 000·0	7·255 7	0·5	0·0	4	3	
Tsai–Hill	20 411·3	47·07	0·5	0·5	4	3	
	530 000·0	7·255 7	0·5	0·0	4	3	
Hoffman	20 327·5	46·88	0·5	0·5	4	3	
	530 000·0	7·255 7	0·5	0·0	4	3	
Tsai–Wu	21 126·8	48·72	0·5	0·5	4	3	
	530 000·0	7·255 7	0·5	0·0	4	3	
Maximum stress (independent)	20 454·9	47·17	0·5	0·5	4	3	
	530 000·0	7·255 7	0·5	0·0	4	3	
Maximum stress	25 056·6	13·52	0·5	0·0	4	3	CC1
	988 380·0	8·096 2	0·5	0·0	2	1	
Maximum strain	25 165·5	13·58	0·5	0·0	4	3	
	988 380·0	8·096 2	0·5	0·0	2	1	
Tsai–Hill	25 049·5	13·52	0·5	0·0	4	3	
	988 380·0	8·096 2	0·5	0·0	2	1	
Hoffman	25 058·9	13·52	0·5	0·0	4	3	
	988 380·0	8·096 2	0·5	0·0	2	1	
Tsai–Wu	25 139·0	13·57	0·5	0·0	4	3	
	988 380·0	8·096 2	0·5	0·0	2	1	
Maximum stress (independent)	25 079·1	13·53	0·5	0·0	4	3	
	1 091 251·4	8·375 3	0·5	0·4	4	3	

(continued)

TABLE 7—contd.

(i) Laminate I $(-45°/45°/-45°/45°)_t$

Failure criterion	$\bar{q}_f$	$\bar{w}_c^f$	$\bar{x}$	$\bar{y}$	k	s	Support condition
Maximum stress	31 304·8[b]	60·72	0·5	0·5	4	3	SS1
	1 153 117·0[c]	10·74	0·4	0·2	4	3	
Maximum strain	31 619·2	61·33	0·1	1·0	4	3	
	1 521 513·0	11·78	0·3	0·2	4	3	
Tsai–Hill	31 738·0	61·56	0·5	0·5	4	3	
	1 187 706·0	10·84	0·6	0·8	4	3	
Hoffman	31 391·7	60·89	0·5	0·5	4	3	
	1 176 179·4	10·81	0·4	0·2	4	3	
Tsai–Wu	33 293·0	64·58	0·1	1·0	4	3	
	1 405 642·9	11·47	0·4	0·2	4	3	
Maximum stress	33 339·3	64·67	0·5	0·5	1	1	
(independent)	1 298 596·9	11·17	0·2	0·4	4	3	
Maximum stress	28 644·9	18·60	0·4	0·0	1	1	CC1
	653 268·0	8·127 9	0·4	0·0	1	1	
Maximum strain	32 336·7	20·99	0·4	0·0	1	1	
	780 873·3	8·638 2	0·6	1·0	1	1	
Tsai–Hill	28 988·7	18·82	0·4	0·0	1	1	
	666 468·0	8·184 2	0·4	0·0	1	1	
Hoffman	28 708·6	18·64	0·4	0·0	1	1	
	660 000·0	8·156 5	0·4	0·0	1	1	
Tsai–Wu	31 272·6	20·30	0·4	0·0	1	1	
	743 267·1	8·493 6	0·6	1·0	1	1	
Maximum stress	31 514·5	20·46	0·4	0·0	1	1	
(independent)	924 158·9	9·148 5	0·4	0·0	1	1	

[a] Quarter-plate analysis for cross-ply lay-ups (5 × 5 uniform mesh); full-plate analysis for all other types of lay-up (10 × 10 uniform mesh).
[b] Small deflection analysis results.
[c] Large deflection analysis results.
NB: Multiple failure locations are not indicated in this table in order to save space, but the same patterns as described in the Appendix apply.

deflections are so large that neither small nor large deflection plate theory is valid. Indeed even the small deflection results for the four thickest laminates are not valid because the deflections far exceed the limit of one quarter of the plate thickness. The reason why so many of the results in Tables 7 and all of the results in Tables 3–6 are not valid is that the plate geometry, which was originally chosen by Reddy and Pandey,[15] is such that the plates are extremely slender (see Table 2). Consequently, only for the thicker plates does failure initiate in the classical large deflection regime. For the thinner plates failure initiates in the large deflection–large rotation regime and the Mindlin laminated plate theory used here is not valid under these circumstances.

It is clear from the results presented that the present DR small deflection initial failure analysis, which has been corroborated by an independent exact analysis, does not support the conclusion implied by Reddy and

Pandey's finite element analysis, viz. that the Tsai–Hill failure criterion grossly underestimates the initial failure pressure and predicts the incorrect failure location.

8 CONCLUDING REMARKS

An exact small deflection and two approximate small and large deflection analyses have been developed for the initial failure response of laminated rectangular plates subjected to transverse pressure loading. The exact analysis is of the Navier-type and the approximate analysis uses a finite-difference implementation of the DR method. The exact analysis has been used to verify the small deflection DR analysis and both analyses have been used to demonstrate that the Tsai–Hill failure criterion may be used with at least the same confidence as other failure criteria for initial flexural failure analysis. Finally, the two approximate analyses have been used to investigate, through small and large deflection results comparisons, the effects of membrane action on the initial flexural failure response, i.e. such action in very thin plates leads to a significant increase in the initial failure pressure and a substantial reduction in the associated maximum deflection.

9 POSTSCRIPT

Although our previous experience and the results of this study suggest that Reddy and Pandey's lack of confidence in the Tsai–Hill criterion for the flexural failure analysis of laminated plates is unjustified, it should, in fairness, be pointed out that only nine different lay-ups have beeen analysed. Ideally many more configurations require analysis to prove the point conclusively.

The following additional remarks, which are essentially comments on or criticisms of Reddy and Pandey's finite element analysis, are intended to shed light on the issue.

(a) The choice of plate slenderness (width to thickness ratio) was unusually large—corresponding to a very thin plate. It is known that Mindlin plate elements may suffer from 'shear locking', especially when used to analyse very thin plates. Could this phenomenon have affected the finite element analysis results? 'Shear locking' may also manifest itself when finite differences are used but steps were taken in the DR analysis to eliminate this effect.

(b) In the finite element analysis stresses were evaluated at either the centre (four-node) or the quarter points (nine-node) of the elements. Thus stresses were not evaluated at the centre or at points along the plate centrelines (the actual failure locations)—only at the nearest Gauss point. In the DR analysis stresses were computed at these points as well as the plate corners and along the edges.

(c) Except for the results obtained with the Tsai–Hill criterion, the finite element results indicate that failure usually occurs in or near the loaded plate surface. This is surprising! The DR results indicate, in general, failure in the loaded surface at the plate corners or failure in the unloaded surface at the centre. Surely, according to the maximum stress (independent) criterion and small deflection theory, centre failure must initiate simultaneously in or near the loaded and unloaded surfaces of a symmetrically laminated plate because the strengths Y_t and Y_c are equal. This was the case with the DR analyses (also see Appendix).

(d) All of the finite element analyses are quarter-plate analyses. Reddy and Pandey do not indicate what membrane edge and centreline conditions were used for their antisymmetrically laminated plate analyses. The DR results show that the initial failure response depends on these conditions and that it is sometimes necessary to analyse the whole plate.

(e) The finite element initial failure results for cross-ply and antisymmetric angle-ply laminates are not corroborated by the exact small deflection results whereas the DR results are.

(f) An iterative scheme of incrementing loads and displacements is used in the small deflection finite element analysis to determine the initial failure pressure. By contrast, one DR analysis coupled with a simple scaling procedure is used in the present study. For truncated tensor polynomial failure criteria containing both linear and quadratic stress terms it has been found that care is needed in the choice of the magnitude of the pressure for the single DR analysis, otherwise the scaling procedure leads to an erroneously large value for the initial failure pressure. Is this feature, in some way, also inherent in Reddy and Pandey's iterative scheme? The Tsai–Hill and maximum stress (independent) failure criteria are insensitive to the choice of pressure for the DR analysis because the former contains only quadratic and the latter only linear stress terms. In this respect both criteria are superior to the others.

(g) In general, the DR initial failure pressures, failure locations and associated maximum deflections are consistent. The finite element failure pressures are less consistent and are generally higher than the DR values. It is not at all clear why such low failure pressures were computed when the Tsai–Hill criterion was used in conjunction with the finite element analysis!

(h) It is not explained what constitutes failure of a lamina in the finite element analysis. In the DR analysis failure arises when one point in either its upper or lower surface ($s = 1$ or 3) satisfies the chosen failure criterion.

ACKNOWLEDGEMENTS

One of us (M.Y.O.) wishes to record his gratitude to the Sudanese Ministry of Higher Education for providing a studentship to carry out this research. Both authors wish to express their appreciation for support from the Department of Engineering. They also wish to thank Mrs Audrey Parker for preparing tracings of the figures.

REFERENCES

1. WHITNEY, J. M., *Structural Analysis of Laminated Anisotropic Plates*. Technomic Publishing Co. Inc., Stamford, Connecticut, 1987.
2. CHIA, C.-Y., *Nonlinear Analysis of Plates*. McGraw-Hill, New York, 1980.
3. TURVEY, G. J., An initial flexural failure analysis of symmetrically laminated cross-ply rectangular plates. *Int. J. Solids and Structures*, 16(5) (1980) 451–63.
4. TURVEY, G. J., Flexural failure analysis of angle-ply laminates of GFRP and CFRP. *J. of Strain Anal.*, 15(1) (1980) 43–9.
5. TURVEY, G. J., A study of the onset of flexural failure in cross-ply laminated strips. *Fibre Sci. and Technol.*, 13(5) (1980) 325–36.
6. TURVEY, G. J., Uniformly loaded, clamped, cross-ply laminated, elliptic plates— an initial failure study. *Int. J. of Mech. Sci.*, 22(9) (1980) 551–62.
7. TURVEY, G. J., The influence of the in-plane boundary condition on the initial flexural failure of cross-ply strips. *J. of Comp. Materials* (Supplement), 14 (1980) 1–14.
8. TURVEY, G. J., Initial flexural failure of square, simply supported, angle-ply plates. *Fibre Sci. and Technol.*, 15(1) (1981) 47–63.
9. TURVEY, G. J., Uniformly loaded, antisymmetric cross-ply laminated, rectangular plates: an initial flexural failure analysis. *Fibre Sci. and Technol.*, 16(1) (1982) 1–10.
10. ADALI, S. and NISSEN, H., Micromechanical initial failure analysis of symmetrically laminated cross-ply plates. *Int. J. of Mech. Sci.*, 29(1) (1987) 83–92.

11. ADALI, S. and JHETAM, G. S., Micromechanical failure analysis of elliptic, cross-ply laminates under flexural loads. In *Composite Structures 4*, ed. I. H. Marshall. Elsevier Applied Science, London, Chapter 12, 1987, pp. 164–74.

12. TURVEY, G. J., Effects of shear deformation on the onset of flexural failure in symmetric cross-ply laminated rectangular plates. In *Composite Structures 4*, ed. I. H. Marshall. Elsevier Applied Science, London, Chapter 11, 1987, pp. 141–63.

13. REDDY, J. N., A simple higher-order theory for laminated composite plates. *J. of Appl. Mech.*, **51**(4) (1984) 745–52.

14. TURVEY, G. J., A study of the effects of imperfections and lay-up antisymmetry on the initial flexural failure response of cross-ply strips. In *ICCM 3*, ed. A. R. Bunsell. Pergamon Press, Oxford, 1980, pp. 291–304.

15. REDDY, J. N. and PANDEY, A. K., A first-ply failure analysis of composite laminates. *Computers and Structures*, **25**(3) (1987) 371–93.

16. DAY, A. S., An introduction to dynamic relaxation. *The Engineer*, **219**(5688) (1965) 218–21.

17. OTTER, J. R. H., CASSELL, A. C. and HOBBS, R. E., Dynamic relaxation. *Proceedings of the Institution of Civil Engineers*, **35**(4) (1966) 633–56.

18. CASSELL, A. C. and HOBBS, R. E., Numerical stability of dynamic relaxation analysis of non-linear structures. *Int. J. for Numer. Methods in Eng.*, **10**(6) (1976) 1407–10.

19. TURVEY, G. J. and OSMAN, M. Y., Elastic large deflection analysis of isotropic rectangular Mindlin plates. (Submitted.)

20. REDDY, J. N., *Energy and Variational Methods in Applied Mechanics*. John Wiley, New York, 1984, pp. 389–401.

21. TURVEY, G. J. and OSMAN, M. Y., Large deflection analysis of orthotropic Mindlin plates. (Submitted.)

APPENDIX: MULTIPLE FAILURE POINTS IN LAMINATED PLATES

It appears, particularly from the results of the small deflection initial failure analyses, that failure may initiate simultaneously at several points within the plate depending on the lay-up and the failure criterion. From the results obtained in this study the following patterns of multi-point failure initiation appear to apply.

A.1 Cross-Ply Laminates

For these lay-ups it appears that corresponding to each point of failure initiation in quadrant (1) (see Fig. A.1(a)) there are symmetrically situated points in quadrants (2), (3) and (4), i.e. the failure points have coordinates $[\alpha, \beta]$, $[(a - \alpha), \beta]$, $[(a - \alpha), (b - \beta)]$ and $[\alpha, (b - \alpha)]$.

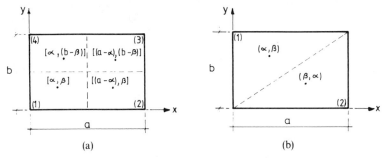

FIG. A.1.   Plan of plate showing multiple failure points. (a) Cross-ply laminates; (b) angle-ply laminates.

A.2 Angle-Ply Laminates

For these lay-ups it appears that corresponding to each point of failure initiation in triangle (1) (see Fig. A.1(b)) there is an antisymmetrically situated point in triangle (2), i.e. the failure points have coordinates (α, β) and (β, α).

A.3 Special Case

For a symmetric laminate failing in the small deflection regime according to the maximum stress (independent) criterion failure arises at points which are symmetrically situated with respect to the plate mid-plane in or near the plate faces.

5

Multilayer Theory for Delamination Analysis of a Composite Curved Bar Subjected to End Forces and End Moments

WILLIAM L. KO and RAYMOND H. JACKSON

NASA Ames Research Center, Dryden Flight Research Facility, PO Box 273, Edwards, California 93523, USA

ABSTRACT

A composite test specimen in the shape of a semicircular curved bar subjected to bending offers an excellent stress field for studying the open-mode delamination behavior of laminated composite materials. This is because the open-mode delamination nucleates at the midspan of the curved bar. The classical anisotropic elasticity theory was used to construct a 'multilayer' theory for the calculations of the stress and deformation fields induced in the multilayered composite semicircular curved bar subjected to end forces and end moments. The radial location and intensity of the open-mode delamination stress were calculated and were compared with the results obtained from the anisotropic continuum theory and from the finite element method. The multilayer theory gave more accurate predictions of the location and the intensity of the open-mode delamination stress than those calculated from the anisotropic continuum theory.

NOTATION

a	Inner radius of semicircular curved bar
a_i	Outer radius of ith layer of semicircular curved bar
a_m	Mean radius of semicircular curved bar, $\frac{1}{2}(a+b)$
A, B, C, D	Arbitrary constants associated with F for loading case of end forces P
A', B', C', D'	Arbitrary constants associated with F for loading case of end moments M

$\bar{A}, \bar{B}, \bar{C}, \bar{D}$	Arbitrary constants associated with F for loading case of end moments M for isotropic materials
b	Outer radius of semicircular curved bar
e	Loading axis offset
E_L	Modulus of elasticity of single ply in fiber direction
E_r	Modulus of elasticity in r direction
E_T	Modulus of elasticity of single ply transverse to fiber direction
E_θ	Modulus of elasticity in θ direction
$E41$	Quadrilateral membrane element
F	Airy stress function
G_{LT}	Shear modulus of single ply
$G_{r\theta}$	Shear modulus associated with r–θ system
h	Width of semicircular curved bar
i	Index associated with ith layer, $i = 1, 2, 3, \ldots, N$
k	Anisotropic parameter, $\sqrt{E_\theta/E_r}$
M	Applied end moment
N	Total number of laminated layers
P	Applied end force
r	Radial distance
r_D	Radial location of σ_D
r_m	Radial location of $(\sigma_r)_{max}$
r'_m	Radial location of $(\sigma'_r)_{max}$
r_0	Radial location of zero σ_θ
u_r	Displacement in r direction
u_θ	Displacement in θ direction
x, y	Rectangular Cartesian coordinates
β	Anisotropic parameter,
	$$\sqrt{1 + (E_\theta/E_r)(1 - 2v_{r\theta}) + E_\theta/G_{r\theta}}$$
$\gamma_{r\theta}$	Shear strain in r–θ plane
δ	Composite ply thickness
ε_r	Strain in r direction
ε_θ	Strain in θ direction
θ	Tangential coordinate
v_{LT}	Poisson ratio of single-ply composite
$v_{rz}, v_{zr}, v_{r\theta}, v_{z\theta}, v_{\theta r}, v_{\theta z}$	Poisson ratios
σ_D	Delamination stress in C-coupon
σ_r	Radial stress

$(\sigma_r)_{max}$	Delamination stress for the case of end forces P, $\sigma_r(r_m, \pi/2)$
$(\sigma_r')_{max}$	Delamination stress for the case of end moments M, $\sigma_r(r_m')$
σ_θ	Tangential stress
$\tau_{r\theta}$	Shear stress
$[\ ]^{(i)}$	Quantity associated with ith layer
$[\ ]_i$	Quantity associated with ith layer

1 INTRODUCTION

One of the major causes of stiffness and strength degradations in laminated composite structures is the delaminations between composite layers. In most engineering applications, laminated composite structures have certain curvatures (for example curved panels and curved beams). If the curved composite structure is subjected to bending that tends to flatten the composite structure, tensile stresses can be generated in the thickness direction of the composites. Also shear stresses could be induced if the bending is not a 'pure' bending. Under normal operations, if the above type of bending occurs cyclically, open-mode delaminations or shear-mode delaminations could nucleate at the sites of peak interlaminar tensile stresses or at the sites of peak interlaminar shear stresses. Continuation of these bending cyclings will cause the delamination zones to grow in size and ultimately cause the composite structures to lose their structural integrity (loss of stiffness and strength) due to excessive delaminations. The type of delamination failure (open-mode or shear-mode) depends on which type of interlaminar strength (tensile or shear) is reached first.

One of the most appealing geometries of a fatigue test coupon for studying the composite delamination phenomenon is the semicircular curved bar shape (C-coupon). When such a test specimen is subjected to end forces (that is nonpure bending), the peak radial stress (tension if the bending tends to increase the radius of the curvature of the curved bar) and the peak shear stress induced in the curved bar will be identical in magnitude but are out of phase in the tangential direction by $\pi/2$.[1] Namely, the peak radial stress is located at the midspan point of the semicircular curved bar, but the peak shear stresses occur at both ends of the semicircular curved bar. The radial distance of both the peak radial and the peak shear stresses are exactly the same.[1] The above nature of the semicircular curved bar offers an excellent situation for studying the initiation and subsequent propagation of delamination zones (open-mode or shear-mode) under cyclic loadings and for studying the fatigue behavior (degradation of stiffness and strength) of multilayered composite materials. In Ref. 1 Ko represented the multilayered composite semicircular bar with an equivalent continuous anisotropic material, and calculated radial locations and intensities of peak radial stresses induced in the curved bar subjected to end forces and end moments. Radial location and intensity of peak radial stress were calculated for different curved bar geometries (ratios of outer and inner radii) and for different degrees of anisotropy. Tolf[2] also conducted stress analyses of curved laminated beams using both continuous and discrete theories. He considered only the 'pure' bending case. In this paper the multilayer theory (discrete theory instead of continuous theory) and the finite element method were used to perform similar delamination analyses of the multilayered semicircular composite curved bar subjected to end forces and end moments. The resulting predictions of locations and intensities of peak radial stresses are compared with the results of the anisotropic continuum theory presented in Ref. 1.

2 COMPOSITE CURVED BAR

Figure 1 shows the geometry of the composite curved bar (C-coupon) for delamination fatigue tests of composite materials. Because finite areas are needed for the load application points both ends of the curved bar must be extended slightly. Thus the C-coupon consists of a semicircular curved region with straight regions at both ends. Under the application of end forces P, the loading axis will have certain offset e from the vertical diameter of the semicircle. Thus the loading condition on the C-coupon is the summation of the following two loading conditions (see Fig. 2): (1) end forces P at the ends of the semicircle and (2) end moment $M = Pe$ at the ends of the semicircle.

Because the interface between $0°$ and $90°$ composite plies has the highest Poisson ratio mismatch in laying up the composite plies for fabricating the C-coupon, it is desirable to place the $90°$ or angle plies at the peak radial stress point to ensure that the delamination will nucleate at that point. Because of this demand the precise location of the peak radial stress point must be known. The following sections will show how to determine the intensities and radial locations of peak radial stresses in the semicircular composite curved bar.

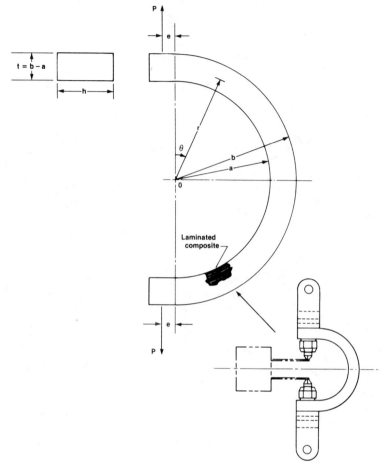

FIG. 1.   Laminated composite curved bar test coupon for fatigue delamination study.

3 ANISOTROPIC CONTINUUM THEORY

For bending a linearly elastic continuous curved bar with cylindrical anisotropy, the Airy stress function F, written in cylindrical coordinate system, takes on the following functional forms:[3]

(a) For end forces P (Fig. 2(B))

$$F = [Ar^{1+\beta} + Br^{1-\beta} + Cr + Dr \ln r] \sin \theta \tag{1}$$

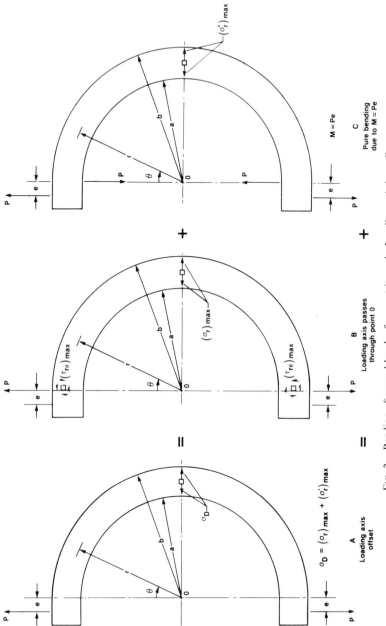

FIG. 2. Bending of curved bar by forces at its ends. Loading axis has offset e.

(b) For end moments M (Fig. 2(C))

$$F = A' + B'r^2 + C'r^{1+k} + D'r^{1-k} \tag{2}$$

where $\{A, B, C, D\}$ and $\{A', B', C', D'\}$ are arbitrary constants that must be determined from boundary conditions, and the two anisotropic parameters β and k are respectively defined as

$$\beta \equiv \sqrt{1 + \frac{E_\theta}{E_r}(1 - 2\nu_{r\theta}) + \frac{E_\theta}{G_{r\theta}}} \tag{3}$$

$$k \equiv \sqrt{\frac{E_\theta}{E_r}} \tag{4}$$

For the isotropic case, $\beta = 2$ and $k = 1$.

The functional form given in eqn (1) is also applicable to the isotropic case ($\beta = 2$). However, the functional format given in eqn (2) cannot be applied directly to the isotropic case by simply setting $k = 1$. For the isotropic case, eqn (2) must be expanded in the neighborhood of $k = 1$ using the relationship $\ln r^{\pm(k-1)} \approx r^{\pm(k-1)} - 1$, where $k - 1 \ll 1$ to the following familiar form:[4]

$$F = \bar{A} + \bar{B}r^2 + \bar{C}r^2 \ln r + \bar{D} \ln r \tag{5}$$

where $\{\bar{A}, \bar{B}, \bar{C}, \bar{D}\}$ are a different set of arbitrary constants.

Stresses in the cylindrically anisotropic body may be expressed in terms of the stress function F as

$$\sigma_r = \frac{1}{r}\frac{\partial F}{\partial r} + \frac{1}{r^2}\frac{\partial^2 F}{\partial \theta^2} \tag{6}$$

$$\sigma_\theta = -\frac{\partial^2 F}{\partial r^2} \tag{7}$$

$$\tau_{r\theta} = -\frac{\partial^2}{\partial r \partial \theta}\left(\frac{F}{r}\right) \tag{8}$$

and the stress–strain relationships for the plane stress case are given by

$$\varepsilon_r = \frac{1}{E_r}\sigma_r - \frac{\nu_{\theta r}}{E_\theta}\sigma_\theta \tag{9}$$

$$\varepsilon_\theta = -\frac{\nu_{r\theta}}{E_r}\sigma_r + \frac{1}{E_\theta}\sigma_\theta \tag{10}$$

$$\gamma_{r\theta} = \frac{1}{G_{r\theta}}\tau_{r\theta} \tag{11}$$

with the reciprocity relationship of

$$\frac{v_{r\theta}}{E_r} = \frac{v_{\theta r}}{E_\theta} \qquad (12)$$

For the plane strain case, E_r, E_θ, $v_{\theta r}$ and $v_{r\theta}$ are simply replaced with

$$\frac{E_r}{1 - v_{zr}v_{rz}}, \frac{E_\theta}{1 - v_{z\theta}v_{\theta z}}, \frac{v_{\theta r}}{1 - v_{z\theta}v_{\theta z}}\left(1 + \frac{v_{rz}v_{\theta z}}{v_{\theta r}}\right) \text{ and}$$

$$\frac{v_{r\theta}}{1 - v_{zr}v_{rz}}\left(1 + \frac{v_{z\theta}v_{rz}}{v_{r\theta}}\right), \text{ respectively.}$$

The strains are related to the displacements through the following formulae:

$$\varepsilon_r = \frac{\partial u_r}{\partial r} \qquad (13)$$

$$\varepsilon_\theta = \frac{1}{r}\frac{\partial u_\theta}{\partial \theta} + \frac{u_r}{r} \qquad (14)$$

$$\varepsilon_{r\theta} = \frac{1}{2}\left(\frac{1}{r}\frac{\partial u_r}{\partial \theta} + \frac{\partial u_\theta}{\partial r} - \frac{u_\theta}{r}\right) \qquad (15)$$

3.1 Stresses

Substitution of eqns (1) and (2) into eqns (6)–(8) yields the following stress equations in terms of the unknown arbitrary constants $\{A, B, D\}$ or $\{B', C', D'\}$:

(a) For end forces P

$$\sigma_r(r, \theta) = [A\beta r^{\beta-1} - B\beta r^{-\beta-1} + D/r] \sin\theta \qquad (16)$$

$$\sigma_\theta(r, \theta) = [A\beta(1 + \beta)r^{\beta-1} - B\beta(1 - \beta)r^{-\beta-1} + D/r] \sin\theta \qquad (17)$$

$$\tau_{r\theta}(r, \theta) = -[A\beta r^{\beta-1} - B\beta r^{-\beta-1} + D/r] \cos\theta \qquad (18)$$

(b) For end moments M

$$\sigma_r(r) = 2B' + C'(1 + k)r^{k-1} + D'(1 - k)r^{-k-1} \qquad (19)$$

$$\sigma_\theta(r) = 2B' + C'k(1 + k)r^{k-1} - D'k(1 - k)r^{-k-1} \qquad (20)$$

$$\tau_{r\theta} = 0 \qquad (21)$$

Notice that the magnitudes of σ_r (eqn (16)) and $\tau_{r\theta}$ (eqn (18)) for the case of end forces P are identical, but are out of phase in the θ direction by $\pi/2$.

3.2 Displacements

Using eqns (9)–(12), (16)–(18) and (19)–(21), the displacements u_r (eqn (13)) and u_θ (eqn (14)) may be integrated to give the following forms, neglecting the rigid body motion terms:

(a) For end forces P

$$
u_r(r, \theta) = \left\{ A r^\beta \left[\frac{1}{E_r} - (1 + \beta) \frac{v_{\theta r}}{E_\theta} \right] + B r^{-\beta} \left[\frac{1}{E_r} - (1 - \beta) \frac{v_{\theta r}}{E_\theta} \right] \right.
$$
$$
\left. + D(\ln r) \left(\frac{1}{E_r} - \frac{v_{\theta r}}{E_\theta} \right) \right\} \sin \theta \tag{22}
$$

$$
u_\theta(r, \theta) = \left\{ A r^\beta \left[\frac{1}{E_r} - \beta(1 + \beta) \frac{1}{E_\theta} - \frac{v_{\theta r}}{E_\theta} \right] \right.
$$
$$
+ B r^{-\beta} \left[\frac{1}{E_r} + \beta(1 - \beta) \frac{1}{E_\theta} - \frac{v_{\theta r}}{E_\theta} \right]
$$
$$
\left. + D \left[(\ln r) \left(\frac{1}{E_r} - \frac{v_{\theta r}}{E_\theta} \right) - \left(\frac{1}{E_\theta} - \frac{v_{\theta r}}{E_\theta} \right) \right] \right\} \cos \theta \tag{23}
$$

(b) For end moments M

$$
u_r(r) = B' \left\{ 2r \left(\frac{1}{E_r} - \frac{v_{\theta r}}{E_\theta} \right) \right\} + C' \left\{ (1 + k) r^k \left(\frac{1}{k} \frac{1}{E_r} - \frac{v_{\theta r}}{E_\theta} \right) \right\}
$$
$$
- D' \left\{ (1 - k) r^{-k} \left(\frac{1}{k} \frac{1}{E_r} + \frac{v_{\theta r}}{E_\theta} \right) \right\} \tag{24}
$$

$$
u_\theta(r, \theta) = B' \left\{ 2r \left(\frac{1}{E_\theta} - \frac{1}{E_r} \right) \right\} \theta \tag{25}
$$

3.3 Delamination Stresses and their Locations

For the continuous (or single-layer) curved bar, the two sets of unknown constants $\{A, B, D\}$ and $\{B', C', D'\}$ can be determined explicitly from the boundary conditions to give closed-form expressions for the stresses (eqns (16)–(21)) and the displacements (eqns (22)–(25)).[1,3] By using the extreme condition $(\partial/\partial r)\sigma_r = 0$, the functional expressions for the delamination stress (maximum value of σ_r) and its radial location were derived in Ref. 1 for both of the aforementioned loading cases.

4 MULTILAYER THEORY

Figure 3 shows the multilayer (N-layers), semicircular curved bar subjected to both end forces P and end moments M. The stress field and displacement field in each layer i ($i = 1, 2, \ldots, N$) for each loading case may be obtained from the results given in Section 3.

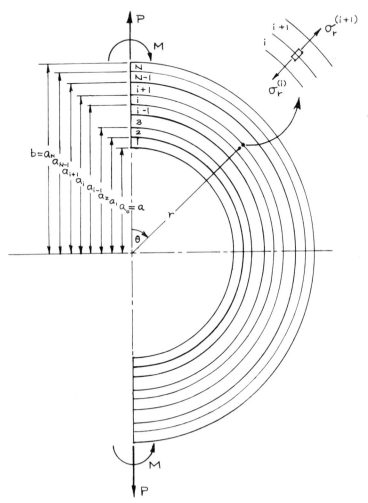

Fig. 3. Bending of laminated anisotropic semicircular curved beam by end forces and end moments.

4.1 Boundary Conditions

At each interface between layers i and $i+1$ $(i=1,2,...,N-1)$ the following boundary conditions for the continuities of stresses and displacements must hold (no sliding between layers):

(a) For end forces P

at $r = a_i$
$$\sigma_r^{(i)}(a_i, \theta) = \sigma_r^{(i+1)}(a_i, \theta) \tag{26}$$

$$\tau_{r\theta}^{(i)}(a_i, \theta) = \tau_{r\theta}^{(i+1)}(a_i, \theta) \tag{27}$$

$$u_r^{(i)}(a_i, \theta) = u_r^{(i+1)}(a_i, \theta) \tag{28}$$

$$u_\theta^{(i)}(a_i, \theta) = u_\theta^{(i+1)}(a_i, \theta) \tag{29}$$

The boundary conditions at the traction-free inner surface $(i-1=0)$ and outer surface $(i=N)$ of the curved bar are:

at $r = a_0 = a$
$$\sigma_r^{(1)}(a, \theta) = 0 \tag{30}$$

$$\tau_{r\theta}^{(1)}(a, \theta) = 0 \tag{31}$$

at $r = a_N = b$
$$\sigma_r^{(N)}(b, \theta) = 0 \tag{32}$$

$$\tau_{r\theta}^{(N)}(b, \theta) = 0 \tag{33}$$

(b) For end forces M

at $r = a_i$
$$\sigma_r^{(i)}(a_i) = \sigma_r^{(i+1)}(a_i) \tag{34}$$

$$u_r^{(i)}(a_i) = |u_r^{(i+1)}(a_i) \tag{35}$$

$$u_\theta^{(i)}(a_i) = u_\theta^{(i+1)}(a_i) \tag{36}$$

at $r = a_0 = a$
$$\sigma_r^{(1)}(a) = 0 \tag{37}$$

at $r = a_N = b$
$$\sigma_r^{(N)}(b) = 0 \tag{38}$$

As mentioned earlier, σ_r and $\tau_{r\theta}$ (eqns (16) and (18)) for the case of end forces P have identical r dependency. Thus, if σ_r satisfies the boundary conditions, $\tau_{r\theta}$ will also satisfy the boundary conditions automatically. Therefore the boundary conditions associated with $\tau_{r\theta}$ (eqns (27), (31) and (33)) are not needed.

For each loading case, each set of the previous boundary conditions will give $2 + 3(N-1) = 3N - 1$ equations for determining the $3N$ unknowns A_i, B_i, D_i $(i=1,2,...,N)$ for the case of end forces P or $3N$ unknowns B_i', C_i', D_i' $(i=1,2,...,N)$ for the case of end moments M.

The last equation needed for each loading case is the condition that the end force P or the end moment M is balanced by the stresses in the curved bar:

(a) For end forces P

$$-P = \sum_{i=1}^{N} \int_{a_{i-1}}^{a_i} \tau_{r\theta}^{(i)}(r,0)\,\mathrm{d}r \qquad \theta = 0 \qquad (39)$$

(b) For end moments M

$$-M = \sum_{i=1}^{N} \int_{a_{i-1}}^{a_i} (r - r_0)\sigma_\theta(r)\,\mathrm{d}r \qquad (40)$$

where the negative signs in front of P and M are to increase the radius of curvature of the curved bar, and r_0 is the unknown radial location where $\sigma_\theta = 0$. For pure bending we have

$$\sum_{i=1}^{N} \int_{a_{i-1}}^{a_i} \sigma_\theta(r)\,\mathrm{d}r = 0 \qquad (41)$$

Therefore (since the r_0 term vanishes) eqn (40) becomes

$$-M = \sum_{i=1}^{N} \int_{a_{i-1}}^{a_i} r\sigma_\theta(r)\,\mathrm{d}r \qquad (42)$$

4.2 Boundary Conditions in Final Forms

After substitution of stress and displacement expressions given respectively in Sections 3.1 and 3.2 into the boundary conditions given in Section 4.1, the following final forms of the boundary conditions are obtained:

(a) For end forces P

for $\sigma_r^{(1)}$ (eqn (30))

$$A_1\beta_1 a^{\beta_1} - B_1\beta_1 a^{-\beta_1} + D_1 = 0 \qquad (43)$$

for $\sigma_r^{(i)}$ (eqn (26))

$$A_i\beta_i a_i^{\beta_i} - B_i\beta_i a_i^{-\beta_i} + D_i - A_{i+1}\beta_{i+1} a_i^{\beta_{i+1}} + B_{i+1}\beta_{i+1} a_i^{-\beta_{i+1}} - D_{i+1} = 0$$

$$(44)$$

for $\sigma_r^{(N)}$ (eqn (32))

$$A_N \beta_N a_N^{\beta_N} - B_N \beta_N a_N^{-\beta_N} + D_N = 0 \qquad (45)$$

for $u_r^{(i)}$ (eqn (28))

$$A_i a_i^{\beta_i} \left\{ \frac{1}{E_r^{(i)}} - (1 + \beta_i) \frac{v_{\theta r}^{(i)}}{E_\theta^{(i)}} \right\} + B_i a_i^{-\beta_i} \left\{ \frac{1}{E_r^{(i)}} - (1 - \beta_i) \frac{v_{\theta r}^{(i)}}{E_\theta^{(i)}} \right\}$$

$$+ D_i (\ln a_i) \left(\frac{1}{E_r^{(i)}} - \frac{v_{\theta r}^{(i)}}{E_\theta^{(i)}} \right) - A_{i+1} a_i^{\beta_{i+1}} \left\{ \frac{1}{E_r^{(i+1)}} - (1 + \beta_{i+1}) \frac{v_{\theta r}^{(i+1)}}{E_\theta^{(i+1)}} \right\}$$

$$- B_{i+1} a_i^{-\beta_{i+1}} \left\{ \frac{1}{E_r^{(i+1)}} - (1 - \beta_{i+1}) \frac{v_{\theta r}^{(i+1)}}{E_\theta^{(i+1)}} \right\}$$

$$- D_{i+1} (\ln a_i) \left(\frac{1}{E_r^{(i+1)}} - \frac{v_{\theta r}^{(i+1)}}{E_\theta^{(i+1)}} \right) = 0 \qquad (46)$$

for $u_\theta^{(i)}$ (eqn (29))

$$A_i a_i^{\beta_i} \frac{\beta_i}{E_\theta^{(i)}} \{(1 + \beta_i) - v_{\theta r}^{(i)}\} - B_i a_i^{-\beta_i} \frac{\beta_i}{E_\theta^{(i)}} \{(1 - \beta_i) - v_{\theta r}^{(i)}\}$$

$$+ D_i \frac{1}{E_\theta^{(i)}} (1 - v_{\theta r}^{(i)}) - A_{i+1} a_i^{\beta_{i+1}} \frac{\beta_{i+1}}{E_\theta^{(i+1)}} \{(1 + \beta_{i+1}) - v_{\theta r}^{(i+1)}\}$$

$$+ B_{i+1} a_i^{-\beta_{i+1}} \frac{\beta_{i+1}}{E_\theta^{(i+1)}} \{(1 - \beta_{i+1}) - |v_{\theta r}^{(i+1)}\} - D_{i+1} \frac{1}{E_\theta^{(i+1)}} (1 - v_{\theta r}^{(i+1)}) = 0 \qquad (47)$$

Equation (47) was obtained by taking the difference between the boundary conditions (28) and (29). This was done because the resulting expression (47) is simpler than using eqn (29):

for P (eqn (39))

$$\sum_{i=1}^{N} [A_i(a_i^{\beta_i} - a_{i-1}^{\beta_i}) + B_i(a_i^{-\beta_i} - a_{i-1}^{-\beta_i}) + D_i(\ln a_i - \ln a_{i-1})] = P \qquad (48)$$

(b) For end moments M

for $\sigma_r^{(1)}$ (eqn (37))

$$2B_1' + C_1'(1 + k_1)a^{k_1 - 1} + D_1'(1 - k_1)a^{-k_1 - 1} = 0 \qquad (49)$$

for $\sigma_r^{(i)}$ (eqn (34))

$$2B_i' + C_i'(1 + k_i)a_i^{k_i - 1} + D_i'(1 - k_i)a_i^{-k_i - 1} - 2B_{i+1}'$$
$$- C_{i+1}'(1 + k_{i+1})a_i^{k_{i+1} - 1} - D_{i+1}'(1 - k_{i+1})a_i^{-k_{i+1} - 1} = 0 \qquad (50)$$

for $\sigma_r^{(N)}$ (eqn (38))

$$2B_N' + C_N'(1 + k_N)b^{k_N - 1} + D_N'(1 - k_N)b^{-k_N - 1} = 0 \qquad (51)$$

for $u_r^{(i)}$ (eqn (35))

$$B_i'\left\{2a_i\left(\frac{1}{E_r^{(i)}} - \frac{v_{\theta r}^{(i)}}{E_\theta^{(i)}}\right)\right\} + C_i'\left\{(1 + k_i)a_i^{k_i}\left(\frac{1}{k_i}\frac{1}{E_r^{(i)}} - \frac{v_{\theta r}^{(i)}}{E_\theta^{(i)}}\right)\right\}$$

$$- D_i'\left\{(1 - k_i)a_i^{-k_i}\left(\frac{1}{k_i}\frac{1}{E_r^{(i)}} + \frac{v_{\theta r}^{(i)}}{E_\theta^{(i)}}\right)\right\} - B_{i+1}'\left\{2a_i\left(\frac{1}{E_r^{(i+1)}} - \frac{v_{\theta r}^{(i+1)}}{E_\theta^{(i+1)}}\right)\right\}$$

$$- C_{i+1}'\left\{(1 + k_{i+1})a_i^{k_{i+1}}\left(\frac{1}{k_{i+1}}\frac{1}{E_r^{(i+1)}} - \frac{v_{\theta r}^{(i+1)}}{E_\theta^{(i+1)}}\right)\right\}$$

$$+ D_{i+1}'\left\{(1 - k_{i+1})a_i^{-k_{i+1}}\left(\frac{1}{k_{i+1}}\frac{1}{E_r^{(i+1)}} + \frac{v_{\theta r}^{(i+1)}}{E_\theta^{(i+1)}}\right)\right\} = 0 \qquad (52)$$

for $u_\theta^{(i)}$ (eqn (36))

$$B_i'\left\{2a_i\left(\frac{1}{E_\theta^{(i)}} - \frac{1}{E_r^{(i)}}\right)\right\} - B_{i+1}'\left\{2a_i\left(\frac{1}{E_\theta^{(i+1)}} - \frac{1}{E_r^{(i+1)}}\right)\right\} = 0 \qquad (53)$$

for M (eqn (42))

$$\sum_{i=1}^{N}[B_i'(a_i^2 - a_{i-1}^2) + C_i'k_i(a_i^{k_i+1} - a_{i-1}^{k_i+1}) - D_i'k_i(a_i^{-k_i+1} - a_{i-1}^{-k_i+1})] = -M$$

$$(54)$$

4.3 Delamination Stresses and their Locations

At exactly which layer the value of σ_r for each loading case will become maximum cannot be predicted until after all the unknown arbitrary constants $\{A_i, B_i, C_i\}$ or $\{B_i', C_i', D_i'\}$ are determined from the appropriate boundary conditions given in Section 4.2. Suppose $(\sigma_r)_{\max}$ (or $(\sigma_r')_{\max}$), the maximum value of σ_r due to end forces P (or end moments M), occurs in the ith layer, then by using the extreme condition $(\partial/\partial r)\sigma_r = 0$ the radial location r_m (or r_m') of $(\sigma_r)_{\max}$ (or $(\sigma_r')_{\max}$) may be calculated from eqn (16) (or eqn (19)) as:

(a) For end forces P

$$r_m = \left[\frac{D_i - \sqrt{D_i^2 - 4A_iB_i\beta_i^2(\beta_i^2 - 1)}}{2A_i\beta_i(\beta_i - 1)}\right]^{1/\beta_i} \qquad (55)$$

(b) For end moments M

$$r'_m = \left[-\frac{D'_i}{C'_i} \right]^{1/2k_i} \tag{56}$$

And the delamination stresses $(\sigma_r)_{max}$ and $(\sigma'_r)_{max}$ for the two loading cases may be written as:

(a) For end forces P

$$(\sigma_r)_{max} \equiv \sigma_r(r_m, \pi/2) = [A_i\beta_i r_m^{\beta_i - 1} - B_i\beta_i r_m^{-\beta_i - 1} + D_i/r_m] \tag{57}$$

(b) For end moments M

$$(\sigma'_r)_{max} \equiv \sigma_r(r'_m) = 2B'_i + C'_i(1 + k_i)(r'_m)^{k_i - 1} + D'_i(1 - k_i)(r'_m)^{-k_i - 1} \tag{58}$$

4.4 Delamination Stress in C-coupon

The delamination stresses $(\sigma_r)_{max}$ (due to P) and $(\sigma'_r)_{max}$ (due to M) do not occur at the same radial locations of the curved bar (see eqns (55) and (56)). Thus the delamination stress σ_D in the C-coupon cannot be constructed by simply summing up $(\sigma_r)_{max}$ and $(\sigma'_r)_{max}$. The value of σ_D must be evaluated at $r = r_D$, where the summation of the radial stress $\sigma_r(r, \pi/2)$ due to P and the radial stress $\sigma_r(r)$ due to M become maximum, Namely,

$$\sigma_D = \overset{\text{Due to } P}{\sigma_r(r_D, \pi/2)} + \overset{\text{Due to } M}{\sigma_r(r_D)} \tag{59}$$

where the value of r_D is calculated from the following extreme condition:

$$(d/dr)[\overset{\text{Due to } P}{\sigma_r(r, \pi/2)} + \overset{\text{Due to } M}{\sigma_r(r)}] = 0 \tag{60}$$

which, after substitution of eqns (16) and (19) and after performing differentiation, becomes

$$A_i\beta_i(\beta_i - 1)r_D^{\beta_i} + B_i\beta_i(\beta_i + 1)r_D^{-\beta_i} - D_i + (k_i^2 - 1)(C'_i r_D^{k_i} + D'_i r_D^{-k_i}) = 0 \tag{61}$$

As will be seen later, the radial location r_D of the delamination stress σ_D in the C-coupon is somewhere between r_m and r'_m (that is $r_m < r_D < r'_m$).

5 NUMERICAL EXAMPLES

The anisotropic continuum theory and the multilayer theory presented respectively in Sections 3 (or Ref. 1) and 4 will now be applied to the delamination analysis of the composite C-coupon. One type of composite

C-coupon under development has the following geometry and ply properties:

Inner radius	$a = 2.1590$ cm (0.85 in)
Outer radius	$b = 2.9724$ cm (1.170 22 in)
Loading axis offset	$e = 0.9525$ cm (0.375 in)
Width	$h = 2.54$ cm (1 in)
Ply thickness	$\delta = 0.01506$ cm (0.005 93 in)
Mean radius	$a_m = (a + b)/2 = 2.5657$ cm (1.010 11 in)
Radii ratio	$(b/a) = 1.3767$

$$E_L = 17.2369 \times 10^{10}\,\text{N/m}^2\ (25 \times 10^6\,\text{lb/in}^2)$$
$$E_T = 0.8274 \times 10^{10}\,\text{N/m}^2\ (1.2 \times 10^6\,\text{lb/in}^2)$$
$$G_{LT} = 0.4137 \times 10^{10}\,\text{N/m}^2\ (0.6 \times 10^6\,\text{lb/in}^2)$$
$$v_{LT} = 0.33$$
$$v_{TL} = 0.01584$$

The afore mentioned C-coupon has 54 composite plies having the stacking sequence of $[0^\circ_{25}/+15^\circ/-15^\circ/-15^\circ/+15^\circ/0^\circ_{25}]$.

5.1 Equivalent Continuum

In order to apply the anisotropic continuum theory, the laminated composite C-coupon will be represented by an equivalent anisotropic continuum having the following effective material properties:

$$E_\theta = 16.3220 \times 10^{10}\,\text{N/m}^2\ (23.6731 \times 10^6\,\text{lb/in}^2)$$
$$E_r = 0.8274 \times 10^{10}\,\text{N/m}^2\ (1.2 \times 10^6\,\text{lb/in}^2)$$
$$G_{r\theta} = 0.4137 \times 10^{10}\,\text{N/m}^2\ (0.6 \times 10^6\,\text{lb/in}^2)$$
$$v_{r\theta} \approx 0.01673$$

Based on these effective material properties, the equivalent continuum representing the C-coupon has the values of anisotropic parameters as $\beta = 7.7151$ and $k = 4.4416$.

5.2 Multilayer System

For the purpose of applying the multilayer theory to the C-coupon, the extended linear regions at both ends will be neglected, and only the semicircular region subjected to two types of loadings (end forces P and end moments M, Fig. 2) will be considered. For simplification, each group of 25 layers of the 0° plies will be represented by one layer of anisotropic continuum, and the center region of 4 layers of $\pm 15^\circ$ angle plies will be represented by another anisotropic continuum. Thus the 54-layer composite will be represented by three layers of anisotropic continua.

(a) 0° plies

The inner $(i = 1)$ and the outer $(i = 3)$ layers have the following effective material properties:

$$E_\theta^{(i)} = E_L = 17.2369 \times 10^{10} \, N/m^2 \, (25 \times 10^6 \, lb/in^2)$$
$$E_r^{(i)} = E_T = 0.8274 \times 10^{10} \, N/m^2 \, (1.2 \times 10^6 \, lb/in^2)$$
$$G_{r\theta}^{(i)} = G_{LT} = 0.4137 \times 10^{10} \, N/m^2 \, (0.6 \times 10^6 \, lb/in^2)$$
$$v_{r\theta}^{(i)} = 0.015\,84$$

which give the values of anisotropic parameters as $\beta_1 = \beta_3 = 7.9272$ and $k_1 = k_3 = 4.5644$.

(b) $\pm 15°$ plies

The center layer $(i = 2)$ has the following effective material properties:

$$E_\theta^{(2)} = 4.8873 \times 10^{10} \, N/m^2 \, (7.0884 \times 10^6 \, lb/in^2)$$
$$E_r^{(2)} = 0.8274 \times 10^{10} \, N/m^2 \, (1.2 \times 10^6 \, lb/in^2)$$
$$G_{r\theta}^{(2)} = 0.4137 \times 10^{10} \, N/m^2 \, (0.6 \times 10^6 \, lb/in^2)$$
$$v_{r\theta}^{(2)} \approx 0.055\,90$$

which give the values of the anisotropic parameters as $\beta_2 = 4.2498$ and $k_2 = 2.4304$.

The $3N$ $(N = 3)$ boundary conditions for the aforementioned three-layer laminated system may be written in matrix form for solving the $3N$ unknown constants A_i, B_i, D_i $(i = 1, 2, 3)$, for the case of end forces P, or for solving the other set of $3N$ unknown constants B_i', C_i', D_i' $(i = 1, 2, 3)$, for the case of end moments M.

6 FINITE ELEMENT ANALYSIS

To verify the solution accuracies of the analysis for which the actual loading condition of the C-coupon was represented with the superposition of two loading cases of the semicircular curved bar (see Fig. 2), finite element stress analysis was performed on the semicircular curved bar (under two loading cases) and on the C-coupon using the structural performance and resizing (SPAR) finite element computer program.[5] Figures 4 and 5 respectively show the SPAR finite element models set up for the semicircular curved bar and the C-coupon. Because of symmetry with respect to the x axis, only the half span of the semicircular curved bar and the C-coupon were modeled. Both systems were first reduced to three-layer

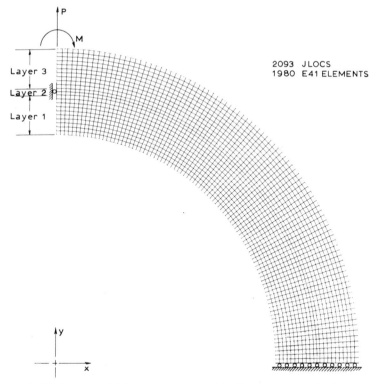

FIG. 4.   Finite element model for multilayered semicircular curved bar under bending.

systems as defined in Section 5.2. Then layers 1, 2 and 3 were respectively modeled in 10, 2 and 10 layers of quadrilateral membrane $E41$ elements in the r direction. In the tangential direction, the quarter-circular region ($0 \leq \theta \leq \pi/2$) of the two systems was modeled with 90 $E41$ elements. The extended region ($-e \leq x \leq 0$) of the C-coupon was modeled with 25 layers of $E41$ elements in the x direction.

The $\theta = 90°$ plane for each model was allowed to move freely in the x direction (shown with rollers in Figs 4 and 5) but not in the y direction. At the upper end of each model, only one point lying in the middle surface ($\sigma_\theta = 0$ point was found to be very close to the middle surface of the curved bar) was constrained to move freely in the y direction only (no movement in the x direction). Thus the end of each model could rotate freely (shown with only one roller, Figs 4 and 5). The applied force P and the applied moment M were represented respectively with the distributions of $\tau_{r\theta}(r, 0)'$ and $\sigma_\theta(r)$

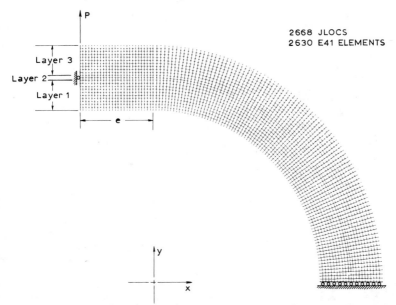

FIG. 5. Finite element model for laminated composite delamination test C-coupon.

obtained from the multilayer analysis. The sizes of the two SPAR models are listed below:

	Semicircular curved bar	C-coupon
JLOCs	2093	2668
$E41$ elements	1980	2530

7 RESULTS

7.1 Semicircular Curved Bar

Figure 6 shows the distributions of σ_r in the $\theta = \pi/2$ plane for the case of end forces P calculated from different theories. The values of $(\sigma_r)_{max}$ and r_m calculated from different theories are indicated in the figure. The values of $(\sigma_r)_{max}$ calculated from different theories are quite close, except its location r_m. The multilayer theory and the finite element method predicted close values of r_m. The $(\sigma_r)_{max}$ site predicted from the anisotropic continuum theory is located slightly closer to the middle surface than the $(\sigma_r)_{max}$ sites predicted from the multilayer theory and SPAR. The $(\sigma_r)_{max}$ site for the isotropic material is closest to the middle surface and is always located between the middle surface and the $(\sigma_r)_{max}$ site predicted from the

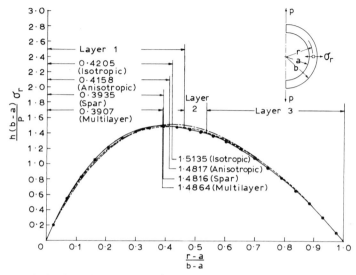

FIG. 6. Distribution of radial stress in $\theta = \pi/2$ plane, curved bar under end forces P. $b/a = 1.3767$. ———, $\beta_1 = \beta_3 = 7.9272$, $\beta_2 = 4.2498$ (multilayer); ---, $\beta = 7.7151$ (anisotropic continuum); ----, $\beta = 2$ (isotropic continuum): ●, SPAR.

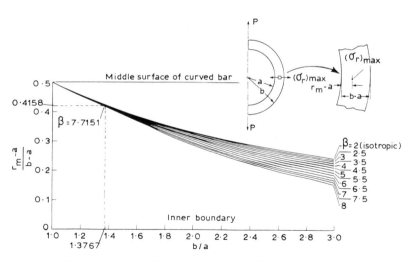

FIG. 7. Plots of locations of $(\sigma_r)_{max}$ as a function of b/a for different values of β.

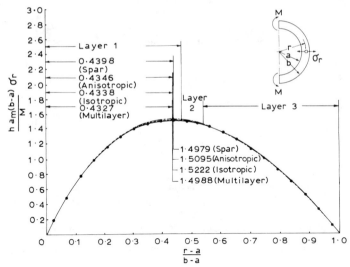

FIG. 8.   Distribution of radial stress, curved bar under end moments M. $b/a = 1.3767$. ———,
$k_1 = k_3 = 4.5644$, $k_2 = 2.4304$ (multilayer); ---, $k = 4.4416$ (anisotropic continuum); ———,
$k = 1$ (isotropic continuum); ●, SPAR.

anisotropic continuum theory. This can be seen more clearly from the plots
shown in Fig. 7.

The distance between the sites of $(\sigma_r)_{max}$ predicted from the multilayer
theory and the anisotropic continuum theory is

$$(r_m)_{\text{anisotropic continuum}} - (r_m)_{\text{multilayer}} = (0.4158 - 0.3907)(b - a)$$
$$= 0.0203 \text{ cm } (0.0080 \text{ in})$$

which is 1.3554 times the single-ply thickness of 0.01506 cm (0.00593 in).

Figure 8 shows the distributions of σ_r for the case of end moments M
calculated from different theories. Unlike the previous case, the values of r'_m
and $(\sigma'_r)_{max}$ calculated from different theories are quite close, showing that
the value of r'_m is quite insensitive to the theory used. The multilayer theory
predicted the shortest distance of r'_m (that is the $(\sigma'_r)_{max}$ site is closest to the
inner boundary of the curved bar). The site of $(\sigma'_r)_{max}$ predicted from the
anisotropic continuum theory always lies between the middle surface and
the $(\sigma'_r)_{max}$ site, based on isotropic theory. This is shown in Fig. 9.

The distance between the sites of $(\sigma'_r)_{max}$ predicted from the multilayer
theory and the anisotropic continuum theory is given below:

$$(r'_m)_{\text{anisotropic continuum}} - (r'_m)_{\text{multilayer}} = (0.4346 - 0.4327)(b - a)$$
$$= 0.001545 \text{ cm } (0.000608 \text{ in})$$

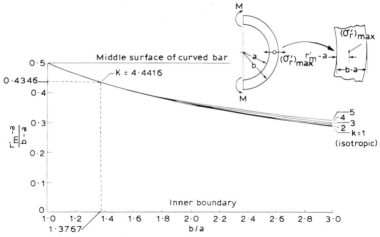

FIG. 9. Plots of locations of $(\sigma_r')_{max}$ as a function of b/a for different values of k.

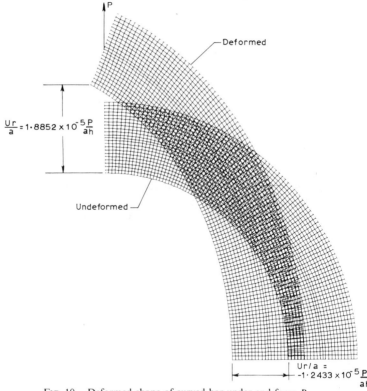

FIG. 10. Deformed shape of curved bar under end force P.

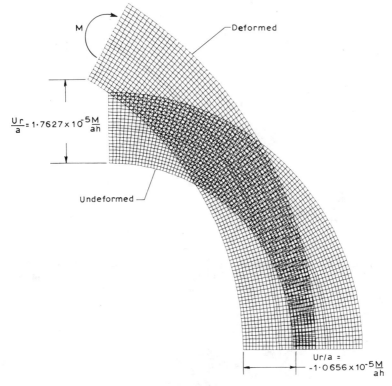

FIG. 11.   Deformed shape of curved bar under end moment M.

which is only 0·1025 times the single-ply thickness of 0·015 06 cm
(0·005 93 in), and is therefore insignificant.

Figures 10 and 11 respectively show the deformed shapes of the
semicircular curved bar subjected to end forces P and end moments M. The
dimensionless radial displacements u_r/a at $\theta = 0$ and $\theta = \pi/2$ are shown in
the figures.

Table 1 summarizes all the values of $(\sigma_r)_{max}$, $(\sigma_r')_{max}$, r_m and r_m' calculated
from different theories.

7.2 C-coupon

In order to determine the radial location r_D and magnitude of
delamination stress σ_D for the C-coupon, the two radial stresses calculated
from the semicircular curved bar due to P and M were summed up. The

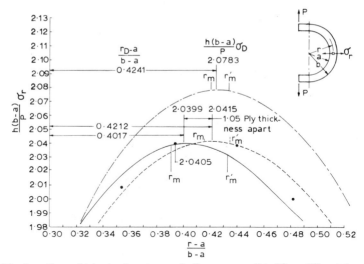

FIG. 12.   Locations of delamination stress σ_D in C-coupon predicted from different theories. ——, Multilayer; – – –, anisotropic continuum; — — —, isotropic continuum; ●, SPAR (C-coupon).

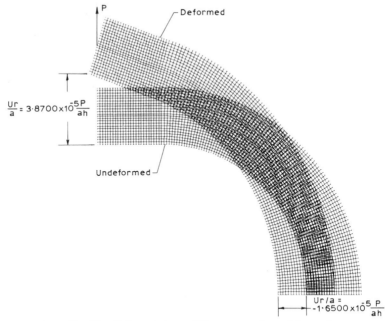

FIG. 13.   Deformed shape of C-coupon under end forces P.

TABLE 1

Intensities and locations of delamination stresses in semicircular curved bar

Theory	End forces P		End moments M	
	Item			
	$\dfrac{h(b-a)}{P}(\sigma_r)_{max}$	$\dfrac{r_m-a}{b-a}$	$\dfrac{ha_m(b-a)}{M}(\sigma_r')_{max}$	$\dfrac{r_m'-a}{b-a}$
Anisotropic continuum	1·481 7	0·415 8	1·509 5	0·434 6
Multilayer theory	1·486 4	0·390 7	1·498 8	0·432 7
Isotropic continuum	1·513 5	0·420 5	1·522 2	0·433 8
SPAR	1·481 6	0·393 5	1·497 9	0·439 8

results are shown in Fig. 12 for multilayer, anisotropic and isotropic cases. Notice that the radial location r_D predicted from all three theories lie between r_m and r_m' but closer to r_m instead of r_m' because the stress contribution due to P is larger than that due to M ($e < a_m$). The distance between the locations of σ_D predicted from multilayer and anisotropic continuum theories is about 1·05 times the ply thickness. The finite element solution data points obtained from the C-coupon model lie in the vicinity of the two curves obtained from the multilayer and anisotropic continuum theories. The intensity and the radial location of σ_D predicted from SPAR (C-coupon) are closest to those predicted from the multilayer theory. Table 2 summarizes the values of σ_D and r_D predicted from different theories.

Figure 13 shows the deformed shape of the C-coupon subjected to end forces P. The dimensionless radial displacement u_r/a at midspan and at the free end are shown in the figure.

TABLE 2

Intensities and locations of delamination stresses in C-coupon

Theory	C-coupon under end forces P	
	$\dfrac{h(b-a)}{P}\sigma_D$	$\dfrac{r_D-a}{b-a}$
Anisotropic continuum	2·041 5	0·421 2
Multilayer theory	2·039 9	0·401 7
Isotropic continuum	2·078 3	0·424 1
SPAR (semicircular curved bar)	2·035 3	0·393 5
SPAR (C-coupon)	2·040 5	0·393 5

8 CONCLUSIONS

The multilayer theory was developed for delamination analysis of a semicircular composite curved bar subjected to end forces and end moments. The difference between the radial locations of the delamination stress (maximum radial stress) predicted from the multilayer theory and from the anisotropic continuum theory was approximately 1·4 times the ply thickness for the case of end forces and about $\frac{1}{10}$ of the ply thickness for the case of end moments. The superposition method (namely, by summing up the two radial stresses induced in the semicircular curved bar subjected to end forces and end moments), used to construct the delamination stress in the C-coupon, gave reasonably accurate intensity of the delamination stress for the C-coupon. The finite element analysis of the C-coupon gave the radial location of the delamination stress in the C-coupon much closer to that predicted from the multilayer theory than from the anisotropic continuum theory.

REFERENCES

1. Ko, W. L., Delamination stresses in semicircular laminated composite bars. NASA TM-4026, 1988.
2. Tolf, G., Stresses in a curved laminated beam. *Fiber Sci. and Technol.*, **19**(4) (1983) 243–67.
3. Lekhnitskii, S. G., *Anisotropic Plates.* Gordon and Breach, New York, 1968.
4. Fung, Y. C., *Foundations of Solid Mechanics.* Prentice–Hall, Englewood Cliffs, New Jersey, 1965.
5. Whetstone, W. D., *SPAR Structural Analysis System Reference Manual, System Level 13A*, Vol. 1. Program Execution, NASA CR-158970-1, 1978.

6

A Study on Longitudinal Compressive Strength of FRP-Composites

XUE YUAN-DE

Department of Engineering Mechanics, Tongji University, Shanghai, People's Republic of China

and

WANG ZHENG-YING

Shanghai GRP Research Institute, Shanghai, People's Republic of China

ABSTRACT

The failure mechanism of FRP-composites under longitudinal compressive stress is studied. Considering the effect of the initial curvature of the fibre and the yielding in shear of matrix, etc., a new formula to predict the FRP longitudinal compressive strength has been derived.

Some ways to improve the compressive testing, including specimen form, loading fixture and transverse supporting fixture, are presented. By numerous tests, compressive properties of various FRP-composites are obtained. Finally, the theoretical prediction of the compressive failure mode is identified with the experimental results and the longitudinal compressive strength can be greatly increased through using a high-modulus fibre, a matrix with high shear strength and improving the interface bonding strength, etc.

INTRODUCTION

In recent years the compressive behaviour of FRP-composites has been systematically investigated by us, for the compressive strength of composites is greatly influenced by some factors, such as the modulus of

199

fibres, the shear strength of the matrix, the debonding strength of the fibre–matrix interface and other technological factors. In some cases the longitudinal compressive strength of composites may be less than its tensile strength, which results in compressive failure or compressive fatigue failure, for instance, for thin-walled composite bending components. Failure modes of uni-composites under longitudinal compressive stress have been investigated and the compressive stress–strain curves of various kinds of FRP-composites experimentally estimated. Finally we succeeded in obtaining a new formula to predict FRP-longitudinal compressive strength and finding some ways to increase FRP-compressive strength, so that it makes it possible to design and manufacture such kinds of composite products, which are loaded by a highly compressive stress, for example, a FRP vaulting pole[1] and a FRP leaf spring.[2]

FAILURE MODE AND ANALYTICAL METHOD

Over the past decades much effort has been made to clarify the failure mechanism of a laminate under longitudinal compressive load and to predict its strength.

As we all know, early in the 1960s, Rosen presented two formulae to predict the internal buckling stress of the fibres embedded in the matrix for two buckling cases. One is the so-called transverse buckling mode and the other the shear instability mode.

But Rosen's formula is not conservative enough. Others suggested a modified factor, for instance, considering the inelastic behaviour of the matrix, Lager and June used a reduction factor of 0·63 instead of the initial elastic modulus in Rosen's formula.

In recent years a new model, named the Kink Band Formation Model, has been presented by Budiansky,[3] Hahn,[4] etc. (Fig. 1). In this model some realistic geometrical and physical characterizations of the composites such as random imperfections of the fibre embedded in the matrix and plasticity in shear of the matrix supporting the fibre are taken into account. They point out that Rosen's formula, in fact, is about initially straight fibres and pure elastic matrix.

This problem is also paid attention to by some investigators in China. Early in the 1970s Professor Zhu[5] pointed out that fibres have some initial curvature, which can cause additional shear stress or tensile stress in the matrix resulting in matrix yielding or interface debonding and finally fibre buckling under compressive stress less than that predicted by Rosen's

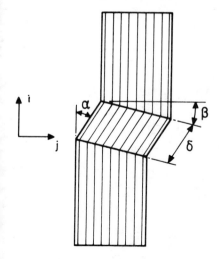

FIG. 1. A schematic kink band geometry.

formula. In 1981 Xue and Wang[6] suggested a modified formula to predict the longitudinal compressive strength of composite laminates, considering the existence of the initial curvature of fibres (caused by fibre misalignment, imperfection of interface bonding, voids, etc.) and the plasticity of matrix in shear. Then Xue *et al.*[7] obtained the value of tensile stress in the fibre matrix interface due to the Poisson's ratio of matrix being larger than that of the fibre. Here we set up a 3-D FEM model composed of a fibre embedded in the matrix as well as assuming that the displacement continuity of two neighbouring elements is constant. Recently Shao and Xue obtained the value of the additional shear stress in the matrix caused by the initial curvature of the fibre measured using the photo-elastic method.

Now we describe the modified formula in brief as follows. Consider a fibre-column subjected to a compressive stress in the fibre direction. The fibre may have initial curvature with the initial deflection of fibre described by:

$$Y = \delta_0 \sin(\pi x/1)$$

where δ_0 is the amplitude of the initial curvature (Fig. 3).

When a compressive load P is applied, the final deflection of the fibre becomes:

$$Y = (\delta_0 + \delta_1)\sin(\pi x/1) = Y_0 + Y_1$$

The differential equation is

$$EJ(d^2 Y/dx^2) = -P(Y_0 + Y_1)$$

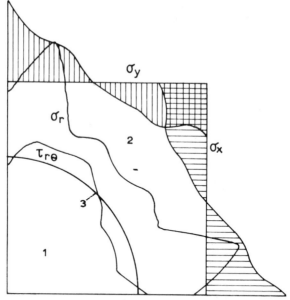

FIG. 2.   Stress distribution at the interface under longitudinal compression. 1, fibre; 2, matrix; 3, interface.

Solving the equation, we can obtain the bending deflection of the fibre

$$Y_1 = \frac{P/P_{cr}}{1 - P/P_{cr}} \sin(\pi x/1)$$

where P_{cr} is the critical stress for a compressive column.
   Hence we can obtain

$$P/P_{cr} = 1/(1 + \delta_0/\delta_1)$$

Further in the case of a fibre embedded in a matrix, we can consider the fibre as a beam on elastic foundation. Suppose m is the fibre buckle mode number and k the elastic coefficient of foundation. In this case the critical stress P_{cr} is

$$P_{cr} = \frac{\pi^2 EJ}{l^2}\left(m^2 + \frac{kl^4}{m^2\pi^2 EJ}\right)$$

The differential equation of bending deflection Y_1 becomes

$$EJ\frac{\mathrm{d}^4 Y_1}{\mathrm{d}X^4} + kY_1 = -P\frac{\mathrm{d}^2}{\mathrm{d}X^2}(Y_0 + Y_1)$$

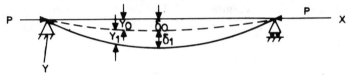

FIG. 3.   A schematic fibre-column with initial curvature under compression.

Solving the equation, we can obtain

$$\delta_1 = \frac{P\dfrac{m^2\pi^2}{l^2}}{EJ\dfrac{m^4\pi^4}{l^4} - P\dfrac{m^2\pi^2}{l^2} + k}\delta_0$$

Also we can obtain

$$P/P_{cr} = 1/(1 + \delta_0/\delta_1)$$

Two different kinds of buckling modes are investigated as follows. For the transverse buckling mode, the buckling stress of the fibre-column σ_{cr} is

$$\sigma_{cr} = 2V_f\sqrt{\frac{V_f E_m E_f}{3(1 - V_f)}}$$

and the maximum tensile strain of the matrix $\varepsilon_{m_{max}}$ is

$$\varepsilon_{m_{max}} = \frac{2\delta_1}{(1 - V_f)t}$$

Hence, considering the existence of the initial curvature of the fibre, δ_0, the failure stress σ^- for laminates in this case is

$$\sigma^- = \frac{2V_f\sqrt{\dfrac{V_f E_m E_f}{3(1 - V_f)}}}{1 + \dfrac{\delta_0}{t}\dfrac{2}{(1 - V_f)\varepsilon_{m_{max}}}}$$

For the shear instability mode, the buckling stress σ_{cr} is

$$\sigma_{cr} = G_m/(1 - V_f)$$

and the maximum shear strain of matrix $\gamma_{m_{max}}$ is

$$\gamma_{m_{max}} = \frac{1}{1 - V_f}\frac{\delta_1}{t}\sqrt{\frac{12G_m}{1 - V_f}\frac{1}{E_f V_f + E_m V_m}}$$

also we can predict the compressive failure stress for laminates in the case of shear instability mode and under consideration of the initial curvature of the fibre

$$\sigma^- = \cfrac{\cfrac{G_m}{1 - V_f}}{1 + \cfrac{\delta_0}{t}\cfrac{1}{\gamma_{m_{max}}}\cfrac{1}{(1 - V_f)V_f}\sqrt{\cfrac{12G_m}{1 - V_f}\cfrac{1}{E_f V_f + E_m V_m}}}$$

where δ_0 equals maximum initial deflection of a curved lamina and t is the thickness of the lamina.

$\gamma_{m_{max}}$ equals maximum shear strain of resin under the combined stress state, i.e. compressive stress and additional shear stress.

The details of this formula and its application can be seen in Ref. 6.

EXPERIMENTAL PROCEDURE AND RESULTS

As is well known, three main difficult problems need overcoming in compressive test methods of FRP thin laminates, i.e. avoiding specimen buckling, preventing load eccentricity and forestalling specimen breakage at its ends. In fact the problems are more complicated than the above-mentioned ones, as we always hope that in a specimen there is a uniform, uni-axial stress region, in which specimen breakage will take place and the strain gauges can be installed. However, it's not easy to do.

In the first, we have tried to measure the compressive properties of composites according to the method of ASTM D3410. This method has many advantages but there are some problems to be solved, so we make some effort to improve this method as follows.

As about 50% of the specimens break at the roots of the tab regions in our experiments (Fig. 4), the D3410 specimen including end tab and adhesive layer is analysed using FEM. The details can be seen in Ref. 7. Results indicate that in the cross-section near the tab region the longitudinal stress is not uniform. The stress concentration factor (SFC) equals 2·0. The value of the transverse compressive stress in this region is about 50% of the average longitudinal stress (Fig. 5). Measurements using microelectric strain gauges (0·2 mm × 0·2 mm) prove that the above-mentioned transverse compressive stress is present.

To eliminate the effect of gripping on the stress state in a rectangular specimen, the gauge region of the specimen should be lengthened, which

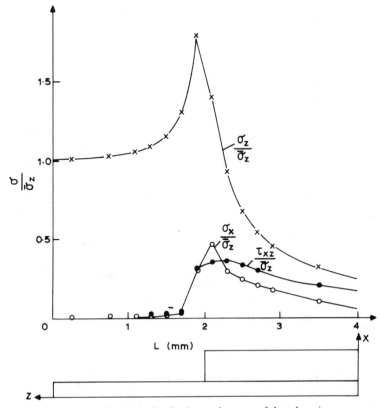

FIG. 4. The stress distribution at the roots of the tab region.

may cause the specimen to buckle when compressed. For the purpose of preventing buckling of a thin laminate specimen, a transverse supporting fixture with springs is needed. The dynamic friction forces can be adjusted by four springs and may be controlled within the range of 2% of the broken load, if polytetrafluoroethylene films are placed between fixture and specimen. The results of measurements using microelectric strain gauges indicate that the transverse supporting forces for preventing the buckling of the specimen are so small that they have nearly no effect on the transverse deformation of the specimen (Fig. 6).

By the way, for preventing load eccentricity and forestalling specimen breakage at its ends, it is necessary to choose an adequate loading fixture. However, the fixture manufactured according to D 3410 seems to be too large and thus heavy for the specimen in dimensions. So a set of new loading

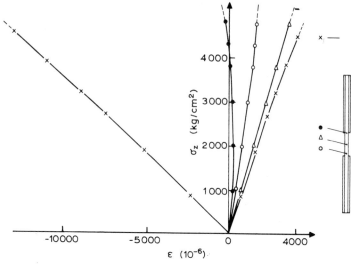

FIG. 5.   Measurements of the strains along the specimen's thickness.

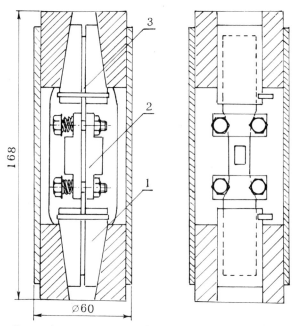

FIG. 6.   Loading and transverse supporting fixture. 1, loading fixture; 2, transverse supporting fixture; 3, specimen.

FIG. 7.   Directly gripped specimen on an electro-servo-hydraulic test machine.

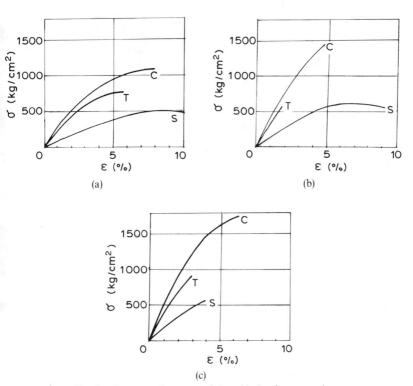

(a)

(b)

(c)

FIG. 8.   Stress–strain curves of three kinds of epoxy resin.

fixtures and transverse supporting fixtures have been designed (Fig. 6). In another way, the hydraulic pressure grips are used to directly grip the specimen if the compressive testing is done on an Instron or MTS Electro-Servo-hydraulic machine (Fig. 7).

Many compressive stress–strain curves of various kinds of composites have been estimated according to previous test methods. We can conclude that C/E_p and $G1/E_p$ have higher longitudinal compressive strengths controlled by fibres buckling, while Kevelar/E_p has a lower one because of lower compressive strength of Kevelar fibres.

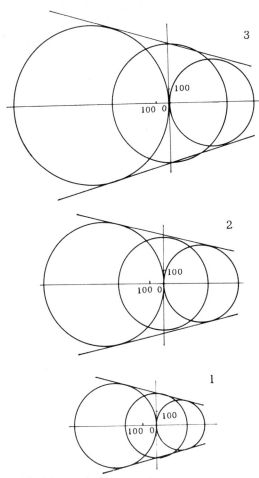

FIG. 9.   Biaxial strength of epoxy resin, controlled by Mohr's criterion.

TABLE 1
Compressive strength and modulus of laminates

Fibre V_f Epoxy resin		E Glass			HM Glass		
		0·36	0·20	0·04	0·36	0·20	0·04
Strength	Type 1	323	227	125	369	285	157
(MPa)	Type 3	491	333	215	610	388	244
Modulus	Type 1	26·9	17·2	10·2	33·0	19·9	10·5
(GPa)	Type 3	26·7	17·6	10·8	36·0	20·0	12·6

In order to clarify the role of the matrix when a laminate is longitudinally compressed, three kinds of epoxy resin are used. Their mechanical properties, including modulus and strength under tensile, compressive and shear stress, are determined and shown in Fig. 8. It is necessary that the shear strength of epoxy cast is measured using a circular bar under torsion and the inelastic behaviour of epoxy matrix must be considered when we calculate it. It is obvious that the resin of type 3 has a much higher shear modulus and shear strength than the other two, despite the fact that its tensile elongation is lower. For these three kinds of epoxy resin we can say that the biaxial strength is controlled by Mohr's failure criterion, shown in Fig. 9, so we can obtain the maximum shear strain of epoxy resin under biaxial stresses, i.e. the longitudinal compressive stress and the additional shear stress caused by the initial fibre curvatures.

The compressive strengths of laminates for three different fibre volume

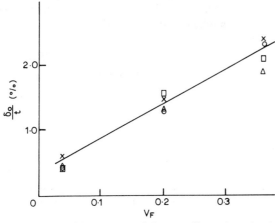

FIG. 10.   Relative initial curvature δ_0/t versus fibre volume fraction curve.

fractions are listed in Table 1. It shows that the strength of laminates made of matrix type 3 is increased by about 50% over that for type 1.

From the experimental results mentioned above, we can also obtain the estimated value of the initial curvatures of fibres for the different fibre volume fractions, shown in Fig. 10. Hence we can predict the theoretical compressive strength of a laminate if we know its fibre volume fraction and matrix biaxial strength, etc.

CONCLUSIONS

Considering the effect of the initial curvature (misalignment, etc.) of fibres and the shear strength of the matrix as well as the tensile strength of the fibre–matrix interface, a new formula to predict the FRP longitudinal compressive strength has been derived.

Some ways to improve the compressive testing method, including specimen form, loading fixture and transverse supporting fixture, are suggested. Finally, the theoretical prediction of the compressive failure mode is identified with the experimental results.

Based on the above-mentioned analysis, there are three ways to increase the longitudinal compressive strength of uni-composites, i.e.

(1)   Using high-modulus fibre, the longitudinal compressive strength of uni-composites is increased, but its strain at breakage isn't increased.

(2)   Using the matrix with high shear strength, e.g. the epoxy resin of type 3, both the longitudinal compressive strength and the strain at breakage of composites can be greatly increased.

(3)   Improving the bonding strength of fibre/matrix interfaces, the interface cracking can be postponed and the FRP longitudinal compressive strength can also be increased.

REFERENCES

1. SHEN KUN-YUAN, WANG ZHENG-YING AND XUE YUAN-DE, Some mechanical problems in the design of a glass fibre reinforced vault-pole. *Mechanics and Practice*, **5**(6) (1983) 2–8.
2. TAO JIANXIN and XUE YUAN-DE, A design study of a composite leaf spring, *Composite Structures—5*. ed. I. H. Marshall. Elsevier Science Publishers, London, pp. 379–91.
3. BUDIANSKY, B., *Micromechanics. Computers and Structures*, **16** (1983) 20.

4. HAHN, H. T., Analysis of kink band formation under compression. *Proceedings of ICCM—6 and ECCM—2,* Vol. 1 (1987), pp. 269–77.
5. ZHU YI-LING, A general review of mechanical properties of composite materials. *Mechanics and Practice,* **2**(1) (1980) 1–6.
6. XUE YUAN-DE and WANG ZHEN-YING, Analysis of GRP compressive strength. *Shanghai Mechanics,* **2**(4) (1981) 37–46.
7. XUE YUAN-DE, TONG JIA-XIAN and WANG ZHENG-YING, A study on compressive properties and testing methods of fiber reinforced laminates. *Jnl Appl. Mech.,* **4**(2) (1987) 77–82.

7

Moisture Management and Artificial Ageing of Fibre Reinforced Epoxy Resins

T. A. COLLINGS

Materials and Structures Department, X32 Building,
Royal Aerospace Establishment, Farnborough, Hants GU14 6TD, UK

ABSTRACT

When attempting to represent the effects of in-service environmental exposure in the airworthiness structural testing of fibre reinforced composite components, it is important that the problems and the limitations associated with moisture conditioning and accelerated ageing are both understood, and allowed for. This paper identifies some of these problems, and makes recommendations for the successful conditioning of structures.

1 INTRODUCTION

For some time the subject of environmental degradation of fibre reinforced plastics (FRP) has been a cause for concern, especially in the airworthiness substantiation of aircraft structures made from such materials. As a consequence there has been a pressing need to quantify the degree of degradation that can occur during a typical service life of an aircraft (usually 25 years), and to make allowance for it in component design and structural testing.

The deterioration that occurs in FRP during the service life is, in general, linked with the level of moisture (water) that is absorbed. This is usually confined to the resin matrix although some absorption by the fibre can take place with aramid fibres. The way in which moisture is absorbed is dependent upon many factors. The factor that features most is that of climatic exposure, that is the severity of exposure to humidity and temperature. Since both humidity and temperature vary considerably,

213

depending upon the geographical location, any decision made on the strength allowance necessary to provide against environmental degradation will, as a consequence, need to consider the intended operational location of an aircraft. Some assessment of the effects of various climates on the total and on the distribution of moisture within a laminate has already been carried out,[1] the result of which is a recommendation to discontinue the use of moisture level as a measure of degradation and to use an alternative criterion, a constant relative humidity (RH) that will produce a representative moisture condition. From this work a recommended simulated worldwide worst environment is suggested. This has been accepted within the UK and within some parts of Europe in collaborative Anglo/European fixed wing and rotary wing projects.

The natural process of moisture absorption in epoxy matrices is normally very slow, and this makes it very difficult to reach an adequate degree of degradation in a structural test element in a practical time. It has been found necessary, therefore, to speed up the moisture diffusion process by employing an accelerated conditioning technique that can ensure a representative level of degradation in a much reduced time. Some attempts at sensible accelerated ageing, in practical timescales, have been reported[2] and are now being considered by the aerospace industry.[3]

During the conditioning of a structure it is convenient to have some measure of the progress of the total moisture uptake and its distribution through the thickness. Normally this is achieved through the use of 'travellers' which are weighed at frequent and regular time intervals. As well as monitoring moisture uptake it is sometimes necessary to be able to predict moisture progress using a moisture diffusion mathematical model. This requires the measurement of the materials moisture constants and is achieved by exposing pretravellers to different environments of steady-state RH and temperature until an equilibrium condition is reached. From these data the necessary relationships between diffusion coefficient and temperature (Arrhenius plot) and RH and moisture equilibrium (M_∞) can be obtained.

Both travellers and pretravellers have for some time now been considered highly desirable for the competent moisture management of complex structures. Until recently, however, little had been done to define the basic requirements needed for a traveller to ensure that it mimicked a structure's response to moisture exposure. This recent work[4] has shown that several important features need to be included, and certain dimensional limits need to be imposed, before a traveller can be accepted as adequate as a structural moisture model.

The aim of this paper is, therefore, to consider the fundamental problems of FRP conditioning and to make recommendations that will provide a route to the successful moisture management and artificial conditioning of FRP.

2 SIMULATED CLIMATIC EXPOSURE

The degree of degradation that occurs in an FRP structure when in service is linked directly with the amount of moisture that is absorbed. In the past the moisture content of a laminate was used to define the extent of any environmental degradation. It is now recognised that this is not really valid since the moisture absorption kinetics of epoxy resins differ widely and also change with physical ageing.[5] An alternative criterion that has now been accepted is to define a constant relative humidity environment that will produce a moisture level that is representative of an FRP structure that has been exposed to a real life environment at a particular geographical location. The variability of a natural environment, that is the daily, monthly or seasonal changes in temperature and humidity, are known[1] to be a major factor in determining both the final moisture equilibrium level deep within a laminate's thickness and the way in which moisture is distributed in the outer surface layers. The results of calculations[1,6] made using climatic data taken from three different locations and assuming Fickian-type moisture diffusion theory (see Fig. 1) show quite clearly the influence that geographical location has on the distribution of moisture within a laminate. Similar calculations on the effect of laminate thickness on moisture distribution are given in Fig. 2. On the evidence of this work it is suggested that the worst worldwide environment might best be simulated by a constant humidity of $84\% + 2\% - 0\%$ or $85\% \pm 1\%$. Further calculations,[6] made using climatic data taken for other geographical locations within Europe, also suggest that, for military aircraft stored out of doors, the worst North European climate is equally as severe as the worst worldwide environment and, therefore, the same RH would be appropriate.

The amount of moisture that can be absorbed in a particular period of time is dependent on the rate at which moisture can be diffused; this is temperature dependent. Thus an average environmental exposure temperature must be specified when simulating a natural climatic exposure since it will determine the maximum thickness of structure that will reach an equilibrium condition at the end of the service life of an aircraft. It will also determine the shape and 'wetness' of the moisture gradient in structures that are too thick to reach an equilibrium condition. For the

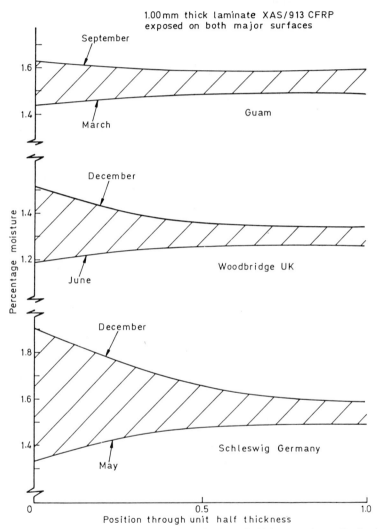

FIG. 1. The effect of geographical location on the through-thickness moisture distribution after reaching an equilibrium condition.

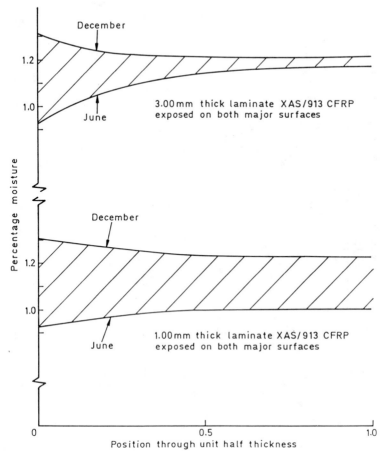

FIG. 2.   The effect of laminate thickness on the through-thickness moisture distribution after reaching an equilibrium condition (Bahrain).

worst worldwide simulated conditioning a mean annual temperature of 26°C is recommended, and for a North European location a mean annual temperature of 10°C is advised.

3 ACCELERATED CONDITIONING

3.1 Elevated Temperature

Perhaps the first logical step towards accelerating moisture uptake is to increase the diffusivity of the resin matrix by raising the temperature of the

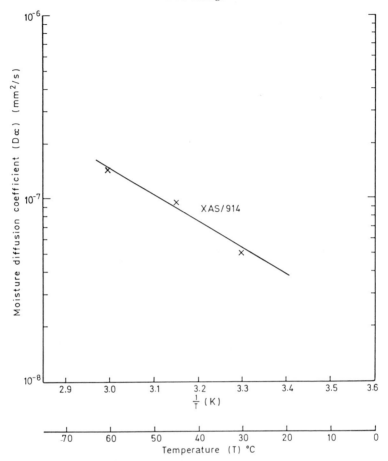

FIG. 3.   Arrhenius plot of moisture diffusion coefficient.

conditioning environment. For example, the diffusion coefficient of the Ciba–Geigy 914 resin system can be increased by a factor of 5 (see Fig. 3) simply by raising the temperature from 20 to 60°C. Even with this increased diffusion factor the processes of moisture absorption are still slow, and ideally further increases in temperature to further accelerate the conditioning would be an advantage.

However, there is a limit, for most epoxy resins, to which temperature can be raised without affecting a change in the mechanism by which moisture is absorbed. Above this critical temperature there is a strong possibility of altering the process of degradation. For this reason it is essential to understand the natural moisture diffusion behaviour for each particular fibre resin system of interest and the maximum safe temperature limit to which this diffusion mechanism can be maintained. For most resin systems of current interest in the UK it has been demonstrated that natural moisture diffusion behaviour can be modelled accurately using Fickian diffusion theory.[7] However, at higher temperatures and for some resin systems at high humidity it has been shown that the Fickian model is no longer valid.

Work by Collings and Copley,[8] and by Whitney and Browning,[9] have demonstrated that a maximum safe temperature exists, above which moisture diffusion deviates from the more classical single phase, described by Fick, to one that is a two-phase Langmuir type.[10] Such a limitation on temperature will therefore be a major consideration in deciding upon an adequate accelerating technique.

3.2 Conditioning to 95% of Equilibrium

Even with an increased conditioning temperature the rate of moisture absorption is still very slow (see Fig. 4), and so a means of obtaining an adequate level of moisture in a much further reduced time is needed. One way of achieving this is through a change in the acceptance criteria, that is to condition to a percentage of the chosen equilibrium condition instead of trying to obtain the full equilibrium state. The advantage to be gained can be readily seen by looking at a typical moisture absorption curve (see Fig. 4). Here it can be seen that the time required to reach a moisture level of about 95% of the full equilibrium condition is relatively short compared with the time taken to reach 100% of equilibrium. For example, calculations show[6] that a 2-mm thick XAS/914 laminate exposed on all surfaces to an environment of 84% RH at 60°C takes 210 days to reach a full equilibrium condition but only 88 days to reach a condition of 95% of equilibrium ($0.95M_\infty$). Thus the final 5% of moisture required to achieve a full equilibrium condition is obtained at the expense of a further 122 days of conditioning. The consequence of this in terms of the calculated through-the-thickness moisture distribution is given in Fig. 5. Clearly a very large saving in time is possible if a 95% of equilibrium can be justified in terms of a non-significant change in the structural strength and stiffness properties. Some assessment of the significance of any strength change between M_∞ and $0.95M_\infty$ has been made[4] by reference to the interlaminar shear strength (ILSS) and compression strength of a composite made from a carbon fibre epoxy resin system. By interpolation of results and assuming a linear relationship between loss of strength or stiffness with increasing moisture

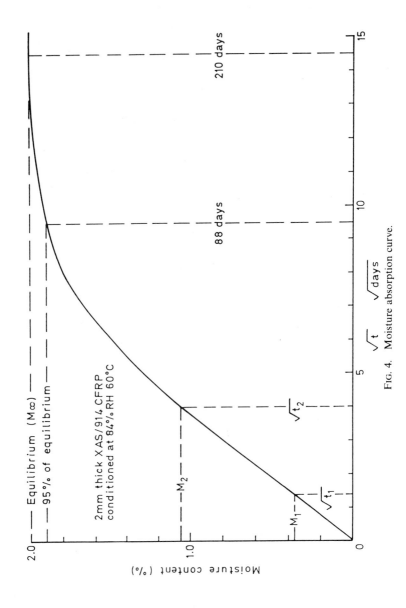

FIG. 4. Moisture absorption curve.

0°XAS/914   0.6 vf

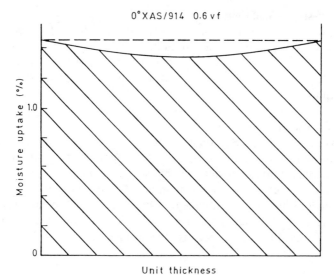

FIG. 5.   Moisture conditioning to 95% of equilibrium.

content it was shown there was a difference of less than 1% in properties between the two conditions. It is suggested, therefore, that a 95% of equilibrium condition is adequate for the conditioning of FRP.

The saving of time and energy costs that can be made using this method will obviously increase for structures of greater thickness. In order to assess the magnitude of the possible time savings, calculations have been carried out on various laminate thicknesses to estimate the time required to reach various states of equilibrium in an environment of 84% RH and 60°C. These are given in Table 1 for laminates made from the XAS/914 fibre resin system.

TABLE 1

Times to reach % of equilibrium when conditioned at 84% RH, 60°C on both sides of an infinite laminate
(0° XAS/914 0·6 vf)

Laminate thickness (mm)	Time to reach % of moisture equilibrium (days)							
	100	99	98	97	96	95	90	85
2	210	138	116	104	95	88	66	53
4	800	550	464	434	378	352	264	213
6	1 850	1 234	1 042	931	850	789	595	480
8	3 300	2 350	1 850	1 662	1 509	1 407	1 056	952

3.3 Multi-stage Conditioning

In order to further accelerate the moisture conditioning process two more, less simple, methods have been investigated. The aim of each is to achieve as quickly as possible a state of full or near moisture equilibrium in the centre plies and then to adjust the distribution through the thickness to that required.

3.3.1 Three-stage conditioning

It has been shown in Section 3.1 that some acceleration of time can be achieved by raising the temperature of the conditioning environment. A further acceleration can be obtained by increasing the relative humidity to the maximum that can be readily maintained, that is 96% RH. Using this approach a mean moisture uptake of 1·46% (equivalent to equilibrium at 84% RH) can be, for example, achieved in a 2-mm thick XAS/914 laminate in 44 days (see Fig. 6), but because the surface plies will be at a higher

2mm thick XAS/914 laminate

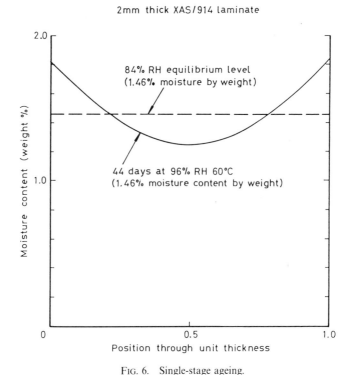

FIG. 6.   Single-stage ageing.

moisture equilibrium level than the centre plies a severe through-the-thickness moisture gradient will exist and the centre plies will be at an abnormally low moisture level. Clearly this is unrepresentative, and what is required is an accelerating method that takes full advantage of the high rates of diffusion given by temperature and humidity, and yet still achieve a reasonably realistic moisture distribution. Such a method has been reported by Collings and Copley,[2] where a three-stage conditioning approach that overcomes the high moisture gradient problem and still achieves considerable time savings was demonstrated.

The three-stage accelerated ageing technique involves applying three different humidities for three different periods at the maximum safe temperature. The first stage exposes the laminate to the maximum practical humidity level that is available, usually 96% RH. In this way the maximum advantage is taken of the short time needed to reach a required moisture level in the centre plies of the laminate. During this first stage the required total moisture level will have been exceeded (see Fig. 7). The second stage removes the excess moisture that lies mostly in the outer plies of the

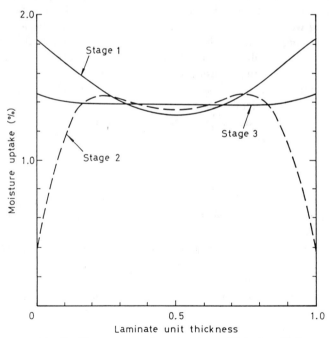

FIG. 7. Three-stage accelerated ageing to 95% of equilibrium.

TABLE 2

Accelerated ageing of an infinite XAS/914 (0·6 vf) laminate exposed on both major surfaces to achieve an equivalent moisture level and distribution of 0·95M$_\infty$ for an 84% RH condition

Laminate thickness (mm)	Time spent conditioning (days)			
	Stage 1 96% RH 60°C	Stage 2 15% RH 60°C	Stage 3 84% RH 60°C	Total time (0·95M$_\infty$)
2	48	2	4	54
4	191	7	11	209
6	428	16	25	469
8	766	25	43	834

laminate by exposure to a low RH environment. During this 'drying' stage the moisture, deeper within the laminate, will continue to diffuse towards the centre plies. The precise value of this RH is not important as long as it is reasonably low (about 15%) for it to have a sensible drying effect. Finally, the third stage uses an RH that will give the desired final moisture level (i.e. 84% for a worldwide worst case). This corrects the moisture level in the outermost plies during which the inner moisture distribution flattens out. Effective use of multi-stage conditioning does, of course, rely upon a mathematical moisture diffusion modelling capability to define the three stages of conditioning. In this paper the finite difference model used in Ref. 6 has been used to calculate the three stages and the total conditioning time for a range of laminate thicknesses made from the XAS/914 fibre resin system. The results are presented in Table 2 for a 60°C conditioning temperature.

3.3.2 Two-stage conditioning

An alternative approach that has been investigated[4] and can achieve an acceptable level of moisture is a two-stage conditioning process. The first stage consists of conditioning at 96% RH at the maximum safe temperature in much the same time scale as that needed for a three-stage method. The second stage uses an RH set at a level that will, given the time, achieve the desired final moisture level. However, if the second stage requires a high RH the drying effect on the outer plies will be slow, therefore the time spent at this condition will need to be longer than that usually required for the second stage of a three-stage method. As a consequence there is essentially no total conditioning time advantage to be gained over

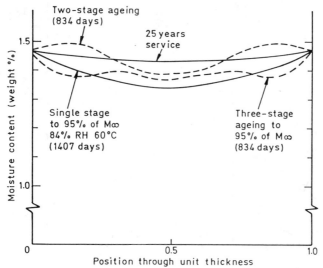

FIG. 8. Moisture distribution in an 8-mm thick structure after exposure on both major surfaces to accelerated ageing methods designed for an 8-mm thick structure.

that of a three-stage method and, because of the reduced drying effect, the two-stage method gives a much higher moisture content and a higher level through-the-thickness moisture; see for example Fig. 8.

The main advantage to be gained from a two-stage method, besides being simpler to apply, is in the conditioning of laminates to high levels of moisture that require low through-the-thickness moisture profiles. This is clearly demonstrated in Fig. 8. Here it can be seen that for a time period of 834 days the two-stage fits the higher level 25 years' service moisture gradient whereas the three-stage fits the lower level single stage 95% of M_∞ moisture gradient.

Clearly both moisture level and through-the-thickness moisture profile will need to be considered before a decision can be made on whether to use a two- or three-stage conditioning method.

4 MULTI-THICKNESS STRUCTURES

A typical range of thicknesses that can be encountered in an aircraft structure is from 2 to 15 mm and, therefore, in service the thinner components will reach an equilibrium level much sooner than the thicker. Over a normal service use of 25 years not all structures can be expected to

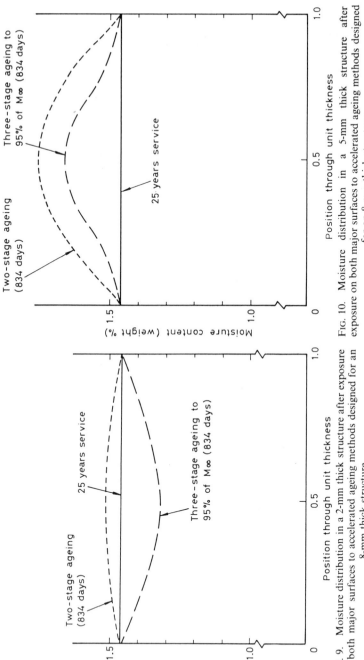

Fig. 9. Moisture distribution in a 2-mm thick structure after exposure on both major surfaces to accelerated ageing methods designed for an 8-mm thick structure.

Fig. 10. Moisture distribution in a 5-mm thick structure after exposure on both major surfaces to accelerated ageing methods designed for an 8-mm thick structure.

reach an equilibrium level; indeed it is known from calculations that any structure greater than 8 mm thick, and exposed on both major surfaces, will not reach a full equilibrium condition. Thus all structures greater than 8 mm thick will contain a through-the-thickness moisture gradient. This raises problems in the accelerated ageing of multi-thickness structures since all thicknesses cannot be accelerated to produce the real-life moisture condition within a shortened time scale using multi-stage conditioning. Thus serious consideration must be given to the number of different structural thicknesses involved for which an optimum conditioning is to be selected. To this end, using a predictive method described by Copley,[6] the results of conditioning using both two- and three-stage ageing have been calculated for a structure made up of thicknesses ranging from 2 to 15 mm thick with conditioning optimised for an 8-mm thick laminate exposed on both major surfaces. The resulting through-the-thickness moisture profiles are given in Figs 8–12.

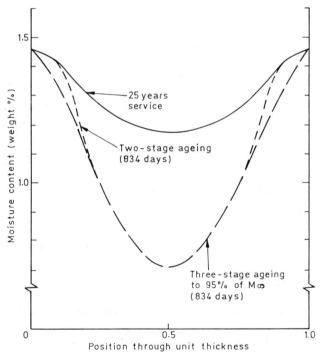

FIG. 11.   Moisture distribution in a 12-mm thick structure after exposure on both major surfaces to accelerated ageing methods designed for an 8-mm thick structure.

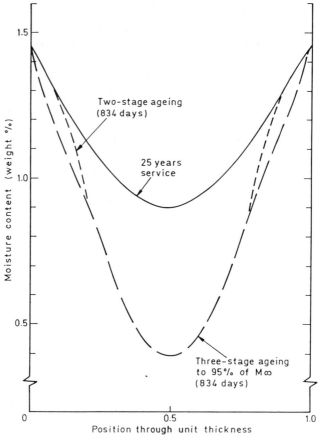

FIG. 12.   Moisture distribution in a 15-mm thick structure after exposure on both major surfaces to accelerated ageing methods designed for an 8-mm thick structure.

5 PREDICTION AND MONITORING OF MOISTURE

5.1 Prediction of Moisture

The ability to predict the diffusion of moisture in an FRP structure is essential for determining the moisture distribution through the thickness of a complex structure after exposure to either steady-state or variable environments of temperature and humidity. It is necessary also for estimating the conditioning time needed to reach a required moisture equilibrium condition and for assessing the ability of accelerated ageing techniques to mimic the process of natural moisture absorption. In order to achieve this, pretravellers are required to enable the necessary moisture constants to be measured. Two constants need to be known: the equilibrium level, which is the maximum amount of moisture that can be absorbed at a given relative humidity, and the diffusion coefficient, which determines the rate of uptake of water.

5.1.1 Equilibrium level

The equilibrium level, M_∞, depends only on the concentration of moisture or relative humidity of the environment, and within limits is generally assumed to be independent of temperature.

5.1.2 Diffusion coefficient

The diffusion coefficient, D, is assumed to depend only on absolute temperature, T, as

$$\log_e D \propto 1/T \tag{1}$$

Although stress can have some effect, it is small and for practical purposes can normally be ignored.

Under conditions of steady-state temperature and humidity the moisture uptake or loss in a composite can be expressed as a percentage of the original dry weight using

$$M = \left(\frac{W_i - W_d}{W_d}\right)100 \tag{2}$$

where M is the percentage moisture uptake or loss, W_d is the original dry weight of the specimen and W_i is the weight of the specimen after a time t_i.

The diffusion coefficient is defined by the equation[11]

$$D = \pi\left(\frac{h}{4M_\infty}\right)^2\left(\frac{M_1 - M_2}{\sqrt{t_1} - \sqrt{t_2}}\right)^2 \tag{3}$$

where M_1 and M_2 are the percentage of water uptake at times t_1 and t_2 respectively, h is the laminate thickness and M_∞ is the moisture equilibrium level for a given relative humidity. The term

$$\left(\frac{M_1 - M_2}{\sqrt{t_1} - \sqrt{t_2}}\right) \tag{4}$$

is the slope of the linear portion of the plot of M against $\sqrt{t}$ (see Fig. 4).

The value of D in eqn (3) is obtained by measuring the moisture absorption of a specimen of finite size and will therefore include moisture diffusion from all six surfaces. To obtain the true one-dimensional diffusion coefficient, D_∞, needed for the calculation of the through-thickness moisture distribution, a correction factor given by Shen and Springer[11] can be used, namely

$$D_\infty = D\left(1 + \frac{h}{b} + \frac{h}{l}\right)^{-2} \tag{5}$$

where b and l are the laminate breadth and length respectively.

5.2 Monitoring of Moisture

During the conditioning of a structure it is necessary to monitor both the rate at which moisture is being absorbed and the total amount of moisture that has been absorbed. This is accomplished by the frequent weighing of a number of relatively small finite sized travellers. As the name implies, travellers should always accompany their parent structure. In this way they will experience an identical environmental exposure history and should be, therefore, fully representative of the moisture uptake of the parent structure.

5.3 Requirements of Pretravellers and Travellers

For all moisture analysis of composites using pretravellers and travellers (hereafter referred to as travellers) it is essential that they should accurately mimic the moisture behaviour of the particular structures under observation. For this to be achieved several points need to be considered, and allowed for, in the design of travellers.

5.3.1 Overall size

Travellers should be of sufficient major surface area to ensure that the moisture absorbed through the edge does not account for a significant proportion of the total moisture uptake. This is important during the early stages of the absorption process, where the proportion of moisture absorbed through the edges accounts for a larger proportion of the total absorbed moisture than at or near the equilibrium condition. For a 50 × 50 mm traveller, after 3 days of exposure to 60°C and 84% RH, the moisture absorbed through the edges would account for 7% of the total, whereas for 88 days of exposure the edges would account for only 2%.

Standardisation of travellers is recommended in order to preserve some capability for the direct interpretation of data collected from several sources. The recommended sizes for travellers have therefore been taken from those that have seen the most regular use, that is 50 × 50 mm and 75 × 75 mm.

TABLE 3
Effect of thickness on the moisture diffusion coefficient
(0° XAS/914 0·6 vf exposed at 96% RH, 60°C)

Laminate thickness (mm)	Diffusion coefficient (mm²/s)	
	D	D_∞
1·03	$3·56 \times 10^{-7}$	$3·53 \times 10^{-7}$
2·02	$3·71 \times 10^{-7}$	$3·57 \times 10^{-7}$
3·07	$3·92 \times 10^{-7}$	$3·70 \times 10^{-7}$
4·11	$4·43 \times 10^{-7}$	$4·10 \times 10^{-7}$
5·23	$4·96 \times 10^{-7}$	$4·51 \times 10^{-7}$

5.3.2 Thickness

It has been shown by Collings and Copley,[8] for laminates of different thickness but identical lay-up, that the moisture diffusion coefficient increases significantly with laminate thickness. Table 3 shows, for example, an increase of 43% in the D_∞ for an increase in thickness of 1 to 5 mm. The use of a wide variety of thicknesses in an aircraft structure must give rise to practical and economic problems in the total moisture management of such a structure since the moisture constants for all thicknesses cannot be expected to be determined. Clearly only a representative number of thicknesses can be used to both predict and monitor moisture uptake. Travellers should, therefore, be selected to give a good representation of the total range of thicknesses used, and to rely on interpolation to assess the moisture behaviour of intermediate thicknesses. For structures that see primarily moisture absorption from one major surface only it is recommended that the traveller thickness is double that of the structure being mimicked. Alternatively one major surface can be sealed against moisture ingress by facing with metal. In the former, some error can occur due to the difference in D_∞ for the thickness involved. In the latter, moisture ingress through the edge of the adhesive layer, needed for bonding the metal to the laminate, will contribute a small moisture error. Care should be taken to ensure the original weight of the traveller is established before sealing takes place in order to provide a dry datum.

5.3.3 Lay-up

It has been shown[12] (see Table 4 and Fig. 13), for identical lay-ups and thickness and identical proportions of 0°/90° plies, that the D_∞ can be significantly altered by the way in which the 0° and 90° plies are distributed

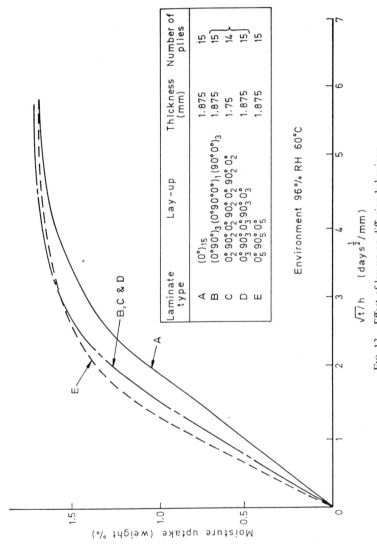

FIG. 13.   Effect of lay-up on diffusion behaviour.

TABLE 4
Effect of lay-up on diffusion coefficient
(XAS/914 exposed at 96% RH, 60°C)

Laminate type	Lay-up	Thickness (mm)	Number of plies	M_∞ (%)	D (mm^2/s)	D_∞ (mm^2/s)
A	$(0°)_{15}$	1·875	15	1·66	$2·37 \times 10^{-7}$	$2·05 \times 10^{-7}$
B	$(0°90°)_3(0°90°0°)_1(90°0°)_3$	1·875	15 ⎫			
C	$0_2°90_2°0_2°90_2°0_2°90_2°0_2°$	1·75	14 ⎬	1·70	$3·75 \times 10^{-7}$	$3·2 \times 10^{-7}$
D	$0_3°90_3°0_3°90_3°0_3°$	1·875	15 ⎭			
E	$0_5°90_5°0_5°$	1·875	15	1·68	$4·4 \times 10^{-7}$	$3·8 \times 10^{-7}$

through-the-thickness. Since laminate lay-up can vary considerably throughout an aircraft structure, both lay-up and its distribution through-the-thickness must be considered in the choice of travellers.

6 MONITORING OF EQUILIBRIUM CONDITION

During the moisture management of a structure it is essential to be able to assess the progress of moisture ingress in order to determine when an adequate condition has been reached. The decision on when to stop conditioning does, however, present a problem since moisture level alone, as determined by a traveller, cannot be relied upon to give an indication of the proximity to an equilibrium condition. This is due to a phenomenon known as 'physical ageing'[5] which occurs in most epoxy resin systems, the result of which is to reduce the total moisture absorption capacity of the matrix so that the moisture equilibrium level is in some doubt. To avoid this problem conditioning was usually allowed to continue until an acceptably low rate of change of moisture, δM (percentage by weight), was reached, normally 0·005% between weekly weighings. This method was considered to be acceptable until recently when further theoretical studies[3,4] revealed some severe limitations. The results of the investigations showed that for laminates up to 8 mm thick errors of > 25% can occur between the actual and assumed conditioning level. Some results of the actual percentage of equilibrium obtained using a δM of 0·005%, for travellers up to 8 mm thick, are given in Tables 5 and 6. In order to achieve a desired percentage of equilibrium it will be necessary to use a modified value of δM, the magnitude of which will depend upon the thickness of traveller used. Calculated values of the δM% (together with actual changes in laminate

TABLE 5
The effect of thickness of a 50 × 50 mm traveller on the percentage of equilibrium reached in an infinitely large structure using a $\delta M \leq 0.005\%$ between weekly weighings (conditioned at 84% RH, 60°C)

Traveller thickness (mm)	Time to reach a moisture change of $\delta M \leq 0.005\%$/week (days)	Equivalent moisture content in an infinitely large structure (%)	Equilibrium % reached in an infinitely large structure (%)
2	108	1·417	97·4
4	274	1·322	90·9
6	406	1·173	80·7
8	500	1·018	70·0

TABLE 6
The effect of thickness of a 75 × 75 mm traveller on the percentage of equilibrium reached in an infinitely large structure using a $\delta M \leq 0.005\%$ between weekly weighings (conditioned at 84% RH, 60°C)

Traveller thickness (mm)	Time to reach a moisture change of $\delta M \leq 0.005\%$/week (days)	Equivalent moisture content in an infinitely large structure (%)	Equilibrium % reached in an infinitely large structure (%)
2	120	1·429	98·3
4	290	1·337	92·0
6	430	1·196	82·2
8	550	1·059	72·8

TABLE 7
δM (%) required between weekly weighings plus laminate weight change in grams for a 50 × 50 mm 0° XAS/914 0·6 vf traveller (conditioned at 84% RH, 60°C)

% of equilibrium required (infinite structure)	Specimen thickness (mm)	δM (%) between weekly weighings	Actual weight change in a 50 × 50 mm traveller (grams) between weekly weighings
95	2	0·014	0·001 1
	4	0·002	0·000 31
	6	0·000 28	0·000 066
	8	0·000 04	0·000 013

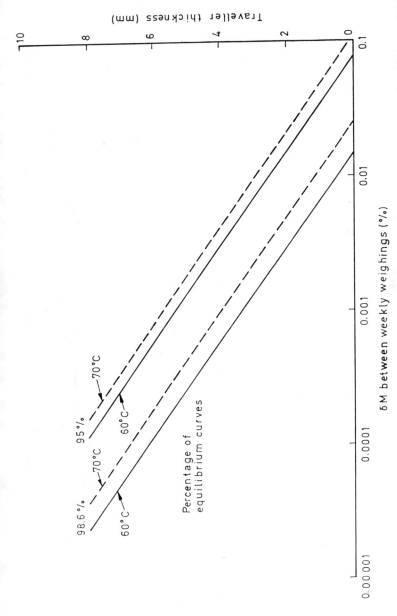

FIG. 14. Change of δM with traveller thickness (50×50 mm $0°$ XAS/914 0·6 vf) to achieve a constant percentage of equilibrium at 84% RH.

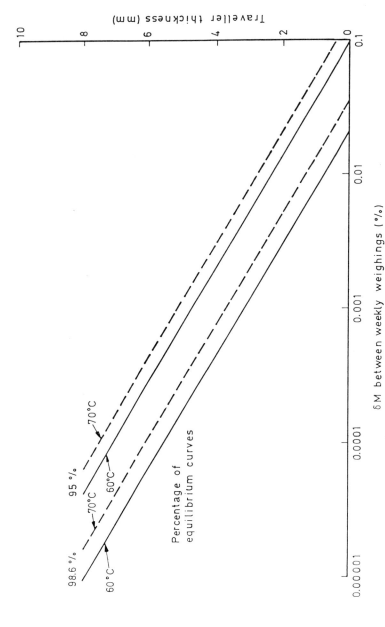

FIG. 15. Change of δM with traveller thickness (75×75 mm $0°$ XAS/914 0·6 vf) to achieve a constant percentage of equilibrium at 84% RH.

TABLE 8
δM (%) *required between weekly weighings plus laminate weight change in grams for a 75 × 75 mm 0° XAS/914 0·6 vf traveller (conditioned at 84% RH, 60°C)*

% of equilibrium required (infinite structure)	Specimen thickness (mm)	δM (%) between weekly weighings	Actual weight change in a 75 × 75 mm traveller (grams) between weekly weighings
95	2	0·015	0·002 65
	4	0·003	0·001 06
	6	0·000 58	0·000 31
	8	0·000 115	0·000 08

weight) necessary to achieve a true 95% of the 84% RH equilibrium level are given in Tables 7 and 8 and Figs 14 and 15 for traveller thicknesses up to 8 mm for both of the recommended sizes of travellers.

The conclusions drawn from this work on the use of weekly weighings cannot, because the weight changes are too small to measure accurately, be recommended for travellers > 3·0 mm thick (sized 50 × 50 mm) and for travellers > 3·4 mm thick (sized 75 × 75 mm). For travellers of greater thickness it is suggested that the initial moisture uptake is used to check the moisture diffusion coefficient and then to calculate the time required to reach the necessary moisture condition.

7 RECOMMENDATIONS

The fundamental problems that arise in the moisture conditioning of FRP have been considered and discussed above. Recommendations for a successful route to their moisture management and artificial conditioning are summarised below.

(i) Decide upon a steady-state RH level and mean temperature to satisfy the worst variable environmental condition for the particular geographical locations of operation of the structure, the worldwide worst condition being represented by an RH of 84% + 2% − 0% (or 85% ± 1%) and a temperature of 26°C.

(ii) Environmentally expose until 95% of the equilibrium condition is reached.

(iii) Acceleration of the conditioning process can be made: (a) by increasing the environmental temperature to the maximum allowable for the matrix system without altering the moisture diffusion behaviour from that obtained during natural exposure, or (b) by using a multi-stage accelerated conditioning by taking advantage of both high temperature and RH whilst still keeping within the Fickian limits.

(iv) Measure both D_∞ and M_∞ for the particular composite material being used by the use of pretravellers.

(v) Use travellers to monitor moisture uptake.

(vi) Traveller size should be either 50×50 mm or 75×75 mm.

(vii) All pretravellers and travellers should be identical in thickness, lay-up orientation and distribution through-the-thickness to that of the structure being monitored.

(viii) The minimum practical level of δM between weekly weighings to monitor an acceptable level of conditioning is 0·005% by weight. This is only advised for laminates of <3 mm for 50×50 mm travellers and $<3·4$ mm for travellers of 75×75 mm.

(ix) For a traveller of thickness greater than those in (viii) calculated times to reach an acceptable level of conditioning should be used.

REFERENCES

1. COLLINGS, T. A., The effect of observed climatic conditions on the moisture equilibrium level of fibre reinforced plastics. *Composites*, **17**(1) (January 1986) 33–41.
2. COLLINGS, T. A. and COPLEY, S. M., On the accelerated ageing of CFRP. *Composites*, **14**(3) (July 1983) 180–8.
3. Environmental effects in the testing of composite structures for aircraft. Report to Composite Research Advisory Group BAe-MSM-R-GEN-0603, 1987.
4. COLLINGS, T. A. (unpublished). Royal Aircraft Establishment, Farnborough, UK.
5. KONG, E. S. W. and ADAMSON, M. J., Physical ageing and its effects on the moisture absorption of amine-cured epoxies. *Polymer Communications*, **24** (June 1983).
6. COPLEY, S. M., A computer program to model moisture diffusion and its application to accelerated ageing of composites. RAE Technical Report 82010, 1982, Royal Aircraft Establishment, Farnborough, UK.
7. CRANK, J., *The Mathematics of Diffusion*, 2nd edn. Clarendon Press, Oxford, 1975.
8. COLLINGS, T. A. and COPLEY, S. M. (unpublished). Royal Aircraft Establishment, Farnborough, UK.
9. WHITNEY, J. M. and BROWNING, C. E., Some anomalies associated with moisture diffusion in epoxy matrix composite materials. ASTM STP 658, American Society for Testing and Materials, Philadelphia, PA, USA, 1978.

10. BONNIAU, P. and BUNSELL, A. R., A comparative study of water absorption theories applied to glass epoxy composites. *J. Comp. Mater.*, **15** (May 1981) 272–93.
11. SHEN, C. H. and SPRINGER, G. S., Moisture absorption and desorption of composite materials. *J. Comp. Mater.*, **10** (January 1976) 2–20.
12. COLLINGS, T. A. and STONE, D. E. W., Hygrothermal effects in CFC laminates: damaging effects of temperature, moisture and thermal spiking. *Comp. Struct.*, **3** (1985) 341–78.

8

Environmental Effects on the Behaviour of Laminated Shells

P. PEGORARO and M. TOURATIER

Laboratoire de Génie de Production,
ENIT–BP1629, 65016 Tarbes Cedex, France

ABSTRACT

A higher-order shear deformation theory of elastic shells is developed for laminated shells, taking into account environmental effects on the behaviour of the shell such as the following: moisture, temperature, prestress and obviously normal charge at the shell surface. The shell model is based on a displacement field in which the displacements of the middle surface are expanded as cubic functions of the thickness coordinate, and the transverse displacement is assumed to be constant through the thickness. So, the distribution of shear deformation is parabolic and there is no need to use shear correction factors. Some numerical results are presented.

INTRODUCTION

The aim of this work is to model some moderately thick structures in the form of shells in order to simulate the behaviour of composite structures exposed to environmental effects. As a matter of fact, the use of polymer matrix composites in naval structural applications requires a careful evaluation of the effects of environmental exposure on the material's properties. The influences of the common environmental mechanisms of moisture and temperature have been found to significantly degrade the matrix-dominated mechanical properties. In order to evaluate the influence of moisture and temperature upon the structural characteristics of such a structure, analytical investigations are performed. A general shell type was chosen for this analysis. Such a structure is representative of skin panels found in submarine chasers for instance. The loading applied to this shell was a compressive load in the middle surface, with normal pressure on the top of the shell. The shells chosen were composite shells in glass–epoxy or glass–polyester laminates. Because of the laminated structure, and consequently the importance of shear deformation effects, this study concerns only moderately thick shells. For classical theories of thin shells, see, for example Refs 1–3.

To evaluate the influences of moisture and temperature in the stress distribution, the global behaviour of the structure was predicted from a homogenization technique in each laminate. For classical fibre composites, the mechanical properties (viscoelastic properties), coefficients of thermal expansion and hygroscopic expansion, are determined by a concentric cylindrical model (and a wave propagation model for the transverse Young's modulus). This choice results from model comparisons of homogenization described in Refs 4 and 5.

The shell is modelled by a displacement approach and a higher-order shear deformation which allows us to take boundary conditions into account on the top and bottom surfaces of the shell and the parabolic distribution of the transverse shear stresses. The theoretical model is obtained by assuming that the displacement components in a shell can be expanded to a series of powers of the thickness coordinate. Then we assume that the transverse normal strain is negligible and we take boundary conditions into account on the inner and outer surfaces of the shell (the tangential surface forces are zero). This allows us to reduce the model to one with only five generalized displacements as in Refs 6–8. But here, the strains are exactly computed from the gradient operator of the displacement field, and consequently, the strains are different from those obtained successively by Bhimaraddi,[6] then Soldatos[7] and Reddy and Lui.[8] Other approaches of higher-order shell theories have been published,[9–11] but these theories contain more than five generalized displacements. One plate theory[12] has been constructed with *only* five generalized displacements and taking into account normal transverse deformation, higher-order shear deformation, and thickness-stretch effects. However, this theory requires second-order derivatives in the displacement field, and the development of finite elements upon this theory involves some difficulties. The advantage of all these models is that they contain the same dependent unknowns as in the first-order shear deformation theory, but account for parabolic distribution of the transverse shear strains through the thickness of the shell, and there is no need to use shear correction factors which are dependent on material properties.

The numerical simulations with this model use the classical diffusion equation to take the hygrothermal behaviour of the material into account, the hygrothermal behaviour having been introduced in the shell model by the constitutive law. All computations are obtained without explicit development of the displacement equilibrium equations by using some linear operators whose associated matrices are programmed. So, the analytical computations are greatly simplified. Numerical results are presented by Navier type solutions for the following: plates, cylindrical panels, spherical shells under simply-supported boundary conditions.

THE THEORETICAL MODEL OF THE SHELL

Let r, s, z denote the orthogonal curvilinear coordinates for the shell, Fig. 1, such that the r- and s-curves are lines of curvature on the midsurface $z = 0$; and z-curves are straight lines perpendicular to the surface $z = 0$. For particular geometries, such as for cylindrical and spherical shells, the lines of principal curvature coincide with the coordinate lines. The principal radii of curvature of the middle surface of the shell are denoted by R_1 and R_2, see Fig. 1.

In the orthogonal curvilinear coordinates chosen for the shell, the displacement field is represented by the following form:

$$\mathbf{U}(r, s, z) = \mathbf{U}_0(r, s) + z\mathbf{\Omega}_1(r, s) + z^2\mathbf{\Omega}_2(r, s) + z^3\mathbf{\Omega}_3(r, s) \tag{1}$$

where

$$\mathbf{U}_0^{\mathrm{T}} = \{u \quad v \quad w\} \qquad \mathbf{\Omega}_i^{\mathrm{T}} = \{\omega_{ri} \quad \omega_{si} \quad 0\} \qquad i = 1, 2, 3 \tag{2}$$

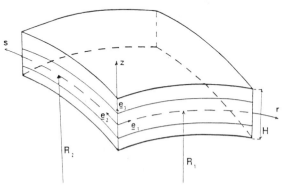

FIG. 1.   The geometry of the shell.

The expressions (1), (2) for the displacement field are dictated by the following considerations:

    (i) the normal transverse strain is zero,
    (ii) the final model must have five independent generalized displacements as in the first order shear deformation theory,
   (iii) the distribution of shear deformation must be parabolic.

So, we need to compute strains in order to satisfy the hypotheses (i)–(iii). For this, the gradient linear operator in orthogonal curvilinear coordinates is computed and its representative matrix in the e_i basis is:

$$\text{Grad } \mathbf{U} = \begin{bmatrix} \dfrac{R_1}{z+R_1}\dfrac{\partial U_1}{\partial r} + \dfrac{U_3}{z+R_1} & \dfrac{R_2}{z+R_2}\dfrac{\partial U_1}{\partial s} & \dfrac{\partial U_1}{\partial z} \\[2mm] \dfrac{R_1}{z+R_1}\dfrac{\partial U_2}{\partial r} & \dfrac{R_2}{z+R_2}\dfrac{\partial U_2}{\partial s} + \dfrac{U_3}{z+R_2} & \dfrac{\partial U_2}{\partial z} \\[2mm] \dfrac{R_1}{z+R_1}\dfrac{\partial U_3}{\partial r} - \dfrac{U_1}{z+R_1} & \dfrac{R_2}{z+R_2}\dfrac{\partial U_3}{\partial s} - \dfrac{U_2}{z+R_2} & \dfrac{\partial U_3}{\partial z} \end{bmatrix} \quad (3)$$

In (3), we have noted

$$U_i(r,s,z) = e_i \cdot \mathbf{U}(r,s,z) \quad (4)$$

In order to take the prestressed effects into account, the Green–Lagrange strains are computed by

$$\varepsilon(\mathbf{U}) = \tfrac{1}{2}(\text{Grad } \mathbf{U} + \text{Grad}^T\mathbf{U} + \text{Grad}^T\mathbf{U} \cdot \text{Grad } \mathbf{U}) \quad (5)$$

and only the following non-linear terms are retained: w^2; $(\partial w/\partial \alpha)^2$, $\alpha = r, s$. Then, since the normal transverse stress is zero, the boundary conditions on the top and bottom surfaces of the shell are

$$\sigma_{1z} = \sigma_5 = 0 \quad \text{and} \quad \sigma_{2z} = \sigma_4 = 0 \quad \text{at} \quad z = \pm H/2 \quad (6)$$

It is easy to show that (6) becomes, for a monoclinic symmetry of a homogeneous material

$$E_{1z} = E_5 = 0 \quad \text{and} \quad E_{2z} = E_4 = 0 \quad \text{at} \quad z = \pm H/2 \quad (7)$$

where E_4 and E_5 are the transverse shear strains, computed from (5).

Finally the assumptions (i)–(iii) with (1)–(7) involve the following form for the displacements in the middle surface:

- U_0 and Ω_1 are unchanged (displacements assumed independent),

- $\Omega_2^T = \left\{ \dfrac{1}{3R_1}\left(\omega_r + \dfrac{\partial w}{\partial r} - \dfrac{u}{R_1}\right); \dfrac{1}{3R_2}\left(\omega_s + \dfrac{\partial w}{\partial s} - \dfrac{v}{R_2}\right); 0 \right\}$

- $\Omega_3^T = \left\{ \dfrac{-4}{3H^2}\left(\omega_r + \dfrac{\partial w}{\partial r} - \dfrac{u}{R_1}\right); \dfrac{-4}{3H^2}\left(\omega_s + \dfrac{\partial w}{\partial s} - \dfrac{v}{R_2}\right); 0 \right\}$

$$(8)$$

In these equations we have introduced the following notations:

$$\omega_r = \omega_{r1} \qquad \omega_s = \omega_{s1} \qquad \bar{3} = \frac{1}{4}\left(12 - \frac{H^2}{R_1^2}\right) \qquad \tilde{3} = \frac{1}{4}\left(12 - \frac{H^2}{R_2^2}\right) \qquad (9)$$

The strains contain the terms $R\alpha/(z + R\alpha)$; $\alpha = 1, 2$. We assume that $z/R\alpha \ll 1$ and then, the E_{ij} components of ε are expressed in a power series of z by using the displacement field (1), (2) with (8), on restricting approximations to the third order (limited expansion). So, the generalized strain vector has the form

$$\mathbf{E} = \mathbf{D}(\mathbf{U}) \qquad (10)$$

where

$$
\begin{aligned}
\mathbf{E}^T &= \{\varepsilon_0^T \quad \varepsilon_1^T \quad \varepsilon_2^T \quad \varepsilon_3^T \quad \Gamma_0^T \quad \Gamma_1^T \quad \Gamma_2^T \quad \Gamma_3^T\} \\
\varepsilon_l^T &= \{E_1^l \quad E_2^l \quad E_6^l\} \qquad\qquad l = 0,1,2,3 \\
\Gamma_l^T &= \{E_4^l \quad E_5^l\} \qquad\qquad\quad l = 0,1,2,3 \\
E_\gamma &= E_\gamma^0 + z E_\gamma^1 + z^2 E_\gamma^2 + z^3 E_\gamma^3 \qquad \gamma = 1,2,6,4,5
\end{aligned}
\qquad (11)
$$

and

$$\mathbf{U}^T = \{u \quad v \quad w \quad \omega_r \quad \omega_s\} \qquad (12)$$

The components of the operator $\mathbf{D}$ are listed in the Appendix. The static equilibrium equations of the shell are deduced from the virtual power principle; that is, given the virtual velocity field $\mathbf{U}^*$ which has the same form as $\mathbf{U}$, the body forces $\mathbf{f}$, the surface forces $\mathbf{p}$; find $\mathbf{U}$ such that, for all $\mathbf{U}^*$, we have[13]

$$0 = -\int_A \mathbf{E}^{*T}\hat{\sigma}\, dA + \int_A \mathbf{U}^{*T}\mathbf{f}\, dA + \int_\Gamma \mathbf{U}^{*T}\mathbf{p}\, dT \qquad (13)$$

(A is the middle and Γ the boundary of the shell Ω). In (11) $\mathbf{E}^*$ is computed by the operator $\mathbf{D}$, that is:

$$\mathbf{E}^* = \mathbf{D}(\mathbf{U}^*) \qquad \mathbf{U}^{*T} = \{u^* \quad v^* \quad w^* \quad \omega_r^* \quad \omega_s^*\} \qquad (14)$$

Otherwise, $\hat{\boldsymbol{\sigma}}$ is the generalized stress vector defined by

$$\hat{\boldsymbol{\sigma}}^{\mathrm{T}} = \{\hat{\mathbf{N}}^{\mathrm{T}} \quad \hat{\mathbf{M}}^{\mathrm{T}} \quad \hat{\mathbf{O}}^{\mathrm{T}} \quad \hat{\mathbf{P}}^{\mathrm{T}} \quad \hat{\mathbf{Q}}^{\mathrm{T}} \quad \hat{\mathbf{K}}^{\mathrm{T}} \quad \hat{\mathbf{R}}^{\mathrm{T}} \quad \hat{\mathbf{T}}^{\mathrm{T}}\} \tag{15}$$

with, $\hat{\boldsymbol{\sigma}} = \boldsymbol{\sigma} + \boldsymbol{\sigma}^\sigma + \boldsymbol{\sigma}^\theta + \boldsymbol{\sigma}^{\mathrm{h}}$; $\boldsymbol{\sigma}^\sigma$, $\boldsymbol{\sigma}^\theta$, $\boldsymbol{\sigma}^{\mathrm{h}}$ are initial stress, thermal stress, hygrometric stress respectively, and

$$\{\hat{\mathbf{N}}^{\mathrm{T}} \quad \hat{\mathbf{M}}^{\mathrm{T}} \quad \hat{\mathbf{O}}^{\mathrm{T}} \quad \hat{\mathbf{P}}^{\mathrm{T}}\} = \int_{-H/2}^{H/2} (1, z, z^2, z^3) : \begin{Bmatrix} \hat{\sigma}_1 \\ \hat{\sigma}_2 \\ \hat{\sigma}_6 \end{Bmatrix} \mathrm{d}z$$

$$\{\hat{\mathbf{Q}}^{\mathrm{T}} \quad \hat{\mathbf{K}}^{\mathrm{T}} \quad \hat{\mathbf{R}}^{\mathrm{T}} \quad \hat{\mathbf{T}}^{\mathrm{T}}\} = \int_{-H/2}^{H/2} (1, z, z^2, z^3) : \begin{Bmatrix} \hat{\sigma}_4 \\ \hat{\sigma}_5 \end{Bmatrix} \mathrm{d}z \tag{16}$$

where $\hat{\sigma}_1 = \hat{\sigma}_{rr}$; $\hat{\sigma}_2 = \hat{\sigma}_{ss}$; $\hat{\sigma}_6 = \hat{\sigma}_{rs}$; $\hat{\sigma}_4 = \hat{\sigma}_{rz}$; $\hat{\sigma}_5 = \hat{\sigma}_{sz}$ are local stresses.

From (13), we can show that equilibrium equations of the shell are of the form

$$\varepsilon(\boldsymbol{\sigma}) = \mathbf{F} \tag{17}$$

The equilibrium linear operator is composed of two parts which are deduced from the $\mathbf{D}$ operator

$$\varepsilon = \mathbf{D}_{\mathrm{m}}^{\mathrm{T}} - \mathbf{D}_b^{\mathrm{T}} \tag{18}$$

where

$\mathbf{D}_{\mathrm{m}}$ = part of $\mathbf{D}$ devoted to the membrane effects
$\mathbf{D}_{\mathrm{b}}$ = part of $\mathbf{D}$ devoted to the bending effects

and are listed in the Appendix.

Note that eqn (17) is linear, because the prestressed effect is assumed constant in (13). Also that $\mathbf{F}$ contains the applied forces: normal pressure $\mathbf{p}$ on the top surface of the shell; initial stress $\boldsymbol{\sigma}^\sigma$ limited to vector $\mathbf{N}^\sigma$ (membrane): $\boldsymbol{\sigma}^\theta$ and $\boldsymbol{\sigma}^{\mathrm{h}}$. To analyse the behaviour of the shell we need to define $\boldsymbol{\sigma}$, $\boldsymbol{\sigma}^\theta$ and $\boldsymbol{\sigma}^{\mathrm{h}}$; according to $\mathbf{U}$, $T - T_0$ and $h - h_0$ respectively (T = temperature distribution; h: moisture distribution). These quantities are solutions of diffusion equations.

HYGROTHERMO-ELASTIC BEHAVIOUR LAW

Since the object of this study is the environmental effects on the behaviour of the shell, the local constitutive law introduced is:

$$\hat{\boldsymbol{\sigma}} - \boldsymbol{\sigma}^\sigma = \mathbf{Q}' \varepsilon(\mathbf{U}) + \boldsymbol{\sigma}^\theta + \boldsymbol{\sigma}^{\mathrm{h}} \tag{19}$$

where

$$\boldsymbol{\sigma}^{\mathrm{T}} = \{\sigma_1 \quad \sigma_2 \quad \sigma_6 \quad \sigma_4 \quad \sigma_5\}$$

the normal transverse stress being zero, the matrix $\mathbf{Q}'$ has the following form in the r, s, z axis:

$$\mathbf{Q'} = \begin{bmatrix} \mathbf{Q'_p} & \mathbf{O} \\ \mathbf{O} & \mathbf{Q'_t} \end{bmatrix} \qquad \mathbf{Q'_p} = \begin{bmatrix} Q'_{11} & Q'_{12} & Q'_{16} \\ Q'_{12} & Q'_{22} & Q'_{26} \\ Q'_{16} & Q'_{26} & Q'_{66} \end{bmatrix} \qquad \mathbf{Q'_t} = \begin{bmatrix} Q'_{44} & Q'_{45} \\ Q'_{45} & Q'_{55} \end{bmatrix} \quad (20)$$

where

$$\mathbf{Q'} = \mathbf{\Phi}^{-1}\mathbf{Q}\mathbf{\Phi}$$

$$Q_{ij} = C_{ij} - \frac{C_{i3}C_{3j}}{C_{33}} \quad \text{for } ij = 11,\ 12,\ 22,\ 66 \quad (21)$$

$$Q_{\alpha\beta} = C_{\alpha\beta} \quad \text{for } \alpha\beta = 44,\ 55$$

and Φ is the rotational matrix around the z axis. All other terms in $\mathbf{Q}$ are zero, because of the hypothesis of unidirectional fibre composites.

In eqn (21), the coefficients $C_{\alpha\beta}$ are the three-dimensional elastic moduli, which have been obtained by homogenization. The technique used is the same as[14] for a transversely isotropic material (unidirectional fibre composite) in an arbitrary layer of the shell, but we have extended this method to compute the thermal and hygrometric expansion coefficients and have computed the transverse Young's modulus by a technique of wave propagation,[15] because of the good result obtained by the latter method. Finally, the constitutive law considered in each layer of the shell and in the symmetrical material axis x, y, z (x parallel to the fibres, $y0z$ plane) is

$$\begin{Bmatrix} E_x \\ E_y \\ 2E_{xy} \\ 2E_{yz} \\ 2E_{zx} \end{Bmatrix} = \begin{bmatrix} \dfrac{1}{E_L} & -\dfrac{\nu_{LT}}{E_L} & 0 & 0 & 0 \\[2mm] -\dfrac{\nu_{LT}}{E_L} & \dfrac{1}{E_T} & 0 & 0 & 0 \\[2mm] 0 & 0 & \dfrac{1}{G_{TT}} & 0 & 0 \\[2mm] 0 & 0 & 0 & \dfrac{1}{G_{LT}} & 0 \\[2mm] 0 & 0 & 0 & 0 & \dfrac{1}{G_{LT}} \end{bmatrix} \begin{Bmatrix} \sigma_x \\ \sigma_y \\ \sigma_{xy} \\ \sigma_{yz} \\ \sigma_{zx} \end{Bmatrix}$$

$$+ \begin{Bmatrix} \alpha_L \\ \alpha_T \\ 0 \\ 0 \\ 0 \end{Bmatrix}(T - T_0) + \begin{Bmatrix} \beta_L \\ \beta_T \\ 0 \\ 0 \\ 0 \end{Bmatrix}(h - h_0) \quad (22)$$

where $T - T_0$ is the temperature distribution and $h - h_0$ the hygrometric distribution, which are dependent on r, s, z in general. These parameters are determined by means of diffusion equations for temperature and moisture. The predicted coefficients of (22) are listed in the Appendix. From (19) and (16) we can write the constitutive equations in the following form:

$$^1\hat{\sigma} - {}^1\sigma^\sigma = {}^1\mathbf{A}\,{}^1\mathbf{E} - {}^1\sigma^\theta - {}^1\sigma^h \tag{23}$$

where the index 1 marks layer 1 of the shell, and

$$^1\mathbf{A} = \begin{bmatrix} {}^1\mathbf{A} & {}^1\mathbf{B} & {}^1\mathbf{D} & {}^1\mathbf{C} & & & & \\ {}^1\mathbf{B} & {}^1\mathbf{D} & {}^1\mathbf{C} & {}^1\mathbf{E} & & \mathbf{0} & & \\ {}^1\mathbf{D} & {}^1\mathbf{C} & {}^1\mathbf{E} & {}^1\mathbf{E} & & & & \\ {}^1\mathbf{C} & {}^1\mathbf{E} & {}^1\mathbf{F} & {}^1\mathbf{G} & & & & \\ & & & & {}^1\mathbf{A}' & {}^1\mathbf{B}' & {}^1\mathbf{D}' & {}^1\mathbf{C}' \\ & \mathbf{0} & & & {}^1\mathbf{B}' & {}^1\mathbf{D}' & {}^1\mathbf{C}' & {}^1\mathbf{E}' \\ & & & & {}^1\mathbf{D}' & {}^1\mathbf{C}' & {}^1\mathbf{E}' & {}^1\mathbf{F}' \\ & & & & {}^1\mathbf{C}' & {}^1\mathbf{E}' & {}^1\mathbf{F}' & {}^1\mathbf{G}' \end{bmatrix} \tag{24}$$

where

$$^1\mathbf{X} = \begin{bmatrix} {}^1X_{11} & {}^1X_{12} & {}^1X_{16} \\ {}^1X_{12} & {}^1X_{22} & {}^1X_{26} \\ {}^1X_{16} & {}^1X_{26} & {}^1X_{66} \end{bmatrix} \qquad {}^1\mathbf{X}' = \begin{bmatrix} {}^1X'_{44} & {}^1X'_{45} \\ {}^1X'_{45} & {}^1X'_{55} \end{bmatrix} \tag{25}$$

$$^1\mathbf{X} = {}^1\mathbf{A}, {}^1\mathbf{B}, \ldots, {}^1\mathbf{G}; \quad \text{idem for } {}^1X' \text{ dashed}$$

Next

$$({}^1\mathbf{A}, {}^1\mathbf{B}, {}^1\mathbf{D}, {}^1\mathbf{C}, {}^1\mathbf{E}, {}^1\mathbf{F}, {}^1\mathbf{G}) = \int_{h_{l-1}}^{h_l} (1, z, z^2, z^3, z^4, z^5, z^6)\,{}^1\mathbf{Q}'_p\, \mathrm{d}z$$

$$\tag{26}$$

$$({}^1\mathbf{A}', {}^1\mathbf{B}', {}^1\mathbf{D}', {}^1\mathbf{C}', {}^1\mathbf{E}', {}^1\mathbf{F}', {}^1\mathbf{G}') = \int_{h_{l-1}}^{h_l} (1, z, z^2, z^3, z^4, z^5, z^6)\,{}^1\mathbf{Q}'_t\, \mathrm{d}z$$

Thus, we have for N layers

$$\hat{\sigma} = \sum_{l=1}^{N} l_{\hat{\sigma}} \tag{27}$$

This constitutive law (27) with (23) allows us to define eqn (17) according to U and **p**, $\mathbf{N}^\sigma$. $T - T_0$, $h - h_0$.

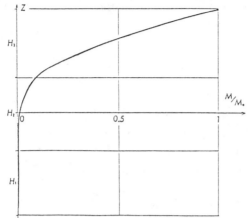

FIG. 2. Moisture distribution with $D_t = 2 \times 10^{-1}\,\text{mm}^2$ and $h_0 = 100\%$.

Finally, eqn (17) becomes

$$\varepsilon(\sigma) = \varepsilon(\mathbf{AD}(\mathbf{U})) = \mathbf{F} \tag{28}$$

where $\mathbf{F}$ is dependent on the following: normal distributed load p; initial stresses N_r^σ, N_s^σ, N_{rs}^σ (components of $\mathbf{N}^\sigma$); temperature $T - T_0$; moisture distribution $h - h_0$ (Fig. 2); body forces $\mathbf{f}$.

NUMERICAL RESULTS

The partial differential equation system (28) has a Navier-type solution for simply-supported shells whose map in the $x_1 x_2$-plane is a rectangle, provided the lamination scheme is of antisymmetric cross-ply or symmetric cross-ply type; that is if the following stiffnesses are zero:

$${}^1X_{16} = {}^1X_{26} = {}^1X_{45}' = 0 \qquad \forall\, 1 \in \{1, 2, \ldots, N\} \tag{29}$$

For a simply supported shell, the boundary conditions are:

$$u(r, 0) = u(r, l) = v(0, s) = v(L, s) = 0$$
$$w(r, 0) = w(r, l) = w(0, s) = w(L, s) = 0$$
$$\omega_r(r, 0) = \omega_r(r, l) = \omega_s(0, s) = \omega_s(L, s) = 0$$
$$N_s(r, 0) = N_s(r, l) = N_r(0, s) = N_r(L, s) = 0 \tag{30}$$
$$M_s(r, 0) = M_s(r, l) = M_r(0, s) = M_r(L, s) = 0$$
$$O_s(r, 0) = O_s(r, l) = O_r(0, s) = O_r(L, s) = 0$$
$$P_s(r, 0) = P_s(r, l) = P_r(0, s) = P_r(L, s) = 0$$

In these equations, L and l denote lengths along the x_1- and x_2-directions, respectively. The following solutions satisfy the boundary conditions (30)

$$u = \sum_{m,n=1}^{\infty} a_{mn} \cos \lambda_m r . \sin \lambda_n s$$

$$v = \sum_{m,n=1}^{\infty} b_{mn} \sin \lambda_m r . \cos \lambda_n s$$

$$w = \sum_{m,n=1}^{\infty} c_{mn} \sin \lambda_m r . \sin \lambda_n s \qquad (31)$$

$$\omega_r = \sum_{m,n=1}^{\infty} d_{mn} \cos \lambda_m r . \sin \lambda_n s$$

$$\omega_s = \sum_{m,n=1}^{\infty} e_{mn} \sin \lambda_m r . \cos \lambda_n s$$

where $\lambda_m = m\pi/L$ and $\lambda_n = n\pi/l$.

TABLE 1

Comparison with Refs 17 and 18 for non-dimensionalized deflection and stresses in a three-layer $(0°/90°/0°)$ rectangular laminate under sinusoidal load $(H_i = H/3;$ $l/L = 3)$

L/H	Source	$\bar{w}$	$\bar{\sigma}_1$	$\bar{\sigma}_2$	$\bar{\sigma}_4$	$\bar{\sigma}_5$	$\bar{\sigma}_6$
4	Pagano	2·82	1·10	0·119	0·033 4	0·387	0·028 1
	Reddy	2·641	1·036	0·103	0·034 8	0·272	0·026 3
	Present	2·643	1·036	0·104	0·027 4	0·382	0·026 5
10	Pagano	0·919	0·725	0·043 5	0·015 2	0·420	0·012 3
	Reddy	0·862	0·692	0·039 8	0·017 0	0·286	0·011 5
	Present	0·864	0·693	0·039 9	0·012 7	0·429	0·011 6
20	Pagano	0·610	0·650	0·029 9	0·011 9	0·434	0·009 3
	Reddy	0·594	0·641	0·028 9	0·013 9	0·288	0·009 1
	Present	0·595	0·641	0·028 9	0·010 2	0·437	0·009 1

The transverse load p is expressed by a double Fourier expansion:

$$p(r,s) = \sum_{m,n=1}^{\infty} p_{mn} \sin \lambda_m r \sin \lambda_n s \tag{32}$$

The hygroscopic nature of the shell is approximated by means of the classical diffusion equation. Moisture absorption and desorption are prevented on one surface of the shell, while the other surface is exposed to the environment. After a long period of exposure, the moisture concentration in the shell approaches the concentration of the environment, then an appropriate answer might be to restrict the region of interest to $0 \le \xi \le \tilde{H}/2$. The solution corresponding to these boundary conditions for absorption is given by eqn[16] (33) where $H = \tilde{H}/2$

$$h(\xi, t) = h_0 \left\{ 1 - \sum_{n=0}^{\infty} p_n \cos q_n(\xi + H/2) \right\} \tag{33}$$

with (n is an integer):

$$p_n = \frac{4}{\pi} \frac{(-1)^n}{2n+1} \exp(-q_n^2 Dt) \qquad q_n = (2n+1)\pi/2H \tag{34}$$

We use an equivalent solution for thermal distribution. In (34) t is time and D the diffusion coefficient.

TABLE 2

Non-dimensionalized deflections and stresses of cross-ply spherical shell under uniform load for a three-layer $(0°/90°/0°)$: $L/H = 10$; $R_1 = R_2 = R$ and $m = n = 19$ in Ref. 31

R/L	Source	$\bar{w}$	$\bar{\sigma}_1$	$\bar{\sigma}_2$	$\bar{\sigma}_4$	$\bar{\sigma}_5$	$\bar{\sigma}_6$
3		—	—	—	—	—	—
	Present	10·585	0·897	0·453	0·176	0·570	0·006 8
5	Reddy	10·332	—	—	—	—	—
	Present	10·811	0·918	0·452	0·181	0·580	0·022 0
10	Reddy	10·752	—	—	—	—	—
	Present	10·909	0·926	0·447	0·183	0·593	0·033 6
10^{30}	Reddy	10·899	—	—	—	—	—
	Present	10·942	0·926	0·439	0·184	0·595	0·045 1

TABLE 3

Non-dimensionalized deflection and stresses of cross-ply spherical shell under uniform load for a five-layer ($0°/90°/0°/90°/0°$): $L/H = 100$; $R_1 = R_2 = R$; $m = n = 19$ in (31); the stresses $\bar\sigma_1$ to $\bar\sigma_6$ are respectively computed at $z = H/2, H/5, 0, 0, H/2$

R/L	$\bar\sigma_1$	$\bar\sigma_2$	$\bar\sigma_4$	$\bar\sigma_5$	$\bar\sigma_6$	$\bar w$
3	0·329	0·220	0·088	0·169	0·069	2·345
5	0·555	0·357	0·153	0·294	0·063	4·127
10	0·785	0·489	0·222	0·427	0·030	6·013
10^{30}	0·899	0·541	0·261	0·501	0·036	7·081

Finally, the partial differential system (28) with (29)–(34) is reduced to the following algebraic linear system:

$$\mathbf{L} - \mathbf{a}_{mn} = \mathbf{G}_{mn} \qquad (35)$$

where $\mathbf{a}^T = \{a_{mn} \quad b_{mn} \quad c_{mn} \quad d_{mn} \quad e_{mn}\}$ are the amplitudes of the generalized displacements $u, v, w, \omega_r, \omega_s$, corresponding to the $\lambda_m \lambda_n$ terms. To program the matrix $\mathbf{L}$ we use a three-dimensional array. The algebraic system (35) is solved by the Gauss direct method. Tables 1, 2 and 3 give some numerical results which have been obtained with the present model. The distribution of stresses is computed by integration of the local equilibrium equations. All numerical computations are obtained without programming of explicit

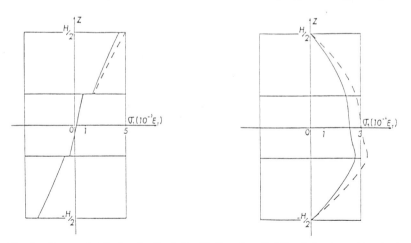

FIG. 3.   Stresses distribution for the shell with $R_1 = R_2 = R$; $R/L = 5$; for the problem in the Table 2. ---, Hygroscopic effect superposed on pressure effect.

equations, but by using the **D** and ε linear operators with their associated matrices. So, the analytical computations are considerably simplified.

In Tables 1 and 2 we have noted:

$$\bar{w} = \left(\frac{wH^3 E_T}{q_0 L^4}\right) \times 10^2 \qquad w = w\left(\frac{L}{2}, \frac{l}{2}\right)$$

$$\bar{\sigma}_1 = \frac{H^2}{q_0 L^2}\sigma_1\left(\frac{L}{2}, \frac{l}{2}, \frac{H}{2}\right) \qquad \bar{\sigma}_2 = \frac{H^2}{q_0 L^2}\sigma_2\left(\frac{L}{2}, \frac{l}{2}, \frac{H}{6}\right)$$

$$\bar{\sigma}_4 = \frac{H}{q_0 L}\sigma_4\left(\frac{L}{2}, 0, 0\right) \qquad \bar{\sigma}_5 = \frac{H}{q_0 L}\sigma_5\left(0, \frac{l}{2}, 0\right)$$

$$\bar{\sigma}_6 = \frac{H^2}{q_0 L^2}\sigma_6\left(0, 0, \frac{H}{2}\right)$$

The material properties are:

$$E_L = 25E_T \qquad G_{LT} = G_{TL} = 0.5E_T \qquad G_{TT} = 0.2E_T$$
$$\nu_{LT} = \nu_{TL} = 0.25 \qquad q_0 = 10^{-2}E_T$$

CONCLUSION

Methods of analysis have been developed for analysing the stresses in moderately thick laminated shells subjected to combined elevated temperature, absorbed moisture, normal pressure and prestressed effects. The analyses developed are based upon the classical diffusion equation and a new model of moderately thick shell. This new model contains more deformation effects than in some other[6-8] higher-order shear deformation theories of shells. The exact solutions obtained from this approximate theory should serve as bench mark results for the finite element approximations.

Moreover, the great advantage of these higher-order shear deformation theories is that they account for parabolic variation of transverse shear strains through thickness and therefore no shear correction factors are needed in these theories. Finally, the theory contains the same generalized displacements as in the first-order shear deformation theory.

The numerical simulations for hygrothermal effects, stability analysis, hygrothermo-viscoelastic behaviour are in progress, as well as some experimental simulations in hygrothermal environments and the homogenization of woven composites.

ACKNOWLEDGEMENTS

This work is financed by the 'STCAN, Direction de la Construction Navale, Département Matériaux et Structures Navales'. The authors thank M. Terrail.

REFERENCES

1. NAGHDI, P. M., A survey of recent progress in the theory of elastic shells. *Appl. Mech. Rev.*, **9** (1956) 365–8.
2. BERT, C. W., Analysis of shells. In *Analysis and Performance of Composites*. ed. L. J. Broutman. Wiley, New York, 1980, pp. 207–58.
3. KOITER, W. T., A consistent first approximation in the general theory of thin elastic shells. *Proceedings of the Symposium on the Theory of Thin Elastic Shells*. Delft, Aug. 1959, pp. 24–8.
4. DATTA, S. K. and LEDBETTER, H. M., Elastic constants of fiber-reinforced boron-aluminium: observation and theory. *Int. J. Solids Struct.*, **19** (1983) 885–94.
5. TOURATIER, M., Mécanismes de reinforcement des composites à matrice métallique. Contrat DRET No. 87/1163 to be published in 1989.
6. BHIMARADDI, A., A higher-order theory for free vibration analysis of circular cylindrical shells. *Int. J. Solids Struct.*, **20** (1984) 623–30.
7. SOLDATOS, K. P., Stability and vibration of thickness shear deformable cross-ply laminated non-circular cylindrical shells. *Proceedings of the 1986 Pressure Vessels and Piping Conference and Exhibition*, Chicago, July 1986, pp. 23–34.
8. REDDY, J. N. and LIU, F. C., A higher-order shear deformation theory of laminated elastic shells. *Int. J. Enging Sci.*, **23** (1985) 319–30.
9. CHENG, S., Accurate fourth order equation for circular cylindrical shells. *J. Engng Mechanics Division (ASCE)*, (1972) 641–56.
10. WIDERA, G. E. O. and LOGAN, D. L., Refined theories for nonhomogeneous anisotropic cylindrical shells. *J. Engng Mechanics Division (ASCE)*, (1980) 1053–74 and 1075–90 (part I and part II).
11. REISSNER, E., On a certain mixed variational theorem and on laminated elastic shell theory. *Lecture Notes in Engineering 28*, Springer-Verlag, Berlin, 1987, pp. 21–7.
12. TOURATIER, M., A refined theory for thick composite plates. *Mech. Res. Comm.*, **15** (1988) 229–36.
13. GERMAIN, P., *Mécanique–Ellipses*, Ecole Polytechnique, Paris, 1986.
14. WHITNEY, J. M. and RILEY, M. B., Elastic stress–strain properties of fiber-reinforced composite materials. Air Force Materials Laboratory Report AFML-TR-65-238, Dec. 1965.

15. BOSE, S. K. and MAL, A. K., Elastic waves in a fiber-reinforced composite. *J. Mech. Phys. Solids*, **22** (1974) 217–29.
16. PIPES, R. B., VINSON, J. R. and CHOU, T-W., On the hygrothermal response of laminated composite systems. *J. Comp. Mater.*, **10** (1976) 129–48.
17. REDDY, J. N., A simple higher-order theory for laminated composite plates. *J. Appl. Mech.*, **51** (1984) 745–52.
18. PAGANO, N. J., Exact solutions for composite laminates in cylindrical bending, *J. Comp. Mater.*, **3** (1969).

APPENDIX

```
C***************************************************************
C                                                              *
C       COMPONENTS OF THE OPERATOR D                           *
C                                                              *
C                                                              *
C       Dm=D(I,J,K) IF  1 <I <12 ,ELSE Dm=0                    *
C                                                              *
C       Db=D(I,J,K) IF 13 <I <20 ,ELSE Db=0                    *
C                                                              *
C       K=1 : Constant terms                                   *
C       K=2 : First derivative / r                            *
C       K=3 : First derivative / s                            *
C       K=4 : Second derivative / rs                          *
C       K=5 : Second derivative / rr                          *
C       K=6 : Second derivative / ss                          *
C                                                              *
C***************************************************************
        SUBROUTINE DEQ(R,RP,H,D,E)
        DOUBLE PRECISION R,RP,H
        DOUBLE PRECISION D(20,5,6),E(5,20,6)
        DOUBLE PRECISION R1,R2,R3,R4,H1,H2,P1,P2,P3,P4,C1,C2
        R1=1.D0/RP
        R2=1.D0/(RP*RP)
        R3=1.D0/(RP*RP*RP)
        R4=1.D0/(RP*RP*RP*RP)
        P1=1.D0/R
        P2=1.D0/(R*R)
        P3=1.D0/(R*R*R)
        P4=1.D0/(R*R*R*R)
        H1=1.D0/H
        H2=1.D0/(H*H)
        C1=3*4.D0/(12.D0-(H*H/(R*R)))
        C2=3*4.D0/(12.D0-(H*H/(RP*RP)))
        D(14,1,1)=-P1
        D(16,1,1)=P2-2.D0*C1*P2
        D(18,1,1)=12.D0*C1*P1*H2-(1.D0-C1)*P3
        D(20,1,1)=-4.D0*C1*P2*H2+(1.D0-C1)*P4
        D(13,2,1)=-R1
        D(15,2,1)=R2-2.D0*C2*R2
        D(17,2,1)=12.D0*C2*H2*R1-(1.D0-C2)*R3
        D(19,2,1)=-4.D0*C2*R2*H2+(1.D0-C2)*R4
        D(14,3,1)=1.D0
        D(16,3,1)=2.D0*C1*P1-P1
        D(18,3,1)=-12.D0*H2*C1+(1.D0-C1)*P2
        D(20,3,1)=4.D0*C1*P1*H2-(1.D0-C1)*P3
        D(13,4,1)=1.D0
        D(15,4,1)=2.D0*C2*R1-R1
        D(17,4,1)=-12.D0*C2*H2+(1.D0-C2)*R2
        D(19,4,1)=4.D0*C2*R1*H2-(1.D0-C2)*R3
        D(1,5,1)=P1
        D(2,5,1)=R1
```

```
D(4,5,1)=-P2
D(5,5,1)=-R2
D(7,5,1)=P3
D(8,5,1)=R3
D(10,5,1)=-P4
D(11,5,1)=-R4
D(1,1,2)=1.D0
D(4,1,2)=-P1
D(7,1,2)=P2-C1*P2
D(10,1,2)=(C1-1.D0)*P3+4.D0*C1*H2*P1
D(3,2,2)=1.D0
D(6,2,2)=-P1
D(9,2,2)=P2-C2*R2
D(12,2,2)=(4.D0*R1*H2+R2*P1)*C2-P3
D(4,3,2)=1.D0
D(7,3,2)=(C1-1.D0)*P1
D(10,3,2)=-4.D0*C1*H2+P2-C1*P2
D(6,4,2)=1.D0
D(9,4,2)=C2*R1-P1
D(12,4,2)=P2-4.D0*C2*H2-C2*P1*R1
D(14,5,2)=1.D0
D(16,5,2)=D(16,3,1)
D(18,5,2)=D(18,3,1)
D(20,5,2)=D(20,3,1)
D(3,1,3)=1.D0
D(6,1,3)=-R1
D(9,1,3)=R2-C1*P2
D(12,1,3)=4.D0*C1*P1*H2+C1*R1*P2-R3
D(2,2,3)=1.D0
D(5,2,3)=-R1
D(8,2,3)=(1.D0-C2)*R2
D(11,2,3)=4.D0*C2*R1*H2+(C2-1.D0)*R3
D(6,3,3)=1.D0
D(9,3,3)=C1*P1-R1
D(12,3,3)=R2-4.D0*C1*H2-C1*P1*R1
D(5,4,3)=1.D0
D(8,4,3)=(C2-1.D0)*R1
D(11,4,3)=(1.D0-C2)*R2-4.D0*C2*H2
D(13,5,3)=1.D0
D(15,5,3)=D(15,4,1)
D(17,5,3)=D(17,4,1)
D(19,5,3)=D(19,4,1)
D(9,5,4)=C2*R1+C1*P1
D(12,5,4)=(C1+C2)*(P1*R1-4.D0*H2)
D(7,5,5)=C1*P1
D(10,5,5)=-4.D0*C1*H2-C1*P2
D(8,5,6)=C2*R1
```

```
C*************************************************************
C                                                            *
C    HOMOGENIZATION                                          *
C                                                            *
C    -M  :  matrix                                           *
C    -F  :  fiber                                            *
C    -1  :  longitudinal                                     *
C    -2  :  transverse                                       *
C    V-  :  volume fraction                                  *
C    Y-  :  Young's moduli                                   *
C    G-  :  shear moduli                                     *
C    P-  :  Poisson's ratio                                  *
C    A-  :  coefficient of thermal expansion                 *
C    H-  :  coefficient of moisture expansion                *
C                                                            *
C*************************************************************
```

```
PM=YM/2.D0/GM-1.D0
PF=YF/2.D0/GF-1.D0
PTM=1.D0-PM-2.D0*PM*PM
PTF=1.D0-PF-2.D0*PF*PF
VM=1.D0-VF
D=PTM*VF*YF+PTF*VM*YM+YF*(1.D0+PM)
Y1=VF*YF+VM*YM+(2.D0*(PF-PM)**2*VF*YF*VM*YM)/D
```

```
A1=(VF*YF*AF+VM*YM*AM-(2.D0*VF*YF*VM*YM*(PF-PM)*((1.D0+PF)*AF-
1(1.D0+PM)*AM)/D))/Y1
G1=GM*(GF+GM+(GF-GM)*VF)/(GF+GM-(GF-GM)*VF)
P1=PM-(2.D0*(PM-PF)*(1.D0-PM**2)*YF*VF)/D
BM=YM/(1.D0-2.D0*PM)/3.D0
G2=GM+2.D0*VF*(BM+GM)*(GF-GM)/(BM+(BM/GM+2.D0)*(VF*GM+VM*GF))
T2=VF*(PTF/YF+2.D0*YF*(P1/Y1)**2)+VM*(PTM/YM+2.D0*YM*(P1/Y1)**2)+
1VF*VM/D*(4.D0*((P1/Y1)**2)*YF*YM*(PF-PM)**2-((PTM*YF-PTF*
1YM)**2)/YF/YM)
YZ=2.D0/(T2+5.D-1/G2)
A2=-((VF*YF*(2.D0*A1-AF)+VM*YM*(2.D0*A1-AM))*P1/Y1-AM*(1.D0+PM)*VM
1-(1.D0+PF)*AF*VF+VM*VF/D*(((1.D0+PM)*AM-(1.D0+PF)*AF)*(PTM*YF+PTF*
1YM+2.D0*YF*YM*P1/Y1*(PF-PM))+4.D0*A1*(PF-PM)**2*YF*YM*P1/Y1))
H1=(VM*YM*HM-(2.D0*VF*YF*VM*YM*(PF-PM)*(1.D0+PM)*(-AM)/D))/Y1
H2=-(P1/Y1*(VF*YF*2*H1+VM*YM*(2.D0*H1-HM))-HM*(1.D0+PM)*VM+VF/D*VM
1*(((1.D0+PM)*HM*(PTF*YM+PTM*YF+2.D0*YF*YM*P1/Y1*(PF-PM)+4.D0*H1*(PF
1-PM)**2*YF*YM*P1/Y1)))
```

9

Weight Change and Mechanical Properties of GRP Panel in Hot Water

Hiroyuki Hamada, Atsushi Yokoyama,
Zen-ichiro Maekawa and Tohru Morii
*Faculty of Textile Science, Kyoto Institute of Technology,
Matsugasaki, Sakyo-ku, Kyoto 606, Japan*

ABSTRACT

This study deals with the effects of water environment on the mechanical properties of glass-fibre reinforced plastics (GRP) panels. After GRP panels are immersed in hot water (80°C) for 3000 h, they are weighed, and tensile, bending and acoustic emission (AE) tests are performed. Ultimate strength and elastic moduli decrease in proportion to increases of weight due to water absorption; however, elastic moduli eventually reach a constant value. It is clear that these phenomena are caused by debonding at the fibre/matrix interface, which is observed by AE amplitude distribution histograms and scanning electron micrographs of the fracture surface.

INTRODUCTION

Recently the demand for glass-fibre reinforced plastics (GRP) structural materials in corrosive environments has surged, since GRP has many advantages over metal as regards specific strength and modulus and corrosion resistance. In particular, GRP panels have often been used as containers for water and for chemical tanks. In wet environments the weight of GRP panels is changed by means of water absorption and matrix dissolution, and as a result a reduction of mechanical properties is induced. Therefore it is important to estimate the effects of the environment, especially water, on the mechanical properties. Several studies have shown the effects of absorbed water on the mechanical properties of GRP.[1-3]

259

Absorbed water may lead to a number of undesirable effects such as polymer degradation, debonding at fibre/matrix interface, etc. However, most investigators have dealt separately with the data of the weight changes and mechanical properties, and very few have clarified the relation between them.[4]

The first objective of this investigation is to determine the relation between weight changes and mechanical properties of GRP panels. The second objective is to clarify the change of the fracture mechanism due to water absorption. The estimation of the fracture mechanism is performed by acoustic emission (AE) technique and scanning electron micrographs.

EXPERIMENTAL PROCEDURE

GRP used in this study are E-glass-fibre reinforced thermosetting plastics. The reinforcement is continuous E-glass-fibre mat and the matrix is orthophthalic-type unsaturated polyester resin. The square-shaped panels are fabricated by a compression moulding method. The length of each panel is 300 mm, and specimens are cut out from them. The compositions of the specimens are summarized in Table 1. A coating of silicon resin is applied at the edge surface to prevent water absorption.

The water immersion tests are conducted at 80°C with the instrument as shown in Fig. 1. The hot water is circulated by a pump to assure uniform temperature. The periods of immersion in this experiment are also shown in Table 1.

The rates at which water is absorbed by the specimens are determined by weighing the specimens (80 mm L × 15 mm W × 3 mm T) after immersion

TABLE 1
Compositions of the specimens and immersion test condition

Specimen	Tensile test	Bending test	AE test
Resin matrix	Orthophthalic type Unsaturated Polyester resin		
Reinforcement	E-Glass-fibre chopped strand mat		
Filler	Calcium carbonate		
Catalyst	Methyl ethyl ketone peroxide		
Accelerator	Naphthenic cobalt		
Moulding method	Compression moulding		
Number of plies	4	4	3
Temperature of hot water (°C)	80		
Immersion time (h)	0, 3, 10, 30, 100, 300, 1000 and 3000		

FIG. 1.   Instrument for immersion in hot water.

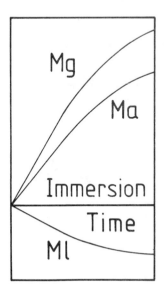

FIG. 2.   Schematic diagram of the net weight gain (M_g), the weight loss (M_l) and the apparent weight gain (M_a).

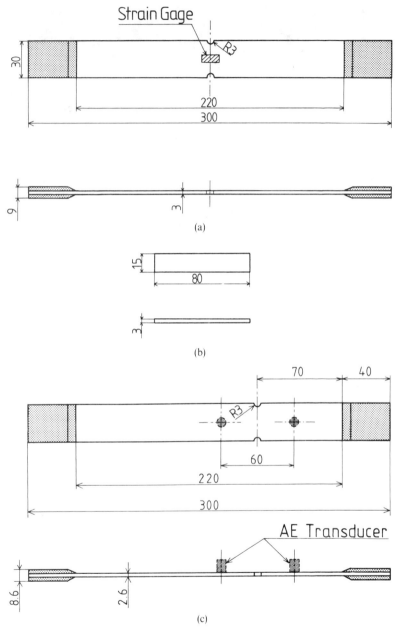

Fɪɢ. 3.　Specimen configuration: (a) Tensile test; (b) bending test; (c) AE test.

time for 3, 10, 30, 100, 300 and 1000 h. It is well known that degradation of the resin matrix and interface, and change in the specimen weight, may arise during immersion in water. This weight change is considered to be produced by the increase of the absorbed water in the specimen, and of the dissolved matrix in the water. The former increases the specimen weight and the latter decreases it, as shown in Fig. 2. In this figure M_g shows the net weight gain by means of the absorbed water and is given by

$$M_g = (W_w - W_d)/W_o$$

where W_w is the weight of the wet specimen after immersion test, W_d is that of the dried specimen and W_o is the original dry specimen weight. M_l shows the weight loss by means of the matrix dissolution and is given by

$$M_l = (W_o - W_d)/W_o$$

Therefore the apparent weight gain, M_a, can be given by

$$M_a = M_g - M_l = (W_w - W_o)/W_o$$

Many investigators have determined the weight change by means of M_a only.

The tensile tests are carried out after immersion tests by the Instron universal testing machine (type 4206) at a crosshead speed of 1 mm/min at room temperature in order to examine the rigidity reduction and the residual strength of the dry specimen. The residual bending elastic modulus and strength are measured using three-point bending with 45-mm span and at a crosshead speed of 2 mm/min. The acoustic emission (AE) test is conducted by an AE apparatus (Shimadzu SAE 1000A) under tensile loading monitoring the damage propagation of GRP. Figure 3 shows the configurations of the tensile, bending and AE test specimens. Moreover, in order to consider the fracture mechanisms of GRP under wet conditions, the fracture surfaces are observed after the tensile tests through a scanning electron microscope (SEM).

RESULTS AND DISCUSSION

Weight Change

Figure 4 shows the relationship between the apparent weight gain (M_a) and the square root of immersion time. The initial part of these plots is linear and M_a reaches the peak value at 100 h; however, after that M_a decreases remarkably. Figures 5 and 6 show the variation in the net weight

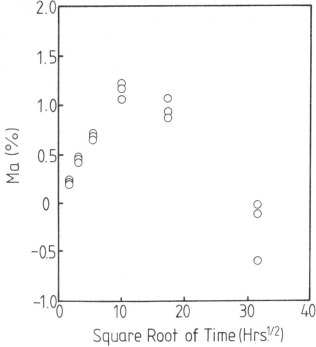

FIG. 4.   Change of the apparent weight gain (M_a).

gain (M_g) and the weight loss (M_1) with the square root of immersion time respectively. M_g increases linearly until 100 h. After this time M_g shifts from the initial straight line and gradually approaches the constant saturation level. The variation in M_g indicates typical Fickian diffusion behaviour. The increases of M_1 scarcely occur at 3 h; however, after this time M_1 is linearly increased.

Therefore it is considered that the mechanism of the weight changes is different before and after 100 h.

- Until 100 h water absorption and matrix dissolution occur simultaneously.
- After 100 h the rate of water absorption is gradually decreased; however, matrix dissolution continuously occurs at a constant rate.

Effects of Water Absorption on Mechanical Properties

Figure 7 shows the relationship between the tensile elastic modulus and M_g. M_g after immersion for 3000 h is calculated by the equation which

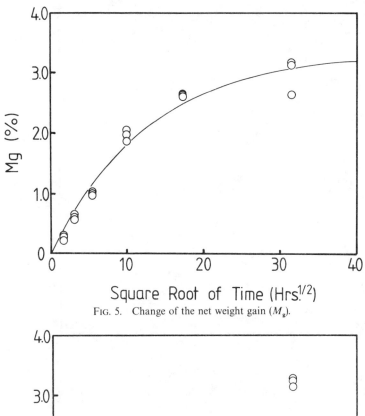

FIG. 5.   Change of the net weight gain (M_g).

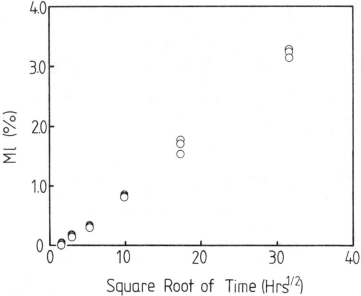

FIG. 6.   Change of the weight loss (M_1).

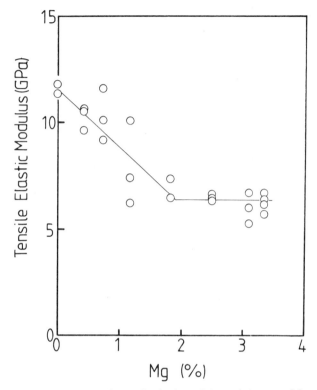

FIG. 7.   Relation between the tensile elastic modulus and the net weight gain (M_g).

approximates to M_g's shown in Fig. 5. The tensile elastic modulus is linearly decreased below approximately 1·9% (before 100 h), at which time M_g shifts from the initial straight line, as shown in Fig. 5, and above that it keeps a constant value. It is well known that the tensile elastic modulus of the composite material is represented by the rule of mixtures.[5] If it is assumed that the tensile elastic modulus of a glass fibre is not decreased due to water absorption, the decrease of the tensile elastic modulus of the composite below 1·9% of M_g may be caused by the decrease of that of the matrix due to water absorption. In Fig. 8 the relationship between the tensile strength and M_g is shown. The tensile strength is linearly decreased against M_g. This tendency differs from the tensile elastic modulus shown in Fig. 7. Therefore it seems that absorbed water has far greater effect on the tensile strength than on the tensile elastic modulus. Phenomenologically it seems as if the

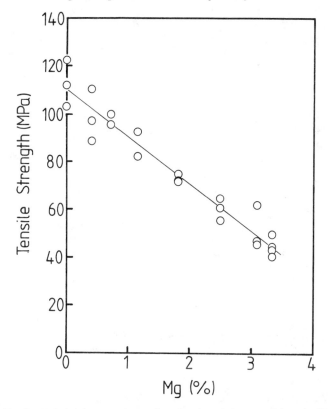

FIG. 8. Relation between the tensile strength and the net weight gain (M_g).

tensile strength is decreased due to increase of M_g. However, the reductions in mechanical properties are caused by fibre/matrix interfacial degradation, etc., in addition to the plasticization in the matrix due to water absorption, so that degradation behaviour cannot be due to the effect of water absorption only. Consequently the causes of the reductions in the tensile elastic modulus and strength are discussed in the next section using an acoustic emission technique and scanning electron micrographs.

Figure 9 shows the relationship between the bending elastic modulus and M_g. The bending elastic modulus is linearly decreased below approximately 1·2%, which corresponds to immersion for 100 h, and above that it keeps almost constant. The M_g at which the bending elastic modulus reaches a constant value is different from that at which the tensile elastic modulus

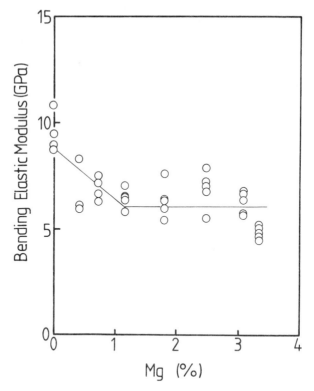

FIG. 9.   Relation between the bending elastic modulus and the net weight gain (M_g).

does. It seems that this difference arises since the specimen is subjected to only tensile load in the tensile test, while in the bending test it is subjected to compressive load as well as tensile load. In Fig. 10 the relationship between the bending strength and M_g is shown. The reduction of the bending strength with M_g shows a similar tendency to that of the tensile strength, since tension failure ultimately occurs at the lower side of the specimen in bending.

From the above-mentioned results tensile and bending properties can be calculated by the following equations if the reductions in mechanical properties are considered to be due only to water absorption.

Tensile elastic modulus (E_t)

$$E_t (\text{GPa}) = -2{\cdot}73 \times M_g (\%) + 11{\cdot}63 \qquad (0 < M_g < 1{\cdot}92\%)$$
$$E_t (\text{GPa}) = 6{\cdot}40 \qquad (M_g > 1{\cdot}92\%)$$

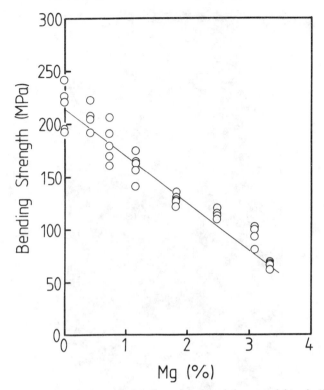

Fig. 10. Relation between the bending strength and the net weight gain (M_g).

Tensile strength (σ_t)

$$\sigma_t\,(\text{MPa}) = -19\cdot6 \times M_g\,(\%) + 110\cdot4$$

Bending elastic modulus (E_b)

$$E_b\,(\text{GPa}) = -2\cdot40 \times M_g\,(\%) + 8\cdot86 \qquad (0 < M_g < 1\cdot15\%)$$
$$E_b\,(\text{GPa}) = 6\cdot09 \qquad\qquad\qquad\quad (M_g > 1\cdot15\%)$$

Bending strength (σ_b)

$$\sigma_b\,(\text{MPa}) = -42\cdot5 \times M_g\,(\%) + 215\cdot1$$

AE Test Results and Observation of Fracture Surfaces

Figure 11 shows the scanning electron micrographs of the fracture surfaces for the original dry specimen and the specimen immersed for 300 h.

FIG. 11.   SEM photographs of the fracture surfaces; immersion time was (a) 0 h.

FIG. 11—*contd.* (b) 300 h.

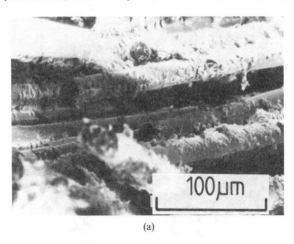

(a)

(b)

FIG. 12.   SEM photographs of the fibre surfaces; immersion time was (a) 0 h and (b) 300 h.

Tight fibre bundles are observed in the former (Fig. 11(a)), while loose fibre bundles are observed in the latter (Fig. 11(b)). Loose fibre bundles occur after 100 h. Accordingly this phenomenon corresponds to the variation of the tensile elastic modulus with M_g; that is, loose fibre bundles at the fracture surface are scarcely observed until 100 h, before which the tensile elastic modulus is decreased, and are observed after 100 h when the tensile elastic modulus remains constant.

Figure 12 shows SEM photographs of the fibre surfaces at the fracture surface. A large quantity of resin matrix adhering on the fibre surfaces is observed in the original dry specimen (Fig. 12(a)), as compared with the specimen immersed for 300 h (Fig. 12(b)). It is considered that Fig. 12(b) indicates the pull-out of the fibre owing to fibre/matrix interfacial degradation.

The amplitude distribution histograms of AE events are shown in Fig. 13. In the original dry specimen (Fig. 13(a)) most of the events are low-(0–2 mV) and high-range (above 6 mV) intensities with relatively few events of middle-range intensities (2–6 mV). The amplitude distribution histogram for the specimen immersed for 30 h (Fig. 13(b)) displays almost the same tendency as that for the original dry specimen; however, the events with high-range amplitude are decreased and those with low- and middle-range amplitude are slightly increased. The amplitude distribution for the specimens immersed for 300 and 3000 h (Fig. 13(c) and (d)) indicates few events with high-range amplitude. It is generally said that the events with middle- and high-range amplitude are due respectively to the fibre/matrix interfacial debonding and the fibre breaking. Therefore the increase of events with middle-range amplitude and the decrease of those with high-range amplitude represent an increase of interfacial debonding and a decrease of fibre breaking respectively.

The debonding at the interface affects the tensile elastic modulus as the number of events with middle-range amplitude hardly vary after 100 h when the tensile elastic modulus remains constant. The increase of debonding at the interface is caused by interfacial degradation. However, debonding at the interface caused by loading is slightly increased due to water immersion since the increase of events with middle-range amplitude is low in Fig. 13(b), (c) and (d) compared with Fig. 13(a). Accordingly it seems that debonding at the interface in the fibre bundles, that is loose fibre bundles as shown in Fig. 11(b), is caused by water penetration before loading. After 100 h fibre bundles in composite completely loosen. As a result interfacial degradation has very little effect on the tensile elastic modulus after 100 h and the tensile elastic modulus remains constant.

Figure 14 shows the variation of the fracture strain with M_g. The fracture strain remains almost constant below 1·9% of M_g (before 100 h) and is remarkably decreased above 1·9% of M_g (after 100 h). The variations of the fracture strain and the tensile elastic modulus are quite reversed. Therefore the mechanisms of the reduction in the tensile strength are different below and above 1·9% of M_g. The reduction in the tensile strength below and above 1·9% is caused by the plasticization in the matrix and interfacial degradation respectively.

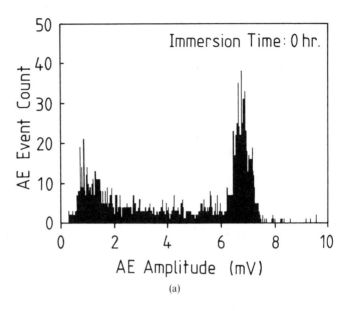

(a)

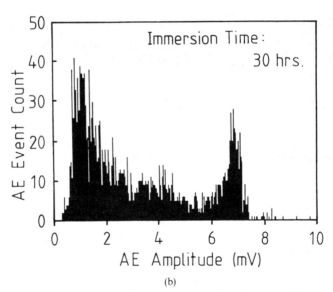

(b)

FIG. 13.   Amplitude distribution histograms; immersion time was (a) 0 h and (b) 30 h.

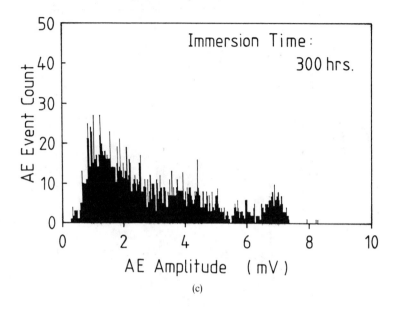

(c)

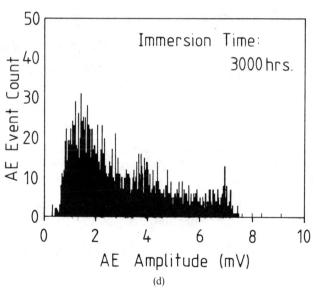

(d)

FIG. 13—*contd.*   (c) 300 h and (d) 3000 h.

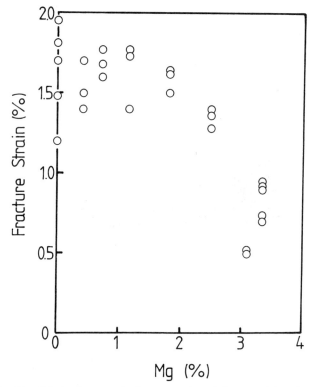

FIG. 14.   Relation between the fracture strain and the net weight gain (M_g).

CONCLUSIONS

The effects of water environment on mechanical properties of GRP panels are investigated. Specific conclusions drawn from this study may be summarized as follows:

1. Debonding at the fibre/matrix interface in the fibre bundle is caused by water penetration after 100 h, and as a result the tensile elastic modulus remains constant.
2. The tensile strength below 1·9% of M_g is decreased by plasticization in the matrix and that above 1·9% of M_g is related to the interfacial degradation.
3. Absorbed water greatly affects mechanical properties, especially tensile and bending strength, and the relations between mechanical

properties and the net weight gain (M_g) are given by equations of the first degree.

ACKNOWLEDGEMENTS

The authors would like to thank Mr Takahiro Hirano (Sekisui Koji Co. Ltd) and Mr Sadaki Mori (Kyoto Institute for Technology) for helpful discussion.

REFERENCES

1. APICELLA, A., MIGLIARESI, C., NICODEMO, L., NICOLAIS, L., IACCARINO, L. and ROCCOTELLI, S., Water sorption and mechanical properties of a glass-reinforced polyester resin. *Composites*, **13** (1982) 406–10.
2. ELLIS, B. and FOUND, M. S., The effects of water absorption on a polyester/chopped strand mat laminate. *Composites*, **14** (1983) 237–43.
3. MISAKI, T. and IWATSU, T., Effect of boiling water aging on strength and fracture properties of chopped strand mat polyester laminates. *J. Appl. Polym. Sci.*, **30** (1985) 1083–93.
4. SHEN, C. H. and SPRINGER, G. S., Effects of moisture and temperature on the tensile strength of composite materials. In *Environmental Effects on Composite Materials*, ed. G. S. Springer. Technomic, Westport, 1981, pp. 79–93.
5. AGARWAL, B. D. and BROUTMAN, L. J., *Analysis and Performance of Fiber Composites*. John Wiley, New York, 1980.

10

Analysis of Nonlinear Forced Vibration of Anisotropic Composite Unsymmetrically Laminated Plates

LI HUA and SHEN DARONG

Department of Engineering Mechanics,
Wuhan University of Technology,
Wuhan, Hubei, People's Republic of China

ABSTRACT

Unsymmetrical angle-ply laminating plates of anisotropic composites are representative in dynamics of plates. After making an analysis of the complete solution analysis process (including calculating examples) of nonlinear flexural forced vibration of plates under the harmonic force, the authors obtain the amplitude–frequency relationship of anisotropic unsymmetrical angle-ply laminating plates under loads of different magnitude, and the effect of the change of the plying layers of laminating plates upon the amplitude–frequency relationship of nonlinear flexural forced vibration. The solution analysis result thus obtained is reliable and the conclusion drawn from the result is representative. What has been done so far is undoubtedly a meaningful attempt in the matter of nonlinear dynamics of composite anisotropic laminating plates and will be of great help in the further research of other nonlinear dynamics of anisotropic laminating plates.

INTRODUCTION

Great concern has been shown for the problem of nonlinear dynamics of composites. Up till now, research into the nonlinear vibration of anisotropic laminated plates has only covered nonlinear free vibration and shock. The study presented in this paper has, however, been directed to the nonlinear flexural forced vibration (NFFV) of anisotropic unsymmetrical angle-ply laminated plates (AUAPLP), which is undoubtedly a great help

279

to further development of nonlinear dynamics of anisotropic laminated plates in other respects.

AUAPLP is chosen as an object of study due to its representative characteristics, for example, the complicated governing equation of vibration and the dynamic characteristics which basically contain those of the nonlinear vibration of the laminated plates laid by various typical methods. The analysis and study of the NFFV of the AUAPLP show that a basic idea can be obtained of the main dynamic characteristics of other anisotropic laminated plates in terms of the NFFV.

PROBLEM AND SOLUTION ANALYSIS PROCESS

The AUAPLP we consider here is a rectangular elastic thin plate with the same thickness all over which satisfies Kirchhoff's hypothesis. In the right handed cartesian system of coordinates x, y, z, the plate thickness, h is directed to the z axis, the position of the rectangular plate is $0 \leq x \leq a$ $0 \leq y \leq b$, the x, y axes are in the plate mid-plane and the effect of the moment of inertia and of the body force is ignored. The constitutive relations of the plate can be written as follows:

$$\begin{Bmatrix} N \\ M \end{Bmatrix} = \begin{bmatrix} A & B \\ B & D \end{bmatrix} \begin{Bmatrix} \varepsilon^{\circ} \\ \kappa \end{Bmatrix} \tag{1}$$

thus

$$\begin{Bmatrix} \varepsilon^{\circ} \\ M \end{Bmatrix} = \begin{bmatrix} A^* & B^* \\ -(B^*)^{\mathrm{T}} & D^* \end{bmatrix} \begin{Bmatrix} N \\ \kappa \end{Bmatrix} \tag{2}$$

in which

$$A^* = A^{-1} \qquad B^* = -A^{-1}B \qquad D^* = D - BA^{-1}B \tag{3}$$

The expressions for the nondimensional general governing equations of the NFFV of the AUAPLP are given by

$$\begin{cases} \bar{A}_{22}^* F_{,\zeta\zeta\zeta\zeta} + \lambda^2(2\bar{A}_{12}^* + \bar{A}_{66}^*)F_{,\zeta\zeta\eta\eta} + \lambda^4 \bar{A}_{11}^* F_{,\eta\eta\eta\eta} \\ \quad + \lambda(\bar{B}_{61}^* - 2\bar{B}_{26}^*)W_{,\zeta\zeta\zeta\eta} + \lambda^3(\bar{B}_{62}^* - 2\bar{B}_{16}^*)W_{,\zeta\eta\eta\eta} \\ \quad = \lambda^2(W_{,\zeta\eta}^2 - W_{,\zeta\zeta}W_{,\eta\eta}) \\ \bar{D}_{11}^* W_{,\zeta\zeta\zeta\zeta} + 2\lambda^2(\bar{D}_{12}^* + 2\bar{D}_{66}^*)W_{,\zeta\zeta\eta\eta} + \lambda^4 \bar{D}_{22}^* W_{,\eta\eta\eta\eta} \\ \quad - \lambda(\bar{B}_{61}^* - 2\bar{B}_{26}^*)F_{,\zeta\zeta\zeta\eta} - \lambda^3(\bar{B}_{62}^* - 2\bar{B}_{16}^*)F_{,\zeta\eta\eta\eta} + \lambda^4 W_{,\tau\tau} \\ \quad = \lambda^2(W_{,\zeta\zeta}F_{,\eta\eta} + W_{,\eta\eta}F_{,\zeta\zeta} - 2W_{,\zeta\eta}F_{,\zeta\eta}) + \dfrac{\lambda^4 b^4}{A_{22}h^3} q(\zeta, \eta, \tau) \end{cases} \tag{4}$$

where the nondimensional parameters are defined as

$$
\begin{cases}
\zeta = \dfrac{x}{a} & \eta = \dfrac{y}{b} & \lambda = \dfrac{a}{b} \\[2ex]
W = \dfrac{w}{h} & F = \dfrac{\psi}{A_{22}h^2} & \tau = \dfrac{t}{b^2}\left(\dfrac{A_{22}h^2}{\rho}\right)^{1/2} \\[2ex]
\bar{A}^*_{ij} = A_{22}A^*_{ij} & \bar{B}^*_{ij} = \dfrac{B^*_{ij}}{h} & \bar{D}^*_{ij} = \dfrac{D^*_{ij}}{A_{22}h^2} & (i,j = 1,2,6)
\end{cases}
\tag{5}
$$

here, A^*_{ij}, B^*_{ij}, D^*_{ij} are elements of the corresponding matrices in eqn (3).

S4, one of the boundary conditions of the four edges simply supported is given by

$$
\begin{cases}
W = W_{,\zeta\zeta} = F_{,\eta\eta} = F_{,\zeta\eta} = 0 & \zeta = 0, 1 \\
W = W_{,\eta\eta} = F_{,\zeta\zeta} = F_{,\zeta\eta} = 0 & \eta = 0, 1
\end{cases}
\tag{6}
$$

The simultaneous equations composed of eqns (4) and (6) just constitute the basic problem of the NFFV of the AUAPLP that the paper studies. Take trial functions for the above-mentioned problem as follows:

$$
\begin{cases}
W(\zeta, \eta, \tau) = f(\tau) \sin \pi\zeta \sin \pi\eta \\
F(\zeta, \eta, \tau) = f^*(\tau) X_1(\zeta) Y_1(\eta) \\
\qquad = f^*(\tau)[ch\alpha_1\zeta - \cos\alpha_1\zeta - v_1(sh\alpha_1\zeta - \sin\alpha_1\zeta)] \\
\qquad \times [ch\alpha_1\eta - \cos\alpha_1\eta - v_1(sh\alpha_1\eta - \sin\alpha_1\eta)]
\end{cases}
\tag{7}
$$

If $\alpha_1 = 4.730\,04$, $v_1 = 0.982\,502$, the trial functions (7) satisfy the boundary condition (6).[3] Substituting (7) into the general governing eqn (4) results in

$$
\begin{cases}
[(\alpha_1^4\bar{A}^*_{22} + (\alpha_1\lambda)^4\bar{A}^*_{11})X_1(\zeta)Y_1(\eta) + \lambda^2(2\bar{A}^*_{12} + \bar{A}^*_{66})X''_1(\zeta)Y''_1(\eta)]f^*(\tau) \\
\quad = [(\lambda\pi^2)^2 \cos((\zeta+\eta)\pi)\cos((\zeta-\eta)\pi)]f^2(\tau) \\
\qquad + [\pi^4\lambda((\bar{B}^*_{61} - 2\bar{B}^*_{26}) + \lambda^2(\bar{B}^*_{62} - 2\bar{B}^*_{16}))\cos\pi\zeta\cos\pi\eta]f(\tau) \\[1ex]
[\lambda^4 \sin\pi\zeta \sin\pi\eta]f''(\tau) \\
\qquad + [\pi^4(\bar{D}^*_{11} + 2\lambda^2(\bar{D}^*_{12} + 2\bar{D}^*_{66}) + \lambda^4\bar{D}^*_{22})\sin\pi\zeta\sin\pi\eta]f(\tau) \\
\quad = \dfrac{(\lambda b)^4}{A_{22}h^3} q(\zeta, \eta, \tau) + [\lambda(\bar{B}^*_{61} - 2\bar{B}^*_{26})X_1^{(3)}(\zeta)Y_1(\eta) \\
\qquad + \lambda^3(\bar{B}^*_{62} - 2\bar{B}^*_{16})X'_1(\zeta)Y_1^{(3)}(\eta)]f^*(\tau) - (\lambda\pi)^2[(X_1(\zeta)Y''_1(\eta) \\
\qquad + X''_1(\zeta)Y_1(\eta))\sin\pi\zeta\sin\pi\eta + 2X'_1(\zeta)Y'_1(\eta)\cos\pi\zeta\cos\pi\eta]f^*(\tau)f(\tau)
\end{cases}
\tag{8}
$$

Using Galerkin's method and taking $\sin \pi\zeta \cdot \sin \pi\eta$ as a weighting function, eqns (8) are weighted average all over the plate. Calculation and arrangement yield respectively:

1. The solution analysis form of the compatible equation of the forced vibration in the nonlinear system of the AUAPLP under the principle of Galerkin's method when taking $\sin \pi\zeta \cdot \sin \pi\eta$ as a weighting function

$$f^*(\tau) = -\frac{4}{3} \frac{(\lambda\pi)^2}{\alpha_1^4(\bar{A}_{22}^* + \lambda^4 \bar{A}_{11}^*)C_1(\alpha_1, v_1) + \alpha_1^4 \lambda^2(2\bar{A}_{12}^* + \bar{A}_{66}^*)C_2(\alpha_1, v_1)} f^2(\tau) \tag{9}$$

simplifies to

$$f^*(\tau) = C^*(\alpha_1, v_1)f^*(\tau) \tag{10}$$

in which

$$\begin{cases} C_1(\alpha_1, v_1) = \left[\frac{\pi(1 + ch\alpha_1)}{\alpha_1^2 + \pi^2} + \frac{1 - \cos(\alpha_1 - \pi)}{2(\alpha_1 - \pi)} + \frac{\cos(\alpha_1 + \pi) - 1}{2(\alpha_1 + \pi)} \right. \\ \qquad\qquad \left. - \frac{v_1 \pi sh\alpha_1}{\alpha_1^2 + \pi^2} + \frac{v_1 \sin(\alpha_1 - \pi)}{2(\alpha_1 - \pi)} - \frac{v_1 \sin(\alpha_1 + \pi)}{2(\alpha_1 + \pi)} \right]^2 \\ C_2(\alpha_1, v_1) = \left[\frac{\pi(1 + ch\alpha_1)}{\alpha_1^2 + \pi^2} + \frac{\cos(\alpha_1 - \pi) - 1}{2(\alpha_1 - \pi)} + \frac{1 - \cos(\alpha_1 + \pi)}{2(\alpha_1 + \pi)} \right. \\ \qquad\qquad \left. - \frac{v_1 \pi sh\alpha_1}{\alpha_1^2 + \pi^2} - \frac{v_1 \sin(\alpha_1 - \pi)}{2(\alpha_1 - \pi)} + \frac{v_1 \sin(\alpha_1 + \pi)}{2(\alpha_1 + \pi)} \right]^2 \end{cases} \tag{11}$$

2. The solution analysis form of the Duffing equation of the forced vibration, in the nonlinear system of the AUAPLP under the principle of Galerkin's method, when taking $\sin \pi\zeta \cdot \sin \pi\eta$ as a weighting function (the load is $q(\zeta, \eta, \tau) = q_0 \cos(\theta(\rho/A_{22}h^2)^{1/2}b^2)\tau)$:

$$f''(\tau) + \left(\frac{\pi}{\lambda}\right)^4 [\bar{D}_{11}^* + 2\lambda^2(\bar{D}_{12}^* + 2\bar{D}_{66}^*) + \lambda^4 \bar{D}_{22}^*]f(\tau)$$

$$- 4\left(\frac{\alpha_1}{\lambda}\right)^4 [\lambda(\bar{B}_{61}^* - 2\bar{B}_{26}^*) + \lambda^3(\bar{B}_{62}^* - 2\bar{B}_{16}^*)]$$

$$\times C_3(\alpha_1, v_1)C^*(\alpha_1, v_1)f^2(\tau)$$

$$+ 4\left(\frac{\alpha_1\pi}{\lambda}\right)^2 C^*(\alpha_1, v_1)C^{**}(\alpha_1, v_1)f^3(\tau)$$

$$= \frac{16q_0 b^4}{A_{22}\pi^2 h^3} \cos\left(\theta\left(\frac{\rho}{A_{22}h^2}\right)^{1/2} b^2\right)\tau \tag{12}$$

in which

$$
\left\{
\begin{aligned}
C_3(\alpha_1, v_1) &= \left[\left(\frac{\pi sh\alpha_1}{\alpha_1^2 + \pi^2} - \frac{v_1\pi(1 + ch\alpha_1)}{\alpha_1^2 + \pi^2} \right)^2 - \left(\frac{\sin(\alpha_1 + \pi)}{2(\alpha_1 + \pi)} - \frac{\sin(\alpha_1 - \pi)}{2(\alpha_1 - \pi)} \right. \right. \\
&\quad \left. \left. + \frac{v_1(1 - \cos(\alpha_1 - \pi))}{2(\alpha_1 - \pi)} - \frac{v_1(1 - \cos(\alpha_1 + \pi))}{2(\alpha_1 + \pi)} \right)^2 \right] \\
C_4(\alpha_1, v_1) &= \left[\left(\frac{sh\alpha_1}{2\alpha_1} - \frac{\alpha_1 sh\alpha_1}{2(\alpha_1^2 + 4\pi^2)} + \frac{v_1(1 - ch\alpha_1)}{2\alpha_1} - \frac{\alpha_1 v_1(1 - ch\alpha_1)}{2(\alpha_1^2 + 4\pi^2)} \right)^2 \right. \\
&\quad - \left(\frac{\sin\alpha_1}{2\alpha_1} - \frac{\sin(\alpha_1 - 2\pi)}{4(\alpha_1 - 2\pi)} - \frac{\sin(\alpha_1 + 2\pi)}{4(\alpha_1 + 2\pi)} \right. \\
&\quad \left. \left. - \frac{v_1(1 - \cos\alpha_1)}{2\alpha_1} + \frac{\alpha_1 v_1(1 - \cos\alpha_1)}{2(\alpha_1^2 - 4\pi^2)} \right)^2 \right] \\
C_5(\alpha_1, v_1) &= \left[-\frac{2\pi sh\alpha_1}{\alpha_1^2 + 4\pi^2} + \frac{\sin(\alpha_1 - 2\pi)}{2(\alpha_1 - 2\pi)} - \frac{\sin(\alpha_1 + 2\pi)}{2(\alpha_1 + 2\pi)} + \frac{2\pi v_1 ch\alpha_1}{\alpha_1^2 + 4\pi^2} \right. \\
&\quad \left. - \frac{v_1(1 - \cos(\alpha_1 - 2\pi))}{2(\alpha_1 - 2\pi)} + \frac{v_1(1 - \cos(\alpha_1 + 2\pi))}{2(\alpha_1 + 2\pi)} \right]^2 \\
C^{**}(\alpha_1, v_1) &= 2C_4(\alpha_1, v_1) + \tfrac{1}{2}C_5(\alpha_1, v_1)
\end{aligned}
\right. \tag{13}
$$

So eqn (12) can be simplified as

$$
f''(\tau) + Pf(\tau) + Qf^2(\tau) + Rf^3(\tau) = \frac{16q_0 b^4}{A_{22}\pi^2 h^3} \cos\left(\theta\left(\frac{\rho}{A_{22}h^2}\right)^{1/2} b^2 \right)\tau \tag{14}
$$

where

$$
\left\{
\begin{aligned}
P &= \left(\frac{\pi}{\lambda}\right)^4 [\bar{D}_{11}^* + 2\lambda^2(\bar{D}_{12}^* + 2\bar{D}_{66}^*) + \lambda^4\bar{D}_{22}^*] \\
Q &= -4\left(\frac{\alpha_1}{\lambda}\right)^4 [\lambda(\bar{B}_{61}^* - 2\bar{B}_{26}^*) + \lambda^3(\bar{B}_{62}^* - 2\bar{B}_{16}^*)]C_3(\alpha_1, v_1)C^*(\alpha_1, v_1) \\
R &= 4\left(\frac{\alpha_1\pi}{\lambda}\right)^2 C^*(\alpha_1, v_1)C^{**}(\alpha_1, v_1)
\end{aligned}
\right. \tag{15}
$$

Using $t = b^2(\rho/A_{22}h^2)^{1/2}\tau$, a substitution calculation is made of the Duffing eqn (14) and leads to

$$
f_{,tt} + \frac{PA_{22}h^2}{\rho b^4}f + \frac{A_{22}h^2}{\rho b^4}(Qf^2 + Rf^3) = \frac{16q_0}{h\rho\pi^2}\cos\theta t \tag{16}
$$

From eqn (16) the linear natural frequency, ω_0, in the vibration system of the laminated plates is obtained

$$\omega_0^2 = \frac{PA_{22}h^2}{\rho b^4} = \frac{A_{22}h^2}{\rho b^4}\left(\frac{\pi}{\lambda}\right)^4 [\bar{D}_{11}^* + 2\lambda^2(\bar{D}_{12}^* + 2\bar{D}_{66}^*) + \lambda^4\bar{D}_{22}^*] \quad (17)$$

Finally, the specific solution analysis form of the Duffing equation is derived

$$f_{,tt} + \omega_0^2 f + \frac{A_{22}h^2}{\rho b^4}(Qf^2 + Rf^3) = \frac{16q_0}{h\rho\pi^2}\cos\theta t \quad (18)$$

in the case of the Duffing equation, assuming a first-order approximate solution to be

$$f = A\cos\theta t \quad (19)$$

Under Galerkin's condition and by way of calculation and arrangement, the result is obtained of the following relationship between the forcing frequency and amplitude of vibration for the NFFV of the AUAPLP

$$\left(\frac{\theta}{\omega_0}\right)^2 = 1 + \frac{3}{4}\frac{A_{22}h^2}{\rho b^4}R\left(\frac{A}{\omega_0}\right)^2 - \frac{16}{\pi^2}\frac{q_0}{h\rho\omega_0^2}\frac{1}{A} \quad (20)$$

or in the form

$$\left(\frac{\theta}{\omega_0}\right)^2 = 1 + \frac{3}{4\pi^4}\frac{\lambda^4 R}{[\bar{D}_{11}^* + 2\lambda^2(\bar{D}_{12}^* + 2\bar{D}_{66}^*) + \lambda^4\bar{D}_{22}^*]}A^2$$

$$- \frac{16}{\pi^6}\frac{\lambda^4}{[\bar{D}_{11}^* + 2\lambda^2(\bar{D}_{12}^* + 2\bar{D}_{66}^*) + \lambda^4\bar{D}_{22}^*]}\frac{q_0b^4}{A_{22}h^3}\frac{1}{A} \quad (21)$$

in which

$$A = f_{max} = W_{max} = \frac{W_{max}}{h} \quad (22)$$

CALCULATION AND CONCLUSION

Isotropic Plate

Here, calculation is made of the isotropic square plate, Poisson's ratio of which is $v = 0.3$, with the four edges simply supported, attempting to derive a corresponding result of the NFFV of the isotropic plate from that of the solution analysis with reference to the NFFV of the AUAPLP. By

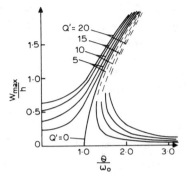

FIG. 1. $v = 0.3$.

comparing the illustrative relationship between amplitude and frequency in Ref. 3 with that obtained here (Fig. 1), it is hoped to prove the reliability of the solution analysis expression used in the AUAPLP in this paper. The comparison shows that although the governing equations of vibration and the way to solve problems are different, the two illustrative relationships, however, coincide in the main (in Fig. 1, the nondimensional load $Q' = q_0 a^4 / E h^4$).

AUAPLP

The unsymmetrically angle-ply laminated (each layer and the axis of the laminated plates criss-cross $+45°/-45°$) square plate here uses graphite–epoxy composites ($E_2/E_1 = 0.025$, $G_{12}/E_1 = 0.015$, $v_{12} = 0.25$) as the material with the boundary condition of the plate four edges S4 simply supported. By computer, the illustrative amplitude-frequency relationship is obtained under different conditions as shown in Figs 2 and 3.

Figure 2(a)–(d) denotes respectively the illustrative amplitude–frequency relationships of the flexural forced vibrations of the AUAPLP with 4 different layers from 2, 4, 6 and 20 under loads of different magnitude (in Fig. 2, the nondimensional load is $Q = q_0 a^4 / A_{22} h^3$).

Figure 3(a)–(e) refers respectively to—when the nondimensional load is $Q = q_0 a^4 / A_{22} h^3 = 0$, 5, 10, 15, 20—the illustrative amplitude–frequency relationships of the flexural forced vibrations of the AUAPLP with different numbers of plying layers.

Discussion and Conclusion

Generally, the first-order approximate equation has yielded the same qualitative result as that of the higher-order approximate equation, so here

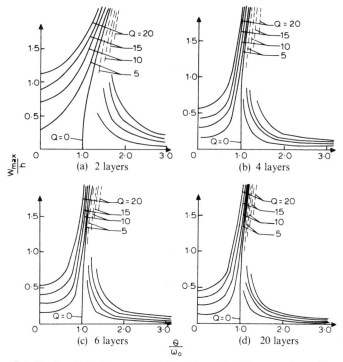

Fɪɢ. 2.   Amplitude–frequency relationships with different numbers of layers.

it is quite enough just to discuss the first-order approximate solution, which is the main harmonic of the whole approximate solution.

As the nonlinear vibration system here is under a simple harmonic force, the main resonance only occurs in the first-order approximate solution and in our calculating examples, the main resonance shows a form of 'jumping'—one jumping to another among the three solutions—in the stable region of the amplitude–frequency relationship curves.

From the amplitude–frequency illustrations of the flexural forced vibration of the AUAPLP in this paper, it is concluded that for a certain given frequency ratio within the stable region, the amplitude decreases rapidly while the anti-deformation increases with the increase of the layers of the plates; for a given amplitude within the weak stable region, however, the frequency ratio decreases and the vibrational system of the composite laminated plates tends towards a lower-order vibration with increase of the layers of the plates.

In addition, from Fig. 3 it is clear that the amplitude–frequency curve of the four-layer plate is rather close to that of the twenty-layer plate.

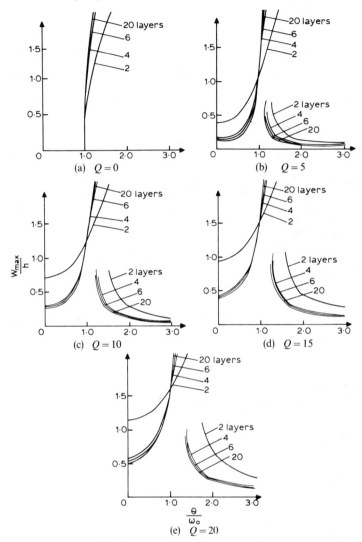

FIG. 3. Amplitude–frequency relationships with different numbers of plies.

Therefore, when the layer number of the composite laminated plate is bigger than or equal to 4, the amplitude–frequency relationship curve becomes stable. In other words, when it is necessary to study the NFFV of the multi-layer composite laminated plates, it is quite enough just to research the four-layer plate and that, to a great degree, aids study of the nonlinear dynamic characteristics of the multi-layer composite plates.

REFERENCES

1. SRIDHAR, S. *et al.*, Part I. Symmetrical response of round plate from the nonlinear resonance of plate for forced response. *J. Sound Vib.* (1975).
2. SATHYAMOORTHY, M., Nonlinear vibrations of plates—a review. *Shock Vib. Dig.* (1983).
3. CHUEN-YUAN CHIA, *Nonlinear Analysis of Plates.* McGraw-Hill, New York, 1980.
4. SATHYAMOORTHY, M., A new method for nonlinear dynamic analysis of composite skew plates. *Shock Vib. Dig.* (1984).
5. SHEBALIN, J. V., Nonlinear dynamics and control of a vibrating rectangular plate. *VA. Rept.* (1983).
6. JONES, R. M., *Mechanics of Composite Material*, translated by Zhu Yiling *et al.* Publishing House of Science and Technology of Shanghai, 1981.
7. QIAN WEICHANG, *Calculus of Variation and Finite Element Method.* Publishing House of Science, China, 1980.
8. NOVOZHINOV, V. V., *Basic Principle of Nonlinear Elasticity*, translated by Zhu Zhaoxiang. Publishing House of Science, China, 1958.
9. Bogolubov, N. N. *et al.*, *Approximate Method of Nonlinear Vibration Theory.* Publishing House of Science and Technology of Shanghai, 1963.
10. WANG LONGFU, *Elastic Theory.* 2nd edn, Publishing House of Science, China, 1984.

11

Free-Vibration and Damping Characterization of Composites

PEARL CHU and OZDEN O. OCHOA

*Mechanical Engineering Department,
Texas A & M University, College Station,
Texas 77843, USA*

ABSTRACT

Many different applications have increased the demand for materials which can reduce fundamental frequencies and obtain high material damping. At present, composite materials offer the widest choice for designing lightweight space structures that can be effectively controlled. This study presents analytical/experimental results for natural frequencies of composite beams and plates. Finite element predictions for the fundamental frequency are checked against the experimental modal analysis results for all specimens. Optimum design parameters are sought for composites in terms of stacking sequence, fiber orientation, aspect ratio and boundary conditions. Effects of damage on damping and vibration is illustrated with the beam specimens.

INTRODUCTION

The increase in designing structures with composite materials can be attributed to their enhanced qualities and to designer's understanding their improved response of each specific use. Naturally, their positive attributes lie in their strength, stiffness, weight reduction, fatigue life, and wear resistance. However, the development of accurate design criteria for the composites' dynamic properties and behavior have become necessary. One of the potential advantages of the structural use of composites is the ability to optimize desired behavior by proper fiber orientations and stacking sequences.[1,2] The information that is available today is inadequate for a

289

designer to choose a composite for a specific application in which the material behavior is defined in advance. The main goal desired through the choice of material is to minimize fundamental frequency or to obtain high material damping. To this end, the present work attempts to identify the characteristics of governing design variables of composites for dynamic response predictions. Even though, statistically, studies of the many different geometries and boundary conditions are needed to determine the dynamic behavior of composites for all applications, the focus will be on the effects of stacking sequences, fiber orientations, and damage on vibrational characteristics.

ANALYSIS

Beams

To determine accurate fundamental frequencies of composites, a shear deformable finite element analysis which includes through-the-thickness

Material:	AS4/3501-6
Layups (Test):	Symmetric $(0_2/\theta/0_2/-\theta)_{2s}$
	Asymmetric $[(0_2/\theta/0_2/-\theta)_s]_2$
(General):	Symmetric $(\pm\theta)_{4s}$
	Asymmetric $(\pm\theta_2)_4$
Material Properties:	$E_{11} = 0.193 \times 10^8$ psi
	$E_{22} = E_{33} = 0.162 \times 10^7$ psi
	$v_{12} = v_{13} = v_{23} = 0.288$
	$G_{12} = G_{23} = G_{13} = 0.102 \times 10^7$ psi

Geometry:

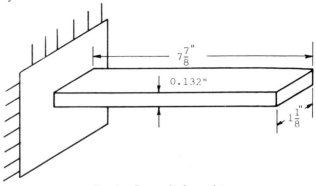

$7\frac{7}{8}$"

0.132"

$1\frac{1}{8}$"

FIG. 1.   Composite beam data.

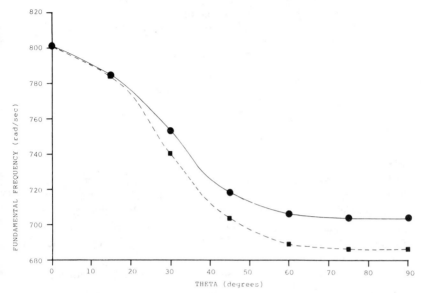

FIG. 2. Fundamental frequency as a function of fiber orientation. —●—, $[0_2/\theta/0_2/-\theta]_{2s}$; —■—, $([0_2/\theta/0_2/-\theta]_s)_2$.

effects is used.[3,4] The details of the composite beam specimens are described in Fig. 1. Two composite beam test layups; namely, symmetric $(0_2/\theta/0_2/-\theta)_{2s}$, and asymmetric $[(0_2/\theta/0_2/-\theta)_s]_2$, of material AS4/3501-6 are studied to determine frequency response to changes in fiber orientation. Data were generated with the FEM analysis with the $\pm\theta$ layers varied from 0 to 90 degrees. Figure 2 shows the results of the data for these test layups of frequency response for both fiber orientation and the comparison for stacking sequence symmetry. This figure illustrates the decrease of fundamental frequency as θ increased, and the consistently higher frequency values for the symmetric layup.

After the test layups were studied, the following angle-ply layups were also analyzed; symmetric $(\pm\theta)_{4s}$, and asymmetric $(\pm\theta_2)4$. The FEM analysis was performed again for these stacking sequences, as composite beams also. The results are presented in Fig. 3. Again, the decrease of fundamental frequency with increasing θ is apparent; however, there is not a significant difference due to symmetry effects.

The effects of varying the sequence of stacking 0, 45, −45 and 90 laminae are illustrated in Table 1. Note the substantial increase in the frequency when the outermost lamina is 0° instead of 90°.

TABLE 1
Quasi-isotropic beam

Stacking sequence	Fundamental frequency (rad/s)
$[\pm 45/0/90]_s$	426·87
$[45/0/-45/90]_s$	523·94
$[0/90/\pm 45]_s$	638·08
$[90/\pm 45/0]_s$	332·35

Plates

The cantilever beam analysis discussed is extended to composite plates. The FEM results are validated for an isotropic case that can be compared to closed-form values. For this purpose, a square plate clamped on all sides is studied. The value determined through the FEM analysis for the fundamental frequency was 2145 rad/s, while the value obtained from classical plate theory was 2178 rad/s.

After the isotropic, clamped plate case was confirmed, test layups, $[\pm 18_n/90_{2n}]_s$, $1 \le n \le 5$, were studied to determine frequency response to changes in stacking sequence and plate geometry. The five

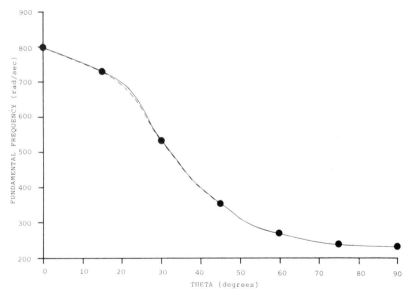

FIG. 3. Fundamental frequency as a function of fiber orientation. —●—, $[\pm\theta]_{4s}$; —■—, $[\pm\theta_2]_4$.

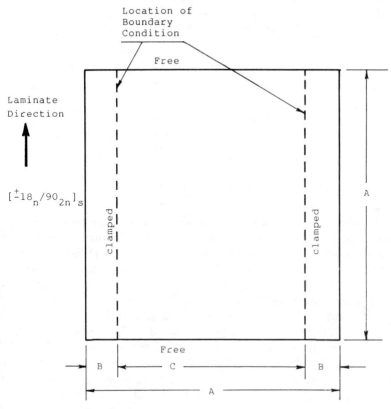

FIG. 4. Composite plate data.

Dimensions

Plate size	A (in)	B ($= 1/8A$) (in)	C (in)	n
2×2 in	2	0·25	1·5	1
4×4 in	4	0·5	3	2
6×6 in	6	0·75	4·5	3
8×8 in	8	1	6	4
10×10 ft	10	1·25	7·5	5

plates analyzed are described in Fig. 4, and the fundamental frequency results are tabulated in Table 2. Note that the scaling effect of the geometry is reflected very well in the magnitudes of the frequency. Only the largest plate, 10 in × 10 in, seems to deviate from the anticipated result of about 600 Hz.

TABLE 2
Composite plate results

Plate size (in)	Fundamental frequency (rad/s)
2 × 2	36 377
4 × 4	18 223
6 × 6	12 163
8 × 8	9 116
10 × 10	6 176

EXPERIMENTS

In addition to the analytical procedures, experimentation has been performed in order to validate the fundamental frequency results. For composite beams, an accelerometer is mounted on the free end. The output is recorded with a spectrum analyzer which generates a frequency–amplitude plot. The resulting graph indicates where natural frequencies occur through peaks in the plot.

Again, an isotropic specimen was first tested to compare the experimental results with closed-form values of the fundamental frequency. After the experimental results were confirmed, the two test $[0_2/\theta/0_2/-\theta]_{2s}$ layups were studied where $\theta = 45°$ on the beam specimens. Three symmetric and three asymmetric test specimens were available, and their initial condition was determined using both X-ray and C-scan procedures. For both the symmetric and asymmetric layups, comparison of the experimental results and FEM analysis showed a difference within 15%.

Damping characteristics are also examined through experimental procedures. Specific damping capacities are calculated with the resulting frequency–amplitude plots using the half-power method.[5] In addition, the prediction of the effects of damage on frequency response and damping characteristics is being examined. One of the symmetric test specimens was increasingly damaged in stages through a three-point bend test. The load and deflection information of each damage stage was plotted during the bend test. After each damage stage, the experimental procedure was re-performed to demonstrate the changes in the frequency and specific damping capacity response. X-rays of the specimen were also taken after each stage to show the amount of induced damage. The fundamental frequency consistently reduced and damping increased after each damage stage as shown in Table 3.

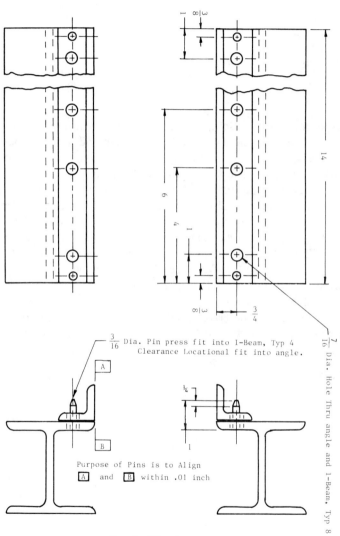

FIG. 5. Vibration test fixture.

TABLE 3
Damage results

	Fundamental frequency (rad/s)	Specific damping capacity
Stage 1	43·36	0·229
Stage 2	39·68	0·263
Stage 3	38·63	0·192
Stage 4	28·14	0·704

Experiments for the square plate geometry have also been performed to analyze both fundamental frequency response and mode shapes. A free-free plate was first set up by placing a 6 in × 6 in composite plate on a large piece of foam. A 9 in × 9 in grid was drawn on the plate, and an accelerometer was mounted on the plate's center point. The plate was then excited using an impulse hammer similar to the cantilever beam procedures. Data were recorded for successive trials where the hammer's impact was initiated at each of the grid's node points. All of these trials were then collectively analyzed to determine mode shapes and fundamental frequencies. For the free-free case, the mode shapes looked as expected from previous isotropic plate experiments.

After the free-free case was completed with the expected results, the clamped plate cases described previously in Fig. 4 were tested. The 6 in × 6 in was clamped on two opposite edges by inserting it between I-beams

Trace A: number 1 (585·064 Hz)
Mode number: 1
Frequency: 585·06 Hz
Damping: 1·61%

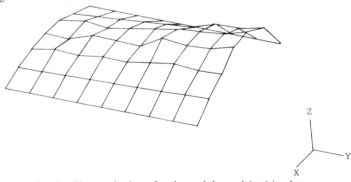

FIG. 6.   First mode shape for clamped-damped 6 × 6 in plate.

Trace A: number 2 (769·378 Hz)
Mode number: 2
Frequency: 769·38 Hz
Damping: 1·57%

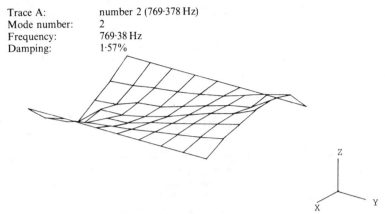

FIG. 7.   Second mode shape of C-C plate.

and angles and bolting to simulate clamped boundary conditions. The apparatus used to clamp the plates is illustrated in Fig. 5. The plate was excited at all grid points with the impulse hammer. The accelerometer was located at the center of the plate. All of the data were recorded, and the frequencies were determined. For the first fundamental frequency, a value of 585 Hz was measured experimentally, compared to 1047 Hz calculated with FEM analysis, approximately a 40% difference. This difference is primarily attributed to the unsuccessful experimental simulation of the clamped boundary conditions. The first three modes of this plate obtained experimentally are illustrated in Figs 6–8.

Trace A: number 3 (1·632K Hz)
Mode number: 3
Frequency: 1·63K Hz
Damping: 1·19%

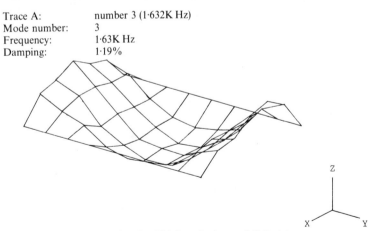

FIG. 8.   Third mode shape of C-C plate.

ACKNOWLEDGEMENTS

The composite plate specimens were provided by Dr R. Bucinell of Hercules Corporation of Magna, Utah and the beam specimens by Drs W. Young and W. Chan of Bell Helicopter Textron, Fort Worth, Texas.

REFERENCES

1. LIAO, D. X., SUNG, C. K. and THOMPSON, B. S., The optimal design of symmetric laminated beams considering damping. *Jnl of Comp. Mater.*, **20** (1986) 485–501.
2. BERT, C. W., Optimal design of a composite-material plate to maximize its fundamental frequency. *Jnl Sound Vib.*, **50**(2) (1977) 229–37.
3. READDY, J. N. and PHAN, N. D., Stability and vibration of isotropic, orthotropic and laminated plates according to a higher-order shear deformation theory. *Jnl Sound Vib.*, **98**(2) (1985) 157–70.
4. OCHOA, O. O., ENGBLOM, J. J. and TUCKER, R., a study of the effects of kinematic and material characteristics on the fundamental frequency calculations of composite plates. *Jnl Sound Vib.*, **101**(2) (1988) 141–8.
5. CRAIG, ROY R. JR, *Structural Dynamics: An Introduction to Computer Methods.* John Wiley, New York, pp. 95–7.
6. WANG, T., *Damping of Laminated Composite Plates.* LTV Aerospace and Defense Presentation.

12

Free Vibration Characteristics of Materially Monoclinic Circular Cylinders

M. Darvizeh

*Department of Mechanical Engineering, Gilan University,
PO Box 401, Rasht, Iran*

and

C. B. Sharma

*Department of Mathematics, University of Manchester (UMIST),
Institute of Science and Technology (UMIST), PO Box 88,
Manchester M60 1QD, UK*

ABSTRACT

The free vibration characteristics (natural frequencies, mode shapes, modal forces and moments) for materially monoclinic cylindrical shells are analysed here using an exact approach. Axial dependence of modal forms is taken in the form of simple Fourier series instead of an exponential dependence used previously. Transverse shear deformation and rotary inertia terms are included in the analysis for a more reliable prediction of response characteristics of such high modulus composite shells. Analytical frequencies obtained from the present study are shown to be in good agreement with some previously published experimental and theoretical results.

NOTATION

a	Shell radius
e_{ij}	Matrix elements
h	Shell thickness
l	Shell length
m	Axial wave number
$M_x, M_\theta, M_{x\theta}$	Moment resultants

299

$\tilde{M}_x^0$ $-\pi M_x(0, \theta)/(l\cos p\theta)$
$\tilde{M}_x^l$ $\pi M_x(l, \theta)/(l\cos p\theta)$
$N_x, N_\theta, N_{x\theta}$ Stress resultants
$\tilde{N}_x^0$ $N_x(0, \theta)/(\cos p\theta)$
$\tilde{N}_x^l$ $N_x(l, \theta)/(\cos p\theta)$
P Circumferential wave number
Q_x Transverse shear force
t Time variable
u, v, w Inplane and radial displacement components
x, θ Axial and circumferential co-ordinates
β_x, β_θ Axial and circumferential rotations respectively
ω Circular natural frequency
Ω Frequency parameter $(= \rho a^2 \omega^2)$

Subscripts and Superscripts
i, j Take the values 1–10
$u, v, w, \beta_x, \beta_\theta$ Denote a variable corresponding to longitudinal modal
               forms
$0, l$ Indicate the values at $x = 0, l$
T   Denotes transpose of a matrix
x, θ Indicate respective directions

INTRODUCTION

Shell type structures in general and their vibration aspects in particular are finding applications in the most diverse branches of modern engineering and technology. The bulk of the literature on shell vibrations deals mainly with homogeneous shells of isotropic material and references to composite laminated shells are rather limited. Some of the most recent references on the latter can be found in an excellent general survey articles by Bert.[1]

In a recent paper Vanderpool and Bert[2] used an iterative but exact approach to calculate resonant vibration frequencies of materially monoclinic, thick wall circular cylindrical shells. The method used is similar to that employed by Forsberg[3] where axial modal dependence is taken in an exponential form. Vanderpool and Bert[2] have obtained theoretical as well as experimental values for natural frequencies. Though the analysis is capable of dealing with arbitrary end conditions a comparison of calculated results with the observed ones was made for the free–free end conditions.

The present paper also provides an exact but iterative approach to analyse various vibration characteristics which include natural frequencies, mode shapes, modal forces and moments for materially monoclinic circular cylinders. Here the axial dependence of modal forms is taken in the form of simple Fourier series in contrast to exponential dependence used in the earlier theoretical works.[2,3] Transverse shear and rotary inertia terms are included in the analysis which are found to be important for the reliable prediction of response characteristics in the case of shells laminated with high modulus composite material.

An appropriate set of modal forms is assumed for the base problem of a freely supported shell with no axial constraints. The derivatives of displacement functions involve the differentiation of Fourier series which is legitimised by Stokes' transformation. This transformation is also essential for releasing those geometric boundary conditions that have been forced due to the nature of the assumed set of modal forms but which violate the actual boundary conditions required for a given problem. The constraints imposed to satisfy the boundary conditions result in a general eigenvalue problem for the base problem. By deleting unwanted rows and columns of the coefficient matrix of the base problem (a direct consequence of the Stokes transformation) one can study the influence of boundary conditions on various vibration characteristics. The ensuing eigenvalue problems are amenable to a simple solution procedure. The interative procedure employed here is on the lines similar to that given earlier.[3] Eigenvalues provide the natural frequencies and the eigenvectors are used to calculate mode shapes, modal forces and moments.

As has been pointed out[2] a materially monoclinic shell has one less plane of symmetry than a specially orthotropic one. It can be characterised by an off-axis helical wound construction where the helices are in a single direction. It provides an extreme practical case because of the complex coupling between surface normal and shear actions. It is this type of materially monoclinic shell which has been chosen to test the validity of the present method and the accuracy of the algebraic manipulations involved. The free vibration characteristics analysed for this case are the natural frequencies, mode shapes, modal forces and moments. A comparison of natural frequencies calculated by the present method is made with those available[2] by calculation and observation and is found to be quite good.

ANALYTICAL FORMULATION

Basic equations given by Love's thin shell theory are in terms of circumferential and axial coordinates θ and x respectively. Details of analysis and the derivation of equations of motion are known[4] along with the expressions for stress and moment resultants. Also since the filaments are wound helically along the shell length, the system is described as materially monoclinic and the stress–strain relations are taken to be the same as before.[2] These relations were obtained first by deriving the components of the stiffness matrix for an orthotropic system and then performing a rotation about the z-axis to yield the stiffness matrix for the materially monoclinic case considered here.

Modal Displacements and Axial Dependence

The displacement and rotation components are assumed to have the following form,[4]

$$u = \psi_u(x)\cos p\theta \qquad v = \psi_v(x)\sin p\theta \qquad w = \psi_w(x)\cos p\theta$$

$$\beta_x = \psi_{\beta_x}(x)\cos p\theta \qquad \beta_\theta = \psi_{\beta_\theta}(x)\sin p\theta \tag{1}$$

Here, p denotes the circumferential wave number and a harmonic time dependence of circular frequency ω has been assumed. In most exact analytical approaches[2,3] functions given by ψ values representing the axial dependence of modal forms are taken to be an exponential form. In this paper these are assumed to be in the form of simple Fourier series on the lines similar to those indicated earlier.[5] The functions $\psi_u(x)$ and $\psi_{\beta_x}(x)$ are given as Fourier cosine series whereas $\psi_v(x)$, $\psi_w(x)$ and $\psi_{\beta_\theta}(x)$ are in the form of Fourier sine series. The fundamental period for all the ψ values is taken to be $2l$.

$$\psi_u(x) = A_{0p} + \sum_{m=1}^{\infty} A_{mp}\cos(m\pi x/l)$$

$$\psi_v(x) = \sum_{m=1}^{\infty} B_{mp}\sin(m\pi x/l)$$

$$\psi_w(x) = \sum_{m=1}^{\infty} C_{mp}\sin(m\pi x/l) \tag{2}$$

$$\psi_{\beta_x}(x) = D_{0p} + \sum_{m=1}^{\infty} D_{mp}\cos(m\pi x/l)$$

$$\psi_{\beta_\theta}(x) = \sum_{m=1}^{\infty} E_{mp}\sin(m\pi x/l)$$

where $A_{mp}, B_{mp}, C_{mp}, D_{mp}, E_{mp}$ are the Fourier coefficients.

It is clear that the Fourier sine series always gives zero values to v, w and β_θ unless one specifies these values[4]

$$v_0 = \psi_v(0) \qquad w_0 = \psi_w(0) \qquad \beta_\theta^0 = \psi_{\beta_\theta}(0)$$
$$v_l = \psi_v(l) \qquad w_l = \psi_w(l) \qquad \beta_\theta^l = \psi_{\beta_\theta}(l) \tag{3}$$

These end values are utilised when one uses Stokes' transformation to differentiate a Fourier series involved in the derivatives of displacement functions.

Derivation of Eigenvalue Problem (Arbitrary End Conditions)

The general formulation given here is capable of handling the vibration problem for a cylindrical shell with arbitrary end conditions. Here, we choose the shell with freely supported ends with no tangential constraint as the base problem. The boundary conditions for such a shell are given by

$$u = N_{x\theta} = Q_x = \beta_x = M_{x\theta} = 0 \quad \text{at} \quad x = 0, l \tag{4}$$

In general, stress resultants are not symmetric, but since we are dealing here with a thin shell, we can assume that $N_{x\theta} = N_{\theta x}$. It is observed that for this base problem none of the ten end conditions given by eqn (4) are satisfied by the assumed mode shapes given by eqns (1) and (2) on a term-by-term basis. Thus, there result constraint conditions to satisfy these boundary conditions by the use of Stokes' transformation. These then lead to the following eigenvalue problem to be solved.

$$(e_{ij})(\tilde{N}_x^0 \tilde{N}_x^l \tilde{M}_x^0 \tilde{M}_x^l v_0 v_l w_0 w_l \beta_\theta^0 \beta_\theta^l)^{\mathrm{T}} = (0) \tag{5}$$

The meanings of these symbols are given in the Notation.

As an illustration and to compare the analytical results with those of Vanderpool and Bert[2] we consider a cylindrical shell with free–free boundary conditions. These are written as follows:

$$N_x = N_{x\theta} = Q_x = M_x = M_{x\theta} = 0 \quad \text{at} \quad x = 0, l \tag{6}$$

It is quite obvious from the assumed modal forms (1), (2) that the tangential displacement v, radial displacement w, and rotation β_θ are identically zero at the ends $x = 0, l$, due to the Fourier sine series expressions in (2). Therefore a releasing procedure is required to remove these unwanted geometrical end conditions. By specifying v and w separately at the two ends and also releasing β_θ at both the ends the relevant eigenvalue problem can be derived from the enforcement of the natural boundary conditions, i.e. forcing $N_{x\theta}$, Q_x and M_x to zero value at both the ends. This is achieved[5] by retaining only the rows and columns associated with the end values v_0, v_l,

w_0, w_l, β_θ^0 and β_θ^l, in the general eigenvalue equation (5) which will now reduce to a simpler form given by

$$(e_{ij})(v_0 v_l w_0 w_l \beta_\theta^0 \beta_\theta^l)^\mathrm{T} = (0) \tag{7}$$

where $i,j = 5, 6, 7, 8, 9, 10$.

The solution of the eigenvalue problem involved follows an approach similar to that given by Forsberg.[3] Starting from some initial estimate for the frequency parameter we can iterate to find its value to a desired accuracy by making the value of the frequency determinant go to zero. The iteration procedure will rapidly converge if good initial estimates are available. The frequency parameters which are calculated easily for some other sets of boundary conditions, are used for this purpose with the signs of increments chosen accordingly.

DISCUSSION OF RESULTS

A single layer monoclinic cylindrical shell is considered here as an example and the results are compared with some available theoretical and experimental results.[2] Physical properties and shell geometrical parameters are the same as given by Vanderpool and Bert.[2]

Frequency Variation

Table 1 gives a comparison of natural frequencies calculated by the present analysis with some previously calculated and observed values.[2] The

TABLE 1

Comparison of theoretical and experimental frequencies (Hz) for a monoclinic cylinder with free–free end conditions

Circumferential wave number (p)	Natural frequencies					
	Axial wave number = 1			Axial wave number = 2		
	Present theory	Previous theory[2]	Experiment[2]	Present theory	Previous theory[2]	Experiment[2]
1	985	—	—	1 775	—	—
2	1 498	1 514	1 490, 1 580, 1 890	2 068	2 117	2 400
3	4 138	4 211	4 100, 4 290	4 205	4 272	3 610
4	6 759	7 869	—	6 831	7 963	—
5	11 131	12 320	10 150	11 341	12 420	11 100

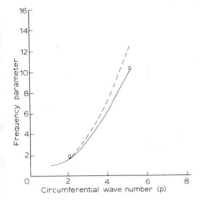

FIG. 1.   Variation of frequency parameter with circumferential wave number for a free–free materially monoclinic cylindrical shell,[2] ($m = 1$; ———, present; - - - -, Ref. 2; $\bigcirc$, experimental).

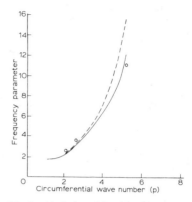

FIG. 2.   Variation of frequency parameter with circumferential wave number for a free–free materially monoclinic cylindrical shell,[2] ($m = 2$; ———, present; - - - -, Ref. 2; $\bigcirc$, experimental).

results correspond to circumferential wave number, $p = 1$ to $p = 5$ and the axial wave number $m = 1, 2$. No results for comparison were available for $p = 1$ for which case the minimum natural frequency is seen to correspond. For other lower circumferential wave numbers ($p = 2, 3$) it is seen that the agreement with the previous results[2] is very good. For the circumferential wave numbers $p = 4, 5$ the results from the present analysis agree well with the experimental results[2] but appear to underestimate the theoretical results.[2] This may be due to various reasons, viz. the way in which the boundary conditions are applied, differences in the use of thin shell theories, etc.

Figures 1 and 2 give a graphical account of the frequency variation with the circumferential wave number, p, and the axial wave number, m, along with the comparison with the corresponding ones given previously.[2] This representation presents a kind of overall picture of frequency variation.

Mode Shapes and Modal Stress Resultants

As an example to illustrate the versatility and the usefulness of the present analysis, mode shapes and modal stress resultants are given in Fig. 3 for a particular case. This gives a kind of quantitative variation of these variables. Since there can be no unique basis to discuss the mode shapes and modal forces we have chosen the maximum radial deflection amplitude to be unity for our purpose here. The graphs exhibit mode shapes and modal forces corresponding to a minimum natural frequency which corresponds

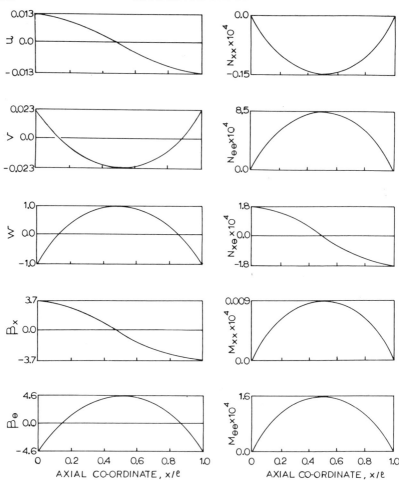

FIG. 3.   Axial variation of mode shapes, modal forces and moments for a free–free materially monoclinic cylinder[2] ($p = 1$, $\Omega = 177\cdot5$); shell geometry: $h/a = 0\cdot16$, $l/a = 3\cdot21$.

to $p = 1$. This is just an illustration and one needs to consider many cases corresponding to various parameters to be able to reach some useful inferences and conclusions.

CONCLUDING REMARKS

The exact analytical method presented in this paper can be used to calculate the free vibration characteristics of thin circular cylindrical shells for an arbitrary choice of end conditions. Although a particular example of a free–free materially monoclinic shell is chosen as an illustration the method can be used for shells with a wide range of shell material and geometrical parameters. Frequency spectrum as well as variation of mode shapes and modal stress resultants can be studied using this method for a desired choice of shell parameters.

REFERENCES

1. BERT, C. W., Advances in dynamics of composite structures. In *Composite Structures 4*, ed. I. H. Marshall. Elsevier Applied Science, London and New York, 1987, pp. 1–17.
2. VANDERPOOL, M. E. and BERT, C. W., Vibration of a materially monoclinic, thick wall circular cylindrical shell. *AIAA J.*, **19** (1981) 634–41.
3. FORSBERG, K., Influence of boundary conditions on the modal characteristics of thin cylindrical shells. *AIAA J.*, **2** (1964) 2150–7.
4. DARVIZEH, M., Free vibration characteristics of orthotropic thin circular cylindrical shells. PhD thesis, University of Manchester, 1986.
5. GREIF, R. and MITTENDORF, S. C., Structural vibration and Fourier series. *J. Sound Vib.*, **48** (1976) 113–22.

13

Resonant Response of Orthotropic Cylindrical Shell Panels

C. B. SHARMA and ISA A. AL-KHAYAT*

*Department of Mathematics,
University of Manchester Institute of Science and Technology (UMIST),
Manchester M60 1QD, UK*

ABSTRACT

An analytical study is carried out to study the vibration characteristics of laminated composite circular cylindrical panels employing Sanders' equations of motion for thin shells. Transverse shear deformation and rotary inertia terms are included in the analysis because of their obvious importance in the case of composite panels. The modal forms are assumed to have the axial dependence in the form of a simple Fourier series. Various aspects studied here involve the effects of change in geometrical parameters of the panel and variation in the directions of orthotropy, etc., on frequencies, mode shapes, modal forces and moments. Analytical results for panels with freely-supported as well as clamped-free curved edges are shown to agree very well with some previous results for the isotropic case.

1 INTRODUCTION

The study of mechanical behaviour of a thin orthotropic circular cylindrical panel is now an established subject. Laminated curved plates form important structural elements in the aerospace industry applications where weight savings are of paramount importance. The advent of advanced fibre-reinforced composite materials such as boron/epoxy and graphite/epoxy has resulted in a dramatic increase in the use of laminated fibre-reinforced plates and other structural components. In the literature

---

*On leave from the Arabian Gulf University, PO Box 26621, Adlya, Bahrain.

one can find approximately a hundred published papers dealing with the vibration of isotropic, homogeneous flat as well as curved plates. Whereas for laminated circular cylindrical panels one can find approximately only about a dozen relevant references. An extensive review of previous work has been given by Leissa.[1] More recent surveys by Leissa[2] and Bert[3] also include a number of papers devoted to vibratory response of such plates. In a recent paper Baharlou and Leissa[4] have developed an approximate procedure to analyse natural frequencies and buckling loads for laminated composite plates having arbitrary edge conditions.

The main aim of the present paper is to study the overall response of thin orthotropic circular cylindrical panels. Transverse shear deformation and rotary inertia terms are included in the analysis because of their importance in the case of composite shells. The modal forms are assumed to have an axial dependence in the form of simple Fourier series[5] rather than the usual exponential axial dependence. Various important aspects of the problem are studied. These include the effects of the inclusion of shear deformation and rotary inertia terms and changes in panel geometrical and material parameters on the frequencies, mode shapes, modal forces and moments. The pattern of behaviour of panel response and associated natural frequencies are studied as a function of changes in the directions of orthotropy. The influence of boundary conditions on the vibration characteristics of composite panels is also included in this study. It is shown that the analytical results obtained by the present analysis are in good agreement with some available theoretical results for an isotropic panel.

2 THEORETICAL CONSIDERATIONS

The governing equations of motion for the free vibration and stress–strain relations including the effect of transverse shear and rotary inertia terms are taken to be the same as given by Hsu *et al.*[6] These are based on the celebrated Sanders' thin shell theory.

2.1 Modal Forms

The displacement and rotation components for an orthotropic circular cylindrical panel can be given by the following relations.

$$u = \psi_u(x) \sin(n\pi\phi/\theta) \sin \omega t \qquad v = \psi_v(x) \cos(n\pi\phi/\theta) \sin \omega t$$

$$w = \psi_w(x) \sin(n\pi\phi/\theta) \sin \omega t \qquad (1)$$

$$\beta_x = \psi_{\beta_x}(x) \sin(n\pi\phi/\theta) \sin \omega t \qquad \beta_\phi = \psi_{\beta_\phi}(x) \cos(n\pi\phi/\theta) \sin \omega t$$

where n is the circumferential wave number and ω is the circular natural frequency. Functions ψ_u, ψ_v, ψ_w, ψ_{β_x}, ψ_{β_ϕ} represent the axial dependence of the respective modal forms.

2.1.1 Axial dependence of modal forms

The functions representing the axial dependence of modal forms are given in the following Fourier series forms:

$$\psi_u(x) = \sum_{m=1}^{\infty} A_{mn} \sin(m\pi x/l)$$

$$\phi_v(x) = B_{0n} + \sum_{m=1}^{\infty} B_{mn} \cos(m\pi x/l)$$

$$\psi_w(x) = C_{0n} + \sum_{m=1}^{\infty} C_{mn} \cos(m\pi x/l) \tag{2}$$

$$\psi_{\beta_x}(x) = \sum_{m=1}^{\infty} D_{mn} \sin(m\pi x/l)$$

$$\psi_{\beta_\phi}(x) = E_{0n} + \sum_{m=1}^{\infty} E_{mn} \cos(m\pi x/l)$$

where A_{mn}, B_{mn}, C_{mn}, D_{mn} and E_{mn} denote the Fourier coefficients. The sine series always give zero values at the end points unless one specifies the affected terms, i.e.

$$\psi_u(0) = u_0 \qquad \psi_u(l) = u_l$$
$$\psi_{\beta_x}(0) = \beta_x^0 \qquad \psi_{\beta_x}(l) = \beta_x^l \tag{3}$$

These values will appear when one uses Stokes' transformation for differentiating Fourier series in the set of displacement functions.

2.2 Derivation of the Eigenvalue Problem

The analysis presented here is capable of analysing the panel vibration characteristics where curved edges can have arbitrary end conditions. In this paper individual problems of panels with freely-supported and clamped-free boundary conditions are deduced as particular cases of ensuring general theoretical considerations. As a base problem for the present investigation a panel with simply-supported ends with no axial constraint along the curved edges is considered. The boundary conditions for such a panel are given as follows:

$$N_x = v = w = M_x = \beta_\phi = 0 \quad \text{at } x = 0, l$$
$$u = w = \beta_x = N_\phi = M_\phi = 0 \quad \text{at } \phi = 0, \theta \tag{4}$$

One can notice that for this problem none of the ten boundary conditions at the curved edges are satisfied by the assumed modal displacement forms (2) on a term-by-term basis. The constraint conditions to satisfy these boundary conditions will lead to the following eigenvalue problem

$$[f_{ij}][\tilde{N}^0_{x\phi}, \tilde{N}^l_{x\phi}, \tilde{Q}^0_x, \tilde{Q}^l_x, \tilde{M}^0_{x\phi}, \tilde{M}^l_{x\phi}, u_0, u_l, \beta^0_x, \beta^l_x] = [O] \quad (i,j = 1, 2, \ldots, 10) \tag{5}$$

2.2.1 FSNT–FSNT end conditions

The boundary conditions for the panel freely supported with no tangential constraint at the curved edges while the straight edges are kept simply supported are given as follows:

$$u = \hat{N}_{x\phi} = \hat{Q}_x = \beta_x = M_{x\phi} = 0 \quad \text{at } x = 0, l$$
$$u = w = \beta_x = N_\phi = M_\phi = 0 \quad \text{at } \phi = 0, \theta \tag{6}$$

It is clear that the modal displacement forms (2) satisfy the above boundary condition on a term-by-term basis. The frequency equation for the FSNT panel is directly obtained as a five by five determinantal equation from the equations of motion.

The elements of this frequency determinant are no longer an infinite series but reduce to single algebraic terms. By solving this frequency equation all the frequencies are given by the eigenvalues. The corresponding eigenvectors are then used to calculate the mode shapes, modal forces and moments for single layer curved panels.

2.2.2 Clamped-free boundary conditions

The non-symmetric set of boundary conditions of clamped-free curved edges panel have the form:

$$u = v = w = \beta_x = \beta_\phi = 0 \quad \text{at } x = 0$$
$$N_x = \hat{N}_{x\phi} = \hat{Q}_x = M_x = M_{x\phi} = 0 \quad \text{at } x = l \tag{7}$$

The conditions $v = 0$, $w = 0$ and $\beta_\phi = 0$ at the end $x = 0$ must be enforced by the release of the quantities $N^0_{x\phi}$, Q^0_x and $M^0_{x\phi}$. At $x = l$ the conditions

$N_x = M_x = 0$ must be enforced with the release of end displacement u_l and rotation β_x^l. The retaining of the rows and columns in the general eigenvalue equation (5) associated with the boundary values $\tilde{N}_{x\phi}^0$, $\tilde{Q}_x^0$, $\tilde{M}_{x\phi}^0$, u_l, β_x^l leads to the following eigenvalue problem,

$$[f_{i,j}][\tilde{N}_{x\phi}^0, \tilde{Q}_x^0, \tilde{M}_{x\phi}^0, u_l, \beta_l^x]^T = [O] \qquad (i,j = 1, 3, 5, 8, 10) \qquad (8)$$

Frequencies are given by the eigenvalues of the problem. To solve the determinantal frequency equation we adopt an iterative procedure. Initial estimates in this case of a panel with clamped-free curved edges are provided by the corresponding values from the analysis of the freely-supported case. This way the number of steps required for convergence is kept down. Once the eigenvalues are found eigenvectors are calculated and utilised in evaluating the mode shapes, modal forces and moments.

3 DISCUSSION OF RESULTS

As an example, a single-layer orthotropic circular cylindrical panel is considered in this paper where the elastic constants and geometry are given by

$$C_{22} = 2.068\,427\,1 \times 10^{10}\,\mathrm{N\,m}^{-2}$$
$$C_{11} = 10.9 C_{22}$$
$$C_{12} = 2.757\,902\,8 \times 10^9\,\mathrm{N\,m}^{-2}$$
$$C_{44} = C_{55} = C_{66} = 7.956\,549\,578 \times 10^9\,\mathrm{N\,m}^{-2}$$
$$l = 0.2794\,\mathrm{m}$$
$$b = 0.2286\,\mathrm{m}$$
$$h = 0.12 \times 10^{-2}\,\mathrm{m}$$
$$R = 2.4384\,\mathrm{m},\ 1.2192\,\mathrm{m}$$

All the figures presented herein are drawn for the axial wave number, $m = 1$.

3.1 Freely-Supported Panel with no Tangential Constraint
3.1.1 Single-layer orthotropic panels

The effect of orthotropy on the natural frequencies is shown in Figs 1 and 2. These figures include a panel with five different orthotropies corresponding to exaggerated axial stiffness ($C_{11}/C_{22} = 10.9$), moderate axial stiffness ($C_{11}/C_{22} = 3.3$), isotropic case ($C_{11}/C_{22} = 1$), as well as moderate meridional stiffness ($C_{22}/C_{11} = 3.3$), and exaggerated meridional stiffness ($C_{22}/C_{11} = 10.9$). Figures 1 and 2 correspond to $R = 2.4384\,\mathrm{m}$ and

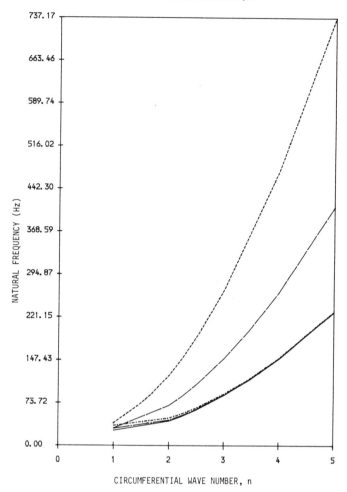

FIG. 1.   Frequency variation with circumferential wave number n for a circular cylindrical panel with FSNT curved edges and variable orthotropy: $-\cdot-\cdot-$, $C_{11}/C_{22} \simeq 10\cdot9$; $-\cdot-$, $C_{11}/C_{22} \simeq 3\cdot3$; $\underline{\quad\quad}$, $C_{11}/C_{22} = 1$; $--$, $C_{22}/C_{11} \simeq 3\cdot3$; $----$, $C_{22}/C_{11} \simeq 10\cdot9$; panel geometry: $l = 0\cdot2794$ m, $b = 0\cdot2286$ m, $h = 0\cdot12 \times 10^{-2}$ m, $R = 2\cdot438$ m.

$1\cdot2192$ m. It is obvious that as C_{11} decreases from its exaggerated value through the moderate to the isotropic case the values of the natural frequencies go up for $n \geq 1$. But for higher values of circumferential wave number, the values of the natural frequency for the three cases converge to each other. By looking at the frequency curves corresponding to panels with moderately and highly exaggerated meridional stiffness one finds that

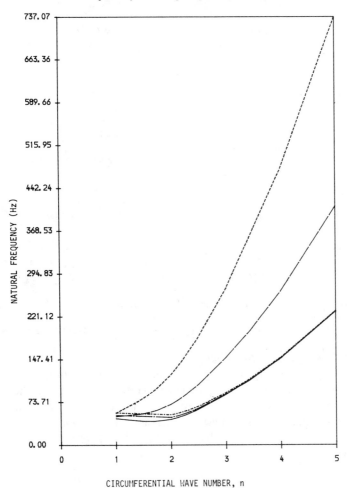

FIG. 2. Frequency variation with circumferential wave number n for a circular cylindrical panel with FSNT curved edges and variable orthotropy: $-\cdot\cdot-$, $C_{11}/C_{22} \simeq 10{\cdot}9$; $-\cdot-$, $C_{11}/C_{22} = 3{\cdot}3$; $---$, $C_{11}/C_{22} = 1$; $--$, $C_{22}/C_{11} \simeq 3{\cdot}3$; $----$, $C_{22}/C_{11} \simeq 10{\cdot}9$; panel geometry: $l = 0{\cdot}2794$ m, $b = 0{\cdot}2286$ m, $h = 0{\cdot}12 \times 10^{-2}$ m, $R = 1{\cdot}2192$ m.

as n, the circumferential wave number, increases, the values of natural frequency are greatly increased. By reducing the value of R from $2{\cdot}4384$ m to $1{\cdot}2192$ m (Figs 1 and 2) there is a noticeable increase in the values of the natural frequencies for $n \leq 2$. For $n > 2$ the natural frequencies converge to each other, i.e. this variation in radius does not affect the values of frequency parameter in this range.

TABLE 1

The effect of shear deformation and rotary inertia terms in the natural frequencies of an isotropic rectangular panel where $l = 0.2794\,m$, $b = 0.2286\,m$, $R = 2.4384\,m$, $h = 0.12 \times 10^{-2}\,m$

(a) *Freely-supported curved ends (m = 1)*

	n						
	1	2	3	4	5	6	7
A	162·914	266·568	543·700	936·620	1 442·234	2 060·274	2 790·700
B	162·907	266·504	543·426	935·804	1 440·300	2 056·330	2 783·469

(b) *Clamped-free curved ends (m = 1)*

	n						
	1	2	3	4	5	6	7
A	87·787	232·196	509·828	900·324	1 404·602	2 022·659	2 753·091
B	87·780	232·137	509·566	899·536	1 402·764	2 018·852	2 746·048

A: Frequency values without shear deformation and rotary inertia terms in the analysis.
B: Frequency values with shear deformation and rotary inertia terms in the analysis.

3.1.2 Single-layer isotropic panel

To check the validity of the present method a comparison with a panel having isotropic material is studied in Table 1(a). It is clear that the effect is very small for lower values of *n*. Whereas for higher values of circumferential wave number *n*, the effect is noticeable where it is shown to lower progressively the values of the natural frequency parameter moderately.

3.1.3 Mode shapes, modal forces and moments

Mode shapes, modal forces and moments of a single-layer orthotropic panel are shown in Fig. 3 (a–j). Since there can be no unique basis for this kind of modal form estimation, we assume the maximum radial deflection amplitude to be unity. Comparison with other available work was not possible since such data are unavailable.

3.2 Clamped-Free Circular Cylindrical Panel

The free vibration characteristics of a panel with clamped-free curved edges are analysed next. Geometrical as well as material properties involved are taken to be the same as in the previous section.

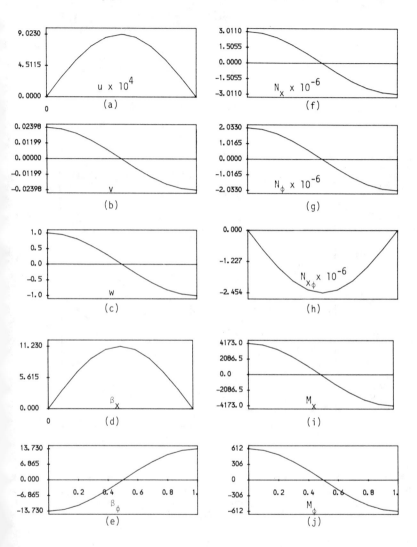

AXIAL COORDINATE, x/ℓ

FIG. 3. Axial variation of mode shapes, modal forces and moments of an orthotropic freely-supported curved circular cylindrical panel ($n = 1$, $\omega = 217$, $C_{11}/C_{22} \simeq 10\cdot9$); panel geometry: $l = 0\cdot2794$ m, $b = 0\cdot2286$ m, $h = 0\cdot12 \times 10^{-2}$ m, $R = 2\cdot4384$ m.

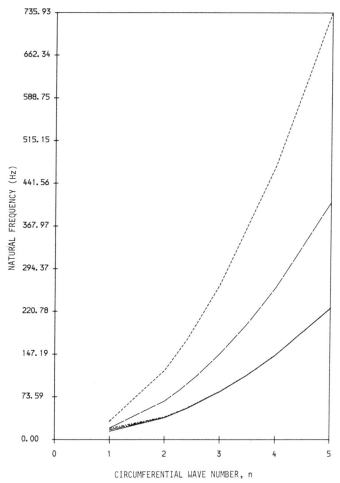

FIG. 4.   Frequency variation with circumferential wave number n for a circular cylindrical panel with clamped-free curved edges and variable orthotropy: — $\cdot$ —, $C_{11}/C_{22} \simeq 10\cdot9$; — $\cdot$ —, $C_{11}/C_{22} \simeq 3\cdot3$; ——, $C_{11}/C_{22} = 1$; – –, $C_{22}/C_{11} \simeq 3\cdot3$; - - - -, $C_{22}/C_{11} \simeq 10\cdot9$; panel geometry: $l = 0\cdot2794$ m, $b = 0\cdot2286$ m, $h = 0\cdot12 \times 10^{-2}$ m, $R = 2\cdot4384$ m.

3.2.1 Frequency spectrum

Variation of natural frequencies with respect to circumferential wave number n is given in Figs 4 and 5 for $R = 2\cdot4384$ m and $R = 1\cdot2192$ m respectively. It is observed from Fig. 4 that for low n values the natural frequency values for exaggerated axial stiffness $(C_{11}/C_{22} = 10\cdot9)$ are higher than the corresponding ones for moderate axial stiffness $(C_{11}/C_{22} = 3\cdot3)$

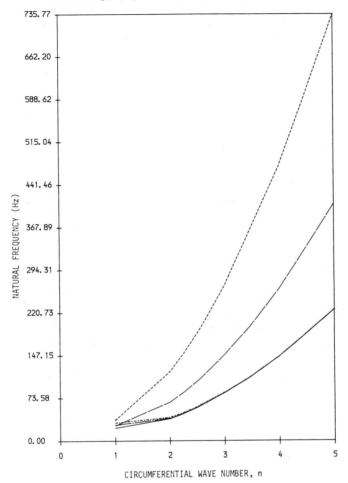

FIG. 5. Frequency variation with circumferential wave number n for a circular cylindrical panel with clamped-free curved edges and variable orthotropy: $-\cdot-\cdot-$, $C_{11}/C_{22} \simeq 10\cdot9$; $-\cdot-$, $C_{11}/C_{22} \simeq 3\cdot3$; ———, $C_{11}/C_{22} = 1$; $--$, $C_{22}/C_{11} \simeq 3\cdot3$; $----$, $C_{22}/C_{11} \simeq 10\cdot9$; panel geometry: $l = 0\cdot2794$ m, $b = 0\cdot2286$, $h = 0\cdot12 \times 10^{-2}$ m, $R = 1\cdot2192$ m.

and the isotropic case ($C_{11}/C_{22} = 1$). As the values of n increase the natural frequencies for the three different cases converge towards each other. The effect of variation in meridional stiffness on natural frequencies is observed by going through the moderate and exaggerated values of C_{22}/C_{11}. It is found that the natural frequencies for these two cases are close to those for isotropic panels as well as the ones with exaggerated axial stiffness for $n = 1$.

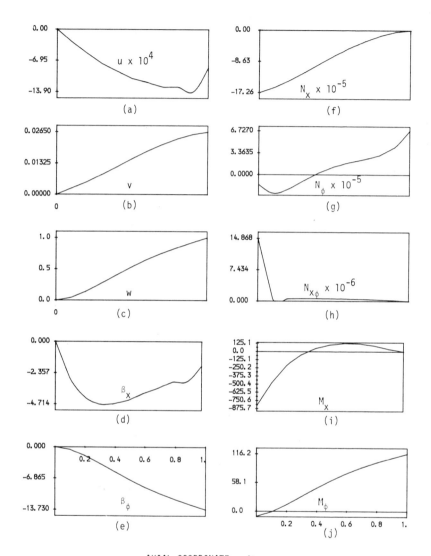

AXIAL COORDINATE, x/ℓ

FIG. 6. Axial variation of mode shapes, modal forces and moments of an orthotropic clamped-free curved circular cylindrical panel ($n = 1$, $\omega = 122$, $C_{11}/C_{22} \simeq 10{\cdot}9$); panel geometry: $l = 0{\cdot}2794$ m, $b = 0{\cdot}2286$ m, $h = 0{\cdot}12 \times 10^{-2}$ m, $R = 2{\cdot}4384$ m.

But as n increases, the values of natural frequencies increase for this case and become much higher than the other two cases. Frequencies in this case also overestimate the corresponding ones for the orthotropic panels considered in the previous section. Table 1(b) shows the effect of including shear deformation and rotary inertia terms. As expected the effect is obvious for higher values of n and hardly noticeable for low values of n.

3.2.2 Mode shapes, modal forces and moments

Mode shapes, modal forces and moments for a single-layer orthotropic clamped-free circular cylindrical panel are presented in Fig. 6 (a–j) and correspond to the minimum values of frequency parameter. The geometrical and material properties are the same as those for the previous case. Once again the modal forms are normalised with respect to the radial displacement w, whose normalised value is taken to be unity. It is observed that there is a significant change in the pattern of mode shapes, modal forces and moments, as compared to the FSNT case as can be expected.

4 CONCLUDING REMARKS

A general solution procedure is presented herein to investigate the free vibration characteristics of single-layer orthotropic circular cylindrical panels. The analysis used here includes the effect of shear deformation and rotary inertia terms. Frequency pattern is shown to be significantly influenced by the change in the direction of orthotropy and change in boundary conditions. Two different cases are obtained by changing the axes of orthotropy of a circular cylindrical panel, one with exaggerated axial stiffness and the other with exaggerated meridional stiffness. Results are given only for some particular cases but the approach can be utilised to calculate the required vibration characteristics of panels with varying geometrical and material parameters. Influence of other sets of boundary conditions can also be studied by using the approach presented herein.

REFERENCES

1. Leissa, A. W., *Vibration of Shells*. NASA SP 288, 1973.
2. Leissa, A. W., Recent studies in plate vibration: 1981–1985. *The Shock and Vibration Digest*, **19** (1987) 10–24.
3. Bert, C. W., Recent advances in dynamics of composite structures. *Composite Structures 4*, **2** (1987) 1–17.

4. BAHARLOU, B. and LEISSA, A. W., Vibration and buckling of generally laminated composite plates with arbitrary edge conditions. *Int. J. of Mech. Sci.*, **29** (1987) 545–55.
5. GREIF, R. and MITTENDORF, S. C., Structural vibrations and Fourier series, *Jnl of Sound and Vibration*, **48** (1976) 113–22.
6. HSU, Y. S., REDDY, J. N. and BERT, C. W., Thermoelasticity of circular cylindrical shells laminated of bimodulus composite materials. *Jnl of Therm. Stress.*, **4** (1981) 155–77.

14

Effect of Impactor Shape on Residual Tensile Strength and Tensile Failure of Carbon/Epoxy Laminates

M. M. STEVANOVIĆ, T. B. STECENKO

Institute of Nuclear Science 'Boris Kidrič'-Vinča,
POB 522, 11001 Belgrade, Yugoslavia

M. C. KOSTIĆ

Lola Institute, POB 802, 11001 Belgrade, Yugoslavia

and

D. B. BRIŠKI-GUDIĆ

Aeronautical Institute, Žarkovo, 11001 Belgrade, Yugoslavia

ABSTRACT

Carbon/epoxy laminate of ($\pm 45/0_2/\pm 45/0/90$)$_{3s}$ lay-up has been subjected prior to a tensile test, to low-velocity drop weight impacts with two impactors of different shape. From tensile tests of unimpacted and impacted coupons residual tensile strengths are determined. Tensile failure of impacted and unimpacted coupons has been studied by observation on optical and scanning electron microscopes. On the basis of these studies an attempt is made to explain the influence of impact with different impactor shapes on tensile failure and tensile strength of tested laminates.

INTRODUCTION

Mechanical performance after impact of carbon/epoxy laminates for structural application, especially of those used for aircraft components, is of particular interest. The impact during manufacture (e.g. from dropped tools) or during service (from different objects or various mass and shape) is the most important source of a wide range of defects and damage which

323

may significantly reduce residual strength. In the previous studies of this problem[1-5] special attention was paid to the impact damage nature, size and extent,[1,2] and to its influence on the residual compressive[2,3] or tensile strength,[3-5] as well as on the fatigue behaviour of carbon/plastic laminates.[5]

This paper discusses the experiments that addressed the effect of impactor shape on residual tensile strength as well as on tensile failure of impacted carbon/epoxy laminates. Low-velocity (energy) impact conditions have been chosen on the way that two used impactors did not produce barely visible impact damage on laminate surfaces. Residual strength after impact was determined in tensile tests, and failure developed in samples during impact and tensile loading was studied by microscopic observation of fracture surfaces.

EXPERIMENTAL

The experiments were carried out on samples cut from plates of $(\pm 45/0_5/\pm 45/0/90)_{3s}$ lay-up. Laminate plates were obtained by curing (600 kPa, 177°C) of unidirectional carbon fibre reinforced epoxy resin prepreg supplied by Hexcel (Belgium). The prepreg reinforcement was Torayca T300 high-strength carbon fibres, while the matrix was made of Hexcel F263 epoxy resin system. The test specimens had the configuration shown in Fig. 1. Tested laminate had density 1500 kg/m³, fibre content 52·0%v and pore content lower than 1%v.

Prior to tensile test, specimens were subjected to the low-velocity impact. The drop weight impacts have been performed at the same energy level (impact energy 21·3 J, drop height 1·267 m, impactor mass 1·71 kg, impactor diameter 0·015 m) with two impactors of different contact shapes: spherical and blunt-tipped.

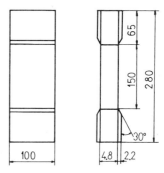

FIG. 1. Test specimen configuration.

TABLE 1
Laminate strength values

Virgin strength (MPa)	Residual strength (MPa)	
	Spherical impactor	Blunt-tipped impactor
515·2 ± 12·5	436·1 ± 45·3	447·1 ± 58·2

During impact the laminate specimens had as supports their tabs and the impactor dropped on the middle point of the gauge length of the test coupon.

The tensile tests were performed with an Amsler Testing Machine of 1000 kN capacity, utilising wedge action grips. From tensile tests of unimpacted and impacted coupons the virgin and residual strength values of tested laminates were determined (Table 1).

Damage which developed in tested unimpacted and impacted samples was studied by observation of tensile failure surfaces on an optical (Figs 2 and 3) as well as on a scanning electron microscope (Figs 4–8). For SEM studies fractured surfaces of failed samples were coated by vacuum evaporation with films of Au–Pd alloys and graphite. Fractured surfaces were examined on a Jeol scanning electron microscope (model JSM-35).

DISCUSSION

From determined tensile strength mean values of unimpacted and impacted samples (Table 1) we deduced strength decreases of tested laminates of 13·2 and 15·3% due to impacts with blunt-tipped and spherical impactors respectively.

These established decreases have to be considered as identical, having in mind the deviations of tensile strength results from their mean values. It indicates that used impactor shapes have no influence on residual strength of examined laminate.

It was also noted that deviation of strength results from mean value or coefficient of variation as a measure of this deviation is much higher in cases of impacted than of unimpacted samples.

By microscopic observation of failure on an optical (at low magnifications) and on a scanning electron microscope we found characteristic failure features in the unimpacted then in the impacted samples and tried to deduce from them the impact influence on failure and mechanical behaviour of examined laminate.

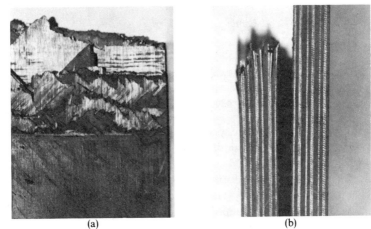

(a)   (b)

FIG. 2.   Tensile failure of unimpacted coupon: (a) transverse fracture; (b) axial fracture. Original magnification × 2·5.

The contour of an overall transverse fracture in unimpacted coupons after tensile tests is of an indented shape (Fig. 2(a)). Transverse fracture across 90° ply is situated mainly in the matrix. Transverse fracture through ± 45° plies meanders down 45° and − 45° directions. By shear failure in 1–2 plane of these plies (Fig. 4(b)) and cutting the fibres under 90° angle the fracture propagates from one ply to another (Fig. 5(a)). Transverse fracture

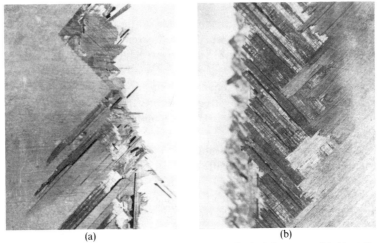

(a)   (b)

FIG. 3.   Tensile failure of impacted coupons: (a) spherical impactor; (b) blunt-tipped impactor. Original magnification × 5.

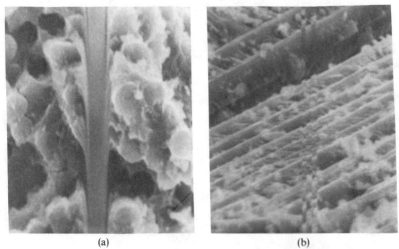

(a)   (b)

FIG. 4.   Transverse fracture of unimpacted coupon: (a) 0° ply; (b) 45° ply. Original
magnification × 1500.

in 0° ply (Fig. 4(a)) follows the fracture in neighbouring ±45° plies (Figs 2 and 5(a)). It is because the failure appears first in ±45° plies and then in 0° plies. Inclined transverse fracture in 0° ply induced the important contribution of pull-out failure mechanism (Fig. 4(a)).

All the unimpacted coupons from the tensile tests have a main axial decohesion going through the 90° midplane ply (Fig. 2(b)). Besides the axial

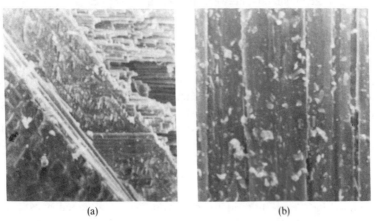

(a)   (b)

FIG. 5.   Axial fracture of unimpacted coupon: (a) interlayer decohesions; (b) main axial
decohesion. Original magnification: (a) × 200; (b) × 600.

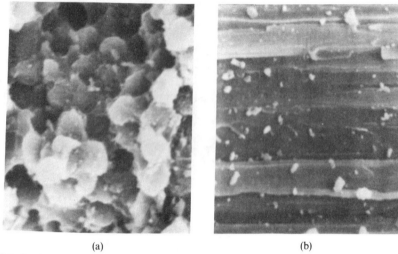

(a)   (b)

FIG. 6.   Transverse fracture of coupon impacted by spherical impactor; (a) 0° ply; (b) 90° ply.
Original magnification × 1500.

component of unimpacted coupons' tensile failure is made of short distance delaminations in the region of transverse fracture (Figs 3 and 5(a)). These delaminations arise between all the present plies as well as in the 0° ply (Fig. 4(a)). The main axial intralaminar crack extending through the matrix of the 90° midplane ply (Fig. 5(b)) breaks the fibres of this ply and traverses into the interply surface between the 90° and 0° plies.

The tensile failure of both groups of impacted (by spherical and blunt-ipped impactors) samples (Fig. 3) is similar to that of unimpacted coupons; ontours of overall macrofracture are practically identical. However, the ilure details, i.e. the microstructure of the failure surfaces, especially of al ones, are somewhat different. Impacted samples possess the same sile failure components as unimpacted ones, but the density of particular ks is higher in impacted samples.

ie profile of impacted sample transverse fracture (Fig. 3) has the same nts as that of unimpacted ones (Fig. 2). Nevertheless, the impacted ns show on the external 45° plies the numerous cracks extending transverse fracture in the fibre direction (Fig. 3).

icrographs of transverse fracture surfaces in present plies of impacted ples (Figs 6 and 7) are identical to those of unimpacted ones (Fig. 4). ne half of the examined impacted coupons have had a big axial crack ugh the 90° midplane ply. Its microfractograph is identical to that of an

Fig. 7. Transverse fracture of coupon impacted by blunt-tipped impactor: (a) all the present plies; (b) 45° ply. Original magnification: (a) × 300; (b) × 1500.

Fig. 8. Axial fracture of coupon impacted by spherical impactor: (a) decohesion between 0° and 45° plies; (b) decohesion between 90° and 0° plies. Original magnification × 750.

(a)   (b)

FIG. 9.   Axial fracture of coupon impacted by blunt-tipped impactor: (a) decohesion between
45° and −45° plies; (b) decohesion between 90° and 0° plies. Original magnification ×750.

unimpacted coupon (Fig. 5(b)). Other numerous axial decohesions in
impacted samples are always interlayer ones (Figs 7(a), 8 and 9). They are
created by crack propagation through the fibre–matrix interface, so they
show micrographs characteristic for interlaminar shear failure (Figs 8 and
9). In unimpacted samples the axial decohesions are located in the matrix
mainly, and they do not show many characteristic figures in the matrix. This
is due to the fact that the axial decohesions in the impacted samples are
initiated by the impact. The impact is the reason for intense delaminations
in the region of transverse failure as well as for dislocation of some plies
from the primary orientation, all of this by induction of stresses and by
creation of cracks which disturb the symmetrical lay-up of the tested
laminate.

The main transverse failure of impacted coupons was not necessarily
located on the impact place. This is due to the large area of impact damage
because of the fact that during the impact tests the laminates to be impacted
had as supports their tabs, not the basic steel plate with a hole in front of the
impactor target.

CONCLUSIONS

Residual tensile strength and tensile failure of carbon fibre/epoxy resin
laminates, subjected prior to tensile tests to impact with two different

impactor contact shapes, have been studied. The results indicate that impactor shapes used have no influence on the residual tensile strength.

The tensile failure of samples impacted by spherical and blunt-tipped impactors is very similar. Contours of their overall macrofracture are practically identical to that of unimpacted coupons. However, the failure details, i.e. microstructure of failure surfaces, especially of axial ones in impacted and unimpacted samples, are different.

From strength results and performed microscopic observations of tensile failure it follows that impact influence on residual tensile strength as well as on tensile failure of tested laminates is by induction of internal stresses in interfaces and interlayers, by initiation of interface and interlayer cracks, as well as by breakage of some fibres in load-bearing plies, all of this in large enough areas around the impacted place.

REFERENCES

1. LABOR, J. D., Impact damage effects on the strength of advanced composites. In *Non-destructive Evaluation and Flaw Criticality for Composite Materials*, ASTM STP 696, 1979, pp. 172–84.
2. JONES, R. and BAKER, A. A., Compressive impact damaged graphite/epoxy laminates. In *Composite Structures—3*, ed. I. H. Marshall. Elsevier Applied Science, London, 1985, pp. 402–15.
3. BISHOP, S. M. and DOERY, G., Effect of damage on the tensile and compressive performance of carbon fibre laminates. AGARD, Conference on Characterisation Analysis and Significance of Defects in Composite Materials, AGARD-CP-355, 1983.
4. BISHOP, S. M., The mechanical and impact performance of advanced carbon-fiber reinforced plastics. In *Developments in the Science and Technology of Composite Materials*, ECCM, First Conference on Composite Materials and Exhibition, Bordeaux, 1985, pp. 222–7.
5. RAMKUMAR, R. L., Effect of low velocity impact damage on the fatigue behavior of graphite/epoxy laminates. In *Long-term Behavior of Composites*, ASTM STP, ed. T. K. O'Brien. American Society for Testing and Materials, Philadelphia, 1983, pp. 116–35.

15

Transverse Impact of Filament-Wound Pipes

K. L. AINSWORTH and K. E. EVANS

*Department of Materials Science and Engineering, University of Liverpool,
PO Box 147, Liverpool L69 3BX, UK*

ABSTRACT

Glass fibre reinforced filament-wound pipes have been tested for their transverse impact properties. Tests have been conducted at velocities up to 30 m s⁻¹ using both full-length simply-supported and end-cradled pipe support conditions. Measurements have been made of residual mechanical properties after impact. Good correlation has been found between the effects of impact and equivalent pipes tested under quasi-static load conditions up to equivalent damage energies. Finite element analysis has also been used to model the deformation of the pipe under load and good agreement has been achieved.

INTRODUCTION

Filament-wound GRP pipes are used in applications, such as in the off-shore oil industry, for their high strength to weight ratio, corrosion resistance and ease of handling.[1] One important area of practical interest is the degree of incidental damage due to low velocity impacts (up to 30 m s⁻¹) that such pipes can withstand without losing integrity or suffering fluid leakage. Low velocity impacts are of importance since they arise from situations such as accidental dropping of the pipe itself or of hand tools which lead to localised areas of damage that are often difficult to detect and can cause premature failure.[2,3] It is, therefore, important to assess the extent of such damage and its effect on pipe properties. This work examines

333

this problem and considers the possibility of correlations between quasi-static and impact tests. Such correlations would allow simple static tests to be used to predict impact damage performance. Finite element modelling techniques have also been used to examine the detailed stress and strain distributions in the pipes under loading.

EXPERIMENTAL METHODS

The specimens under test were complete pipes of length 500 mm, external diameter 162 mm and wall thickness 6 mm. They were filament-wound with a winding angle of ± 55° from E-glass with an epoxy resin matrix (Epikote 828).

Two loading geometries were used for both static and impact tests. These were (a) simply- (or flat) supported along the pipe length or (b) simply-supported in 81 mm radius semicircular cradles (50 mm wide) at the pipe ends. In both cases the load was applied by a solid cylinder of length 150 mm and diameter 25 mm applied at the mid-point transverse to the axis of the pipe. The two geometries were used to model support conditions found in realistic pipe applications.

The following tests have been conducted.

(a) Static Tests

All static tests were performed on an Instron 1185 mechanical testing machine set up in compression mode with a crosshead speed of 5 mm/min. The two support conditions will be considered separately.

(i) Flat support static tests

Tests were initially carried out to full damage to establish the major features of the load/displacement curves. Once this had been achieved, a series of tests to points along the load/displacement curve were carried out to observe the delamination patterns obtained at various stages on the load/displacement curve and to relate the effects of varying amounts of damage to residual properties.

From the load/displacement curve it is possible to obtain peak load, deflection to peak load, initial gradient, and energy both to peak load and to test end. The latter is used as a basis for performing equivalent impact tests, which will be discussed later.

(ii) Cradled static tests

These were conducted with the pipe supported in semicircular cradles, so that the test geometry resembled a three-point bend with a 400-mm span. Again the intention is to consider the development of damage throughout the load/displacement curve.

(b) Impact Tests

All impact tests to date were performed on a large drop hammer rig using a 38·65-kg tup with a maximum drop height of 8 m. Both multiple and single bounce test results are available. The majority of results presented here are for multiple bounce tests. As before, the two support conditions are considered separately.

(i) Flat support impact tests

These were conducted to mirror the static tests by using the static energy to test end to determine the required drop height. In order to obtain impact data which would be comparable with static data (e.g. peak load, initial gradient, etc.), the first bounce of the multiple bounce impacts was instrumented using laser-doppler velocimetry.[4] This provides a direct method of measuring velocity (and hence acceleration) as a function of time. From this plots can be obtained, for each specimen, of velocity/time, energy/time, force/time, energy/displacement and force/displacement.

(ii) Cradled impact tests

Again tests were performed equivalent to the static tests in energy. A full range of impact energies have not yet been completed.

(c) Examination of the Effects of Damage

Two main methods were used to examine the effects of the damage caused by the static and impact tests, one to observe the extent of damage, the other to quantify its effects on pipe performance.

(i) Photograph techniques

This is a very simple technique which consists of inserting a strong light source inside the pipe to examine the damaged surfaces. This causes the delamination patterns formed to be very clearly illuminated and allows quantitative measurements of the extent of the damage to be made. The patterns were recorded photographically for further analysis.

(ii) Residual property tests

After the amount of damage to each specimen had been recorded, all specimens, whether initially statically tested or impacted, were retested on the Instron 1185. Using the same geometries as in the original tests, the pipes were retested to failure to measure their residual properties. Finally, the pipes were re-examined using the backlighting technique to see what effects the retests had on delamination growth.

EXPERIMENTAL RESULTS

(a) Static Tests

(i) Flat supported static tests

Figure 1 shows an example of the static load/displacement curve for the flat supported specimens. The curve has two distinct regions, indicating a two-part failure process. The first region represents elastic behaviour up to a first, discernible failure point. After this the load continues to increase but at a slower rate, marked by a less steep gradient with progressive damage until a second major failure occurs. Although the test is accompanied by much tearing, it is not catastrophic and, despite the huge deformations involved (up to 55 mm), the pipe recovers nearly all of its original shape. Numerical data extracted from the load/displacement curves are presented in Table 1.

On inspection after test the damage can be seen to be very localised. The basic upper surface pattern consists of an indentation mark (or crush crater), gel coat cracking and, for tests to full damage, regions where the tup nose entering has caused the resin to shear away. This damage is of a similar nature for all the tests conducted with the only real difference being the degree of severity. The delamination patterns are discussed later.

(ii) Cradled static tests

The first thing to note is that despite the change of geometry the failure process follows the same two-part pattern as in the flat cases, although it

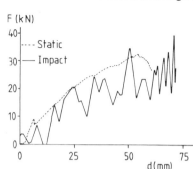

FIG. 1. Comparison of impact and static force (F)/displacement (d) curves for specimens IT_{10} and T_{10}.

TABLE 1
Static flat supported data

Specimen number	When test stopped	Peak load (kN)	Deflection (mm)	Gradient (N/mm × 10³)	Energy (J)	Energy to test end (J)
P_1	First failure	7·28	6·50	1·450	—	20·84
P_3	15 kN	15·00	20·25	0·688	—	167·65
P_5	20 kN	20·00	25·00	0·720	267·10	370·98
T_{10}	Full damage	32·50	54·25	0·789	1 064·48	1 246·73
T_{11}	Full damage	27·30	40·75	0·871	652·26	763·60
T_{12}	Full damage	31·90	55·25	1·081	1 066·32	1 121·46

occurs at a much lower load (see Table 2). There are two interesting features to note. The first is that first failure occurs at around the same peak load (6·4 kN as opposed to 6·8 kN) in both the impact and static cases. The second is that although the cradled pipes have a higher gradient initially after the first failure the gradients agree rather well. The significance of these points will be discussed later.

One important thing to note is that under test the close-fitting cradles do not appear to allow significant whole-pipe flexure. On inspection the damage sustained in the tests is similar to that seen in the flat tests only less severe.

(b) Impact Tests

(i) Flat supported impact tests

Figure 1 shows an example of the impact load/deflection curve for flat specimens. It should be noted that similar curves were obtained for all five specimens. The impact curve follows a similar pattern to its equivalent static curve, i.e. an overall increase in force (F) with displacement (d), but also includes a large vibratory effect caused by specimen oscillation. It

TABLE 2
Cradled static data

Specimen number	Peak load (kN)	Deflection (mm)	Gradient (N/mm × 10³)	Energy (J)	Energy to test end (J)
C_1	20·50	32·00	0·870	429·24	620·79
C_2	18·70	26·00	0·752	300·78	374·30
C_3	20·75	37·00	0·830	520·81	571·98

TABLE 3

Impact data

Specimen number	Drop height (m)	Velocity (m/s)	Tup mass (kg)	Peak load (kN)	Peak load deflection (mm)	Peak load energy (J)	Max. deflection (mm)	Tup energy (J)
IP_1	0·06	1·08	38·65	30·47	5·44	17·64	6·00	22·75
IP_3	0·44	3·04	38·65	26·00	17·78	149·15	19·63	166·84
IP_5	0·98	4·45	38·65	48·24	26·07	374·72	26·07	371·60
IT_{11}	2·01	6·25	38·65	34·94	52·74	744·76	52·74	762·16
IT_{10}	3·30	7·89	38·65	39·35	71·11	1 170·73	71·85	1 251·31

appears that, for the full damage curve presented, the static test has a higher gradient (i.e. stiffness) than the impact test. Impact data obtained are presented in Table 3. Once more, despite large deflections observed in the tests (over 70 mm), almost all the deformation is recoverable.

On inspection after the tests the damage is very similar to that previously obtained only more severe than even the flat static case.

(ii) Cradled impact tests

As has already been stated, the only results available so far for this test type arise from instrumented tests to full damage. The same type of damage pattern is obtained for the cradled impact tests as for the cradled static tests, i.e. less severe than in the flat pipe case.

(c) Examination of the Effects of Damage
(i) Photographic technique

The power of this very simple technique can be seen by reference to Figs 2 and 3. Figure 2 shows how the delamination pattern builds up. The example given is for the flat supported static case, but the damage builds up in the same way for both static and impact tests in either of the two support conditions considered. Figure 2(a) shows the damage after the first failure. Here the local crush indentation mark is visible and delamination has been

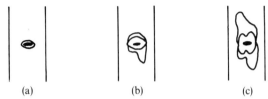

(a)   (b)   (c)

FIG. 2.   Schematic representation of delamination development. (a) Specimen after initial damage; (b) specimen after loading to 15 kN; (c) specimen at test end.

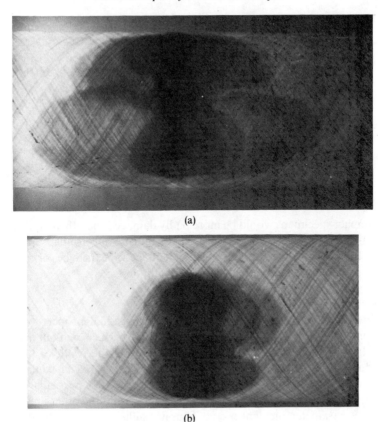

FIG. 3. Photograph of pipe delamination obtained by backlighting for (a) flat supported pipe (IT_{11}) and (b) cradled pipe (IC_2).

initiated. Figure 2(b) shows damage at 15 kN, which is half the load to failure. Further delaminations have occurred and these continue to grow until failure is reached, indicated by Fig. 2(c). It is possible to distinguish different layers delaminating by this method alone.

Although the process is identical for all four test types, the degree of severity and actual delamination patterns do vary, as can be seen in Fig. 3. This compares full damage impacts on flat supported pipes (Fig. 3(a)) and cradled pipes (Fig. 3(b)). It can be clearly seen that the extent of delamination in the flat supported pipe is much greater than that in the cradled pipe. This observation also extends to static tests (see Table 4). It was originally expected that, due to multiple bounce tests being performed,

TABLE 4
Residual properties for static flat supported tests

Specimen number	Delaminated area (m²)	Peak load (kN)	Gradient (N/mm × 10³)	Deflection (mm)	Energy (J)
P_1	0·007 1	17·85	0·671	30·50	311·17
P_3	0·031 1	18·05	0·548	30·25	264·82
P_5	0·110 8	19·50	0·360	33·50	282·46
T_{10}	0·222 3	17·80	0·235	58·38	450·48
T_{11}	0·149 1	18·50	0·303	51·50	462·04
T_{12}	0·214 5	18·60	0·225	54·00	403·30

the impact specimens would be more badly damaged than their equivalent quasi-static tests. However, this is seen to be the case only for flat supported tests. In the cradled pipe cases the area of damage appears to be very similar whether the pipes were impacted or statically tested.

(ii) Residual property tests

The numerical information obtained from residual property tests is presented in Table 4. So far this analysis has only been carried out in detail for the static flat supported tests. Figure 4 shows the result of plotting residual gradient (S_r) and deflection (d_r) against the area of damage (A) caused by the initial tests. It should be noted that in this particular case a linear relationship exists between area of damage and the other important damage parameter, incident energy. Thus it is sufficient to consider the area of damage plots alone since incident energy plots would be of the same form.

One final thing to note is the effect of these retests on the delamination growth. In the pipes originally partially damaged, new delaminations are formed upon retest. If, however, pipes originally taken to full damage are examined after retest, no new delaminations are revealed.

ANALYTICAL AND COMPUTER MODELLING

Modelling of the experimental situation up to first failure (i.e. to the end of elastic behaviour) was undertaken for both flat and cradled specimens to obtain detailed stress and strain distributions in the pipe. It is hoped that it will be possible to correlate these with the observed damage patterns.

TABLE 5
Comparison of modelling with experiment

Test condition	Method of displacement	Deformation	Displacement (mm)
Unsupported	Experimental	P_1	6·00
		P_3	5·50
		T_{11}	5·50
		T_{12}	6·50
	Analytical solution		3·40
	Computer solution		4·87
Cradled	Experimental	C_1	3·88
		C_2	3·50
		C_3	3·75
	Computer solution		3·68

(a) Modelling the Flat Supported Pipe

For an initial comparison with experimental work a simple analytical model[5] was considered. This consisted of a system where two equal and opposite point loads were applied at the pipe mid-point, allowing the deflection δ under the load P to be calculated as shown below:

$$\delta = 6\cdot5 P/Et(R/t)^{3/2}(L/R)^{-3/4}$$

where P is the load to first failure (6·75 kN in this case), E is the modulus of elasticity taken as 25 GPa, t the wall thickness, R the inner pipe radius and L the pipe length. The displacement predicted is shown in Table 5.

The second form of modelling was by finite element analysis using the NISA2 package.[6] The elements used were four-noded laminate shell elements and the material properties are listed in Table 6. The model pipe

TABLE 6
Unidirectional material properties for FEM

Material property	Value for composite layers	Value for resin layers
Longitudinal Young's modulus	$E_x = 3\cdot86 \times 10^{10}$ Pa	$E_x = 4 \times 10^9$ Pa
Transverse Young's modulus	$E_y = 8\cdot27 \times 10^9$ Pa	$E_y = 4 \times 10^9$ Pa
Poisson ratio	$v_{xy} = 2\cdot60 \times 10^{-1}$	$v_{xy} = 3\cdot30 \times 10^{-1}$
	G_{xy}	G_{xy}
Shear moduli	$G_{yz} = 4\cdot14 \times 10^9$ Pa	$G_{yz} = 1\cdot5 \times 10^9$ Pa
	G_{xz}	G_{xz}

constructed was 500 mm long with an external diameter of 162 mm and a wall thickness of 6 mm. Isotropic gel coat layers 1 mm thick were used with the remaining 4 mm thickness constructed as composite layers 0·5 mm thick and oriented at ± 55°. The model was then restrained along its bottom line in x and y to model the support conditions. Additionally, the top centre node was restrained in x and z to prevent rotation. The next step was to apply the load to first failure (6·75 kN) to the model. In order to recreate the experimental situation as closely as possible, this was initially applied as a line load in the x direction (i.e. transverse to the pipe axis). Then, to simulate the circular shape of the nose, the load was further subdivided parabolically in the z direction. The model was run successfully and the displacement predicted is shown in Table 5, giving excellent agreement with experiment.

(b) Modelling the Cradled Pipe

In this case there is no analytical situation so only finite element modelling was undertaken. The same model was used as in the flat pipe case with the following modifications. Firstly, the restraint conditions were altered to simulate the cradles. Secondly, the load applied was altered to 6·39 kN, the value of first failure for the cradled pipes. This model was also successfully run and the displacement predicted is shown in Table 5, again in good agreement with experiment.

DISCUSSION

(a) Static Tests

The first thing to note is that complete pipe sections exhibit a two-part failure process independent of support conditions. An initial elastic region is followed by failure due to resin gel coat yielding and delamination initiation. The curve then continues with a less steep gradient with progressive delamination and crush damage. No final catastrophic failure process is observed since the remainder of the pipe continues to support a load even after major damage, well beyond any practical design limit, such as a weepage criterion, has occurred.

Comparing the two support conditions used, the cradled pipes, surprisingly, are initially stiffer, resulting in less energy being required to arrive at the load necessary for delamination initiation. After this has occurred, all pipes are of similar stiffness but the cradled specimens fail at lower loads for less deflection and with less energy. This is believed to be due to the cradles restricting the pipe transverse deformation.

(b) Impact Tests

The oscillatory behaviour of pipes under impact is caused by a combination of a dynamic elastic response and the process of failure. At low velocities it is known that elastic behaviour dominates and so, in an attempt to predict this dynamic behaviour more quantitatively, a simple mass-spring model was employed.[7] This regards the pipe as a massless spring being impacted by a mass m, in this case the 38·65-kg tup. From this approximation the force F at any time t can be calculated using

$$F = V_o \sqrt{mk} \, \sin{(\sqrt{k/m} \; t)}$$

where V_o is the impact velocity and k the pipe stiffness obtained from the average gradient of the force/displacement plot. Reference to Table 7 shows that considering how simple the model is, it gives a very good first approximation to behaviour characterised by the maximum force and the half period, in three of the five cases. In fact, the model only breaks down at very low velocities (i.e. specimen IP_1). It should be noted that IP_5 has an uneven cross-section and this is the probable reason for poor prediction here.

(c) Examination of the Effects of Damage

The simple backlighting technique used here has proved very effective. It allows the delamination patterns to be seen very clearly and, since it only requires a strong light source, is thus a technique which can easily be used in the field. The information yielded by this technique has allowed the load/deflection curve to be more fully interpreted, e.g. it is now known that first failure is the delamination initiation point (Fig. 2(a)). The technique allows delamination patterns to be compared for the static and impact cases, and this can be quantified using image analysis techniques.

TABLE 7
Predictions from the mass-spring model

Specimen number	Impact velocity (m/s)	Pipe stiffness (N/m × 10⁶)	Maximum force (N) Predicted	Maximum force (N) Actual	Half period (ms) Predicted	Half period (ms) Actual
IP_1	1·08	2·80	11 200	30 470	11·65	14·00
IP_3	3·04	1·47	22 850	26 000	16·10	14·87
IP_5	4·45	1·87	37 900	48 240	14·35	10·62
IT_{11}	6·25	0·66	31 600	34 940	24·10	22·20
IT_{10}	7·89	0·58	37 300	39 350	25·70	21·74

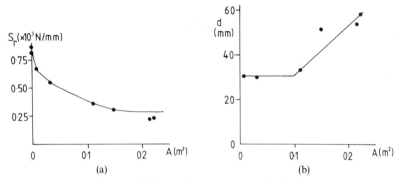

FIG. 4. Variation of (a) residual gradient and (b) residual displacement with area delamination for statically tested pipes.

The other form of damage analysis considered was measurement of residual properties and, in the static flat supported case, good correlations are found between area of damage and both residual gradient and residual deflection. Figure 4(a) shows that pipe stiffness falls continuously with increasing area of damage whilst Fig. 4(b) reveals a damage threshold for deformation. Below an area of $0.09 \, \text{m}^2$, increasing the area of damage has no effect on the residual deflection, i.e. the pipes are essentially undamaged. Above it, increasing the area of damage results in a higher deformation to failure. This ties in with the observation that delamination growth, which requires increasing deformations, is occurring in this region. Another point which should be noted is the effect of first failure. Observation of the pipe at this point reveals very little superficial damage but reference to Table 4 reveals that it has a significant effect on residual properties, especially peak load.

(d) Analytical and Computer Modelling

This has proved very successful (Table 5). The analytical solution, considering how simple it is, provides a very good first approximation in the flat supported case and finite element modelling has improved on this. The agreement is especially good in the case of the cradled pipe where no analytical techniques are available.

CONCLUSION

The work so far completed on transverse pipe deformation has shown good correlations between static and impact tests. It is possible to correlate the area of damage to the resultant residual mechanical properties. Finite element analysis has successfully modelled the deformation of the pipes in both the flat and cradled support geometries and is being used to relate the overall stress and strain states of the pipes to the delamination envelopes.

Further work is in progress to examine single bounce impact tests so as to provide a quantitative correlation between identical energy quasi-static and impact tests. Further mass-spring models are also being used to examine oscillation effects under impact.

ACKNOWLEDGEMENTS

The authors wish to acknowledge the support of Shell Research Ltd and the SERC through the provision of a CASE award studentship. In particular, the authors wish to acknowledge the assistance of Paul Marks at Shell Research Ltd, Thornton.

REFERENCES

1. COOPER, L. T., Development, manufacture and applications of glass reinforced plastic pipes. *Anti-corrosion Methods and Materials*, **25** (1978) 3–10.
2. CAPRINO, G., CRIVELLI VISCONTI, I. and DI ILIO, A., Composite materials response under low velocity impact. *Comp. Struct.*, **2** (1984) 261–71.
3. SHARMA, A. V., Low velocity impact tests on fibrous composite sandwich structures. In *Test Methods and Design Allowables for Fibrous Composites*, ASTM STP 734, ed. C. C. Chamis. American Society for Testing and Materials, Philadelphia, 1981, pp. 54–70.
4. HODGKINSON, J. M., VLACHOS, N. S., WHITELAW, J. H. and WILLIAMS, J. G., Drop weight impact tests with the use of laser-doppler velocimetry. *Proc. R. Soc.*, **A379** (1982) 133–47.
5. ROARK, R. and YOUNG, L., *Formulas for Stress and Strain* (1975), Example 8, p. 495.
6. NISA2, Engineering Mechanics Research Corporation, 1707 W. Big Beaver, Troy, MI 48084, USA.
7. SJÖBLOM, P. O., HARTNESS, J. T. and CORDELL, T. M., On low-velocity impact testing of composite materials. *J. Comp. Mater.*, **22** (Jan. 1988) 30–52.

16

Dynamic and Impact Behaviour of Composite Plates

C. DOAN and P. HAMELIN

Centre d'Etudes et de Recherches sur les Matériaux Composites—CERMAC, INSA, Université Claude Bernard (UCB) Bât 304, 20, Avenue Albert Einstein, 69621 Villeurbanne Cedex, France

ABSTRACT

An experimental analysis of dynamic and impact tests applied on composite laminated plates using polymer matrices and glass fibres is presented. For each plate for different conditions of loading (energy level, side and geometry of impactor) and for different dynamic conditions of targets (viscoelastic, orthotropic properties; edge's conditions; damage...), the relationships between load and deflection can be drawn by using accelerometers, optical methods for measuring the force and the displacements. This allows analysis of failure modes and crack propagation. The vibration and indentation tests are necessary to understand well the impact behaviour of composite plates. A method of predicting the dynamic and impact behaviour of composite plates is proposed. It must be noted that the impact response of composite structures depends on their mechanical properties, especially edge conditions and viscoelasticity. A good agreement between experimental data and theoretical values is obtained.

INTRODUCTION

With the increase in application of composite materials to structures, especially aeronautical, the necessity of studying their behaviour under impact conditions has grown. For example, aircraft can be subjected to two different ranges of impact: high-velocity (in service: hail, birds, etc.) and low-velocity (in ground service: dropping tools, etc.). Both ranges are important.

In general, the analytical modelling of impact phenomena is greatly complicated by structural effects. In this work, an attempt to predict the impact response of composite plates is presented. At first the modelling of the dynamic behaviour of viscoelastic composite plates is studied, because of the importance of structural effects.

An experimental analysis of the problem is shown. The effects of both pressure on edges and damage are investigated.

EXPERIMENTAL TEST

Impact Machine

In order to understand the impact response of composite plates a simple machine is used. Specimens (A) are rectangular plates of 160–388 mm with clamped boundaries. Impact is provided by a drop-impactor (B), of different heads. This means the shape and the mass of the head can be varied (Fig. 1). A cylinder (E) with holes was used to localise the impact and vary its velocity.

During the impact, the signals from the accelerometer (H) clamped on top of the impactor, are being registered as a function of time. The acceleration–time data allows force–time plotting. Double integration of

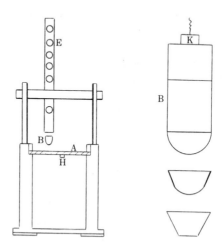

FIG. 1.   Set-up for measuring the impact response of plates.

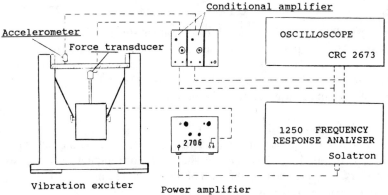

FIG. 2. Scheme for measuring the plate dynamic characteristics.

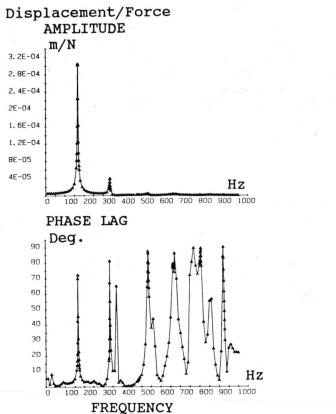

FIG. 3. Dynamic characters of viscoelastic plates versus frequency.

the test data provides central displacement–time data. Another accelerometer (H) can be placed to measure the deflection in another point of the plate. The end of the impact was defined as being the moment when the impactor leaves the plate; this is when the force is zero. The deflection at the end of the impact was found to be non-zero.

Dynamic Measurement

The impact response of plates depends on their dynamic behaviour, which is dictated by their natural shapes. Before impact testing it was necessary to determine the vibration of the plates. The method of vibration measurement is shown in Fig. 2.

The specimens were of the same type as in the impact test. The plates were given a sinusoidal load by the vibration exciter. This was commanded by the frequency response analyser. The electrical signals of the force and deflection from the accelerometer, and the force transducer are transmitted to the conditional amplifiers and then they go to the frequency response analyser. By the use of the FFT (Fast Fourier Transport) processor, the analyser gave immediately the force or the deflection magnitudes, their ratio and their phase lag.

Typical variations of force/deflection amplitude and phase lag with frequency are shown in Fig. 3. The peaks show natural frequencies.

ANALYSIS

The mechanical behaviour of composite plates subjected to a concentrated load is studied in two stages; at first analysis of the free vibrations of plates, and secondly, determination of the contact force.

Dynamic Behaviour of Visoelastic Composite Plates

After the viscoelastic-elastic principle of correspondence,[8-10] we have the stiffness matrix complex as follows:

$$[Q^*]^{(1,2)} = \begin{bmatrix} \dfrac{E_1^*}{1 - v_{12}^* v_{21}^*} & \dfrac{E_1^* v_{21}}{1 - v_{12}^* v_{21}^*} & 0 \\ \dfrac{E_2^* v_{12}^*}{1 - v_{12}^* v_{21}^*} & \dfrac{E_2^*}{1 - v_{12}^* v_{21}^*} & 0 \\ 0 & 0 & G^* \end{bmatrix} \tag{1}$$

The constitutive relation as the starting point of the method can be written based upon the well known classic laminated plate theory (Whitney and Pagano[14]) as follows:

$$\begin{Bmatrix} N^* \\ M^* \end{Bmatrix} = \begin{bmatrix} A^* & B^* \\ B^* & D^* \end{bmatrix} \begin{Bmatrix} \varepsilon^* \\ K^* \end{Bmatrix} \tag{2}$$

where the stiffness reduced matrices are defined as:

$$(A^*, B^*, D^*) = \sum_k [Q^*]_k^{(x,y)}(h_1, h_2, H_3) \tag{3}$$

with

$$h_1 = z_{k+1} - z_k \qquad h_2 = \tfrac{1}{2}(z_{k+1}^2 - z_k^2) \qquad h_3 = \tfrac{1}{3}(z_{k+1}^3 - z_k^3) \tag{4}$$

The stiffness matrix in each lamina is obtained from its viscoelastic modulus. The latter can be predicted from the dynamic properties of the constituent materials (Doan et al.[3,4]).

Equation of Motion of Laminates

Let q_j be a set of observed variables (displacements) and Q_i be a set of generalised forces defined by the following virtual work (L) condition:

$$Q_i = \partial L / \partial q_i \tag{5}$$

From the Lagrange equation of motion

$$\frac{\mathrm{d}}{\mathrm{d}t}\left(\frac{\partial T}{\partial \dot{q}_i}\right) - \frac{\partial T}{\partial q_i} - \frac{\partial V}{\partial q_i} = Q_i \tag{6}$$

the equation for a fixed rectangular plate of length a, width b and of thickness h, is obtained

$$q_{mn} + \alpha^2 q_{mn} = \beta_{mn} Q_{mn} \tag{7}$$

where the deflection is given by

$$W = \sum_{m=1}^{\infty} \sum_{n=1}^{\infty} Q_{mn}(t)\left(1 - \cos\frac{2m\pi}{a}x\right)\left(1 - \cos\frac{2n\pi}{b}y\right) \tag{8}$$

Natural frequencies are:

$$\alpha_{mn}^2 = \frac{16\pi^4}{9\rho hab}\left[D_{11}\frac{m^4}{a^4} + D_{22}\frac{n^4}{b^4} + (2D_{12} + 4D_{66})\frac{m^2 n^2}{a^2 b^2}\right] \tag{9}$$

The potential energy is

$$V = 2ab\pi^4 \left[3\frac{D_{11}}{a^4} \sum \sum m^4 q_{mn}^2 + 3\frac{D_{22}}{b^4} \sum \sum n^4 q_{mn}^2 \right.$$
$$\left. + \frac{2D_{12} + 4D_{66}}{a^2 b^2} \sum \sum m^2 n^2 q_{mn}^2 \right] \quad (10)$$

The kinetic energy is

$$T = \frac{9\rho abh}{8} \sum \sum q_{mn}^2 \quad (11)$$

and kinetic coefficient is

$$\beta_{mn} = \frac{4}{9\rho hab} \quad (12)$$

The solution for q_{mn} is then

$$q_{mn} = A_{mn} \cos \alpha t + B_{mn} \sin \alpha t + \frac{\beta}{\alpha} \int_0^t Q(s) \sin \alpha(t - s) \, ds \quad (13)$$

where A_{mn} and B_{mn} are determined from initial conditions. If the plate is motionless at initial time, $t = 0$ ($w = 0$). This means all A_{mn} and B_{mn} will equal 0.

For example, the point (γ, η) is excited with a harmonic force, written $p = p_0 \cos \omega t$. The generalised force will be:

$$Q_{mn} = P_0 \cos \omega t \left(1 - \cos \frac{2m\pi}{a} \gamma \right) \left(1 - \cos \frac{2n\pi}{b} \eta \right) \quad (14)$$

Hence we have the following solution:

$$q_{mn} = \left[\frac{P_0 \beta}{\alpha_{mn}^2 - \omega^2} (\cos \omega t - \cos \alpha_{mn} t) \right] \left(1 - \cos \frac{2m\pi}{a} x \right) \left(1 - \cos \frac{2n\pi}{b} y \right)$$
$$\times \left(1 - \cos \frac{2m\pi}{a} \gamma \right) \left(1 - \cos \frac{2n\pi}{b} \eta \right) \quad (15)$$

Because of viscous damping, all the energy associated with free oscillations will be dissipated after some time, at which point only forced vibrations will be present. The deflection becomes

$$W^* = P_0 \beta 4 \left[\sum_m \sum_n \frac{1}{\alpha_{mn}^{*2} - \omega^2} \left(1 - \cos \frac{2m\pi}{a} x \right) \right.$$
$$\left. \times \left(1 - \cos \frac{2n\pi}{b} y \right) \left(1 - \cos \frac{2m\pi}{a} \gamma \right) \left(1 - \cos \frac{2n\pi}{b} \eta \right) \right] \quad (16)$$

where α_{mn}^* is determined from the linear viscoelasticity correspondence principle.

Since the deflection is written $w = w_0 \cos(\omega t + \delta)$ both the magnitude and the phase shift can be obtained as follows:

$$W_0 = (W'^2 + W''^2)^{1/2} \qquad \mathrm{tg}\,\delta = -W''/W' \tag{17}$$

where W' and W'' are real and imaginary parts of W^*.

The viscous properties give rise to a dissipation of energy, which can be expressed as $\omega V_d/2\pi$ over a unit of time (V_d is the energy dissipated after a period). It follows:

$$V_d = -2ab\pi^4\left[3\frac{D_{11}''}{a^4}\sum\sum m^4 q_{mn}''^2 + 3\frac{D_{22}''}{b^4}\sum\sum n^4 q_{mn}''^2\right.$$
$$\left. + \frac{2D_{12}'' + 4D_{66}''}{a^2 b^2}\sum\sum m^2 n^2 q_{mn}''^2\right] \tag{18}$$

with

$$q_{mn}''^2 = \left[\left(\frac{4P_0\beta}{\alpha_{mn}^{*2} - \omega^2}\right)\left(1 - \cos\frac{2m\pi}{a}\gamma\right)\left(1 - \cos\frac{2n\pi}{b}\eta\right)\right]^2 \tag{19}$$

Calculation of Contact Force

The dynamic response of the plate due to the impact is transient in nature. The deflection at any point on the plate can be expanded in terms of the mode shapes of natural vibration. For the effect of local deformation at the point of contact, the Hertz law is assumed

$$P = K\alpha^{2/3} \tag{20}$$

Referring to Fig. 4, we have the indentation α as

$$\alpha = W_0 - W \tag{21}$$

where W_0 denotes the displacement of the mass m_0

$$W_0 = v_0 t - \frac{1}{m_0}\int_0^t P(r)(t-r)\,dr \tag{22}$$

v_0 is the initial velocity of the mass and w is the deflection of the plate at the place (γ, η).

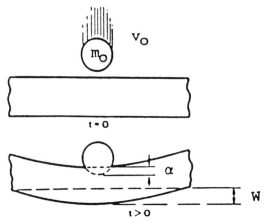

FIG. 4. Transverse impact of plates. Impact velocity: $v_0 = \sqrt{2gh}$, h is the drop-height.

Substituting (8), (13), (22) and (23) into (20) we obtain the non-linear integral equation for the contact force

$$(P(t)/K)^{2/3} = v_0 t - \frac{1}{m_0} \int_0^t P(r)(t - r)\,dr$$

$$- \sum_m^\infty \sum_n^\infty \frac{\beta}{\alpha_{mn}} \left[\int_0^t P(r) \sin \alpha_{mn}(t - r)\,dr \right] R_{mn} \quad (23)$$

which can only be solved by numerical methods.

Interaction Impactor–Composite Materials Target

The mechanical behaviour of plates subjected to impact depends directly on the interaction between the impactor and the composite plates. This means the indentation of the impactor into the plate. The famous elastic indentation law (20) was derived by Hertz.

Theoretically, K is determined as (Greszczuk[20])

$$K = \left[\frac{16}{3\pi(K_1' + K_2')} \right] \left(\frac{C_R}{S^3} \right)^{1/2} \quad (24)$$

where C_R and S are terms which take into account the geometries of the impactor and the target. K_1 and K_2 are parameters which take into account the elastic properties.

If the target and the impactor are made of transversely isotropic material K_2 can be derived from the results of Conway

$$K_2 = \frac{\sqrt{A_{22}}[(\sqrt{A_{11}A_{22}} + G_{zr})^2 - (A_{12} + G_{zr})^2]^{1/2}}{2\pi\sqrt{G_{zr}}(A_{11}A_{22} - A_{12}^2)} \tag{25}$$

where

$$A_{11} = E_z(1 - v_r)\beta \qquad \beta = \frac{1}{1 - v_r - 2v_{zr}^2\delta}$$

$$A_{22} = \frac{E_r\beta(1 - v_{zr}^2\delta)}{1 + v_r} \qquad \delta = \frac{E_r}{E_z} \tag{26}$$

$$A_{12} = E_r v_{zr}\beta$$

and E, G and v are Young's modulus, the shear modulus and Poisson's ratio respectively of the target. While r and z denote the implane and the thickness direction.

In our case, where the impactor is spherical with radius R, we have

$$K_1 = \frac{1 - v^2}{\pi E} \tag{27}$$

where E and v are Young's modulus and Poisson's ratio of the isotropic impactor.

If the target is orthotropic, K_2 is modified very little because it depends essentially on the target properties in the impact direction.

An empirical formula is proposed for E_r and v_r as follows:

$$E_r^{-1} = \frac{\cos^4 \alpha}{E_L} + \frac{\sin^4 \alpha}{E_T} + \frac{1}{4}\left(\frac{1}{G_{LT}} - \frac{2v_{LT}}{E_L}\right)x \sin^2 2\alpha$$

$$v_r = \frac{E_r}{E_L}\left[v_{LT} - \frac{1}{4}\left(1 + 2v_{LT} + \frac{E_L}{E_T} - \frac{E_L}{G_{LT}}\right)x \sin^2 2\alpha\right] \tag{28}$$

where L is the orthotropic direction, α is the orientation of the orthotropic axes.

We study the case where the target is a laminated plate and the impactor is spherical. We have $C_R = R/2$ and $S = 2$, therefore:

$$K = 4\sqrt{R}/3\pi(K_1 + K_2) \tag{29}$$

The contact area is circular with the radius

$$a = \left[\frac{3\pi R}{4}P(K_1 + K_2)\right]^{1/3} \tag{30}$$

The pressure distribution over the contact area is

$$q_{x,y} = q_0 \left[1 - \frac{x^2}{a^2} - \frac{y^2}{a^2} \right]^{1/2} \tag{31}$$

where q_0 is the surface pressure at the centre $(x = y = 0)$ of the contact area

$$q_0 = 3P/2\pi a^2 \tag{32}$$

The above coefficient of indentation is derived from the contact between a sphere and a half-space. It is uncertain for laminated composites.

This explains why the following experimental procedure was used. The experimental set-up is shown in Fig. 5.

The strain rate effect influences the P–α relation (Caprino *et al.*[15]). A quick camera was used to measure the indentation during the impact. However, the difficulties encountered in doing this may have affected the validity of our results.

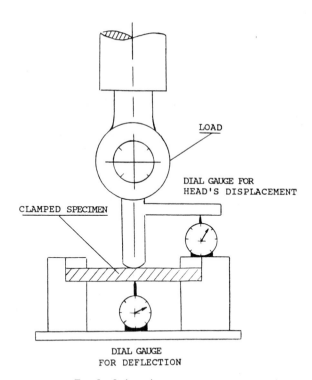

FIG. 5.   Indentation test set-up.

RESULTS AND DISCUSSION

Results of Measurement of the Dynamic Behaviour of Plates

Figures 6–12 show the variation of both the stiffness (deflection/force) amplitude and the loss factor, tg δ, versus frequency. These graphs apply to glass/epoxy fabric composite plates (300 × 300 × 3 mm) with 12 plies of 40% of fibre (except where indicated).

The first predicted natural frequency is 170 Hz whilst the experimental

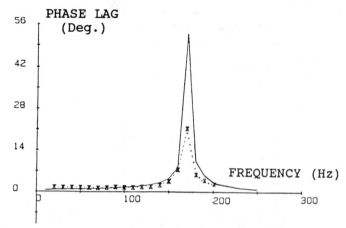

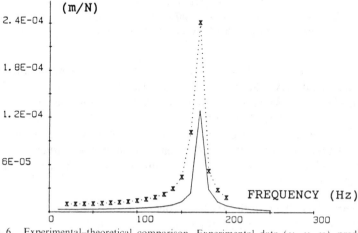

Fig. 6.   Experimental–theoretical comparison. Experimental data (× – × – ×), prediction curve (——).

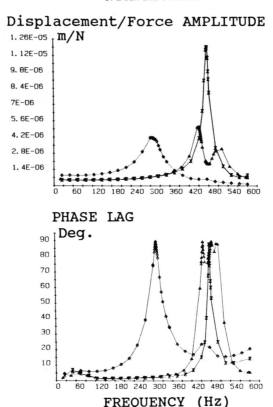

FIG. 7.   Effects of the pressure on the plate edges. Tightening by hand (◇-◇-◇), tightening each screw by spanner until 0·6 m kg (▲-▲-▲) and tightening each screw by spanner until 1·5 m kg (—✕—).

value is 168 Hz. An excellent correlation between experimental points and the theoretical curve can be seen in Fig. 6 for the loss factor.

The effect of pressure on the plate ends on the dynamic behaviour for a glass/epoxy composite plate (300 × 300 × 8·8 mm) with 32 plies of 40% of fibre is shown in Fig. 7. One can form, here, the following conclusions; the pressure exerted on the plate edges decreases their natural frequency, and at a certain value of pressure, the curve of stiffness versus frequency displays a disturbed zone.

Influence of Damage to the Dynamic Behaviour of Plates

During the impact there are three main damage modes which may occur at the same time or in sequence. The potential occurrence of multiple

Force/Displacement AMPLITUDE

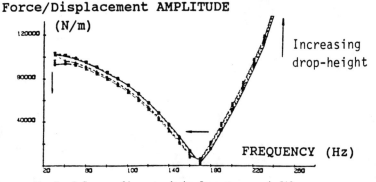

FIG. 8.   Influence of impact velocity. Impactor mass is 510 g.

damage modes can make the classification of the critical failure modes, in post-test inspection, difficult.

We have studied the effects of the damage level on the dynamic behaviour of plates. We have tested the same plates and effected the following procedures:

—caused impact by the dropping of balls of different masses from different heights;
—created a hole and increased its diameter.

Figure 8 shows the measurement of the plate's dynamic characters with a ball of mass 510 g falling from heights of 60, 120, 160 and 200 cm. The plate stiffness was slightly decreased when the impact was of a higher velocity. That becomes more clear if the ball is heavier (Fig. 9).

Force/Displacement AMPLITUDE

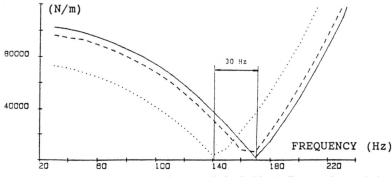

FIG. 9.   Effects of the impactor mass. Impact velocity is 6·2 m/s. For non-damaged plate (——), for impact energy of 43·8 J (·····), and for impact energy of 9·9 J (– – –).

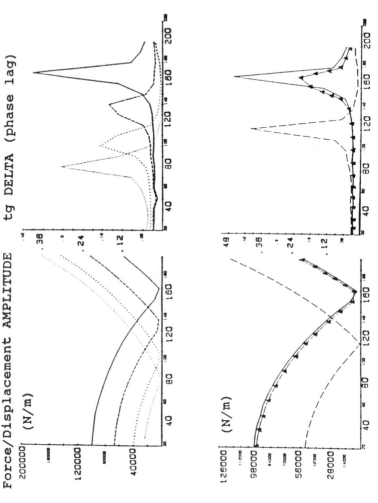

FIG. 10. Effect of slip at the plate ends during impact. For non-damaged plate (———), for damaged plate remaining in the apparatus (– – –), for damaged plate after taking and replacing in the apparatus (▲–▲–▲).

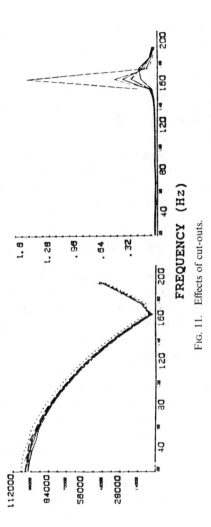

FIG. 11. Effects of cut-outs.

One can draw the conclusion, from the above facts, that the degree of damage is strictly dependent on the height and mass of the balls, i.e. the energy of the impact.

The plates were removed from the apparatus and damaged in many places. One can see, referring to Fig. 10, an important variation in the dynamic behaviour of the plates.

After taking a damaged plate and replacing it in the apparatus its dynamic characteristics were measured. Similar results, as for the non-damaged plates were obtained (Fig. 10). This signifies that the plates could not return to their initial state after impact. One concludes there is a slip at the plate ends during the impact.

The dynamic characteristics of a plate with a circular hole of diameter from 2 to 40 mm, are shown in Fig. 11. They are not significantly changed. However, the theoretical results obtained shows that a clear variation is found, if the hole has a diameter from $a/3$, where a is the plate dimension.

CONCLUSION

A scheme of experimental analysis of dynamic and impact behaviour of composite plates is presented. The method is complicated. However, it is necessary to apply dynamic and impact tests to composite materials because of their use and structural complexity. The analytical method presented in this paper provides a means of computing the dynamic characteristics of composites response to impact. The latter is strongly dependent on the former.

This study allows us to analyse failure modes and crack propagation and verify our different criteria of failure, our different expressions which approach failure modes of the plate (as shear mode, bending mode), localisation effects and damage propagation.

It must be noted that the predicted values of the dynamic characteristics agree well with the experimental results. It is necessary to mention here that the dynamic behaviour of composite plates depends very much on the edge pressure effects, whilst the damage is not so important. The law of indentation is more complicated than the one proposed by Hertz. This is explained by the structural and rate-dependent effects.

BIBLIOGRAPHY

1. ALAM, N. and ASNANI, N. T., Vibration and damping analysis of fiber-reinforced composite material plates. *J. Comp. Mater.* (1986).

2. BERT, C. W., Optimal design of a composite materials plate to maximise its fundamental frequency. *J. Sound Vib.* (1977).

3. DOAN, C., GERARD, J. F., HAMELIN, P., MERLE, G. and XIE, M., Prediction of viscoelastic properties of glass/ether-amide block copolymer composite materials. *Comp. Sci. Technol.*, **34** (1989) 1–15.

4. DOAN, C. and HAMELIN, P., Comportement dynamique de plaques stratifies orthotropes douees de viscoelasticite, *Annales de l'Institut Technique du Batiment et des Travaux Publics*, No. 471 (1989) 142–7.

5. FERRY, J. D., *Viscoelastic Properties of Polymers, 3rd edn*. John Wiley, New York, 1970.

6. FLUGGU, W., *Viscoelasticity, 2nd edn*. Springer-Verlag, Berlin, 1975.

7. HAMELIN, P., DE WILDE, W. P. and NARMON, B., *Introduction to the Structural Behavior of Fiber Reinforced Composite Material*. CERMAC-INSA Lyon et VUB-STRU, Vitgaven, 1985.

8. HASHIN, Z., Elastic moduli of heterogenous materials. *J. Appl. Mech.* (1962) 143–50.

9. HASHIN, Z., Dynamic behavior of viscoelastic composites. *Int. J. Solid Struct.*, 1970.

10. HASHIN, Z., Complex moduli of viscoelastic composites. *Int. J. Solid Struct.*, 1970.

11. RAJAMAMI, A., and PRABHAKARAN, R., Dynamic response of composite plates with cut-outs. *Journal of Sound and Vibration*, 4(54) (1977) 549–76.

12. TIMOSHENKO, S. P., *Vibration Problems in Engineering*. Van Nostrand, New York, 1955.

13. TSAI, S. W., HANN, H. T., *Introduction to Composite Materials*. Technomic Pub. Co. Westport, CI, 1980.

14. WHITNEY, J. M. and PAGANO, N. J., Shear deformation in heterogenous anisotropic plates. *J. Appl. Mech.* (1970), 1031–6.

15. CAPRINO, G., CRIVELLI-VISCONTI, T. and DI-ILLO, A., Composite materials response under low velocity impact. *Comp. Struct.*, **2**(3) (1984) 261–71.

16. CHEN, J. K. and SUN, C. T., Analysis of impact response of buckled composite laminates. *Comp. Struct.*, **3**(2) (1985) 97–118.

17. DOBYNES, A. L. and PORTER, T. R., A study of structural integrity of graphite composite structures subjected to low velocity impact. *Poly. Eng. Sci.*, **21**(8) (1981) 493–8.

18. DOBYNES, A. L., Analysis of simple-supported orthotropic plates. *AIAA Journal*, **19** (1981).

19. GOLDSMITH, W., *Impact*. Edward Arnold, London, 1960.

20. GRESZCZUK, L. B., Damage in composite materials due to low velocity impact. In *Impact Dynamics*. John Wiley, New York 1982, pp. 55–94.

21. LAL, K. M., Low velocity transverse impact behavior of 8 ply graphite-epoxy laminates. *Journal Rein. Plast. Comp.*, **2** (1983) 216–316.

22. LAL, K. M., Residual strength assessment of low velocity impact damage of graphite-epoxy laminates. *Journal Rein. Plast. Comp.*, **2** (1983) 226–39.

23. LEE, J. D., DU, S. and LIEBOWITZ, H., Three dimensional finite element and dynamic analysis of composite laminates subjected to impact. *Comput. Struct.*, **19**(5/6) (1984) 807–13.

24. LORD RAYLEIGH, On the production of vibration by forces of relatively long duration, with application to the theory of collisions. *Phil. Mag.*, **11** (1906).

25. MOON, F. C., A critical survey of wave propagation and impact in composite materials. NASA Report CR-121226, 1973.

26. SUN, C. T. and LAI, R. I. S., Exact and approximate analysis of transient wave propagation in an anisotropic plate. *AIAA Journal* (Oct. 1974) 1415–17.

27. SUN, C. T. and CHEN, J. K., On the impact of initially stressed composite laminates. *J. Comp. Mater.*, **19** (Nov. 1981) 490–504.

17

Some Geometric and Component Interaction Effects during the Axial Collapse of Glass/Polyester Tubes at Force Levels Compatible with Rail Vehicle Design

JOHN F. KELLY

British Railways Board, Research Department, Railway Technical Centre, London Road, Derby DE2 8UP, UK

ABSTRACT

Glass/polyester tubes can be triggered into stabilised axial collapse at force levels at least up to 2 MN whilst obtaining energy density levels up to some 58 kJ/kg. The geometric arrangement of the tube influences the triggering force, energy density and debris flow during collapse although multi-tube arrangements, in equilateral triangular arrays, do not appear to affect performance. These devices can offer commercial and technical advantages in rail vehicle applications.

1 INTRODUCTION

The effectiveness of a structure in absorbing energy during controlled collapse is basically a function of shape and material type. If the structure has primarily a loading requirement, the energy-absorbing capability is usually a compromise. However, with a dedicated energy absorber, designed to protect the main structure and any payload, much more latitude is available to optimise the choice of shape and material against the geometric and financial constraints of the application. Energy absorption during structural collapse on impact is a requirement across the complete spectrum of passenger transportation, although the extent to which this can be provided is very much conditioned by the mode. The materials of construction most commonly used are aluminium or steel, and it is the

John F. Kelly

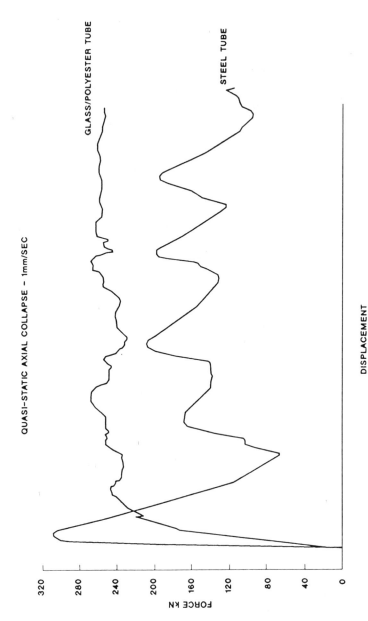

Fig. 1. Typical axial collapse characteristics for metal and composite tubes.

plasticity of these which dominates the energy absorption in such structures during severe impact when extensive bending of the structure occurs.

Within the range of practical structural metallic shapes the most effective shape for energy absorption is a cylinder in axial collapse creating symmetrical bending. The main disadvantage of this approach to energy absorption is that high initiation forces are required and the force is of a cyclical nature as the collapse progresses. The general characteristic then is of a high peak/mean force ratio together with a considerable oscillation about the mean value.

Small composite cylinders, manufactured by a variety of methods and materials, have been shown to be particularly effective at absorbing energy when triggered into a stabilised axial collapse.[1,2] Energy densities are much greater than could be expected with metals and both the initiation characteristics and the stabilised collapse are more favourable than metals for structural protection (Fig. 1).

2 BACKGROUND TO APPLICATION

The considered application for the energy-absorbing tubes is rail vehicles, either as a component within the vehicle interconnectors or part of the end of vehicle structures. The maximum compressive force to be resisted by a British Rail passenger vehicle, without permanent deformation, is 2 MN. Therefore any device would be required to 'trigger' and maintain stabilised collapsing forces below this value. Whilst it would be impractical to absorb vehicle impact energy at the higher speeds, previous damage analysis of passenger vehicles[3] has indicated that there is an economic case, based on repair costs, up to velocities of 14 m/s.

3 TEST SPECIMENS

In order to maintain reasonable quality control of the dimensions of the samples and the resin/glass ratio during manufacture, it was decided to utilise the filament winding technique. The materials of construction were dictated by the economic constraints of the application and against this background continuous filament 'E' glass fibres were selected together with a polyester resin of the type Crystic 491. The range of geometric dimensions of the test specimens was selected against the likely installation envelope available on the vehicle and an approximation of achievable energy

FIG. 2. Typical sample geometry before and after collapse.

densities previously reported.[2] Earlier workers[4] had suggested that a chamfer on the end of a composite tube was sufficient to trigger a stable collapse regime. It was therefore decided to machine a 45° chamfer on one end of all the test specimens (Fig. 2). The target glass/resin ratio for all the samples was 75%.

4 TEST VARIABLES

From the foregoing the test variables selected were:

Impact velocity—quasi-static	1 mm/s
Impact velocity—dynamic	4–14 m/s
Impact energy	up to 300 kJ
Tube inner diameter	5–300 mm
Tube wall thickness	10–20 mm
Tube length/diameter ratio	1:1 and 2:1

5 TEST EQUIPMENT

The testing was carried out to compare the quasi-static and dynamic modes over the variables listed above.

5.1 Quasi-static

The quasi-static testing was conducted on a Dartec 2000-kN Universal Testing Machine (Fig. 3). The force/deflection relationship of the test

FIG. 3. Dartec 2000-kN test machine.

FIG. 4. British Rail drop weight test machine.

specimen was recorded on a two-axis plotter and the test velocity, which is the velocity of the closing plattens, was kept constant at 1 mm/s throughout the tests. The machine had been previously calibrated to British Standard 1610, Class 1, indicating an error of less than 1%.

5.2 Dynamic

The dynamic testing was conducted on the British Rail Dropped Weight Test Facility (Fig. 4), which has a maximum capacity of 300 kJ and a maximum drop height of 10 m. The load cell was calibrated at the National Engineering Laboratory and the general accuracy of the installed instrumentation is better than $\pm 2\%$ with the exception of the values of acceleration, which were $\pm 7\%$. Timing slots were attached to the tup which, with the aid of a laser, enabled the position and velocity to be determined with an interval of 5 mm (Fig. 5).

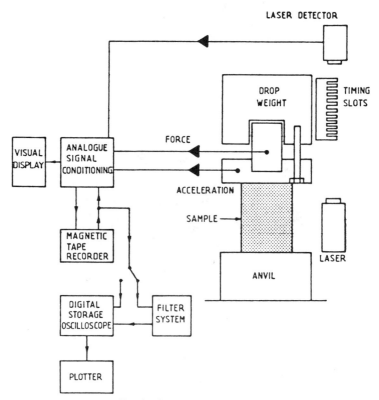

FIG. 5.   Instrumentation layout.

The dynamic tests were conducted using a mass of 1 and 3 t; the drop height was varied from 0·8 to 9·0 m, resulting in an impact approach velocity within the range 3·8–13·3 m/s. Due to the presence of high-frequency components in the force recordings the effect of filtering was investigated and it was decided that filtering at 1024 Hz was most acceptable. The main effect of filtering is to reduce the force peaks with very little effect on the energy levels.

6 TEST RESULTS

6.1 Geometric Tests

As reported in Ref. 5, the 45° taper is very successful at initiating the collapse at these high force levels over the range of tube geometries tested, in both a quasi-static and dynamic mode. The indications are that as the L/D of the tube is increased the collapse will still initiate at the tapered face, although shear failures will occur remote from this which considerably reduce the absorbed energy. For this reason dynamic tests were not undertaken on tube L/D ratios greater than 1. The effect of wall thickness/diameter ratio on energy density is unclear. From the quasi-static tests the energy density remains virtually constant against TL/D^2 down to a value of 0·05 although such a correlation was not apparent from the dynamic tests. From Fig. 6 it can be seen that the energy density has a tendency to reduce above some 4 m/s, being more pronounced with the smaller tubes, whilst at the 300 mm diameter the energy density is virtually unaffected by velocity.

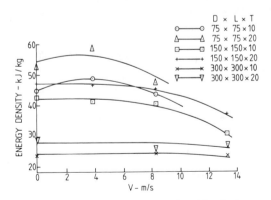

FIG. 6. Variation of energy density with velocity.

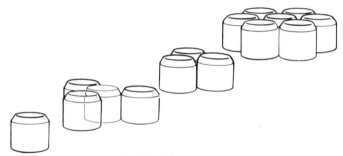

FIG. 7.   Tube arrays.

6.2 Tube Interaction during Collapse

The interaction during collapse was examined using the 75 mm internal diameter tubes with a 10 mm wall thickness. The test geometry chosen, from the initial variables, was thought to be the one most likely to be used in a multi-tube rail installation. The test velocity was selected in order that a single tube could absorb the tup energy, without having to provide additional rig facilities.

Four tube arrays were investigated, as illustrated in Fig. 7, involving one, three, four and seven tubes; as can be seen the three-tube array, whilst being kept equilateral, was given two spacings. Figure 8 shows the force characteristics obtained filtered at 1024 Hz. The single-tube test at 3·8 m/s gave a characteristic which had been repeated numerous times during the test programme, the onset of collapse was progressive and, as it stabilised,

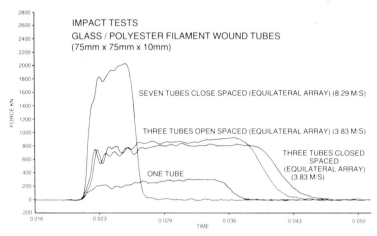

FIG. 8.   Comparative force characteristics.

FIG. 9. Seven-tube array prior to impact test.

the mean force was 278 kN. The three-tube array test, conducted with and without spacing at 3·8 m/s, gave very similar characteristics, as shown in Fig. 8. It was concluded from this test that the effect of spacing was very marginal with this array and the stabilised mean force was 852 and 820 kN respectively. Based on the conclusion of the three-tube tests it was decided to conduct the seven-tube array test without spacing at 8·39 m/s, as shown in Fig. 9 prior to test, which gave a stabilised mean force of 2000 kN.

6.3 Debris Flow during Tube Collapse

In order to obtain some indication of the debris flow at the crush front as a tube collapses, quasi-static tests were carried out and the tube size selected

FIG. 10.   Test piece clamped for resin impregnation.

from the test range was 150 mm internal diameter with a 20 mm wall thickness. The test technique used, with the aid of the Dartec machine, was to crush the first tube for 50 mm. The tube was then removed from the machine, at which point the crushed front of the tube expanded and placed in the clamp, as shown in Fig. 10. The clamp was then adjusted to return the tube to the test length and the tube and clamp were then dipped in a resin bath and subsequently cured. After removing the clamp, the resin-

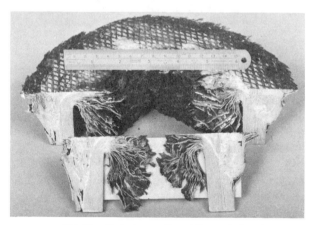

FIG. 11.   Test piece sectioned for analysis.

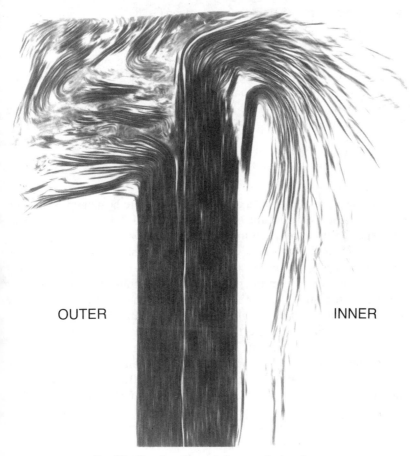

OUTER   INNER

FIG. 12.   Sample radiograph from crushed sections.

impregnated sample was then sectioned for analysis. A 3-mm slice was sawn from the section, from which radiographs were then produced (Fig. 11). This process was repeated on a number of other tubes crushing to 75 and 100 mm respectively.

Typical results from these tests are shown in Fig. 12. During the first 50 mm the intra-wall crack, which is attributed to compression-induced bending as described by Kendall,[6] has extended throughout the test piece and the majority of the crush front is flowing to the outside of the tube. Witness marks on the face of the upper platten of the Dartec machine showed that considerable friction was taking place as the front was being turned. Over the next 50 mm the crush regime is in two distinct phases and has reversed the major flow direction. Whilst more than half the debris is now flowing inwards, shear failures are beginning to occur on the inner wall, with pieces of debris breaking away. On the outer wall discrete sections appear to be buckling and folding in quite an ineffective way. On reaching 150 mm more than half the section is still flowing inwards, with virtually no detachment on the inner face and the outer wall failure regime has advanced considerably ahead of the inner, still buckling and folding.

7 DISCUSSION AND CONCLUSIONS

The geometric tests enabled a crush characteristic to be determined over a range of impact velocities and geometric ratios, which is illustrated in Fig. 6 and reported in more detail in Ref. 5.

The indications are that the energy absorption of multi-tube equilateral triangular arrays during axial collapse is not very sensitive to spacing between the tubes. From the results this demonstrates that the stabilised mean crushing force, for a multi-tube installation, could be predicted by simply factoring from the single-tube result with an error of less than 3%.

The radiographs indicate that the intra-tube crack is produced by compression-induced bending and occurs very early in the compression, proceeding right through the sample. The reason for the reversal of the majority of debris from outside to inside between 50 and 75 mm is not clear, although this could be the process which initiates the buckling and folding regime on the outside of the tube, which is in evidence from 75 mm onwards. Whilst there was evidence of heating of the platten and the sample on completion of the test, attempts to measure this with a thermal imaging camera during the crush were unsuccessful. The intra-wall crack does not appear to dominate the initial part of the crush, although as this progresses, certainly at a crush length of 75 mm, the differing collapse regimes appear to be quite well established and arranged either side of it, together with an asymmetric distribution of the compressive force.

Glass/polyester filament-wound tubes can offer a method of energy absorption, at force levels and within the geometric and financial constraints found within rail vehicle design criteria, which is substantially superior to that found with metals. Whilst this work indicates some aspects of the collapse mechanism over a range of geometries, the investigation was by no means exhaustive and possible applications should be regarded with these conclusions in mind.

ACKNOWLEDGEMENTS

The author would like to thank the Directors of British Railways Board Research Division for permission to publish this work and colleagues within the division who gave invaluable assistance, together with Professor D. Hull of the University of Cambridge for his helpful comments.

REFERENCES

1. COPPA, A. P., NASA TN-D1510, 1962.
2. MAGEE, C. L. and THORNTON, P. H., SAE 780434, 1978.
3. SCHOLES, A., *IMechE Conf. Railway Vehicle Body Structures*, C284/85, Derby, 1985.
4. KIRSH, P. A. and JAHNIE, H. A., *Soc. Auto. Engng* (1981), No. 0148-7191/81/0223-0233502.5.
5. KELLY, J. F., ECCM-3, Bordeaux, 1989.
6. KENDALL, K., *J. Mater. Sci.*, **11** (1976) 1767–9.

18

A Design Study of a Composite Leaf Spring

Tao Jianxin and Xue Yuan-de

*Department of Engineering Mechanics, Tongji University,
Shanghai, People's Republic of China*

ABSTRACT

The design method of a composite leaf spring is presented in this paper. The composite leaf spring manufactured has excellent static and fatigue behavior. The life prediction for specimens subjected to three-point bending fatigue and proof-test which will guarantee a minimum life for a specimen are also discussed. The evaluation methods of impact damage we proposed, especially the single-specimen test, have prospects for engineering applications.

1 INTRODUCTION

The automotive industry is one of the four largest markets for composite materials. A leaf spring is an important component which can be made of composite materials. A composite leaf spring has a lot of advantages such as light weight, comfortable ride and long life. It is also highly resistant to corrosion and possesses damping properties which will greatly reduce vibration.

The lower elastic modulus is significant for a composite leaf spring, because the spring is a flexible component which is required to absorb and release energy. In addition, composite materials can be tailored to meet the design requirements. It is necessary to find a design method for a composite leaf spring.

Since the spring will be subjected to fluctuating load and impact load, fatigue and impact resistance are two important properties to which we must pay great attention in the design studies of a composite leaf spring. How to predict the life of a composite leaf spring subjected to three-point bending fatigue, how to ensure a minimum life for the spring and how to evaluate the properties of impact resistance of the material, etc., all of these problems are required to be solved.

2 DESIGN METHOD

Figure 1 shows the loading state of a leaf spring. It can be taken as a component subjected to three-point bending.

A composite leaf spring should have the same stiffness as the steel spring it replaces. The stiffness, therefore, should be taken as the primary design parameter. A curved beam with equal width and varied thickness was chosen. The ratio of the radius of curvature to the thickness of the spring is so large that it is accurate enough to use straight beam theory.

According to optimum structural design, the normal stress in every section should be equal. The thickness of the beam, therefore, is in a parabolic form along the longitudinal direction, i.e.

$$h(x) = a\sqrt{c_3 - x} \tag{1}$$

where a is a shape parameter which will be determined by using a stiffness-equivalence condition later.

The leaf spring whose thickness follows eqn (1) cannot resist shear load at the supports. The thickness in the two tips must be increased according to the shear stress condition, it will be

$$h_{min} = \mu 3P/4b(\tau) \tag{2}$$

in which h_{min} is the thickness in the tips of the beam, τ is the allowable shear stress of the material and μ is a parameter depending on the cross-sectional contour.[1] In addition, the central portion of the beam must be flat and straight in order that the spring can be tightly mounted on the chassis.

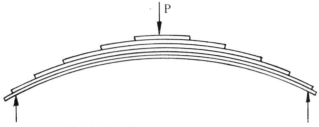

FIG. 1.   Loading state of a leaf spring.

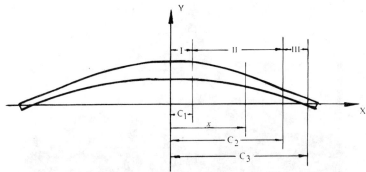

FIG. 2. The profile of a composite leaf spring.

Therefore, the leaf spring should have the profile as shown in Fig. 2. The central portion is rigidly clamped while the spring is mounted on the vehicle, the bending rigidity EI_1 of this portion can be treated as infinite.

Using Castigliano's method, the displacement of the central portion is

$$\Delta = \frac{\partial U}{\partial P} = 2 \int_{c_1}^{c_2} \frac{M\bar{M}}{EI_2} \, dx + 2 \int_{c_2}^{c_3} \frac{M\bar{M}}{EI_3} \, dx \tag{3}$$

where M and $\bar{M}$ are the bending moment under load P and the bending moment under unit force respectively. EI_2 and EI_3 are the rigidities of the portion II and the portion III respectively.

The design criterion becomes

$$k_c = P/\Delta = k_{st} \tag{4}$$

where k_c and k_{st} are the stiffnesses of the composite spring and steel spring replaced respectively.

The shape parameter, a, can be determined by substituting eqns (1) and (2) into eqns (3) and (4).

After determination of a, check the strength condition of normal stress

$$\sigma \le [\sigma] \tag{5}$$

If eqn (5) cannot be satisfied, other materials or other structural forms should be used.

3 MANUFACTURE AND TEST

We used the prepreg manufacturing process to fabricate springs for a commercial light truck. In this process, the prepreg, which is epoxy resin

FIG. 3.   Static test.

reinforced by unidirectional woven E-glass fiber, is laminated to the required profile, then it is cured with hot pressing. This process makes it possible to fabricate the spring into a mono-leaf one with a constant width and a variable thickness. The spring weighs 2·5 kg, being 1/6 the weight of the multi-leaf steel spring replaced. It exhibits good static and dynamic properties superior to the steel spring, the fatigue life in the test rig being over 500 000 cycles without failure. At the time of submission of this paper, the springs were mounted on the trucks being tested in the field. Figures 3, 4 and 5 are the composite leaf springs subjected to static test, fatigue test and field test respectively.

FIG. 4. Fatigue test.

FIG. 5.   Field test.

4   EVALUATION OF MATERIAL PROPERTIES

4.1 Fatigue Property and Life Prediction

To understand the fatigue behavior of the material and to explore the approaches to predict fatigue life, we subjected about 250 specimens to a three-point bending fatigue test in an INSTRON fatigue test machine. The nominal size of the specimens is $180 \times 25 \times 10$ mm (L × W × H). The span is 160 mm. The stress ratio is 0·1. All specimens were tested under constant load at a frequency of 5 Hz. The maximum stress levels (σ/S (S is the strength of the material)) are 0·66, 0·73, 0·77, 0·83 and 0·88.

In the test, the specimens exhibit very good fatigue behavior. Figure 6 shows the curves of stiffness degradation in the three-point bending fatigue test of the material. From Fig. 6 it can be seen that, during fatigue, the stiffness degradation rate of the specimens is small. In the early stages, the damage is first initiated by the occurrence of some tiny broken-fiber points in the compressive surface near the pressing-head. With increase of cycles, these damage points link up and propagate along the longitudinal direction and meanwhile, the damage penetrates gradually into the inside. When the stiffness decreases gradually to about 90% of its initial value, delamination and fiber-failure are initiated on the tensile surface of the specimen resulting in a rapidly decreasing stiffness. Consequently, the final failure occurs. In general, the time from the beginning of the rapid decrease in

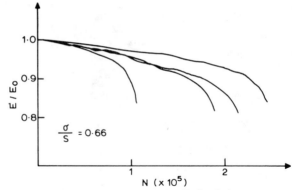

FIG. 6.   Stiffness degradation during fatigue.

stiffness to the final failure occupies no more than 5% of the entire life. The
failure criterion, therefore, can be defined as

$$E_{cr} = 0.9 E_0 \qquad (6)$$

The data from fatigue tests scatter over a wide range, even in the same
stress level, the lives differ probably by several times. So, it will be difficult to
find a rule for stiffness degradation from a single curve of degradation. But,
if all the curves are plotted on the same ordinate, it will be found that the
curves of normalized bending stiffness (E/E_0) versus normalized life (N/N_f)
fall into a narrow region (Fig. 7) indicating that there is a certain
relationship between the degradation of stiffness and life during fatigue.

Damage in the specimens is often initiated on the compressive surface of

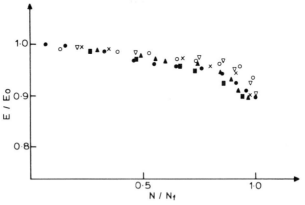

FIG. 7. E/E_0 versus N/N_f.

the central portion. We assume that the damage growth rate is directly proportional to the power of the maximum compressive strain in the specimen, $\varepsilon_{max}(D)$, i.e.

$$dD/dN = A'[\varepsilon_{max}(D)]^n \tag{7}$$

in which A' is a constant, D is the damage parameter which is defined as

$$D = 1 - E/E_0 \tag{8}$$

Notice that $\varepsilon_{max}(D)$ can be expressed as

$$\varepsilon_{max}(D) = \sigma_{max}/E(D) \tag{9}$$

Using eqns (8) and (9), eqn (7) may be rewritten as

$$-\frac{1}{E_0}\frac{dE}{dN} = A\left[\frac{\sigma_{max}}{E/E_0}\right]^n \tag{10}$$

in which $A = A'/E_0^n$.

Integration of eqn (10) gives a relationship between the specified modulus, E/E_0, its corresponding cycle number, N, and maximum stress, σ_{max}

$$E/E_0 = [1 - (n + 1)A\sigma_{max}^n N]^{1(n + 1)} \tag{11}$$

If we define that when

$$E = E_{cr} = 0.9E_0 \qquad N = N_f$$

the expression for fatigue life, N_f, can be obtained as follows

$$N_f = [1 - 0.9^{n + 1}]/(n + 1)A\sigma_{max}^n \tag{12}$$

where n and A are constants. They can be determined by a set of data obtained from the stiffness degradation test.

Table 1 lists the comparison between experimental data and theoretical results by using the proposed equation. It can be seen that the agreement is satisfactory.

TABLE 1
Comparison between experimental data and theoretical results

Applied stress level (σ/S)	0·73	0·77	0·83	0·88
Predicted life	394 154	113 753	26 493	6 779
Experimental fatigue life[a]	295 143	83 143	26 214	5 733

[a] Mean values of 7 specimens.

4.2 Proof-Test

The characteristics of composites have inevitably a certain dispersion due to some defects of the materials themselves and some other defects induced in the manufacturing process. We found that both the strength and the fatigue life of the material can be described by Weibull distributions with two parameters:

strength: $\qquad R_s(S) = \exp\left[-(S/S_0)^{\alpha_s}\right]$ (13)

fatigue life: $\qquad R_f(N) = \exp\left[-(N/N_0)^{\alpha_f}\right]$ (14)

In eqns (13) and (14) S_0 and N_0 are the characteristic strength and characteristic life respectively, α_s and α_f are shape parameters of the strength distribution and life distribution respectively.

The problem we are concerned with is how through a proof-test prior to the springs being allowed to leave the factory to weed out those products whose properties do not satisfy the requirements. In other words, how to determine proof load in the proof-test according to the requirements such as fatigue stress and expected life.

To solve this problem, we adopt the so-called strength–life equal rank assumption,[2] that is, a specimen which has a higher strength also has a higher fatigue life, and vice versa. In mathematical terms this assumption is expressed as

$$R_s(S \mid \sigma) = R_f(N) \tag{15}$$

where $R_s(S \mid \sigma)$ is the strength distribution of the specimens whose strength is above the fatigue stress, σ. $R_f(N)$ is the fatigue life distribution.

Then the relationship between strength and life can be obtained from eqn (15)

$$-\left(\frac{\sigma}{S_0}\right)^{\alpha_s} + \left(\frac{S}{S_0}\right)^{\alpha_s} = \left(\frac{N}{N_0}\right)^{\alpha_f} \tag{16}$$

or eqn (16) can be rewritten as

$$S = S_0\left[\left(\frac{\sigma}{S_0}\right)^{\alpha_s} + \left(\frac{N}{N_0}\right)^{\alpha_f}\right]^{1/\alpha_s} \tag{17}$$

If fatigue stress is σ, the fatigue life expected is $\bar{N}$, then the proof stress in the proof-test should be

$$\sigma_p = S_{N=\bar{N}} = \left[\left(\frac{\sigma}{S_0}\right)^{\alpha_s} + \left(\frac{\bar{N}}{N_0}\right)^{\alpha_f}\right]^{1/\alpha_s} \tag{18}$$

TABLE 2
Results of the proof-test

Applied stress level (σ/S)	Guarantee life $\bar{N}$	Proof stress level (σ/S)	Number of specimens	Failure no. in fatigue
0·73	500 000	1·030	6	1
0·73	50 000	0·883	6	0
0·78	10 000	0·834	6	0
0·78	5000	0·806	6	0

It means that to guarantee the products to have lives greater than N, the products must be subjected to a proof-test to σ_p as shown in eqn (18).

To testify the effectiveness of the proposed method, we conducted four sets of fatigue life tests after proof-test. Table 2 lists the experimental results. It can be seen that the results are very good. Therefore, the proposed method is probably an available method of quality control which can guarantee a minimum life with a high reliability.

4.3 Impact Damage and its Evaluation

A conventional method to evaluate impact properties is the Charpy or Izod impact test. But, because the mechanism of impact damage in composites is more complex than in metals, this method of evaluation may not be satisfactory for providing data of the impact behavior of composites. To evaluate properties of the materials in impact resistance and to obtain quantitatively the criterion used for evaluating the damage, two types of impact test were conducted.[3] One was a multi-specimen test, in which each of the specimens was impacted at different energies to record the $P-t$ (load–time) and $P-\Delta$ (load–displacement) curves as well as degradation of the stiffness. Another type is a single-specimen test, in which only one specimen is impacted at an energy great enough to break it and from $P-t$ and $P-\Delta$ curves of the specimen to determine the impact damage property of the specimen. The typical $P-t$ and $P-\Delta$ curves for the single specimen test are shown in Fig. 8.

In the multi-specimen test, for every specimen, the relationship between the degradation of stiffness and impact energy absorbed by per square of cross-section is investigated. Figure 9 shows the curve of damage D ($D = 1 - E/E_0$) versus energy absorbed by specimen, G. In general, there are two transition points on the curve of $D-G$. When the first transition point C_1 is reached, the damage increases rapidly. At second point C_2, the specimen is in a stage of unstable fracture failure. Therefore, the energy corresponding

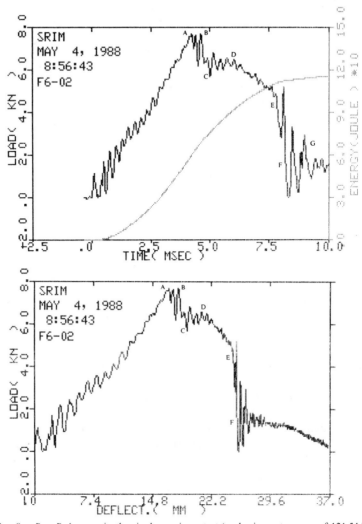

FIG. 8.   *P–t, P–Δ* curve in the single-specimen test (under impact energy of 121·54 J).

to the first transition point, G_c can be defined as the criterion of material impact resistance. If the impact energy exceeds this value, the growth rate of damage will be increased and then dangerous failure will take place. For the material used for the composite leaf spring, G_c is 27·34 J/cm^2, while G_c/G_F is 64·5% (G_F is energy to fracture). It means that 64·5% of impact energy has been absorbed at the critical point in which the ability to bear the impact

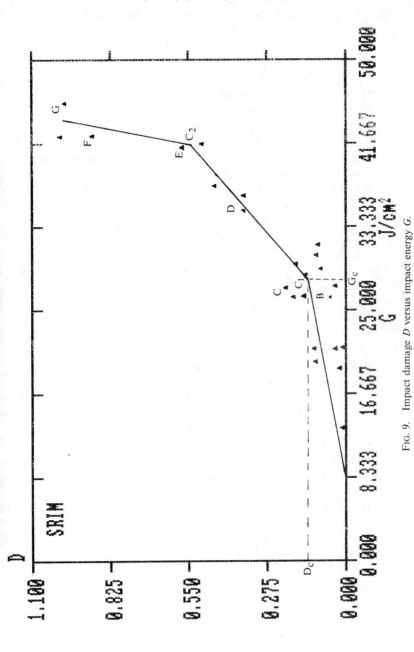

Fig. 9. Impact damage D versus impact energy G.

load has mostly been used. So, the composite leaf spring is of good impact resistance.

It can be seen that the multi-specimen test is a useful way to evaluate the impact resistant property of composite materials. But this method needs lots of specimens. This trouble can be avoided by the single-specimen test. In the single-specimen test, we find that the damage propagation process of the specimen is just similar to the individual damage process for the specimens under different impact energies. The variation rule of stiffness caused by the damage at different stages also agrees with that in the multi-specimen test. In Fig. 9 the points B, C, D, E, F and G are in accordance with the characteristic points in Fig. 8. It means that it is possible to evaluate the impact damage rule by use of a single-specimen test. It only needs to break a specimen using single-specimen test to get $P-t$ and $P-\Delta$ curves of the entire damage process; the critical points can be conveniently obtained from the curves. For instance, the energy corresponding to the characteristic point C in Fig. 8 can be taken as G_c, which agrees with the result obtained from the multi-specimen test. Hence, this method will provide a simple and practical tool for engineering application.

5 CONCLUSIONS

1. The composite leaf spring fabricated for a commercial truck according to the method of stiffness-equivalence and optimum structural design only weighs 2·5 kg, being 1/6 the weight of the multi-leaf steel spring replaced. Its excellent static, fatigue and impact resistance are superior to the steel spring.

2. In the fatigue process, the degradation of bending stiffness is gradual. There exists a certain relationship between the degradation of stiffness and life during fatigue. The stress level has no influence on fatigue failure mode. The life prediction equation developed can give satisfactory results.

3. To guarantee a spring having enough life to withstand the fatigue load, it can be achieved by means of proof-test. The proof load can be well determined by the equation developed.

4. The new methods for evaluating impact damage of composite materials, especially the single-specimen method, have prospect for practical application in engineering. The criterion for assessing impact resistance can be easily obtained using these methods.

5. The properties of fatigue and impact resistance of the material we used are suitable to the needs of the leaf spring.

ACKNOWLEDGEMENTS

The support of the National Natural Science Foundation and Shanghai Steel Leaf Spring Factory are acknowledged. The authors would like to express their great appreciation to Professor Zhu Yiling for encouragement and guidance. Help from Mr Sun Guofang and Mr Lian Hua is also appreciated.

REFERENCES

1. TIMOSHENKO, S. P. and GOODIER, R., *Theory of Elasticity*, 2nd edn. McGraw-Hill, New York, 1951.
2. HAHN, H. T. and KIM, R. Y., Proof testing of composite materials. *J. Comp. Mater.*, **9** (1975) 297.
3. SUN GUOFANG, LIAN HUA, XUE YUAN-DE and TAO JIANXIN, Evaluation of damage to laminated composites, *Proc. 7th Int. Conf. on Composite Materials*, Beijing, 1989.

19

Design Concept for One Piece Panel Vehicular Components in SMC Compression Molding

TSUNEO HIRAI

Department of Mechanical Engineering, Doshisha University,
Imadegawa Sakyo-ku Kyoto 602, Japan

and

MASASHI YAMABE

Central Engineering Laboratory, Nissan Motor Co. Ltd,
1-Natsushima Yokosuka 237, Japan

ABSTRACT

Products molded from fiber reinforced material are subject to several problems during the manufacturing process such as sink marks, warpage and short shots. In an earlier work, the authors used finite element analysis to analyze the in-mold flow behavior of sheet molding compound (SMC) during the compression molding process. The analytical results were then verified through experimentation. This report first proposes an explanation for the mechanism generating sink marks in a ribbed product having a simple shape. Based on that understanding of the mechanism, a method for minimizing sink marks has been devised and confirmed through experimentation to be effective. The results of the present work are now being used to establish a molding concept for a one-piece outer panel.

1 INTRODUCTION

Fiber reinforced plastic (FRP) is used extensively today in manufacturing the outer panels and structural parts of aircraft and transportation vehicles because of its light weight and large degree of molding freedom.

Applications of this material to the automobile can also be seen in the use of SMC (sheet molding compound) and R-RIM (reinforced reaction injection molding) material in the Fierro and Corvette manufactured by General Motors. It is anticipated that FRP will find increased use in the years ahead.

However, there are several issues pertaining to FRP that remain to be solved, including surface quality and paintability problems. Taking SMC as an example, two types of approaches are being taken in an effort to improve its surface quality. Some researchers are approaching this issue from the standpoint of modifying the material composition,[1,2] such as in the case of Class A SMC; other researchers are approaching the problem in terms of the manufacturing conditions and are working to improve the molding methods. Since the optimum molding condition often involves a repeated process of trial and error, it can lead to substantial delays in the development of new parts. In order to shorten the development period, there is an urgent need to establish the techniques for analyzing the molding process by means of numerical analysis.

Some general-purpose software tools, such as MOLDFLOW, have been developed and are widely used at present in carrying out numerical analyses of injection molding.[3] One drawback of these analytical tools, however, is that they only treat the flow of the resin and are unsuitable for heterogeneous materials containing fiber reinforcement. The literature shows few reports of flow behavior analyses for compression molding. One reason for this is assumed to be the difficulty of defining the boundary conditions on account of the movement of the upper mold during material flow.

The authors have carried out a numerical analysis of material flow during the compression molding process and have made clear the relationship between various molding factors and material flow.[4−6] In addition, the accuracy of the analytical results has been confirmed through experimentation. This report first presents an analysis of the mechanism causing sink marks. Based on that analysis, a method for eliminating sink marks by applying back pressure during molding is proposed, and the effectiveness of this technique is discussed.

2 ANALYTICAL FRAMEWORK

A typical compression load curve for SMC compression molding is shown in Fig. 1. The load increases linearly immediately after the material reaches the upper mold. It then shows a gradual rise until it again increases sharply

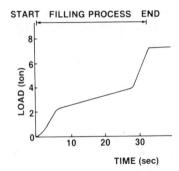

Fig. 1. Typical compression load curve.

just prior to the completion of the filling process. The complexity of the flow behavior as the material fills the mold in compression molding adds to the difficulty of the analysis. The authors have attempted to simplify the analysis by dividing the filling process into three stages.[7] These are initial deformation, flow behavior up to the point where the melt front reaches the mold and the subsequent filling behavior. Each stage is analyzed and modeled using the results of eigenvalue analysis, flow analysis and finally deformation analysis.[8] Through the combined application of these models it is possible to analyze various problems involved in the filling process (Fig. 2).

3 NUMERICAL ANALYSIS PROCEDURE AND FLOW EXPERIMENTS

It was noted in the previous report[7] that the flow behavior just prior to the completion of the filling process was a major cause of sink marks on the

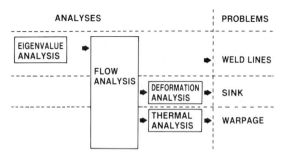

Fig. 2. Analytical framework.

back surface of a rib. It was decided, therefore, to conduct a flow analysis from the onset of filling to the point where the melt front reached the lower mold and to conduct an elastic–plastic analysis after that point. These analyses revealed the stress field of the material and also the reaction force between the material and the mold. A comparison was then made of the analytical results and experimental data.

3.1 Flow Analysis

A detailed explanation of the flow analysis procedure is given in the previous reports.[9-11] This paper will describe the assumptions used in the analysis and the governing equations that were ultimately derived. The heterogeneity of the glass fiber and resin was defined as a quasi-homogeneous anisotropic fluid on the basis of the following two assumptions.

(1) The glass fiber will be oriented in the streamline direction as a result of the material flow during the molding process.

(2) Anisotropic flow resistance will be seen in the axial and perpendicular directions of the glass fiber. This will be treated as anisotropic pseudo-plastic viscosity in each element used in a finite element analysis and will be converted to global coordinates. The Eulerian method will be employed because of the difficulty in carrying out the analysis using the Lagrangian method.

The governing equations are given as the following three expressions:

$$\rho u \frac{\partial u}{\partial x} + \rho v \frac{\partial u}{\partial y} + \frac{\partial p}{\partial x} - \mu_x \left(\frac{\partial^2 u}{\partial x^2} + \frac{\partial^2 u}{\partial y^2} \right) = 0$$

$$\rho u \frac{\partial v}{\partial x} + \rho v \frac{\partial v}{\partial y} + \frac{\partial p}{\partial y} - \mu_y \left(\frac{\partial^2 v}{\partial x^2} + \frac{\partial^2 v}{\partial y^2} \right) = 0$$

$$\frac{\partial u}{\partial x} + \frac{\partial v}{\partial y} = 0$$

Here u and v are the flow velocity in the x and y directions, p the pressure, ρ the material density, and μ_x and μ_y are pseudo-plastic viscosities which take into account the anisotropy of the flow resistance. The constitutive equations are regarded as pseudo-plastic flows and are obtained as the following two expressions which conform to the nth power law:

$$\sigma_{xx} = 2\mu_x \frac{\partial u}{\partial x} - p \qquad \sigma_{xy} = \sigma_{yx} = \mu_y \frac{\partial v}{\partial x} + \mu_x \frac{\partial u}{\partial y}$$

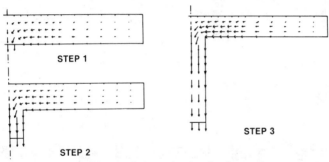

FIG. 3.   Flow analysis results showing flow vectors.

The unsteady state of the material during the flow process was treated using a progressive step-by-step method. Numerical solutions to these partial differential equations were derived using a Galerkin finite element method.

The flow was analyzed from the onset until the melt front reached the lower mold. The flow during that interval was divided into several steps and the flow velocity vector of the material was found for each step (Fig. 3). The results shown in the figure indicate that the flow velocity increased with each step.

3.2 Elastic–Plastic Analysis

The results of the flow analysis were taken as the initial conditions of an elastic–plastic analysis, which was carried out to analyze the nodal reaction force between the upper mold and the material near a rib, as shown in Figs 4 and 5. The nodal reaction force of SMC is shown in Fig. 4 and that of a resin compound (isotropic SMC without glass fiber) is shown in Fig. 5, for each of the steps indicated earlier. These results were obtained at nodal points of the material which came in contact with the upper mold. In the case of SMC, there were nodal points in the back surface of the rib (within a 5 mm distance from the center of the rib) where the reaction force was zero as the filling process proceeded. At locations beyond 5 mm from the center of the rib, high reaction force occurred and it was difficult for the material to flow because it was restrained by the mold. For the resin compound, the reaction force showed a uniform increase, and the force level was low.

3.3 Experimental Setup

In order to verify the analytical results, compression molding experiments were carried out using the molds shown in Fig. 6. Pressure

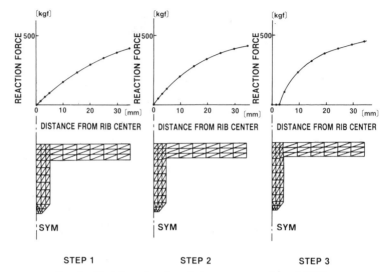

STEP 1   STEP 2   STEP 3

FIG. 4.   Nodal reaction force between upper mold and SMC.

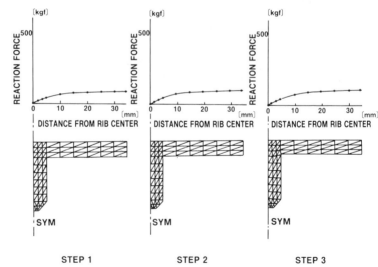

STEP 1   STEP 2   STEP 3

FIG. 5.   Nodal reaction force between upper mold and isotropic resin compound.

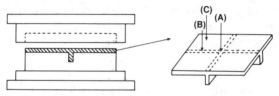

FIG. 6. Experimental setup.

measurements were taken at points A, B and C using pressure sensors that had been embedded in the upper mold in advance. Two types of materials were molded for comparative purposes. One was SMC, the other was a resin compound, i.e. isotropic SMC without any glass fiber.

3.4 Experimental Results

The results measured for SMC (anisotropic flow) and the resin compound (isotropic flow) are presented in Figs 7 and 8, respectively. The horizontal axis represents elapsed time and the vertical axis indicates the pressure level. SMC showed a momentary pressure drop at points A and B just before the completion of the filling process, whereas the pressure rose sharply at point C. These pressure behaviors were not observed for the isotropic resin compound. The experimental data showed good agreement with the results of the elastic–plastic analysis presented earlier. The momentary pressure drop displayed by the anisotropic SMC at points A and B corresponded to the analytical result that indicated zero reaction force between the material and the mold. The sudden pressure rise at point C corresponded to the increase in the hydrostatic pressure component. The experimental and analytical results for the isotropic SMC also showed good agreement.

Surface roughness measurements were made of the back surface of the ribs of the molded parts. The maximum roughness for SMC was 10μ and

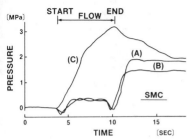

FIG. 7. Pressure measurements for SMC.

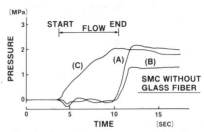

FIG. 8. Pressure measurements for isotropic resin compound.

that for the resin compound was 4μ. These results suggest that there is a correlation between the filling behavior of anisotropic material and the amount of sink marks that occur.

4 SINK MARK MECHANISM AND CORRECTIVE MEASURE

4.1 Mechanism

The mechanism generating sink marks near the ribs can be understood in terms of the four steps outlined below (Fig. 9).

Step 1

The flow action causes the melt front to reach the bottom of the lower mold. The friction that is generated between the material and the mold wall at that time results in a U-shaped melt front.

Step 2

The material begins to fill in the vacant space that has developed between the melt front and the bottom of the mold. In areas away from the rib (indicated by the dotted portion in the figure), the material tends not to undergo deformation because it is strongly restrained by the upper and lower molds. As a result, the hydrostatic pressure component rises.

Step 3

Since the movement of the material in the areas of the dotted portion is restricted, the material near the rib undergoes plastic deformation and fills in the vacant space. There are locations at that time where there is zero reaction force between the upper mold and the material, and the material is not in contact with the upper mold at those places.

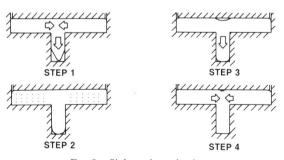

FIG. 9.   Sink mark mechanism.

Step 4

When pressure is then applied and held, the material at those places of noncontact with the mold, which developed in the previous step, undergoes gradual deformation such that it comes in contact with the upper mold. At this time the resin compound fills up the vacant space selectively because of the anisotropic flow resistance. One reason for this is that the peripheral material undergoes elastic recovery, resulting in insufficient molding pressure under ordinary conditions. This leads to the occurrence of sink marks in the molded part.

4.2 Measure for Reducing Sink Marks

Based on the foregoing analysis of the sink mark mechanism, an attempt was made to modify the construction of the mold so as to minimize the occurrence of sink marks. An experimental setup was built and tested which allowed back pressure to be applied to the rib portion only.

4.2.1 Experimental apparatus

As shown in Fig. 10, a hydraulic cylinder has been incorporated in the lower mold which functions to move a plate indicated in the figure as (A). This construction allows pressure to be applied to the back surface of the rib in the direction of the upper mold.

The application of pressure can be understood as follows when seen in side view (Fig. 11). Immediately after pressure is applied, the pressure level of the lower mold, F_b, is greater than that of the upper mold, F_a. Gradually F_a becomes larger than F_b. While the level of F_b is maintained, the plate is then moved, creating a space in which the rib is formed. In this case, since the melt front does not assume a U-shape, it is assumed that the cavities mentioned earlier do not occur.

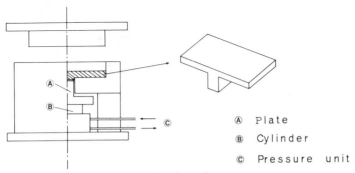

Ⓐ	Plate
Ⓑ	Cylinder
Ⓒ	Pressure unit

Fɪɢ. 10.   Experimental apparatus.

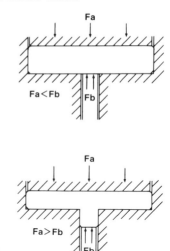

FIG. 11.   Side view of back pressure.

This experimental apparatus was employed to mold test pieces that were 200 mm long, 100 mm wide and 10 mm thick. The rib had a depth of 20 mm and width of 10 mm.

4.2.2 Experimental results

The surface roughness near the rib of a part molded using this experimental apparatus was measured with a surface roughness analyzer (Fig. 12(a)). For the sake of comparison, measurements were also made of a molded part which did not receive any back pressure (Fig. 12(b)). On the measurement, the needle moves with scanning the surface (15 mm length, 20 mm width, scanning pitch 1 mm/each measurement). To observe the sink

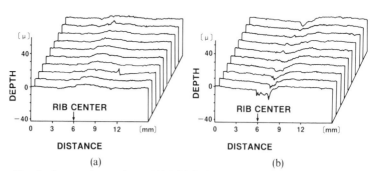

FIG. 12.   Surface roughness of part. (a) Molded with back pressure. (b) Molded without back pressure.

marks only, we have removed a large wave (more than 8 mm wave length), using low pass cut-off filter. Experiments were carried out under several couples of conditions for F_a and F_b; Fig. 12(a) and (b) show the typical case. On Fig. 12(b), in which nonaddition of back pressure, about 12μ depth of sink mark can be measured. On the other hand, with back pressure (Fig. 12(a)), the depth of sink mark was almost 3μ. Compared with these two products, it is evident to realize the sink marks on the surface with nonaddition of back pressure. These results confirmed that the application of back pressure was effective in improving surface quality.

5 CONCLUSION

This paper has described the mechanism generating sink marks and a corrective measure that has been devised for minimizing them. The results indicate that the application of back pressure to the back surface of a rib works to control the shape of the melt front during rib molding and thereby suppresses the occurrence of sink marks. The use of numerical analysis techniques made it possible to explain analytically the relationship between back pressure and the amount of sink marks that occur. The combination of these analytical techniques, the new mold design and structural analysis techniques for ribbed structures should contribute to active application of one-piece SMC outer panels to transportation vehicles.

REFERENCES

1. DERMMLER, K., *Kunststoffe*, **56** (1966) 606.
2. ASAKAWA, K. and FUKUSHIMA, T., *J. Soc. Fiber Reinforced Plastics*, Tokyo (1987) 33–2.
3. ENOMOTO, A. and SATO, S., *J. Soc. Automotive Engineers of Japan* (1984) 38–7.
4. HIRAI, T. and KATAYAMA, T., *Proc. of the 2nd Int. Conf. on Composite Materials*, 1978, p. 1283.
5. YAMABE, M., HIRAI, T., KATAYAMA, T. and FUJIWARA, T., *31st Int. SAMPE Symposium*, 1986, p. 1666.
6. HIRAI, T., KATAYAMA, T., YAMABE, M. and WATANABE, K., *Proc. of Int. Symp. on FRP/CM*, Nanjing, 1988.
7. YAMABE, M., HIRAI, T., KATAYAMA, T. and BAN, Y., *J. Japan Soc. Mechanical Engineers*, **497** (1988) 170.
8. HIRAI, T., KATAYAMA, T. and HAMADA, H., *J. Japan Soc. Technology of Plasticity*, **25** (1984) 287.

9. HIRAI, T. and KATAYAMA, T., *The Science and Engineering Review of Doshisha Univ.*, **18**–1 (1977) 43.
10. HIRAI, T. and KATAYAMA, T., *The Science and Engineering Review of Doshisha Univ.*, **20**–2 (1979) 61.
11. HIRAI, T., Rheology of carbon fiber composite prepreg materials. In *Development in Reinforced Plastics—5*, ed. G. Pritchard. Elsevier Applied Science Publishers, London, 1986, pp. 233–65.

20

A Simple Finite Element Formulation for Three-Dimensional Laminated Composite Plates

A. Miravete* and E. Dueñas

Department of Mechanical Engineering, University of Zaragoza, 50015 Zaragoza, Spain

ABSTRACT

A general finite element formulation for laminated composite plates based on a higher-order shear theory is attempted. The method presented here is simple, accurate and easy to implement in general programs for three-dimensional laminated composite structures. The original second-order equations derived from a higher-order shear theory are transformed into first-order equations by applying a penalty function method. Therefore, a quadrilateral C^0 finite element can be used, making possible the analysis of three-dimensional laminated composite plates by means of an assembly of quadrilateral elements. The conditions of zero transverse shear stresses on the top and bottom faces of the plate are satisfied. Convergence and comparison with other elements are examined.

INTRODUCTION

The classical lamination theory based on the Kirchhoff[1] hypotheses was the first attempt in the development of laminated plate theory. This theory is not computationally efficient from the point of view of simple finite element formulation[2] and is not suitable for composite structures analysis owing to its simplifying assumptions, the most important of which are the neglect of the transverse shear deformations and the transverse normal stresses.

* Presently, Visiting Scientist at the Air Force Materials Laboratory, AFWAL/MLBM, Wright Patterson AFB, Ohio 45433, USA.

Reissner[3] and Mindlin[4] were the first to provide first-order shear deformable theories based on the thin plate assumptions for variation of stresses and displacements through the thickness of the plate, respectively. Studies of low order elements, i.e. three-noded triangles, four-noded and eight-noded quadrilaterals showed violent stress oscillations. Some techniques like reduced and selective integrations,[5 - 8] modified shear strain methods[9] and hybrid and mixed methods,[10] succeeded in generating efficient elements. But, even so, these numerical analyses have certain limitations: the transverse shearing strains are assumed constant through the plate thickness, and a fictitious shear correction coefficient is introduced.

Lo et al.[11,12] and Reissner[13] presented a theory for plates based on an assumed higher-order displacement field. Kant et al.[14] presented, for the first time, a C^0 plate bending finite element formulation of a higher-order theory. This element is a nine-noded Lagrange quadrilateral and has six degrees of freedom per node. Reddy[15] presented a higher-order shear theory in which in-plane displacements are expanded as cubic functions of the thickness coordinate, while the transverse deflection is kept only a function of x and y.

Recently, Pandya and Kant[16] presented a refined higher-order C^0 plate bending element in which the transverse deflection is expressed as a quadratic function of the thickness. This element is a nine-noded Lagrange quadrilateral. H. V. Laksminarayana and Ramani[17] studied some shear flexible triangular laminated composite plate finite elements. G. Prathap and Somashekar[18] presented an eight-noded laminated anisotropic finite element. Tessler[19] presented a higher-order theory in which the transverse deflection is expressed as a quadratic function of the thickness.

The present formulation is based on a higher-order shear theory in which in-plane displacements are expanded as cubic functions of the thickness coordinate. Initially, the transverse deflection is kept only as a function of x and y. A quadrilateral, C^0 finite element is studied. Convergence and comparison with other elements are examined.

HIGHER-ORDER ELEMENT FORMULATION

The higher-order shear theory used here,[15] takes into account the parabolic distribution of the transverse shear stress along the laminate thickness. This requires the use of a displacement field in which the inplane displacements are cubic functions of the thickness coordinate and the transverse deflection is constant along the plate thickness. This definition of the displacement field satisfies the condition that the transverse shear stress be zero on the plate surface and not zero in any other place.

So that the displacement field is given by:

$$u_1(x, y, z) = u(x, y) + z\Psi_y(x, y) + z^2\xi_x(x, y) + z^3\phi_x(x, y)$$
$$u_2(x, y, z) = v(x, y) + z\Psi_x(x, y) + z^2\xi_y(x, y) + z^3\phi_y(x, y) \tag{1}$$
$$u_3(x, y) = w(x, y)$$

where u, v and w are the displacements of a point (x, y) of the midplane, and Ψ_x, Ψ_y are the rotations of the normals to the midplane about the axes x and y, respectively. The functions ξ_x, ξ_y, and ϕ_x and ϕ_y are determined by using the condition that the transverse shear stresses σ_4 and σ_5 are zero on the plate surfaces.

$$\sigma_5(x, y, \pm h/2) = 0 \qquad \sigma_4(x, y, \pm h/2) = 0 \tag{2}$$

For orthotropic plates or plates laminated in orthotropic layers, these conditions are equivalent to the requirement that the corresponding strains be zero on these surfaces.

$$\varepsilon_4 = \frac{\partial u_2}{\partial z} + \frac{\partial u_3}{\partial y} = -\Psi_x(x, y) + 2z\xi_y(x, y) + 3z^2\phi_y(x, y) + \frac{\partial w}{\partial y}$$
$$\varepsilon_5 = \frac{\partial u_1}{\partial z} + \frac{\partial u_3}{\partial x} = -\Psi_y(x, y) + 2z\xi_x(x, y) + 3z^2\phi_x(x, y) + \frac{\partial w}{\partial x} \tag{3}$$

Taking into account that ε_4 and ε_5 are zero on the plate surfaces, the following expressions are obtained:

$$\xi_x = 0 \qquad \xi_y = 0$$
$$\phi_x = -\frac{4}{3h^2}\left(\frac{\partial w}{\partial x} + \Psi_y\right) \qquad \phi_y = -\frac{4}{3h^2}\left(\frac{\partial w}{\partial y} - \Psi_x\right) \tag{4}$$

Introducing eqn (4) into eqn (1), the displacement field becomes:

$$u_1 = u + z\left[\Psi_y - \frac{4z^2}{3h^2}\left(\frac{\partial w}{\partial x} + \Psi_y\right)\right]$$
$$u_2 = v + z\left[-\Psi_x - \frac{4z^2}{3h^2}\left(\frac{\partial w}{\partial y} - \Psi_x\right)\right] \tag{5}$$
$$u_3 = w$$

The strains associated with these displacements are:

$$\varepsilon_1 = \frac{\partial u_1}{\partial x} = \frac{\partial u}{\partial x} + z\frac{\partial \Psi_y}{\partial x} - \frac{4z^3}{3h^2}\left(\frac{\partial^2 w}{\partial x^2} + \frac{\partial \Psi_y}{\partial x}\right)$$

$$\varepsilon_2 = \frac{\partial u_2}{\partial y} = \frac{\partial v}{\partial y} - z\frac{\partial \Psi_x}{\partial y} - \frac{4z^3}{3h^2}\left(\frac{\partial^2 w}{\partial y^2} - \frac{\partial \Psi_x}{\partial y}\right)$$

$$\varepsilon_4 = \frac{\partial u_2}{\partial z} + \frac{\partial u_3}{\partial y} = -\Psi_x - \frac{4z^2}{h^2}\left(\frac{\partial w}{\partial y} - \Psi_x\right) + \frac{\partial w}{\partial y}$$

$$\varepsilon_5 = \frac{\partial u_1}{\partial z} + \frac{\partial u_3}{\partial x} = \Psi_y - \frac{4z^2}{h^2}\left(\frac{\partial w}{\partial x} + \Psi_y\right) + \frac{\partial w}{\partial x}$$ (6)

$$\varepsilon_6 = \frac{\partial u_1}{\partial y} + \frac{\partial u_2}{\partial x} = \frac{\partial u}{\partial y} + z\frac{\partial \Psi_y}{\partial y} - \frac{4z^3}{3h^2}\left(\frac{\partial^2 w}{\partial xy} + \frac{\partial \Psi_y}{\partial y}\right)$$
$$+ \frac{\partial v}{\partial x} - z\frac{\partial \Psi_x}{\partial x} - \frac{4z^3}{3h^2}\left(\frac{\partial^2 w}{\partial xy} - \frac{\partial \Psi_x}{\partial x}\right)$$

where

$$\varepsilon_1 = \varepsilon_1^0 + z(k_1^0 + z^2 k_1^2) \qquad \varepsilon_4 = \varepsilon_4^0 + z^2 k_4^2$$
$$\varepsilon_2 = \varepsilon_2^0 + z(k_2^0 + z^2 k_2^2) \qquad \varepsilon_5 = \varepsilon_5^0 + z^2 k_5^2$$ (7)
$$\varepsilon_6 = \varepsilon_6^0 + z(k_6^0 + z^2 k_6^2)$$

and

$$\varepsilon_1^0 = \frac{\partial u}{\partial x} \qquad\qquad k_1^0 = \frac{\partial \Psi_y}{\partial x}$$

$$\varepsilon_2^0 = \frac{\partial v}{\partial y} \qquad\qquad k_2^0 = -\frac{\partial \Psi_x}{\partial y}$$

$$\varepsilon_4^0 = \frac{\partial w}{\partial y} - \Psi_x \qquad k_1^2 = -\frac{4}{3h^2}\left(\frac{\partial^2 w}{\partial x^2} + \frac{\partial \Psi_y}{\partial x}\right)$$

$$\varepsilon_5^0 = \frac{\partial w}{\partial x} + \Psi_y \qquad k_2^2 = -\frac{4}{3h^2}\left(\frac{\partial^2 w}{\partial y^2} - \frac{\partial \Psi_x}{\partial y}\right)$$ (8)

$$\varepsilon_6^0 = \frac{\partial u}{\partial y} + \frac{\partial v}{\partial x} \qquad k_4^2 = -\frac{4}{h^2}\left(\frac{\partial w}{\partial y} - \Psi_x\right)$$

$$k_5^2 = -\frac{4}{h^2}\left(\frac{\partial w}{\partial x} + \Psi_y\right)$$

$$k_6^0 = \frac{\partial \Psi_y}{\partial y} - \frac{\partial \Psi_x}{\partial x}$$

$$k_6^2 = -\frac{4}{3h^2}\left(2\frac{\partial^2 w}{\partial xy} + \frac{\partial \Psi_y}{\partial y} - \frac{\partial \Psi_x}{\partial x}\right)$$

For a plate of constant thickness h composed of thin layers of orthotropic material, the constitutive equations can be derived as discussed in Ref. 7. Under the assumption that each layer possesses a plane of elastic symmetry parallel to the x–y plane, the constitutive equations for a layer can be written as

$$
\begin{bmatrix} \bar{\sigma}_1 \\ \bar{\sigma}_2 \\ \bar{\sigma}_6 \\ \bar{\sigma}_4 \\ \bar{\sigma}_5 \end{bmatrix} =
\begin{bmatrix}
\bar{Q}_{11} & \bar{Q}_{12} & 0 & 0 & 0 \\
\bar{Q}_{12} & \bar{Q}_{22} & 0 & 0 & 0 \\
0 & 0 & \bar{Q}_{66} & 0 & 0 \\
0 & 0 & 0 & \bar{Q}_{44} & 0 \\
0 & 0 & 0 & 0 & \bar{Q}_{55}
\end{bmatrix}
\begin{bmatrix} \bar{\varepsilon}_1 \\ \bar{\varepsilon}_2 \\ \bar{\varepsilon}_6 \\ \bar{\varepsilon}_4 \\ \bar{\varepsilon}_5 \end{bmatrix}
\tag{9}
$$

where Q_{ij} are the plane-stress-reduced elastic constants in the material axes of the layer, and the bar over the quantities refers to the material axes of the layer. Upon transformation, the lamina constitutive equations can be expressed in terms of stresses and strains in the plate (laminate) coordinate as

$$
\begin{bmatrix} \sigma_1 \\ \sigma_2 \\ \sigma_6 \\ \sigma_4 \\ \sigma_5 \end{bmatrix} =
\begin{bmatrix}
Q_{11} & Q_{12} & Q_{16} & 0 & 0 \\
Q_{12} & Q_{22} & Q_{26} & 0 & 0 \\
Q_{16} & Q_{26} & Q_{66} & 0 & 0 \\
0 & 0 & 0 & Q_{44} & Q_{45} \\
0 & 0 & 0 & Q_{45} & Q_{55}
\end{bmatrix}
\begin{bmatrix} \varepsilon_1 \\ \varepsilon_2 \\ \varepsilon_6 \\ \varepsilon_4 \\ \varepsilon_5 \end{bmatrix}
\tag{10}
$$

where Q_{ij} are the transformed material constants.

Assuming that the only existing plane of symmetry is the plane of the plate, the constitutive equations of the plate become

$$
\begin{bmatrix} N_1 \\ N_2 \\ N_3 \\ M_1 \\ M_2 \\ M_6 \\ P_1 \\ P_2 \\ P_6 \\ Q_1 \\ Q_2 \\ R_1 \\ R_2 \end{bmatrix} =
\begin{bmatrix}
A_{11} & A_{12} & A_{16} & B_{11} & B_{12} & B_{16} & E_{11} & E_{12} & E_{16} & 0 & 0 & 0 & 0 \\
A_{12} & A_{22} & A_{26} & B_{12} & B_{22} & B_{26} & E_{12} & E_{22} & E_{26} & 0 & 0 & 0 & 0 \\
A_{16} & A_{26} & A_{66} & B_{16} & B_{26} & B_{66} & F_{16} & F_{26} & F_{66} & 0 & 0 & 0 & 0 \\
B_{11} & B_{12} & B_{16} & D_{11} & D_{12} & D_{16} & F_{11} & F_{12} & F_{16} & 0 & 0 & 0 & 0 \\
B_{12} & B_{22} & B_{26} & D_{12} & D_{22} & D_{26} & F_{12} & F_{22} & F_{26} & 0 & 0 & 0 & 0 \\
B_{16} & B_{26} & B_{66} & D_{16} & D_{26} & D_{66} & F_{16} & F_{26} & F_{66} & 0 & 0 & 0 & 0 \\
E_{11} & E_{12} & E_{16} & F_{11} & F_{12} & F_{16} & H_{11} & H_{12} & H_{16} & 0 & 0 & 0 & 0 \\
E_{12} & E_{22} & E_{26} & F_{12} & F_{22} & F_{26} & H_{12} & H_{22} & H_{26} & 0 & 0 & 0 & 0 \\
E_{16} & E_{26} & E_{66} & F_{16} & F_{26} & F_{66} & H_{16} & H_{26} & H_{66} & 0 & 0 & 0 & 0 \\
0 & 0 & 0 & 0 & 0 & 0 & 0 & 0 & 0 & A_{44} & A_{45} & D & D \\
0 & 0 & 0 & 0 & 0 & 0 & 0 & 0 & 0 & A_{45} & A_{55} & D & D \\
0 & 0 & 0 & 0 & 0 & 0 & 0 & 0 & 0 & D & D & F_{44} & F_{45} \\
0 & 0 & 0 & 0 & 0 & 0 & 0 & 0 & 0 & D & D & F_{45} & F_{55}
\end{bmatrix}
\begin{bmatrix} \varepsilon_1^0 \\ \varepsilon_2^0 \\ \varepsilon_6^0 \\ k_1^0 \\ k_2^0 \\ k_6^0 \\ k_1^2 \\ k_2^2 \\ k_6^2 \\ \varepsilon_4^0 \\ \varepsilon_5^0 \\ k_4^2 \\ k_5^2 \end{bmatrix}
\tag{11}
$$

where A_{ij}, B_{ij}, etc., are the plate stiffnesses defined by:

$$(A_{ij}, B_{ij}, D_{ij}, E_{ij}, F_{ij}, H_{ij}) = \int_{-h/2}^{h/2} Q_{ij}(1, z, z^2, z^3, z^4, z^6)\,dz \qquad (i,j = 1, 2, 6)$$

$$(A_{ij}, D_{ij}, F_{ij}) = \int_{-h/2}^{h/2} Q_{ij}(1, z^2, z^4)\,dz \qquad (i,j = 4, 5) \tag{12}$$

and N_i, M_i, P_i, Q_i and R_i are the resultant stresses defined by:

$$(N_i, M_i, P_i) = \int_{-h/2}^{h/2} \sigma_i(1, z, z^3)\,dz \qquad (i = 1, 2, 6)$$

$$(Q_1, R_1) = \int_{-h/2}^{h/2} \sigma_4(1, z^2)\,dz \tag{13}$$

$$(Q_2, R_2) = \int_{-h/2}^{h/2} \sigma_5(1, z^2)\,dz$$

Next, this theory is applied to the finite element method for its later implementation.

The finite element selected is a quadrilateral one of four nodes in vertices with linear functions of the type:

$$N_i = (1 + \xi\xi_i)(1 + \eta\eta_i) \qquad (i = 1, 2, 3, 4) \tag{14}$$

The displacement of a point (x, y) of the plate midplane, according to the finite element method is given by

$$u = \begin{bmatrix} u \\ v \\ w \\ \Psi_x \\ \Psi_y \end{bmatrix} = Na^e \tag{15}$$

where N is the shape function matrix of the finite element and a^e is the displacement vector of the nodes of the element e.

The strain vector, deduced from eqns (8) is,

$$\varepsilon = Lu = LNa^e \tag{16}$$

where

$$\varepsilon^T = [\varepsilon_1^0 \varepsilon_2^0 \varepsilon_6^0 k_1^0 k_2^0 k_6^0 k_1^2 k_2^2 k_6^2 \varepsilon_4^0 \varepsilon_5^0 k_4^2 k_5^2] \tag{17}$$

and L is a differential operator of the second order.

This differential operator, which is of the second order and therefore would demand the use of a finite element of continuity C^1, can be obtained from eqns (8) and (16). Two new equations are introduced, by means of the penalty function method, to reduce the derivation order. By achieving this reduction, it will be possible to use a finite element of continuity C^0, with all the advantages that this reduction supposes.

According to this theory[20]

$$\Pi^* = U_s + \alpha \int_\Omega C^T C \, d\Omega + W \tag{18}$$

where C is a linear condition. In this case, there are two second-order derivatives and therefore the following two conditions are implemented:

$$C_1(w, \theta_x) = \frac{\partial w}{\partial x} - \theta_x \qquad C_2(w, \theta_y) = \frac{\partial w}{\partial y} - \theta_y \tag{19}$$

So, we can define

$$\hat{u} = \begin{bmatrix} \theta_x \\ \theta_y \end{bmatrix} \tag{20}$$

The general expression of the generalized displacements and forces is

$$\begin{bmatrix} K_{11} & K_{12} \\ K_{21} & K_{22} \end{bmatrix} \begin{bmatrix} u \\ \hat{u} \end{bmatrix} = \begin{bmatrix} F \\ 0 \end{bmatrix} \tag{21}$$

and, the final expression for the stiffness matrix is

$$K = K_{11} - K_{12}[K_{22}]^{-1} K_{21} \tag{22}$$

Since there are five degrees of freedom, the assembly of elements into a three-dimensional laminated structure can be done directly.[21]

The transverse shear stresses cannot be accurately given by eqn (10) as the continuity condition at the interfaces of any two layers is not satisfied for laminated plates. For this reason, the interlaminar shear stresses are obtained by integrating the equilibrium equations of elasticity for each layer

$$\sigma_4^L\big|_{z=h_{L+1}} = -\sum_{L=1}^{n} \int_{h_{L+1}}^{h_L} \left(\frac{\partial \sigma_1}{\partial x} + \frac{\partial \sigma_6}{\partial y} \right) dz$$

$$\sigma_5^L\big|_{z=h_{L+1}} = -\sum_{L=1}^{n} \int_{h_{L+1}}^{h_L} \left(\frac{\partial \sigma_2}{\partial y} + \frac{\partial \sigma_6}{\partial x} \right) dz \tag{23}$$

NUMERICAL EXAMPLES

Three examples have been analysed:

(a) Square, clamped, uniform loaded plate.
    $a/h = 100$. Material HM Graphite Epoxy. Laminate [0/90/45/
    −45/−45/45/90/0].
(b) Square, simply supported, uniform loaded plate.
    Material HM Graphite Epoxy. Laminate [0/90/0/90/0/90/0/90/0].
(c) Square, simply supported, uniform loaded plate.
    Material HM Graphite Epoxy. Laminate [45/−45/45/−45/45/
    −45/45/−45/45] (Table 1).

TABLE 1
Material properties

Material	HM Graphite Epoxy
E1 (msi)	30
E2 (msi)	0·75
E3 (msi)	0·75
G12 (msi)	0·45
G23 (msi)	0·375
G13 (msi)	0·45
$v_{12} = v_{13} = v_{23}$	0·25

Example (a) In Figs 1 and 2, distributions of interlaminar shear stresses through the plate thickness at A are shown. Figure 1 is obtained by applying the constitutive equations. A continuous distribution is obtained if equilibrium equations are applied (Fig. 2). Figure 3 shows the distribution of interlaminar strains through the plate thickness.

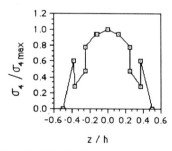

FIG. 1.   Distribution of σ_4 at A. (Constitutive equations.)

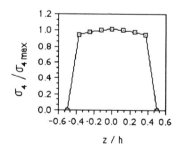

FIG. 2.   Distribution of σ_4 at A. (Equilibrium equations.)

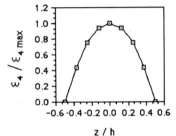

FIG. 3. Distribution of ε_4 at A.

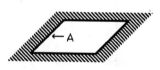

FIG. 4. Localization of point A.

TABLE 2
Accuracy of deflection at B (example (b))

$\dfrac{a}{h}$	$\dfrac{wE^2h^3}{qa^4} \times 10^3$				
	Present	TRIPLT[17]	QUAD4[18]	QUAD8-3[18]	EXACT[22]
10	5·85	5·85	5·84	5·85	5·85
100	4·48	4·48	4·47	4·49	4·49
1 000	4·46	4·45	4·46	4·47	4·47

TABLE 3
Accuracy of stress resultants at B (example (b))

$\dfrac{a}{h}$	$\dfrac{M_x}{qa^2} \times 10^2$				
	Present	TRIPLT[17]	QUAD4[18]	QUAD8-3[18]	EXACT[22]
10	8·43	8·42	8·44	8·42	8·42
100	8·86	8·81	8·88	8·87	8·88
1 000	8·75	8·42	8·88	8·87	8·89

TABLE 4
Accuracy of deflection at B (example (c))

$\dfrac{a}{h}$	$\dfrac{wE^2h^3}{qa^4} \times 10^3$			
	Present	TRIPLT[17]	QUAD4[18]	QUAD8-3[18]
10	3·735	3·732	3·738	3·748
100	2·400	2·400	2·402	2·403
1 000	2·251	2·223	2·389	2·389

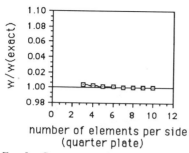

FIG. 5. Convergence of maximum deflection for $a/H = 2$.

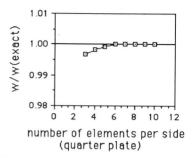

FIG. 6. Convergence of maximum deflection for $a/h = 4$.

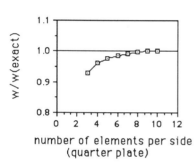

FIG. 7. Convergence of maximum deflection for $a/h = 10$.

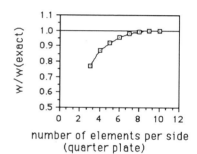

FIG. 8. Convergence of maximum deflection for $a/h = 20$.

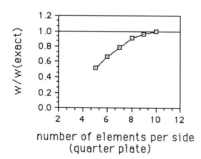

FIG. 9. Convergence of maximum deflection for $a/h = 50$.

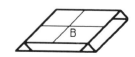

FIG. 10. Localization of point B. (Example B.)

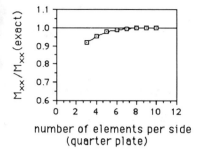

FIG. 11.   Convergence of resultant stress
for $a/h = 2$.

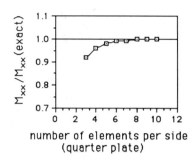

FIG. 12.   Convergence of resultant stress
for $a/h = 4$.

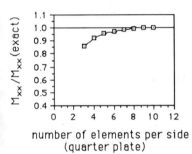

FIG. 13.   Convergence of resultant stress
for $a/h = 10$.

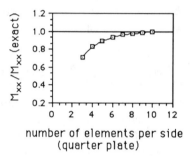

FIG. 14.   Convergence of resultant stress
for $a/h = 20$.

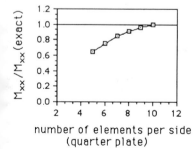

FIG. 15.   Convergence of resultant stress
for $a/h = 50$.

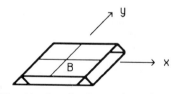

FIG. 16.   Localization   of   point   B.
(Example B.)

TABLE 5
Accuracy of stresses resultants at B (example (c))

$\dfrac{a}{h}$	$\dfrac{M_x}{qa^2} \times 10^2$			
	Present	TRIPLT[17]	QUAD4[18]	QUAD8-3[18]
10	3·47	3·50	3·50	3·47
100	3·58	3·67	3·59	3·60
1 000	3·43	3·45	3·59	3·59

Example (b) Accuracy of deflection and stresses resultants are shown in Tables 2 and 3, respectively. Convergence graphics of maximum deflection are represented in Figs 5–9 for different a/h ratios. Figures 11–15 show convergence of stress resultants for the same laminate.

Example (c) Accuracy of deflection and stresses resultants are shown in Tables 4 and 5, respectively.

CONCLUSIONS

A four-noded quadrilateral, C^0 finite element has been presented. In-plane, bending and interlaminar shear stresses are accounted for. The conditions of zero transverse shear stresses on the top and bottom faces are satisfied. Convergence and comparison with other elements show excellent results. Owing to the fact that this element has five degrees of freedom per node, its implementation in general programs for three-dimensional laminated composite structures is very easy. This formulation can be modified in order to obtain interlaminar normal stresses (σ_z). In this case the transverse deflection is expressed as a quadratic function of the thickness.

REFERENCES

1. TIMOSHENKO, S. P. and WOINOWSKY-KRIEGER, S., *Theory of Plates and Shells.* McGraw-Hill, New York, 1959.
2. COOK, R. D., *Concepts and Applications of Finite Element Analysis.* John Wiley, New York, 1981.
3. REISSNER, E., The effect of transverse shear deformation on the bending of elastic plates. *ASME, J. Appl. Mech.,* **12** (1945) A69–A77.
4. MINDLIN, R. D., Influence of rotatory inertia and shear deformation on flexural motions of isotropic elastic plates. *ASME, J. Appl. Mech.,* **18** (1951) 31–8.

5. ZIENKIEWICZ, O. C., TAYLOR, R. L. and TOO, J. M., Reduced integration techniques in general analysis of plates and shells. *Int. J. Num. Meth. Engng.*, **3** (1971) 275–90.
6. PAWSEY, S. E. and CLOUGH, R. W., Improved numerical integration of thick shell finite elements. *Int. J. Num. Meth. Engng.*, **3** (1971) 545–86.
7. HUGHES, T. J. R., TAYLOR, R. L. and KANOKNUKULCHAL, W., A simple and efficient finite element for plate bending. *Int. J. Num. Meth. Engng.*, **11** (1977) 1529–43.
8. PUGH, E. D. L., HINTON, E. and ZIENKIEWICZ, O. C., A study of quadrilateral plate bending with reduced integration. *Int. J. Num. Meth. Engng.*, **12** (1978) 1059–79.
9. LEE, S. W. and PIAN, T. H., Improvement of plate and shell finite elements by mixed formulations. *AIAA J.*, **16** (1978) 29–34.
10. HUGHES, T. J. R. and TEZDUYAR, T. E., Finite elements based upon Mindlin plate theory, with particular reference to the four-node bilinear isoparametric element. *ASME, J. Appl. Mech.*, **48** (1981) 587–96.
11. LO, K. H., CHRISTENSEN, R. M. and WU, E. M., A high-order theory of plate deformation part I: Homogeneous plates. *ASME, J. Appl. Mech.*, **44** (1977) 663–8.
12. LO, K. H., CHRISTENSEN, R. M. and WU, E. M., A high-order theory of plate deformation part II: Laminated plates. *ASME, J. Appl. Mech.*, **44** (1977) 669–76.
13. REISSNER, E., On transverse bending of plates, including the effects of transverse shear deformation. *Int. J. Solids Struct.*, **11** (1975) 569–73.
14. KANT, T., OWEN, D. R. J. and ZIENKIEWICZ, O. C., A refined higher-order C^0 plate bending element. *Comput. Struct.*, **15** (1982) 177–83.
15. REDDY, J. N., A simple higher-order theory for laminated composite plates. *ASME, J. Appl. Mech.*, **51** (1984) 745–52.
16. PANDYA, B. N. and KANT, T., A refined higher-order generally orthotropic C^0 plate bending element. *Comput. Struct.*, **28** (1988) 119–33.
17. LAKSMINARAYANA, H. V. and RAMANI, T. S., On improving the performance of a shear-flexible triangular laminated composite plate finite element. *Proc. Int. Conf. Com. Mat. and Struct.*, Madras, 1988.
18. PRATHAP, G. AND SOMASHEKAR, B. R., A field-consistent 8-noded laminated anisotropic plate element. *Proc. Int. Conf. Com. Mat. and Struct.*, Madras, 1988.
19. TESSLER, A., An improved higher-order theory for orthotropic plates. *Proc. Rev. Mech. Comp.*, Bal Harbour, Florida, 1988.
20. MIRAVETE, A., *The Finite Element Method Applied to Composites.* Escribano, Zaragoza, 1987.
21. ZIENKIEWICZ, O. C., *The Finite Element Method.* McGraw-Hill, New York, 1967.
22. NOOR, A. K. and MATHERS, M. D., Shear flexible finite element models of laminated composite plates. *NASA TN D-8044,* 1975.

21

Transverse Stress Predictions for Thin-to-Thick Composite Structures: Shear Deformable Finite Element Penalty Formulation

JOHN J. ENGBLOM and JOSEPH P. FUEHNE

*Mechanical Engineering Department, Texas A&M University,
College Station, Texas 77843, USA*

ABSTRACT

A shear deformable finite element formulation for laminated composite plate and shell structures is extended so that thin as well as thick geometries can be studied. This is accomplished by a specially constructed layering of elements in the 'thickness' direction. Adjoining layers are mathematically coupled at common interlaminar boundaries by use of a penalty parameter formulation. Since each element has simply midsurface nodal points at which displacements/rotations are prescribed, constraint equations serve to relate motion at layer midsurfaces to motion at layer interfaces. Continuity in each interface, i.e. between two layers, is represented by a set of three penalty parameters. Two of these parameters provide continuity of interlaminar shear stresses across the interface, while the third parameter provides continuity of interlaminar normal stress. In modelling actual structures, the degree of layering as well as specification of the penalty parameters can be extensively varied to account for changes in geometry, geometric discontinuities and the like.

INTRODUCTION

Design applications, relating to laminated composite plate/shell geometries, typically involve complex geometry and constraint definitions of the type amenable to 'design' analysis only through use of numerical modelling techniques. The finite element method (FEM) has been used with much

419

success in predicting the response of both monolithic and composite material systems. Refined formulations have been developed to characterize the fundamental mechanical behavior of such composites, particularly with respect to quantifying transverse shear and transverse normal stress variations in laminated plate and shell type structures. Transverse effects are especially significant for these material systems because 'interlaminar' strengths are lower than the comparable in-plane values. It should be noted that the transverse stress variation is a function of plate/shell dimensions, lamina orthotropy and laminae fiber orientations.

In the numerical modelling of actual composite structures, the FEM formulation should be capable of treating significant thickness variations. Examples include thickness changes due to ply drop-offs, the use of filler plies and even transitioning from relatively thin plate or shell sections to thick flanges. An additional concern is that of capturing the mechanical behavior at free edges, e.g. in the vicinity of a cut-out in a composite structure. In each of these cases, it is important that the 'interlaminar effects' be well represented because damage accumulation, and ultimately design allowables, are intimately coupled to the through-the-thickness stresses in these geometries. It is apparent that, for geometries of the type described, the FEM approach must provide for the representation of properties on a ply by ply basis for the entire structure.

Whitney[1] showed, in 1969, the importance of transverse shear deformation on the bending of laminated plates. Similarly, Pagano[2,3] demonstrated the deficiencies of classical laminated plate theory (CPT), particularly with regard to thick geometries, by comparing elasticity to CPT solutions for both cylindrical bending and for simply supported plates subjected to sinusoidally varying loads. These works clearly illustrate the need to include transverse shear effects when thick geometries are of interest.

There are alternative methods of including these effects in finite element analysis. Mau *et al.*[4] account for transverse shear effects in their four-noded quadrilateral element by using a hybrid stress method, where an assumed stress distribution is used in the element and an assumed displacement field exists on interelement boundaries. In the displacement field, rotations are independent of midsurface displacements thereby relaxing the Kirchhoff assumption. Spilker and Engelmann[5] use a similar hybrid stress formulation with an eight-noded quadrilateral element.

Higher-order displacement models are also used to model thick composite plates. Typically, terms which are quadratic and/or cubic with respect to the through-the-thickness coordinate, such as those used by

Pandya and Kant,[6] are added to the displacement functions to improve the variation of transverse shear stresses. Using the constitutive equations with a linear displacement model results in constant transverse stresses; however, the addition of higher ordered terms provides a parabolic variation. In a similar method developed by Hinrichsen and Palazotto,[7] the through-the-thickness deformations are modelled by utilizing a cubic spline function. Reddy *et al.*[8] also improved the displacement variation in the thickness direction of a laminate. They summed continuous interpolation functions through the thickness. Each of these formulations give satisfactory results for thick composite plates but are difficult to extend to more complex geometries, e.g. structures having curvature.

It is noted that a Lagrange multiplier approach is an alternative to a penalty parameter approach. Use of penalty functions does not add equations (degrees of freedom) to be solved whereas each Lagrange multiplier adds an equation to the equation set. Chatterjee and Ramnath[9] have used the Lagrange multiplier approach in their efforts to model laminated structures as an assemblage of sublaminates. In this case, a sublaminate by their definition, is equivalent to one layer of elements in the present work. Although Chatterjee and Ramnath do not consider thick plates or shells, Chatterjee does present results for delaminated plates.[10]

The finite element approach developed in this work to represent laminated composite structures, is based on utilizing 'shear deformable' plate/shell element formulations of the type recently reported.[11-14] In these references, middle surface displacements and rotations are prescribed independently for the purpose of relaxing the Kirchhoff hypothesis. Interpolation functions are chosen such that reasonable transverse stress variations can be obtained. Solution of the constitutive equations provides stresses in-the-plane of each ply, while equilibrium equations provide the transverse stresses. Use of the equilibrium equations gives significant improvement over the CPT, without the numerical inefficiencies associated with using a 3-D solid finite element formulation.

It is emphasized that the computational efficiencies achieved in the 'shear deformable' formulations,[11-14] i.e. with regard to improving transverse stress predictions, is extended in the present penalty formulation to thicker geometries. This can be appreciated by considering a particular location on a finite element model, which can be represented as a set of *n* layers. Each layer exhibits independent displacements and rotations, and is comprised of a number of plies which can have different fiber orientations and material properties. It is reasonable to expect, therefore, that the number of layers required in the transverse direction should be very minimal. Furthermore, the equilibrium equations can be integrated very efficiently through-the-thickness of n layers, while allowing the penalty parameters to account for interlaminar stress continuity at layer interfaces. A much more refined finite element mesh would be needed to compute transverse stresses to the same level of accuracy if the more traditional approach, i.e. of using the constitutive equations to determine all stress components, were employed. As a result of solving constitutive and equilibrium equations, computational efficiency is achieved. The penalty formulation provides the added flexibility of allowing delamination to be represented at preselected interfaces.

FINITE ELEMENT FORMULATION

Displacement Field

The shear deformable elements reported upon[11–14] are similar to others in the literature, e.g. Panda and Natarajan,[15] and Ahmad et al.,[16] with the primary exception being the manner by which the transverse stresses are determined. The efficiencies associated with this approach are enhanced in extending the formulation to layering the elements using penalty parameters. Layers are each represented as eight-noded quadrilaterals, as depicted in Fig. 1, each having independently prescribed displacements and rotations. These displacements are defined below as

$$u(x, y, z) = u_0(x, y) + z[n_z\theta_x(x, y) - n_y\theta_z(x, y)]$$
$$v(x, y, z) = v_0(x, y) + z[n_z\theta_y(x, y) + n_x\theta_z(x, y)] \quad (1)$$
$$w(x, y, z) = w_0(x, y) - z[n_x\theta_x(x, y) + n_y\theta_y(x, y)]$$

where u_0, v_0, and w_0 are the midsurface displacements within a layer, and θ_x, θ_y, and θ_z are the midsurface rotations. All of these terms are given in the element cartesian coordinate system. Also shown in Fig. 1 are the curvilinear coordinates ξ, η and ζ, where ξ and η lie in the layer midsurface and ζ defines the transverse (thickness) direction. Curvature is accommodated by definition of surface normals η_x, η_y, and η_z which represent components of the normal to the midsurface in element coordinates. When η_z equals 1 and both η_x and η_y are identically zero, the geometry simply reduces to that of a layered plate. It is noted that in this paper, results are given for plate geometries only. A bi-quadratic interpolation (shape) function is utilized to specify both nodal point location and displacement variation within each layer.

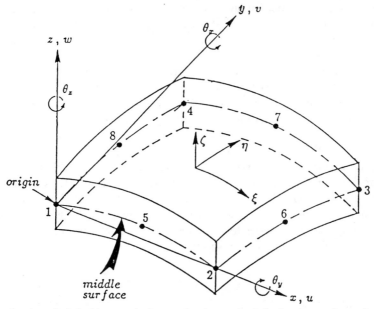

FIG. 1. A typical doubly curved element showing nodes 1–8, the elemental coordinate system, and the six degrees of freedom per element, u, v, w, θ_x, θ_y, θ_z.

Having defined a displacement variation, the stiffness relationship is developed in the traditional finite element sense.[11-14] The displacement field chosen provides a piecewise linear displacement variation in the through-the-thickness direction of the laminate, i.e. assuming that continuity is preserved at the interface between the layers using the penalty parameters.

Stress Variation

Stresses are calculated for sub-layers (plies) within each layer on a ply-by-ply basis; constitutive equations provide definition of stresses within plies, but not in the transverse direction. These stress components are given below for the kth sub-layer as

$$\{\sigma\}^k = [\mathbf{D}]^k \{\varepsilon\}^k \tag{2}$$

In this formulation, if the constitutive equations were used to determine the transverse stresses, a constant stress value would be obtained for each layer. In the simplified case wherein the geometry is modelled with one layer, such that there is no interfacing with penalty parameters, such a constant stress result is not reasonable and can be greatly improved by using the equilibrium equations. The layer interfacing presented in this work gives quite refined transverse stress results without extensive model refinement in the thickness direction.

To illustrate this approach, consider first the variation in the x, y directions, including shear, where the x–y plane is parallel to the plate surface and, therefore, the z axis defines the plate normal direction. Stress variation in the x direction has the functional form

$$\sigma_x = \sigma_x[x, y, xy, x^2, y^2, z(x, y, xy, x^2, y^2)] \tag{3}$$

while the normal stress in the y direction and the shear stress in the x–y plane vary in the same manner. These stresses can be calculated within each sub-layer at each of the Gauss points and approximated by use of a 'smoothing' function. In the absence of body forces, the equilibrium equations can then serve to determine the through-the-thickness stress variation. These equations have the form

$$\frac{\partial \tau_{xz}}{\partial z} = -\left(\frac{\partial \sigma_x}{\partial x} + \frac{\partial \tau_{xy}}{\partial y} \right)$$

$$\frac{\partial \tau_{yz}}{\partial z} = -\left(\frac{\partial \tau_{xy}}{\partial x} + \frac{\partial \sigma_y}{\partial y} \right) \tag{4}$$

$$\frac{\partial \sigma_z}{\partial z} = -\left(\frac{\partial \tau_{xz}}{\partial x} + \frac{\partial \tau_{yz}}{\partial y} \right)$$

where the derivatives on the right hand side of (4) are obtained directly on the basis of the described smoothing functions. The transverse stresses thus are determined by numerically integrating the expressions in (4) in the z direction. The through-the-thickness variations have the form

$$\tau_{xz} = \tau_{xz}[xz, yz, z^2(1, x, y)]$$

$$\tau_{yz} = \tau_{yz}[xz, yz, z^2(1, x, y)] \tag{5}$$

$$\sigma_z = \sigma_z[z^2, z^3]$$

which provides a parabolic transverse stress distribution within each layer, and gives refined stress results for a set of interfacing layers.

Penalty Parameters

Since the referenced elements[11-14] are layered in the thickness direction of the laminate, continuity must be considered at the interface between layers. To demonstrate the present use of penalty functions, consider the

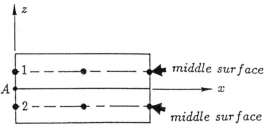

FIG. 2.   Two stacked elements showing nodes 1 and 2 as well as point A, which is a common point on the boundary between the elements.

following example using a simplified displacement field. Figure 2 presents two elements with nodes 1 and 2 and a point A, where the latter point is located at the interface between the two elements. Considering element 1 above, displacement in the x direction is given as

$$u_A^1 = u_{01} + \frac{h_1}{2}\theta_{y1} \tag{6}$$

and, similarly, displacement in the x direction for element 2 below is written

$$u_A^2 = u_{02} - \frac{h_2}{2}\theta_{y2} \tag{7}$$

and, of course, equations of the same type can be written for displacements in the y and z directions. The constraint equation, for motion in the x direction at point A in the interface, then follows

$$u_{01} + \frac{h_1}{2}\theta_{y1} - u_{02} + \frac{h_2}{2}\theta_{y2} = 0 \tag{8}$$

and equations of the same type are written for the other directions, except that the displacements in the Z direction (normal to the plate) are simply set to be equal in the upper and lower layers. This approach to satisfying continuity at the interface results in the constraint equations

$$[C]\{D\} = \{0\} \tag{9}$$

where $[C]$ is a 3×10 matrix, $\{D\}$ is a displacement vector, including all of the translational and rotational terms for the upper and lower layers. To apply these constraint equations, the total potential function is augmented with a penalty term and results in a stiffness matrix modification. The triple matrix product below

$$[C]^T[\alpha][C] \tag{10}$$

is then simply added to the stiffness matrix, where $[C]$ is as previously defined and $[\alpha]$ is a diagonal matrix containing the penalty parameters. These parameters are analogous to springs that serve to keep the interface of adjoining layers together. When the penalty parameters are given large values, the elements are joined as a continuum. Conversely, when these parameters are set equal to zero, a discontinuity exists at the interface.

PRELIMINARY RESULTS

Some of the initial results, obtained using the present formulation for thick plates, have been encouraging. Table 1 presents normalized stresses for a simply supported, square $[0°/90°/0°]$ plate, subjected to a sinusoidally varying pressure loading. The normalized quantities are defined as

$$(\bar{\sigma}_x, \bar{\sigma}_y, \bar{\sigma}_{xy}) = \frac{1}{q_0 S^2}(\sigma_x, \sigma_y, \sigma_{xy})$$

$$(\bar{\sigma}_{xz}, \bar{\sigma}_{yz}) = \frac{1}{q_0 S}(\sigma_{xz}, \sigma_{yz})$$

(11)

where q_0 is the maximum pressure, applied at the center of the plate, and S is the aspect ratio. The elasticity results were derived by Pagano.[3] The

TABLE 1

Normalized stresses for a simply supported square $[0°/90°/0°]$ plate with a sinusoidally varying load

S	Approach	$\bar{\sigma}_x$ $\left(\frac{a}{2},\frac{a}{2},\pm\frac{b}{2}\right)$	$\bar{\sigma}_y$ $\left(\frac{a}{2},\frac{a}{2},\pm\frac{h}{6}\right)$	$\bar{\sigma}_{xy}$ $\left(0,0,\pm\frac{h}{2}\right)$	$\bar{\sigma}_{zz}$ $\left(0,\frac{a}{2},0\right)$	$\bar{\sigma}_{yz}$ $\left(\frac{a}{2},0,0\right)$
	(6 × 6) 1 layer	0·391	0·572	0·045	0·308	0·251
	(6 × 6) 3 layers	0·543	0·411	0·053	0·284	0·201
4	(6 × 6) 5 layers	0·660	0·491	0·057	0·276	0·200
	Elasticity	0·755	0·556	0·051	0·282	0·217
	(6 × 6) 1 layer	0·531	0·189	0·023	0·387	0·095
20	(6 × 6) 3 layers	0·540	0·204	0·023	0·384	0·095
	Elasticity	0·552	0·210	0·023	0·385	0·094
	(6 × 6) 1 layer	0·542	0·167	0·022	0·393	0·083
100	(6 × 6) 3 layers	0·542	0·174	0·021	0·392	0·083
	Elasticity	0·539	0·181	0·021	0·395	0·083

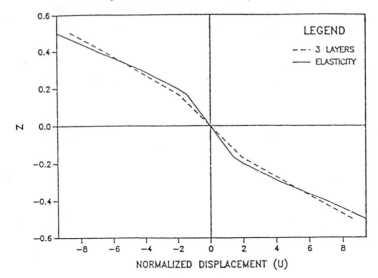

FIG. 3. Normalized displacement through-the-thickness of a plate strip ($S = 10$) subjected to a cylindrical bending load.

orthotropic material properties are $E_L/E_T = 25$, $G_{LT}/G_{TT} = 2\cdot5$, and $v_{LT} = v_{TT} = 0\cdot25$, which are representative of a high modulus graphite epoxy. Consider the plate to have planar dimensions $a \times a$ and total laminate thickness, h. One level of mesh refinement is used in the plane, while the number of layers used in the thickness direction is varied. For an aspect ratio of 4, significant improvement is demonstrated as the layers are increased from 1 to 5. The improvement is not as dramatic for the higher aspect ratios, which is expected since the transverse shear effects are diminished as the plate is thinned down.

A second example studied is the cylindrical bending of a $[0^\circ/90^\circ/0^\circ]$ plate strip. The orthotropic material properties are identical to those used in the previous problem. Figures 3–5 present plots of normalized displacement, normal stress and transverse shear stress through-the-thickness of the plate strip. The displacement and stresses are normalized as follows

$$\bar{\sigma}_x = \frac{\sigma_x(\tfrac{1}{2}z)}{q_0} \qquad \bar{\sigma}_{xz} = \frac{\sigma_{xz}(0, z)}{q_0} \qquad \bar{u} = \frac{E_T u(0, z)}{hq_0} \tag{12}$$

where q_0 is the maximum applied pressure, which occurs at the center of the plate strip, h is the total laminate thickness, and E_T is the elastic modulus transverse to the fiber direction. For the finite element results, a 20-element

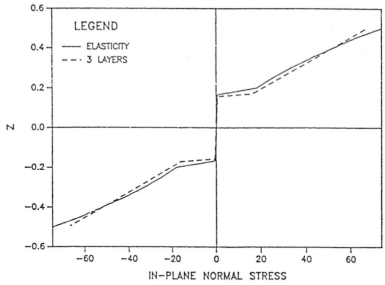

FIG. 4.   Normalized normal stress through-the-thickness of a plate strip ($S = 10$) subjected to a cylindrical bending load.

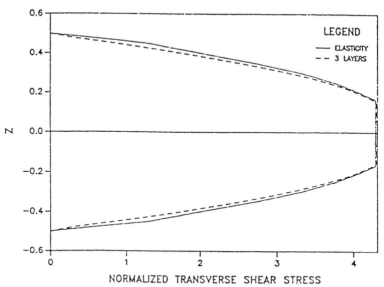

FIG. 5.   Normalized transverse shear stress through-the-thickness of a plate strip ($S = 10$) subjected to a cylindrical bending load.

strip of elements is used in the plane, and 3 layers are used to model in the thickness direction. Results are given for an aspect ratio of 10 and compare well to the elasticity solution given by Pagano.[2]

CONCLUSIONS

Good results have been obtained for the thick plate problems studied. These results have been defined as preliminary at this point because current efforts, focused on improving the satisfaction of equilibrium, should provide greater accuracy in the calculated stresses. Furthermore, the formulation should provide a numerically efficient means of addressing more complex problems, e.g. cutouts, ply drop-offs, and even interior debonds.

REFERENCES

1. WHITNEY, J. M., The effect of transverse shear deformation on the bending of laminated plates. *Jnl Comp. Mater.*, **3** (1969) 534–47.
2. PAGANO, N. J., Exact solutions for composite laminates in cylindrical bending. *Jnl Comp. Mater.*, **3** (1969) 398–311.
3. PAGANO, N. J., Exact solutions for rectangular bidirectional composites and sandwich plates. *Jnl Comp. Mater.*, **4** (1970) 20–34.
4. MAU, S. T., TONG, P. and PIAN, T. H. H., Finite element solutions for laminated thick plates. *Jnl Comp. Mater.*, **6** (1972) 304–11.
5. SPILKER, R. L. and ENGELMANN, B. E., Hybrid-stress isoparametric elements for moderately thick and thin multilayer plates. *Comp. Methods Appl. Mech. Eng.*, **56** (1986) 339–61.
6. PANDYA, B. N. and KANT, T., Finite element analysis of laminated composite plates using a higher-order displacement model. *Comp. Sci. Technol.*, **34** (1988) 137–55.
7. HINRICHSEN, R. L. and PALAZOTTO, A. N., Nonlinear finite element analysis of thick composite plates using cubic spline functions. *AIAA Journal*, **24** (1986) 1836–42.
8. REDDY, J. N., BARBERO, E. J. and TEPLY, J. L., A plate bending element based on a generalized laminate plate theory. *AIAA Paper No. 88-2322* (1988) 937–43.
9. CHATTERJEE, S. N. and RAMNATH, V., Modeling laminated composite structure as assemblage of sublaminates. *Int. J. Sol. Struct.*, **24** (1988) 439–58.
10. CHATTERJEE, S. N., Three- and two-dimensional stress fields near delaminations in laminated composite plates. *Int. J. Sol. Struct.*, **23** (1987) 1535–48.
11. HAMDALLAH, J. M. and ENGBLOM, J. J., Finite element plate formulation including transverse shear effects for representing composite shell structures. *Jnl Rein. Plast. Comp.* (in press).

12. FUEHNE, J. P. and ENGBLOM, J. J., A shear deformable, doubly-curved finite element for the analysis of laminated composite structures. *Comp. Struct.*, **12**, 81–95.

13. ENGBLOM, J. J. and OCHOA, O. O., Finite element formulation including interlaminar stress calculations. *Comput. Struct.*, **23**(2) (1986) 241–9.

14. ENGBLOM, J. J. and OCHOA, O. O., Through-the-thickness stress predictions for laminated plates of advanced composite materials, *Int. Jnl Numer. Meths Engng.*, **21** (1985) 1759–76.

15. PANDA, S. and NATARAJAN, R., Analysis of laminated composite shell structures by the finite element method. *Comput. Struct.*, **14** (1981) 225–30.

16. AHMAD, S., IRONS, B. M. and ZIENKIEWICZ, O. C., Analysis of thick and thin shell structures by curved finite element. *Int. Jnl Numer. Meths in Engng*, **2** (1970) 419–51.

22

Nonlinear Effects on Delamination Characterization of Composites under Multi-axial Loads

DARIO CASTANO and OZDEN O. OCHOA

Texas A & M University, College Station, Texas 77843, USA

ABSTRACT

The paper explores the nonlinear geometric effects on the strain energy release rates of delaminated composite specimens subjected to multi-axial loads. A numerical solution is obtained using an incremental finite element formulation in which large displacements and rotations but small strains are assumed. A Lagrangian formulation is adopted in the development of the finite element model. The stress field obtained is used with the crack closure technique to evaluate the total strain energy release rate as well as the three components: opening, shearing and twisting modes. Comparisons of nonlinear and linear results are presented to illustrate the effects of large deformations on stresses and strain energy release rates.

INTRODUCTION

The ongoing efforts of using composites as primary structural members require a thorough study of their structural/material response and susceptibility to failure. The classical plate theory approach which models symmetric laminates adequately cannot address the complexity that arises in the presence of delamination. At this stage the laminate becomes unsymmetric and the bending–stretching coupling becomes important. In addition to unsymmetric laminates, laminates with a high percentage of 45° plies display bending–stretching coupling which enhances the possibility of

431

a large deformation response. Thus geometrically nonlinear strain–displacement relations are required to model the laminates successfully. A quasi-three-dimensional finite element program developed by Chan and Ochoa[1] for delamination characterization is used as a basis for the implementation of the nonlinear static analysis. The element used is an eight-noded isoparametric element with three degrees of freedom per node; namely three displacements, U_1, U_2 and U_3. The assumed displacement field is modified to incorporate applied uni-axial strain, longitudinal and twisting curvatures. This element is used to model the cross-section of a laminate to observe the free edge effects. Further details are given in Ref. 1.

ANALYSIS

In the following development of the isoparametric finite element formulation it will be assumed that the displacements and rotations of the fibers are large, but the fiber extensions and angle changes between fibers are small. In essence, the stress–strain material property matrix remains constant while the original finite element area undergoes large rigid body displacements and rotations. Consider the motion of a body in a stationary Cartesian coordinate system (Fig. 1), and assume that the solutions for the static and kinematic variables from time 0 to time t have been obtained.

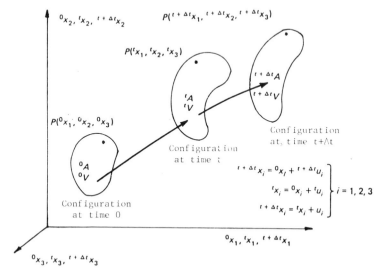

FIG. 1.   Motion of body in stationary Cartesian coordinate system.

Then, to solve for the variables at time $t + \Delta t$, an updated Lagrangian formulation is used.[2]

The equilibrium of the body at time $t + \Delta t$ can be expressed using the principle of virtual displacements as

$$\int_{t+\Delta t_V} {}^{t+\Delta t}\tau_{ij}\delta_{t+\Delta t}e_{ij}{}^{t+\Delta t}\mathrm{d}V = {}^{t+\Delta t}R$$

where ${}^{t+\Delta t}\tau_{ij}$ is the Cauchy stress tensor, $\delta_{t+\Delta t}e_{ij}$ is the variation in the infinitesimal strain tensor, and ${}^{t+\Delta t}R$ corresponds to the virtual work from external forces, body forces, initial strains and so on.

Using the Green–Lagrange strain tensor defined in the updated formulation as

$$^{t+\Delta t}_t\varepsilon_{ij} = \tfrac{1}{2}({}_tu_{i,j} + {}_tu_{j,i} + {}_tu_{k,i}\,{}_tu_{k,j})$$

in which ${}_tu_{i,j} = \partial u_i/\partial {}^tX_j$, then the above virtual displacement equation reduces to

$$\int_{t_V} {}^{t+\Delta t}_tS_{ij}\delta{}^{t+\Delta t}_t\varepsilon_{ij}{}^t\mathrm{d}V = {}^{t+\Delta t}R$$

${}^{t+\Delta t}_tS_{ij}$ is the second Piola–Kirchhoff stress tensor at time $t + \Delta t$ referred to the configuration at time t, and it is defined as

$$^{t+\Delta t}_tS_{ij} = \frac{{}^t\rho}{{}^{t+\Delta t}\rho}\,{}^{t+\Delta t}_tX_{i,m}\,{}^{t+\Delta t}\tau_{mn}\,{}^{t+\Delta t}_tX_{j,n}$$

in which ${}^t\rho/{}^{t+\Delta t}\rho$ represents the ratio of the mass densities, and

$$^{t+\Delta t}_tX_{i,m} = \frac{\partial {}^tX_i}{\partial {}^{t+\Delta t}X_m}$$

Also

$$\delta{}^{t+\Delta t}_t\varepsilon_{ij} = \delta\tfrac{1}{2}({}_tu_{i,j} + {}_tu_{j,i} + {}_tu_{k,i}\,{}_tu_{k,j})$$

The unknown stresses ${}^{t+\Delta t}_tS_{ij}$ can now be incrementally decomposed into known stresses ${}^t_tS_{ij} = {}^t\tau_{ij}$ plus increments in stresses ${}_tS_{ij}$.

The strain ${}^{t+\Delta t}_t\varepsilon_{ij} = {}_t\varepsilon_{ij}$ represents the increment of strain tensor from time t to time $t + \Delta t$ with respect to configuration at time t. ${}_t\varepsilon_{ij}$ can be rewritten as

$$_t\varepsilon_{ij} = {}_te_{ij} + {}_t\eta_{ij}$$

where

$$_te_{ij} = \tfrac{1}{2}({}_tu_{i,j} + {}_tu_{j,i}) \qquad {}_t\eta_{ij} = \tfrac{1}{2}\,{}_tu_{k,i}\,{}_tu_{k,j}$$

Therefore the equation of virtual displacement reduces to

$$\int_{t_V} {}_t S_{ij}\delta_t\varepsilon_{ij}\,{}^t\mathrm{d}V + \int_{t_V} {}^t\tau_{ij}\delta_t\eta_{ij}\,{}^t\mathrm{d}V = {}^{t+\Delta t}R - \int_{t_V} {}^t\tau_{ij}\delta_t e_{ij}\,{}^t\mathrm{d}V$$

Using the approximations ${}_t S_{ij} = {}_t C_{ijrs}\,{}_t e_{rs}$ and $\delta_t\varepsilon_{ij} = \delta_t e_{ij}$, we obtain as an approximate equation of motion

$$\int_{t_V} {}_t C_{ijrs}\,{}_t e_{rs}\delta_t e_{ij}\,{}^t\mathrm{d}V + \int_{t_V} {}^t\tau_{ij}\delta_t\eta_{ij}\,{}^t\mathrm{d}V = {}^{t+\Delta t}R - \int_{t_V} {}^t\tau_{ij}\delta_t e_{ij}\,{}^t\mathrm{d}V \quad (1)$$

where ${}_t C_{ijrs}$ is the tangent material property tensor at time t referred to the configuration at time t. It is also noted that

$${}^t\tau_{ij} = {}_t C_{ijrs}\,{}^t_t\varepsilon_{rs}$$

where ${}^t_t\varepsilon_{rs}$ is the Almansi strain tensor defined as

$${}^t_t\varepsilon_{ij} = \tfrac{1}{2}({}^t_t u_{i,j} + {}^t_t u_{j,i} - {}^t_t u_{k,i}\,{}^t_t u_{k,j})$$

The assumed displacement field from configuration at time t to configuration at time $t + \Delta t$ is given by[1]

$$\begin{aligned}
U_1({}^t X_1, {}^t X_2, {}^t X_3) &= u_1({}^t X_2, {}^t X_3) + (\varepsilon_0\,{}^t X_1 + K^t X_1\,{}^t X_3) \\
U_2({}^t X_1, {}^t X_2, {}^t X_3) &= u_2({}^t X_2, {}^t X_3) + (C^t X_1\,{}^t X_3) \\
U_3({}^t X_1, {}^t X_2, {}^t X_3) &= u_3({}^t X_2, {}^t X_3) + (-C^t X_1\,{}^t X_2 - 1/2K^t X_1^2)
\end{aligned} \quad (2)$$

where u_i are the customary displacements of isoparametric elements, ε_0 is the uniform extension, K is the longitudinal curvature, and C is half the twisting curvature.

The displacements and coordinates of the elements are represented by

$${}^t u_i = \sum_{K=1}^{8} h_K\,{}^t u_i^K \qquad {}^t X_i = \sum_{K=1}^{8} h_K\,{}^t X_i^K \quad (3)$$

for $i = 1, 2, 3$. By substituting the above interpolation functions in the equation of motion, eqn (1), we obtain the following expressions:

(a)
$${}^t\tau_{ij}\delta_t e_{ij} = \delta\hat{u}^{\mathrm{T}}\,{}^t_t B_{\mathrm{L}}^{\mathrm{T}}\,{}^t\hat{\tau}$$
$$\scriptstyle (1\times24)\,(24\times5)(5\times1)$$

where $\delta\hat{u}^{\mathrm{T}}$ is the transform of the virtual nodal displacement vector, ${}^t_t B_{\mathrm{L}}^{\mathrm{T}}$ is the transform of the linear strain–displacement transformation matrix, and ${}^t\hat{\tau}$ is the vector of Cauchy stresses. Notice that in our discussion we use

$$\delta_t e_{ij} = \tfrac{1}{2}\delta({}_t U_{i,j} + {}_t U_{j,i})$$

Also

$${}^t_t\varepsilon_{ij} = \tfrac{1}{2}({}^t_t U_{i,j} + {}^t_t U_{j,i} - {}^t_t U_{k,i}\,{}^t_t U_{k,j})$$

(b) $\quad {}^t\tau_{ij}\delta_t\eta_{ij} = \delta\hat{u}^T \underset{t}{{}^t}B_{NL}^T \; {}^t\tau \; {}^t_t B_{NL} \; \hat{u}$
$\qquad\qquad\quad {\scriptstyle(1 \times 24)(24 \times 6)(6 \times 6)(6 \times 24)(24 \times 1)}$

$$+ \delta u^T {}^t_t B_{NL}^T \begin{Bmatrix} {}^t\tau_{12}(\varepsilon_0 + K^t X_3) + {}^t\tau_{23}K^t X_1 \\ {}^t\tau_{13}(\varepsilon_0 + K^t X_3) + {}^t\tau_{33}K^t X_1 \\ {}^t\tau_{12}C^t X_3 + {}^t\tau_{23}C^t X_1 \\ {}^t\tau_{13}C^t X_3 + {}^t\tau_{33}C^t X_1 \\ {}^t\tau_{12}(-C^t X_2 - K^t X_1) + {}^t\tau_{22}(-C^t X_1) \\ {}^t\tau_{13}(-C^t X_2 - K^t X_1) + {}^t\tau_{23}(-C^t X_1) \end{Bmatrix}$$

where ${}^t_t B_{NL}^T$ is the transform of the nonlinear strain–displacement transformation matrix, ${}^t\tau$ is the matrix of Cauchy stresses, and $\hat{u}$ is the vector of nodal displacements.

(c) $\quad {}_tC_{ijrs}\,{}_te_{rs}\delta_t e_{ij} = \delta\hat{u}^T \underset{t}{{}^t}B_L^T \; {}_tD \; {}^t_t B_L \; \hat{u}$
$\qquad\qquad\qquad {\scriptstyle(1 \times 24)(24 \times 5)(5 \times 5)(5 \times 24)(24 \times 1)}$

$$+ \delta u^T {}^t_t B_L^T \begin{Bmatrix} \bar{Q}_{12}(\varepsilon_0 + K^t X_3) + \bar{Q}_{26}C^t X_3 \\ \bar{Q}_{13}(\varepsilon_0 + K^t X_3) + \bar{Q}_{36}C^t X_3 \\ \bar{Q}_{45}(-C^t X_2) \\ \bar{Q}_{55}(-C^t X_2) \\ \bar{Q}_{16}(\varepsilon_0 + K^t X_3) + \bar{Q}_{66}C^t X_3 \end{Bmatrix}$$

where ${}_tD$ is the material property matrix and the $\bar{Q}$'s are elements in the material matrix.

For unit virtual nodal point displacements, the equations of the nodal displacements take the form

$$\left(\int_{{}^t v} {}^t_t B_L^T \,{}_tD\, {}^t_t B_L \,{}^t dV + \int_{{}^t v} {}^t_t B_{NL}^T \,{}^t\tau\, {}^t_t B_{NL} \,{}^t dV \right)\hat{u}$$

$$= -\int_{{}^t v} {}^t_t B_L^T \begin{Bmatrix} \bar{Q}_{12}(\varepsilon_0 + K^t X_3) + \bar{Q}_{26}C^t X_3 \\ \bar{Q}_{13}(\varepsilon_0 + K^t X_3) + \bar{Q}_{36}C^t X_3 \\ \bar{Q}_{45}(-C^t X_2) \\ \bar{Q}_{55}(-C^t X_2) \\ \bar{Q}_{16}(\varepsilon_0 + K^t X_3) + \bar{Q}_{66}C^t X_3 \end{Bmatrix} {}^t dV \qquad (4)$$

$$-\left(\int_{{}^t v} {}^t_t B_L^T \,{}^t\hat{\tau}\, {}^t dV + \int_{{}^t v} {}^t_t B_{NL}^T \begin{Bmatrix} {}^t\tau_{12}(\varepsilon_0 + K^t X_3) + {}^t\tau_{23}K^t X_1 \\ {}^t\tau_{13}(\varepsilon_0 + K^t X_3) + {}^t\tau_{33}K^t X_1 \\ {}^t\tau_{12}C^t X_3 + {}^t\tau_{23}C^t X_1 \\ {}^t\tau_{13}C^t X_3 + {}^t\tau_{33}C^t X_1 \\ {}^t\tau_{12}(-C^t X_2 - K^t X_1) + {}^t\tau_{22}(-C^t X_1) \\ {}^t\tau_{13}(-C^t X_2 - K^t X_1) + {}^t\tau_{23}(-C^t X_1) \end{Bmatrix} {}^t dV \right)$$

In this analysis it is assumed that there are no external forces or body forces, or residual strains acting directly at the nodes, therefore the forcing parameter is ε_0 (which is incremented during the iteration procedure).

The solution of the nodal displacements for the discrete time $t + \Delta t$ cannot be given directly since all the quantities with the exception of $_tD$ and $\bar{Q}_{ij}$ change with time (we assume that the material properties remain constant). Therefore we iterate until the solution is obtained to sufficient accuracy. The equations used in the modified Newton iteration procedure are

$$_t^t K \Delta \hat{u}^{(i)} = {}^{t+\Delta t}\hat{R} - {}^{t+\Delta t}_{t+\Delta t}\hat{F}^{(i-1)} \tag{5}$$

$$^{t+\Delta t}\hat{u}^{(i)} = {}^{t+\Delta t}\hat{u}^{(i-1)} + \Delta \hat{u}^{(i)} \tag{6}$$

with the initial conditions

$$^{t+\Delta t}\hat{u}^{(0)} = {}^t\hat{u} \qquad {}^{t+\Delta t}\hat{F}^{(0)} = {}^t\hat{F} \tag{7}$$

Using numerical integration, the stiffness matrix $_t^t K$ is evaluated from

$$_t^t K = \sum_{\substack{s=1 \\ t=1}}^{3} W_s W_t ({}_t^t B_L^T \, _t^t D \, _t^t B_L + {}_t^t B_{NL}^T \, {}^t\tau \, _t^t B_{NL}) \det {}^t J \tag{8}$$

Also the forces on the right-hand side of eqn (5) are expressed as

$$^{t+\Delta t}\hat{R} = -\sum_{\substack{s=1 \\ t=1}}^{3} W_s W_t \left({}^{t+\Delta t}_{t+\Delta t} B_L^{(i-1)T} \begin{cases} \bar{Q}_{12}(\varepsilon_0 + K^t X_3) + \bar{Q}_{26} C^t X_3 \\ \bar{Q}_{13}(\varepsilon_0 + K^t X_3) + \bar{Q}_{36} C^t X_3 \\ \bar{Q}_{45}(-C^t X_2) \\ \bar{Q}_{55}(-C^t X_2) \\ \bar{Q}_{16}(\varepsilon_0 + K^t X_3) + \bar{Q}_{66} C^t X_3 \end{cases} \right) \tag{9}$$

$$\times \det {}^{t+\Delta t}J^{(i-1)}$$

It should be noted that $^{t+\Delta t}\hat{R}$ is deformation-dependent, and is therefore evaluated using ε_0, K, C and $^t X_i$ at time $t + \Delta t$, but integrating over the area last calculated in the iteration. The same conditions as above apply to

$$^{t+\Delta t}_{t+\Delta t}\hat{F}^{(i-1)} = \sum_{\substack{s=1 \\ t=1}}^{3} W_s W_t$$

$$\times \left({}^{t+\Delta t}_{t+\Delta t}B_L^{(i-1)T} \, {}^{t+\Delta t}\hat{\tau}^{(i-1)} + {}^{t+\Delta t}_{t+\Delta t}B_{NL}^{(i-1)T} \begin{cases} ^{t+\Delta t}\tau_{12}^{(i-1)}(\varepsilon_0 + K^t X_3) + {}^{t+\Delta t}\tau_{23}^{(i-1)}K^t X_1 \\ ^{t+\Delta t}\tau_{13}^{(i-1)}(\varepsilon_0 + K^t X_3) + {}^{t+\Delta t}\tau_{33}^{(i-1)}K^t X_1 \\ ^{t+\Delta t}\tau_{12}^{(i-1)}C^t X_3 + {}^{t+\Delta t}\tau_{23}^{(i-1)}C^t X_1 \\ ^{t+\Delta t}\tau_{13}^{(i-1)}C^t X_3 + {}^{t+\Delta t}\tau_{33}^{(i-1)}C^t X_1 \\ ^{t+\Delta t}\tau_{12}^{(i-1)}(-C^t X_2 - K^t X_1) + {}^{t+\Delta t}\tau_{22}^{(i-1)}(-C^t X_1) \\ ^{t+\Delta t}\tau_{13}^{(i-1)}(-C^t X_2 - K^t X_1) + {}^{t+\Delta t}\tau_{23}^{(i-1)}(-C^t X_1) \end{cases} \right) \tag{10}$$

$$\times \det {}^{t+\Delta t}J^{(i-1)}$$

Equations (8)–(10) provide a detailed definition of the equilibrium statement which is solved incrementally. An energy convergence criterion is used to terminate the iterations. Convergence is assumed to be reached when

$$\Delta \hat{u}^{(i)\mathrm{T}}({}^{t+\Delta t}\hat{R} - {}^{t+\Delta t}_{t+\Delta t}\hat{F}^{(i-1)}) \le \varepsilon_{\mathrm{E}}[\Delta \hat{u}^{(1)\mathrm{T}}({}^{t+\Delta t}\hat{R} - {}^{t}_{t}\hat{F})]$$

where the energy tolerance ε_{E} is a preset value.[3]

The strain is evaluated at the integration points using the Green–Lagrange strain tensor and the updated Lagrangian formulation, giving

$$^{t+\Delta t}_{t}\varepsilon_{ij} = \tfrac{1}{2}({}_{t}U_{i,j} + {}_{t}U_{j,i} + {}_{t}U_{k,i}{}_{t}U_{k,j})$$

Finally, the stress is obtained from the matrix product of material properties and strain vector. The crack closure technique is implemented with the stress field after convergence to evaluate the strain energy release rates.

RESULTS

Several analytical cases for tension loads are studied. The stacking sequences considered are $[+35/-35/0/90]_s$ and $[45_2/-45^2]_s$. The applied strain for the uni-axial tension is $2000\mu\varepsilon$. Later this is increased to $20\,000\mu\varepsilon$ to highlight the nonlinear effects. The material systems used are AS4/3601-6 and PEEK/AS4.

$[35/-35/0/90]_s$ Laminate

The interlaminar normal stress σ_{zz} and the interlaminar shear stresses σ_{yz} and σ_{xz} are displayed as functions of the width in Figs 2 and 3 respectively. These stress values are for the first lamina, i.e. 90° and they are calculated at the centroid of the elements. There is no difference between the linear and nonlinear results for this strain level of $2000\mu\varepsilon$. The same laminate when subjected to $20\,000\mu\varepsilon$ displays a slight difference in the interlaminar stresses obtained from linear analysis in comparison to those obtained by nonlinear analysis. This can be observed from Figs 4 and 5. Note that the variation of these stresses as functions of the width follows the same trend as those first observed in Figs 2 and 3. Figure 6 illustrates the deformation of the elements which are located at the right edge of the section being modeled. The deformation of the element in the 90° lamina is highlighted.

$[45_2/-45_2]_s$ Laminate

Figures 7 and 8 display the interlaminar normal and shear stresses. Note that σ_{xz} is larger than σ_{yz}. Once again for the applied strain of $2000\mu\varepsilon$ the

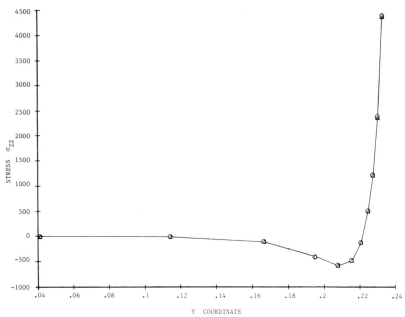

FIG. 2. Nonlinear and linear interlaminar normal stresses at the midplane for a $[35/-35/0/90]_s$ laminate subjected to $\varepsilon_0 = 2000\mu\varepsilon$. ○, Linear; △, nonlinear.

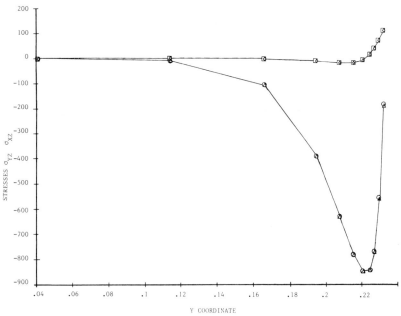

FIG. 3. Nonlinear and linear interlaminar shear stresses at the midplane for a $[35/-35/0/90]_s$ laminate subjected to $\varepsilon_0 = 2000\mu\varepsilon$. ○, σ_{yz} Linear; △, σ_{yz} nonlinear; □, σ_{xz} linear; ◇, σ_{xz} nonlinear.

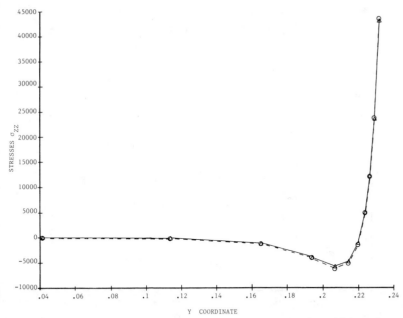

FIG. 4. Nonlinear and linear interlaminar normal stresses at the midplane for a [35/−35/0/90]$_s$ laminate subjected to $\varepsilon_0 = 20\,000\mu\varepsilon$. ○, Linear; △, nonlinear.

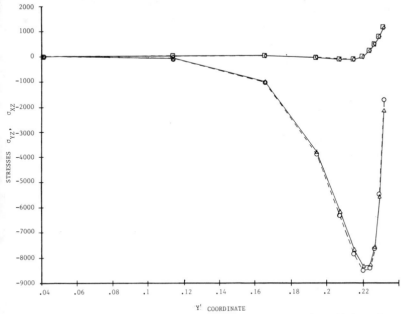

FIG. 5. Nonlinear and linear interlaminar shear stresses at the midplane for a [35/−35/0/90]$_s$ laminate subjected to $\varepsilon_0 = 20\,000\mu\varepsilon$. ○, σ_{yz} Linear; △, σ_{yz} nonlinear; □, σ_{xz} linear; ◇, σ_{xz} nonlinear.

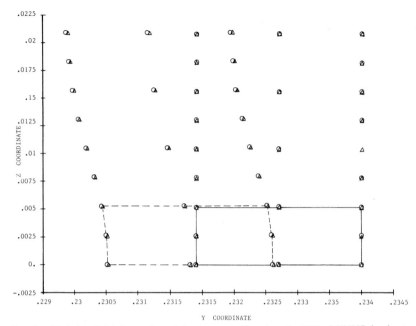

FIG. 6. Global nodal deformation of the element at the edge of a $[35/-35/0/90]_s$ laminate subjected to $\varepsilon_0 = 20\,000\mu\varepsilon$. $\bigcirc$, Linear; $\triangle$, nonlinear.

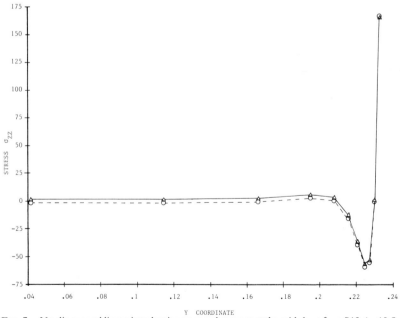

FIG. 7. Nonlinear and linear interlaminar normal stresses at the midplane for a $[45_2/-45_2]_s$ laminate subjected to $\varepsilon_0 = 2000\mu\varepsilon$. $\bigcirc$, Linear; $\triangle$, nonlinear.

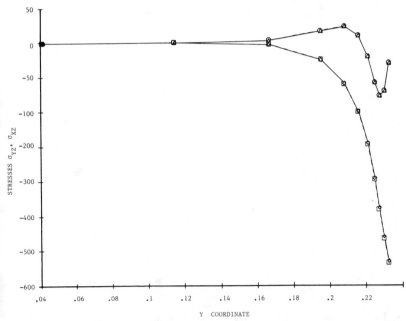

FIG. 8. Nonlinear and linear interlaminar shear stresses at the midplane for a $[45_2/-45_2]_s$ laminate subjected to $\varepsilon_0 = 2000\mu\varepsilon$. $\bigcirc$, σ_{yz} Linear; $\triangle$, σ_{yz} nonlinear; $\square$, σ_{xz} linear; $\lozenge$, σ_{xz} nonlinear.

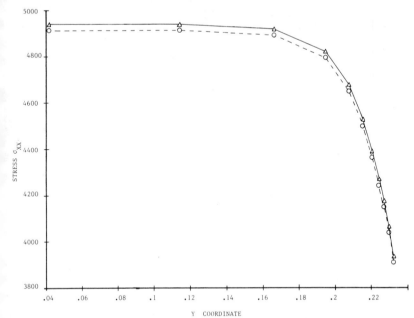

FIG. 9. Nonlinear and linear axial stress at the midplane for a $[45_2/-45_2]_s$ laminate subjected to $\varepsilon_0 = 2000\mu\varepsilon$. $\bigcirc$, Linear; $\triangle$, nonlinear.

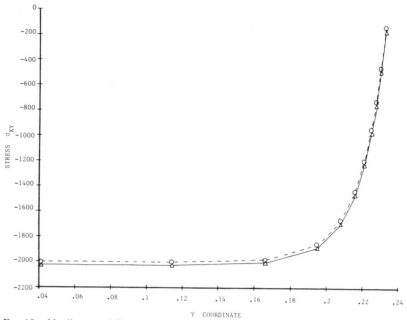

FIG. 10. Nonlinear and linear in-plane stress at the midplane for a $[45_2/-45_2]_s$ laminate subjected to $\varepsilon_0 = 2000\mu\varepsilon$. $\bigcirc$, Linear; $\triangle$, nonlinear.

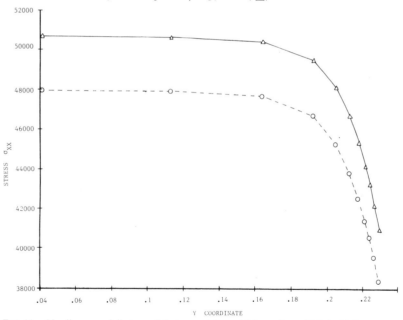

FIG. 11. Nonlinear and linear axial stress at the midplane for a $[45_2/-45_2]_s$ laminate subjected to $\varepsilon_0 = 20\,000\mu\varepsilon$. $\bigcirc$, Linear; $\triangle$, nonlinear.

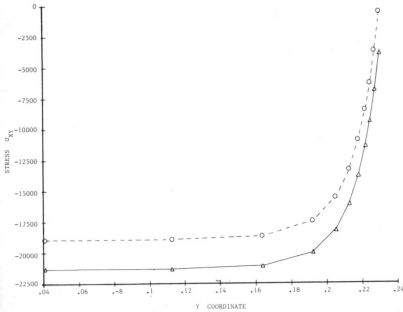

FIG. 12. Nonlinear and linear in-plane shear stress at the midplane for a $[45_2/-45_2]_s$ laminate subjected to $\varepsilon_0 = 20\,000\mu\varepsilon$. $\bigcirc$, Linear; $\triangle$, nonlinear.

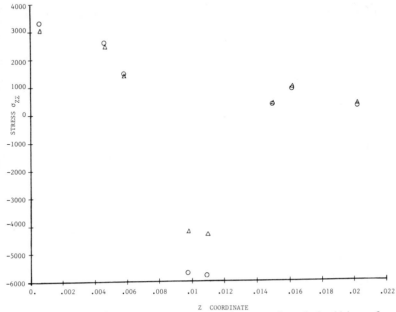

FIG. 13. Nonlinear and linear interlaminar normal stresses through-the-thickness for a $[45_2/-45_2]_s$ laminate subjected to $\varepsilon_0 = 20\,000\mu\varepsilon$. $\bigcirc$, Linear; $\triangle$, nonlinear.

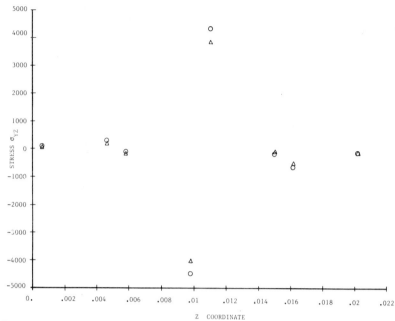

FIG. 14. Nonlinear and linear interlaminar shear stresses through-the-thickness for a $[45_2/-45_2]_s$ laminate subjected to $\varepsilon_0 = 2000\mu\varepsilon$. $\bigcirc$, Linear; $\triangle$, nonlinear.

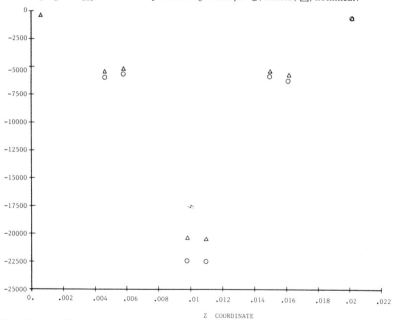

FIG. 15. Nonlinear and linear interlaminar shear stresses through-the-thickness for a $[45_2/-45_2]_s$ laminate subjected to $\varepsilon_0 = 20\,000\mu\varepsilon$. $\bigcirc$, Linear; $\triangle$, nonlinear.

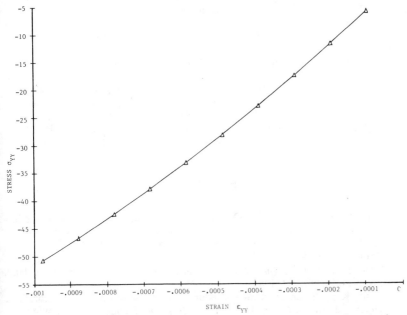

FIG. 16.   In-plane stress as a function of strain in the top 45° lamina near the free edge for a
[45$_2$/−45$_2$]$_s$ laminate subjected to $\varepsilon_0 = 2000\mu\varepsilon$.

difference between linear and nonlinear responses is negligible. However, for the normal stress σ_{xx} and the shear stress σ_{xy}, Figs 9 and 10 illustrate the difference in linear versus nonlinear results.

When this laminate is subjected to $20\,000\mu\varepsilon$, the highest differences in stresses between linear and nonlinear analysis are as depicted in Fig. 11 for σ_{xx} and in Fig. 12 for σ_{xy}. Also note the variation of interlaminar stresses σ_{zz}, σ_{yz} and σ_{xz} as functions of through-the-thickness of the laminate displayed in Figs 13–15. Finally, to illustrate the nonlinearity evaluated by the program, the in-plane stresses σ_{yy} is plotted against the strain ε_{yy} in Fig. 16 for the applied strain ε_0, increasing to $2000\mu\varepsilon$ in 10 steps of $200\mu\varepsilon$. These stress and strain values were evaluated at the centroid of an element near the edge in the top 45° lamina. When the material is changed from AS4/3501-6 to PEEK/AS4, one observes the same trends but for slightly higher magnitudes of stress.

The strain energy release rates are obtained by incorporating a crack of length of eight ply thicknesses at the first interface from the midplane for both stacking sequences, namely between 90/0 and −45/−45 for the laminates studied. The values obtained are given in Table 1.

TABLE 1

Strain energy release rate	$[45_2/-45_2]_s$		$[35/-35/0/90]_s$	
	Linear	Nonlinear	Linear	Nonlinear
G_I	0·000 03	0·000 01	0·542 64	0·542 39
G_{II}	0·004 62	0·004 57	0·046 15	0·044 93
G_{III}	0·050 72	0·049 50	0·000 35	0·000 39
G_{TOT}	0·055 38	0·054 08	0·589 14	0·587 71

REFERENCES

1. CHAN, W. S. and OCHOA, O. O., An integrated finite element model of edge-delamination analysis for laminates due to tension, bending, and torsion loads. AIAA Paper No. 87-0704-CP. *Proc. of 28th AIAA/ASME SDM Conf.,* Monterey, CA, April 6–8, 1987, pp. 27–35.
2. BATHE, K. J., *Finite Element Procedures in Engineering Analysis.* Prentice–Hall, Englewood Cliffs, NJ, 1982.
3. BATHE, K. J., Some practical procedures for the solution of nonlinear finite element equations. *Comp. Meths Appl. Mech. and Engng,* **22** (1980) 59–85.

23

On the Postbuckling Behavior of a Delaminated Thin Cylindrical Shell

Z. Q. CHEN and G. J. SIMITSES

*Georgia Institute of Technology, Atlanta,
Georgia 30332, USA*

ABSTRACT

The buckling and postbuckling behavior of cylindrical shell configurations, in the presence of delaminations, is discussed. Models employed for the buckling analysis of such systems are critically reviewed. Moreover, a review of some similar studies of buckling and postbuckling of rings and liners confined to uniformly contracting circular boundaries is presented. Finally, a simple model is developed for the analysis of a delaminated thin cylindrical film from a contracting cylindrical substrate. Results are obtained by employing this model, and these results seem to be qualitatively reasonable. Suggestions for future studies are offered.

1 INTRODUCTION

Composite structures often contain internal delaminations. Causes of internal delaminations are many and include tool drops, bird strikes, runway debris hits, and manufacturing defects. Moreover, in some cases, especially in the vicinity of holes or close to edges in general, delaminations start because of the development of interlaminar stresses. Several analyses have been reported on the subject of edge delamination and its importance in the design of laminated structures. A few of these works are referenced.[1-4] These and their cited references form a good basis for the interested reader.

Internal delaminations present an important problem especially for laminated structures which are subject to destabilizing loads. The presence

447

of delamination in these situations may cause local buckling and/or trigger global buckling and therefore induce a reduction in the overall load-bearing capacity of the laminated structure. The problem, because of its importance, has received considerable attention in recent years.

A finite element analysis was developed by Whitcomb[5] to analyze a laminated plate with a through-the-width delamination, for pre- and post-buckling behavior. Shivakumar and Whitcomb[6] in 1985 studied the buckling of an elliptic delamination embedded near the surface (thin film analysis) of a thick quasi-isotropic laminate. A one-dimensional analytical model was developed by Chai et al.[7,8] to assess the comprehensive strength of near-surface interlaminar defects in laminated composites. Bottega and Maewal[9,10] considered the dynamics of delamination buckling and growth for circular plates and delaminations. Bottega[11] extended this work to arbitrary-shaped delaminations. Simitses and his collaborators[12-14] investigated the delamination buckling and growth of flat composite structural elements.

Owing to mathematical complexity, only very limited information on the subject of delamination buckling and growth of shells is currently available.

Troshin[15] studied the effect of longitudinal delamination, in a laminar cylindrical shell, on the critical external pressure. The governing equations were solved by the Kutta–Merson method. In another paper,[16] Troshin investigated the delamination stability of triple-layer shells with almost the same method. Sallam and Simitses[17] studied delamination buckling of axially loaded thin cylindrical shells. Moreover, the present authors[18,19] presented results of delamination buckling of pressure loaded cylindrical panels and shells. Finally, as far as delamination growth of shell configurations is concerned, one must refer to the studies of Bottega.[20,21]

In studying the delamination buckling of shell configurations, the investigators[15-19] made several simplifying assumptions. The significance of some of these assumptions in predicting critical conditions is discussed in the next section. Moreover, since knowledge not only of buckling and buckling modes but also of postbuckling response is very much needed in the study of delamination growth of debonded shell configurations, the paper presents a model for postbuckling analysis. The paper also contains a critical review of postbuckling analyses, on similar problems such as those dealing with buckling of circular rings confined to a stationary rigid boundary[22] or a contracting boundary.[23] The work of Bottega is based on the analysis of Ref. 23. The proposed simple model is free of simplifying assumptions and it captures the essential features of the anticipated postbuckling response characteristics.

2 LINEAR BUCKLING ANALYSIS MODELS

As already mentioned in the introduction, several simplifying assumptions were made by the investigators[15-19,24] in dealing with the buckling analysis of delaminated thin complete cylindrical shells and cylindrical panels. This led to a linear eigen-boundary-value problem and the lowest eigenvalue was taken to be a measure of the critical load, while the corresponding eigenfunction represented the buckling mode. In order that we discuss some of the dangers of this linear analysis, let us concentrate on the problem treated by Troshin[15] and Simitses and Chen.[18,19,24] This problem and the solution are briefly described below.

The authors investigated the delamination buckling of laminated cylinders and panels of various constructions (i.e. cross-ply, graphite/epoxy composite construction in Ref. 24) when acted upon by uniform external pressure (see Fig. 1). The geometries employed were such that they covered a wide range of length to radius ratios, and of radius to thickness ratios, as well as panels of various widths. The delaminated region was assumed to have constant width and extended along the entire length of the cylindrical shell or panel. Several parametric studies were performed and reported in connection with this problem. Some of the essential assumptions that led to the solution are listed below.

(i) The shell is free of initial geometric imperfections.

(ii) The geometry is such that regions I, II, III and IV (see Fig. 1) are symmetric with respect to their respective midsurfaces. The implication here is that in the prebuckling state the response is one of membrane state since there is no coupling (because of symmetry) between extension and bending.

(iii) The buckling equations for each region were obtained from the equilibrium equations of a pressure-loaded (uniformly) perfect cylindrical configuration by using the perturbation technique. This, of course, implies infinitesimal changes (in all parameters) from the membrane (primary) state to the bent (buckled) state, including the loading parameter q^i for each region (q^{II} is the pressure exerted on region II at the instant of buckling).

(iv) No provision was made to account for the possibility of contact between regions I and II after buckling. For example, in Refs 18 and 19 it was found that the prevailing mode in some cases was the antisymmetric mode, which implies partial contact between regions I and II along the delamination. Thus, the eigenfunctions

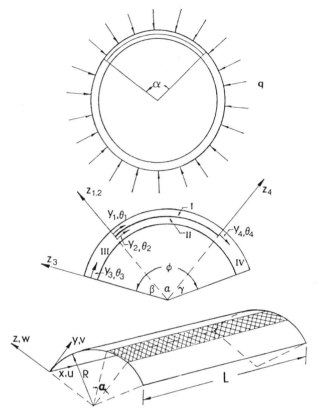

FIG. 1.   Geometry, loading and sign convention.

corresponding to the lowest eigenvalues were taken to represent the buckling modes.

The solution procedure is briefly described for the pressure–load case only.

A separated solution is assumed for the three displacement components with a single trigonometric function (of one full wave) representing buckling response along the cylinder or cylindrical panel, for each region. Substitution into the buckling equations reduces them to a set of ordinary differential equations in the hoop direction.

The solution of the ordinary differential equations is obtained in terms of eight undetermined constants for each region. Thus, the total number of unknowns is 32. The number of boundary and auxiliary conditions is also

32. The auxiliary conditions denote kinematic continuity and local balance of forces and moments at the two edges of the delamination.

Use of the 32 boundary and auxiliary conditions leads to a system of 32 linear, homogeneous, algebraic equations in the 32 unknown constants. A nontrivial solution exists if the determinant of the coefficients vanishes.

A computer program was written in order to obtain critical conditions.

For the case of axial compression the solution procedure is similar to the one described. For details concerning the solution for both load cases, the interested reader is referred to Refs 17–19 and 24.

3 PREVIOUS STUDIES ON SIMILAR PROBLEMS

The only reported studies on the postbuckling behavior of delaminated shells are those reported by Bottega.[20,21] These studies deal primarily with a thin layer (thin film) delaminating from a contracting cylindrical substrate.

In addition, there exist at least two previous studies on problems which are very similar to the one considered herein.

Hsu *et al.*[22] in 1964, studied the postbuckling of a thin ring which is confined to a rigid circular boundary and is subjected to an end-compressive load *P* as shown in Fig. 2. There are some similarities between the model in Ref. 22 and the one in Ref. 18 for an isotropic long cylindrical panel with very thin delaminated layer close to the inner surface (see Fig. 3).

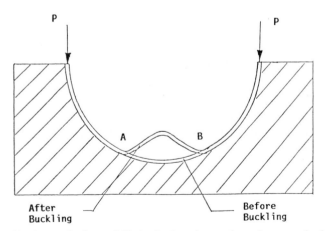

FIG. 2. A thin ring confined to a rigid circular boundary under end compressive load.[22]

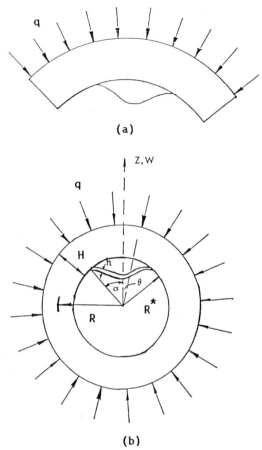

(a)

(b)

Fig. 3.   A long cylinder or panel with a delaminated thin layer subjected to external pressure.

These include:

1.   Segment AB in Fig. 2 is similar to a delaminated thin layer in Fig. 3.
2.   The delaminated layer in Fig. 3 is supposed to be very thin such that the buckling of it does not affect the substrate and therefore the substrate can be considered something similar to the rigid boundary in Fig. 2.

Although the two models are different in loading, boundary conditions etc., the results in Ref. 22 should still provide some insight into the problem of Ref. 18. The authors of Ref. 22 obtained the prebuckling and

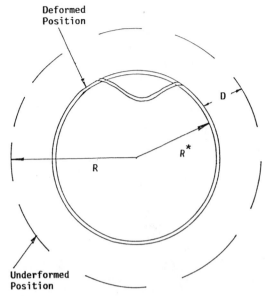

Deformed
Position

Underformed
Position

FIG. 4. A thin ring confined to a uniformly contracting circular boundary.[23]

postbuckling deflections of the thin ring with initial imperfections and concluded that for the perfect configuration the prebuckling deflection branch and the postbuckling deflection branch would meet at infinity, i.e. no critical load could be found.

In 1972, El-Bayoumy[23] investigated the postbuckling behavior of a circular elastic ring confined to a uniformly contracting circular boundary (see Fig. 4). The models shown in Fig. 3 and Fig. 4 would be similar, if one assumed that the circular boundary in Fig. 4 is a ring and relate the contracting displacement, D, to the external pressure. In spite of the similarity, the two models are different because in Fig. 3 only a small segment of a ring debonds from the substrate, while in Fig. 4 the whole ring debonds from the circular boundary. In Ref. 23 the buckled configuration was assumed to consist of two regions, i.e. a detached region and an attached region. The problem was treated as a variational problem with variable end points. Three deflection branches of the crown point, i.e. unbuckled deflection branch, and small and large buckled deflection branches were plotted versus the contracting displacement of the circular boundary. Since the unbuckled deflection branch did not intersect the buckled deflection branch, no limit or bifurcation point could be found. The stability of the various equilibrium branches were discussed using an

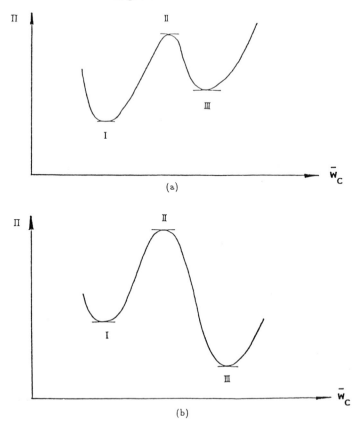

Fig. 5.   Total potential, Π, curves with three static equilibrium points.

energy barrier approach, which showed that the unbuckled and the large buckled deflection branches are stable and the small buckled deflection branch is unstable. The author concluded that the energy barrier decreases with increasing contraction thereby making the ring more sensitive to external disturbance, and that the ability of the ring to sustain external disturbance diminishes as the contraction increases.

Unfortunately, this conclusion[23] is not correct for the reason stated below.

For a given contraction, the total potential, Π, at three equilibrium positions have the relative heights, as shown in Figs 5(a) or 5(b), in which I, II and III correspond to equilibrium positions in the prebuckling branch and small and large buckled branches, respectively. In order to move from the unbuckled equilibrium position (I) to the buckled stable equilibrium position (III), it is necessary to obtain additional total potential energy which is larger than $\Pi_{II} - \Pi_I$, therefore a finite disturbance is needed. Without a finite disturbance, the configuration will stay in its unbuckled state. The conclusion of Ref. 23 that when the difference in energy levels between I and III gets larger it will be easier for the ring to buckle does not make sense. The real situation is that the configuration could not find its way to the buckled state no matter if the energy level at III is higher or lower than that at I, unless there is an imposed finite disturbance. Therefore, for all cases in Ref. 23 and for some cases in Ref. 22, the buckled state of the configuration can never be reached, since no limit or bifurcation points could be found in these cases.[22,23] Furthermore, it is apparent that the models in Refs 22 and 23 need also be modified, in order to carry out a postbuckling analysis, applicable to a delaminated thin cylindrical film.

4 POSTBUCKLING ANALYSIS OF A DELAMINATED THIN CYLINDRICAL FILM

The postbuckling response of a thin film, debonded from a (thin) cylindrical shell, which is subjected to external pressure, is studied. The delaminated region is of constant width and it extends along the entire length of the shell. The length is assumed to be infinite, so the problem reduces to a delamination problem of a thin ring (see Fig. 3(b)). It is further assumed that during the prebuckling state no delamination growth occurs, only the thin film buckles, and its ends move toward the initial center of curvature. The effects of initial imperfection, width of delamination and end rotation upon the postbuckling response are investigated.

4.1 Mathematical Formulation

The ring under investigation is loaded by external pressure and the delaminated layer close to the inner surface of the ring is assumed free of pressure.

It is assumed that the ring is thin and the delaminated layer is extremely thin, i.e. $H/R \ll 1$ and $h/H \ll 1$, where R is the radius of the ring, H and h are the thicknesses of the ring and of the debonded layer, respectively. The stiffness of the delaminated layer is assumed to be small compared to that of the base cylinder so that the buckling of the layer does not affect the prebuckling state (membrane) in the base cylinder. The initial imperfection shape of the thin film is denoted by w_o, which is measured from the perfect cylinder surface to the reference surface of the imperfect one (positive outward). Let w denote the transverse displacement component of 'reference surface' material points and be measured from the undeformed surface (positive-outward), and let v denote the tangential displacement (positive-clockwise).

For the delaminated layer, the extensional strain of material points on the reference surface ε^o, and the change in curvature, κ, can be expressed as:

$$\varepsilon^o = \frac{dv}{ds} + \frac{w}{R} + \frac{1}{2}\left(\frac{dw}{ds}\right)^2 + \frac{dw}{ds}\frac{dw_o}{ds} - \delta_1\left(\frac{v}{R}\right)\left(\frac{dw_o}{ds} + \frac{dw}{ds}\right) + \tfrac{1}{2}\delta_1\left(\frac{v}{R}\right)^2 \quad (1)$$

$$\kappa = -\frac{d^2w}{ds^2} + \delta_1\frac{1}{R}\frac{dv}{ds} \quad (2)$$

where

$$\delta_1 = \begin{cases} 1 \text{ for Sanders' kinematic relations} \\ 0 \text{ for Donnell's kinematic relations} \end{cases}$$

If one uses the variable θ instead of $s(s = R\theta)$, then

$$\varepsilon^o = \frac{1}{R}\frac{dv}{d\theta} + \frac{w}{R} + \frac{1}{2R^2}\left(\frac{dw}{d\theta}\right)^2 + \frac{1}{R^2}\frac{dw}{d\theta}\frac{dw_o}{d\theta}$$

$$+ \delta_1\left[\frac{1}{2}\left(\frac{v}{R}\right)^2 - \frac{v}{R^2}\left(\frac{dw}{d\theta} + \frac{dw_o}{d\theta}\right)\right] \quad (3)$$

$$\kappa = -\frac{1}{R^2}\left(\frac{d^2w}{d\theta^2} - \delta_1\frac{dv}{d\theta}\right) \quad (4)$$

The total potential energy, Π, for the buckled layer is

$$\Pi = U + W \quad (5)$$

where

$$U = \int_{-\alpha}^{\alpha}\int_A \tfrac{1}{2}E(\varepsilon^o + z\kappa)^2 R\,dA\,d\theta \quad (6)$$

$$W = -\int_{-\alpha}^{\alpha}(p_r w + p_\theta v)R\,d\theta \quad (7)$$

are the strain energy and the potential of the externally applied forces, respectively. In eqns (6) and (7), A denotes the cross-sectional area of the thin layer, p_r and p_θ are the distributed load components in the radial and tangential directions, respectively.

The equilibrium equations are derived by employing the principle of the stationary value of the total potential, or

$$\delta\Pi = \int_{-\alpha}^{\alpha} \int_{A} ER(\varepsilon^{\circ} + z\kappa)(\delta\varepsilon^{\circ} + z\delta\kappa)\,\mathrm{d}A\,\mathrm{d}\theta$$

$$- \int_{-\alpha}^{\alpha} (p_r\delta w + p_\theta\delta v)R\,\mathrm{d}\theta = 0 \tag{8}$$

Let N and M denote the hoop load and bending moment, respectively,

$$N = \int_{A} E\varepsilon^{\circ}\,\mathrm{d}A = EA\varepsilon^{\circ} \tag{9}$$

$$M = -\int_{A} Ez^2\kappa\,\mathrm{d}A = -EI\kappa \tag{10}$$

where

$$A = \int_{A} \mathrm{d}A \quad \text{and} \quad I = \int_{A} z^2\,\mathrm{d}A$$

Then substitution of eqns (9) and (10) into eqn (8) yields

$$\int_{-\alpha}^{\alpha} [N\delta\varepsilon^{\circ} - M\delta\kappa - p_r\delta w - p_\theta\delta v]R\,\mathrm{d}\theta = 0 \tag{11}$$

From eqns (3) and (4), one can write the expressions for $\delta\varepsilon^{\circ}$ and $\delta\kappa$:

$$\delta\varepsilon^{\circ} = \frac{1}{R}\frac{\mathrm{d}\delta v}{\mathrm{d}\theta} + \frac{\delta w}{R} + \frac{1}{R^2}\frac{\mathrm{d}\delta w}{\mathrm{d}\theta}\left(\frac{\mathrm{d}w}{\mathrm{d}\theta} + \frac{\mathrm{d}w_\mathrm{o}}{\mathrm{d}\theta}\right)$$

$$+ \delta_1\left[\frac{v}{R^2}\delta v - \frac{\delta v}{R^2}\left(\frac{\mathrm{d}w}{\mathrm{d}\theta} + \frac{\mathrm{d}w_\mathrm{o}}{\mathrm{d}\theta}\right) - \frac{v}{R^2}\frac{\mathrm{d}\delta v}{\mathrm{d}\theta}\right] \tag{12}$$

$$\delta\kappa = -\frac{1}{R^2}\left(\frac{\mathrm{d}^2\delta w}{\mathrm{d}\theta^2} - \delta_1\frac{\mathrm{d}\delta v}{\mathrm{d}\theta}\right) \tag{13}$$

Substitution of eqns (12) and (13) into eqn (11), integration by parts, and requiring continuity at any point of the reference axis lead to the following equilibrium equations:

$$\frac{\mathrm{d}N}{\mathrm{d}\theta} + \frac{N}{R}\left[\left(\frac{\mathrm{d}w}{\mathrm{d}\theta} + \frac{\mathrm{d}w_\mathrm{o}}{\mathrm{d}\theta}\right) - v\right]\delta_1 - \frac{1}{R}\frac{\mathrm{d}M}{\mathrm{d}\theta}\delta_1 + p_\theta R = 0 \tag{14}$$

$$-N + \frac{1}{R}\left\{\frac{\mathrm{d}w}{\mathrm{d}\theta}\left[N\left(\frac{\mathrm{d}w}{\mathrm{d}\theta} + \frac{\mathrm{d}w_\mathrm{o}}{\mathrm{d}\theta}\right)\right] - vN\delta_1\right\} - \frac{1}{R}\frac{\mathrm{d}^2M}{\mathrm{d}\theta^2} + p_r R = 0 \tag{15}$$

The boundary conditions at $\theta = \pm \alpha$ are:

Either   Or

$$N - \frac{M}{R}\delta_1 = 0 \qquad \delta v = 0$$

$$M = 0 \qquad \delta\frac{dw}{d\theta} = 0 \qquad (16)$$

$$\frac{N}{R}\left(\frac{dw}{d\theta} + \frac{dw_o}{d\theta}\right) - \frac{Nv}{R}\delta_1 - \frac{1}{R}\frac{dM}{d\theta} = 0 \qquad \delta w = 0$$

By letting $\delta_1 = 0$, eqns (14) and (15) become Donnell-type governing equations. Moreover, for the chosen model $p_\theta = p_r = 0$. Then,

$$\frac{dN}{d\theta} = 0 \qquad (17)$$

$$-N + \frac{1}{R}\frac{d}{d\theta}\left[N\left(\frac{dw}{d\theta} + \frac{dw_o}{d\theta}\right)\right] - \frac{1}{R}\frac{d^2 M}{d\theta^2} = 0 \qquad (18)$$

For the configuration shown in Fig. 3, the boundary conditions become

$$v(\alpha) = v(-\alpha) = 0 \qquad w(\alpha) = w(-\alpha) = w^p \qquad \frac{dw(\alpha)}{d\theta} = \frac{dw(-\alpha)}{d\theta} = 0 \qquad (19)$$

where w^p is the uniform transverse contraction of the ring, which is assumed to occur during the primary path. The expression for w^p is given by Ref. 25

$$w^p = \frac{qR^2}{EA_r} \qquad (20)$$

where A_r is the cross-sectional area of the ring.

4.2 Solution Procedure

In solving the governing equations, eqns (17) and (18), with the boundary conditions given by eqn (19), the Galerkin method is employed.

Let the trial functions become

$$\bar{w} = \sum_{m=1,3,5,\dots} A_m\left(1 + \cos\frac{m\pi\theta}{\alpha}\right) + \bar{w}^p \qquad (21)$$

$$\bar{v} = \sum_{n=1,2,3,\dots} B_n \sin\frac{n\pi\theta}{\alpha} \qquad (22)$$

where

$$\bar{w} = \frac{w}{R} \qquad \bar{v} = \frac{v}{R} \quad \text{and} \quad \bar{w}^p = \frac{w^p}{R}$$

are the nondimensionalized displacement components. Apparently, all boundary conditions are satisfied by the trial functions.

The initial imperfection shape function is assumed as

$$\bar{w}_0 = \frac{w_0}{R} = e\left(1 + \cos\frac{\pi\theta}{\alpha}\right) \tag{23}$$

where the constant e is an imperfection amplitude parameter.

Substitution of eqns (21)–(23) into eqns (3) and (4), and then into eqns (9) and (10), yields

$$N = EA\Bigg[\sum_{n=1,2,\ldots} B_n \frac{n\pi}{\alpha}\cos\frac{n\pi\theta}{\alpha} + \sum_{m=1,3,\ldots} A_m\left(1 + \cos\frac{m\pi\theta}{\alpha}\right)$$

$$+ \bar{w}^p + \frac{1}{2}\left(\sum_{m=1,3,\ldots} A_m \frac{m\pi}{\alpha}\sin\frac{m\pi\theta}{\alpha}\right)^2$$

$$+ \left(\sum_{m=1,3,\ldots} A_m \frac{m\pi\theta}{\alpha}\sin\frac{m\pi\theta}{\alpha}\right)\left(e\frac{\pi}{\alpha}\sin\frac{\pi\theta}{\alpha}\right)\Bigg] \tag{24}$$

$$M = -\frac{EI}{R}\left(\sum_{m=1,3,\ldots} A_m\left(\frac{m\pi}{\alpha}\right)^2\cos\frac{m\pi\theta}{\alpha}\right) \tag{25}$$

By substituting eqns (24) and (25) into the left hand side of eqns (17) and (18), by writing them as $\Phi_1(A_m, B_n)$ and $\Phi_2(A_m, B_n)$, respectively, and by employing the Galerkin procedure one obtains

$$\int_{-\alpha}^{\alpha} \Phi_1(A_m, B_n)\sin\frac{i\pi\theta}{\alpha}\,d\theta = 0 \qquad (i = 1, 2, 3, \ldots)$$

$$\int_{-\alpha}^{\alpha} \Phi_2(A_m, B_n)\left(1 + \cos\frac{j\pi\theta}{\alpha}\right)d\theta = 0 \qquad (j = 1, 3, 5, \ldots) \tag{26}$$

which leads to a system of cubic algebraic equations. For a one-term solution, we have two unknowns (A_1, B_1) and two equations. For a two-term solution, the number of unknowns would be four (A_1, A_3, B_1, B_2) and so would the number of equations.

4.3 Numerical Results and Modification of the Model

The postbuckling deflection of a delaminated thin layer close to the inner surface of a cylinder is computed. The effect of the delamination width, α, as well as the initial imperfection parameter, e, on the postbuckling behavior is investigated.

The results are presented graphically as plots of the deflection at the crown point $(\theta = 0)$, $\bar{w}_c$, versus the end transverse displacement, $\bar{w}^p$, where

$$\bar{w}_c = \frac{w(0) - w^p}{R} \tag{27}$$

The deflections at the crown point of a thin layer for various α-values are shown in Fig. 6. For every α, there are two branches of solution shown in Fig. 6. In branch (a), $|\bar{w}_c|$ decreases as $|\bar{w}^p|$ increases, while in branch (b), $|\bar{w}_c|$ increases with $|\bar{w}^p|$. In fact, another branch of solution, which corresponds to $\bar{w}_c > 0$, can be found in solving the system of cubic equations. This

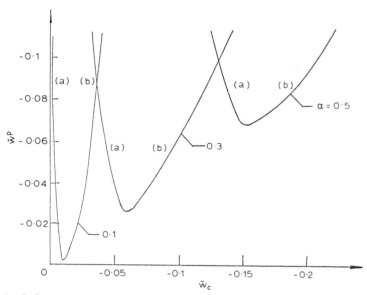

FIG. 6.   Deflection at the crown point, $\bar{w}_c$, versus the end transverse displacement, $\bar{w}^p$, for different delamination widths.

branch is not plotted because it is unrealistic ($\bar{w}_c > 0$ implies that the thin film tends to go upward and penetrate into the ring). This last branch corresponds to the unbuckled or prebuckling solution.

The postbuckling behavior of a thin film with $\alpha = 0.3$ and various values for the initial imperfection parameter, e, is studied and the results are plotted in Fig. 7. It is seen that for smaller $|e|$ values ($e = -0.0001$ or -0.01) there exist two branches of postbuckling response and a prebuckling one (the vertical axis). On the other hand, for larger $|e|$ values ($e = -0.03$ or -0.05) only one branch can be found. Using the stability criterion, it can be shown that the prebuckling equilibrium positions and the positions corresponding to branch (a) are unstable. No bifurcation or limit point exists. For the reasons mentioned in Section 3, the configuration cannot move into the postbuckling state unless there exists a finite disturbance. Therefore, the model derived here is also weak for postbuckling analysis of a delaminated thin layer.

Next, the model is modified by considering an end rotation, which was

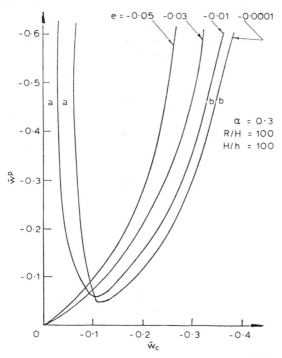

FIG. 7. Deflection at the crown point, $\bar{w}_c$, versus the end transverse displacement, $\bar{w}^p$, for various imperfection amplitudes.

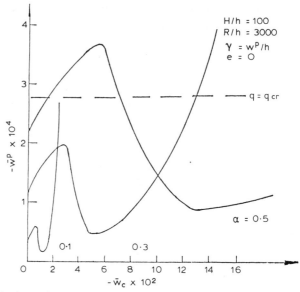

FIG. 8.   Deflection at the crown point, $\bar{w}_c$, versus the end transverse displacement, $\bar{w}^p$, for the end rotation model.

neglected hitherto. One can argue from physical considerations that as the ring moves inward there is contraction in the hoop direction which implies some small relative rotation. Therefore, it is assumed that there exists some end rotation, γ, during postbuckling. Then, the last boundary condition in eqns (19) changes to

$$\frac{1}{R}\frac{dw(\alpha)}{d\theta} = -\frac{1}{R}\frac{dw(-\alpha)}{d\theta} = \gamma \qquad (28)$$

Furthermore, let the trial functions for the transverse displacement be

$$\bar{w} = \sum_{m=1,3,\ldots} A_m\left(1 + \cos\frac{m\pi\theta}{\alpha}\right) + \bar{w}^p - \frac{2\alpha}{\pi}\gamma\cos\frac{\pi\theta}{2\alpha} \qquad (29)$$

This $\bar{w}$ together with $\bar{v}$ in eqn (22), satisfy all boundary conditions. By employing the Galerkin procedure one obtains a system of nonlinear algebraic equations, which can be solved by Newton's method. The example taken here is the postbuckling behavior of a thin film of various widths ($\alpha = 0\cdot1$, $0\cdot3$ and $0\cdot5$) with an end rotation of $\gamma = w^p/h$. The results are shown graphically in Fig. 8. It is clearly seen that the delaminated thin layer is subject to limit-point instability and snap-through buckling is predicted, by the modified model.

4.4 Conclusions

The real situation during delamination buckling is much more complicated than the one described in the models. It has already been mentioned that the linear model, used to compute buckling load, does not take into account (a) the contacting of delamination surfaces during and after buckling and (b) the finite jump of the pressure during buckling. In the linear model it is assumed that the external pressure is distributed to the two parts separated by the delamination (Parts I and II, see Fig. 1) in proportion to the thicknesses. It is true in the prebuckling state, but in the postbuckling state the whole area (for symmetric mode) or part of the area (for antisymmetric mode) of Part II will be free of pressure. In order to fully understand the delamination buckling problem of shells and panels, it is necessary to carry out nonlinear postbuckling analyses. The only other papers on the subject are due to Bottega[20,21] and the two papers[22,23] that have some similarity to the delaminated thin film buckling. These last two works are not closely connected to the present problem for the following reason: the buckled thin part and the thick boundary are two distinctly different parts and the deflection of the thin part does not affect the rigid boundary (see Figs 2 and 4), while in the present problem the base cylinder is definitely influenced by the deflection of the delaminated layer.

In this paper, a nonlinear postbuckling model of a thin cylindrical film was derived and then modified. This model is more closely connected to the true behavior, but it still has some flaws. These are:

1. A reasonable trend was obtained only by considering an end rotation in the modified model. But this end rotation has not been related to pre- and postbuckling load parameters and response parameters.
2. In the model, the surface of the thin film is considered free of pressure, which is true for the symmetric mode in the postbuckling state, but not true in the prebuckling state.
3. Some limit points shown in Fig. 8 correspond to $\bar{w}^p$ values which, when related to the external pressure, q, give q-values which are too large. In some cases they are even larger than the buckling load for the global structure.

Needless to say that further and more detailed studies need be performed in order to acquire complete and full understanding of delamination buckling and growth of laminated shells.

REFERENCES

1. KACHANOV, L. M., Separation failure of composite materials. *PMM*, **5** (1976) 812–15.
2. WILLIAMS, G. J. *et al.*, Recent developments in the design, testing and impact damage-tolerance of stiffened composite panels. NASA TM 80077, Washington, DC, April, 1979.
3. WILKINS, D. J. *et al.*, Characterizing delamination growth in graphite-epoxy. In *Damage in Composite Materials* ed. K. L. Reifsnider. ASTM STP 775, 1982, pp. 168–83.
4. WANG, S. S., Edge delamination in angle-ply composite laminates. *AIAA J.*, **22**(2) (1984) 256–64.
5. WHITCOMB, J. D., Finite element analysis of instability related delamination growth. *J. Comp. Mater.*, **15** (1981) 403–26.
6. SHIVAKUMAR, K. N. and WHITCOMB, J. D., Buckling of a sublaminate in a quasi-isotropic composite laminate. *J. Comp. Mater.*, **19** (1985) 2–18.
7. CHAI, H., BABCOCK, C. D. and KNAUSS, W. G., One dimensional modelling of failure in laminated plates by delamination buckling. *Int. J. Sol. Struct.*, **17** (1981) 1069–83.
8. CHAI, H., KNAUSS, W. G. and BABCOCK, C. D. Observation of damage growth in compressively loaded laminates. *Exp. Mech.*, **23** (1983) 329–37.
9. BOTTEGA, W. J. and MAEWAL, A., Delamination buckling and growth in laminates. *J. Appl. Mech.*, **50**(1) (1983) 184–9.
10. BOTTEGA, W. J. and MAEWAL, A., Dynamics and delamination buckling. *Int. J. Non-Lin. Mech.*, **18**(6) (1983) 449–63.
11. BOTTEGA, W. J., A growth law for the propagation of arbitrary shape delaminations in layered plates. *Int. J. Sol. Struct.*, **19**(11) (1983) 1009–17.
12. SIMITSES, G. J., SALLAM, S. and YIN, W. L., Effects of delamination of axially loaded homogeneous laminated plates. *AIAA J.*, **23**(9) (1985) 1437–44.
13. SALLAM, S. and SIMITSES, G. J., Delamination buckling and growth of flat, cross-ply laminates. *Comp. Struct.*, **4** (1985) 361–81.
14. YIN, W. L., SALLAM, S. and SIMITSES, G. J., Ultimate axial load capacity of a delaminated beam-plate. *AIAA J.*, **24**(1) (1986) 123–8.
15. TROSHIN, V. P., Effects of longitudinal delamination in a laminar cylindrical shell on the critical external pressure. *Mech. Comp. Mater.*, **17**(5) (1985) 563–7.
16. TROSHIN, V. P., Use of three-dimensional model of filler in stability problems for triple-layer shells with delaminations. *Mekhanika Kompozitnykh Materialov*, **4** (1983) 657–62.
17. SALLAM, S. and SIMITSES, G. J., Delamination buckling of cylindrical shells under axial compression. *Comp. Struct.*, **7** (1987) 83–101.
18. SIMITSES, G. J. and CHEN, Z. Q., Delamination buckling of pressure-loaded thin cylinders and panels. In *Composite Structures 4* ed. I. H. Marshall. **1**, Elsevier Applied Science Publishers, London, 1987, pp. 294–306.
19. SIMITSES, G. J. and CHEN, Z. Q., Buckling of delaminated, long, cylindrical panels under pressure. *Comput. Struct.*, **28**(2) (1988) 173–84.
20. BOTTEGA, W. J., On thin film delamination growth in a contracting cylinder. *Int. J. Solids Struct.*, **24**(1) (1988) 13–26.

21. BOTTEGA, W. J., On delamination of thin layers from cylindrical surfaces. In *Proceedings of 29th AIAA/ASME/ASCE/AHS Structures, Structural Dynamics and Materials Conference*, Part 1, AIAA Publications, 1988, pp. 351–8.

22. HSU, P. T., ELKON, J. and PIAN, T. H. H., Note on the instability of circular rings confined to a rigid boundary. *J. Appl. Mech.*, **31**(3) (1964) 559–62.

23. EL-BAYOUMY, L., Buckling of a circular elastic ring confined to a uniformly contracting circular boundary. *J. Appl. Mech.*, **39**(3) (1972) 758–65.

24. CHEN, Z. Q. and SIMITSES, G. J., Delamination buckling of pressure-loaded, cross-ply, laminated, cylindrical shells. *ZAMM*, **68**(10) (1988) 491–9.

25. SIMITSES, G. J., *Elastic Stability of Structures*. Prentice-Hall Inc., Englewood Cliffs, NJ, 1976 (reprinted R. E. Krieger Publishing Co., Melbourne, FL, 1985), Chap. 7.

24

Buckling and Vibration Analysis of Composite Prismatic Plate Structures

V. Peshkam

Mouchel Advanced Composite Analysis Design,
L. G. Mouchel & Partners, West Hall, Parvis Road,
West Byfleet, Weybridge, Surrey KT14 6EZ, UK

and

D. J. Dawe

Department of Civil Engineering, University of Birmingham,
PO Box 363, Birmingham B15 2TT, UK

ABSTRACT

The prediction of buckling stresses and natural frequencies of prismatic plate structures made of composite laminated material is considered, using the finite strip method. Different analysis procedures are outlined, dependent upon whether the plate structure is of finite length or is 'long' and on whether shear deformation plate theory or classical plate theory is used. Features of the analysis include a broad description of material properties, the presence of applied shear stress, accommodation of eccentric connections between plate flats and the use of multi-level substructuring techniques. Developed computer programs are used to solve a number of example problems.

INTRODUCTION

Plate assemblies of rectangular planform and of longitudinally-invariant cross-section, such as box sections or reinforced panels, are important structural components in the realms of aeronautical engineering, ship-building and civil engineering. Often, such structures are made of fibre-reinforced composite laminates which have orthotropic or generally

467

anisotropic material properties, particularly in situations where considerations of weight-saving are of paramount importance. The use of composite laminates has specific implications for the analysis of plate structures since it is quite possible that through-thickness shear (and related) effects will significantly influence the out-of-plane properties of component plate flats: in this regard it is noted that shear deformation plate theory (SDPT) takes account of through-thickness shear effects, of course, whilst classical plate theory (CPT) does not.

In the design of prismatic plate structures there exists a clear need for accurate and economical analysis procedures to predict critical buckling stress levels (including the presence of applied shear stress) and natural frequencies of vibration. Evidence of this need is the considerable effort invested over the past two decades in the development of finite strip, or folded plate, methods of analysis which have an attractive blend of accuracy, economy and ease of modelling. These methods can be broadly categorised as to whether the developed properties of an individual strip are based on the direct solution of a set of governing differential equations or on the application of potential energy or virtual work principles, whether the analysis is of the single-term type (for 'long' structures, i.e. of notionally infinite length) or of the multi-term type (for structures of finite length), and whether the out-of-plane properties of a strip are based on SDPT or CPT. Discussion of these points and reference to earlier studies is available elsewhere[1,2] and will not be repeated here.

In the present work we bring together and up-date in summary form some recent developments in the subject area which collectively extend significantly the scope, power and efficiency of the finite strip analysis of prismatic plate structures of metallic or composite construction. Strip properties are based on a potential energy approach, plate structures may be long or of finite length, analyses employ both SDPT and CPT in turn, and a number of other sophisticated features is included. The types of problem that can be accommodated are those of free vibration, of buckling under the action of either a live stress system or a combination of a dead stress system of fixed magnitude and a live stress system, and of vibration in the presence of a dead stress system.

TYPES OF ANALYSES AND PROGRAMS

Basic Equations

A typical component plate flat which forms part of a prismatic plate structure is shown in Fig. 1. The plate flat may be subjected to the in-plane

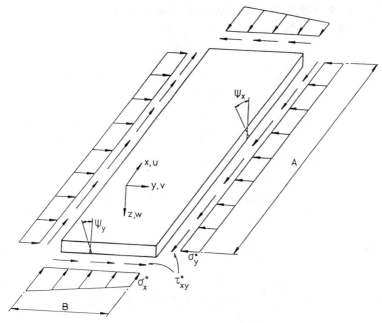

FIG. 1. A component plate flat.

stress system shown, leading to buckling, or it may be undergoing harmonic motion whilst vibrating in a natural mode with circular frequency p, or both these influences may be present.

In SDPT[3-5] the displacement at a general point in the plate flat depends upon five fundamental quantities at the corresponding point on the middle surface, $z = 0$: these are the middle-surface translational displacements, u, v and w, plus the rotations of the middle-surface normal along the x and y axes, ψ_x and ψ_y. All displacement quantities are in fact perturbation quantities, representing changes that occur at the instant of buckling or during vibration about a datum position.

In the finite strip analysis a plate flat is modelled with one or (usually) more plate strips. For an individual plate strip the elastic stiffness matrix, $\mathbf{k}$, geometric stiffness matrix, $\mathbf{k_g}$, and consistent mass matrix $\mathbf{m}$ are developed from expressions for the changes in strain energy, in potential energy of applied in-plane stresses, and in kinetic energy that occur during the perturbation associated with buckling or vibration. The expressions used in the present approach, in the context of SDPT, are of very general form and are now recorded for a laminate of thickness h.

The strain energy per unit middle surface area is

$$
\begin{aligned}
\mathrm{d}U = \frac{1}{2}\Bigg[\!\!\!\!\!\! &\; A_{11}\!\left(\frac{\partial u}{\partial x}\right)^2 + A_{22}\!\left(\frac{\partial v}{\partial y}\right)^2 + A_{66}\!\left(\frac{\partial v}{\partial x}+\frac{\partial u}{\partial y}\right)^2 \\
&+ 2A_{12}\frac{\partial u}{\partial x}\frac{\partial v}{\partial y} + 2A_{16}\frac{\partial u}{\partial x}\!\left(\frac{\partial u}{\partial y}+\frac{\partial v}{\partial x}\right) + 2A_{26}\frac{\partial v}{\partial y}\!\left(\frac{\partial u}{\partial y}+\frac{\partial v}{\partial x}\right) \\
&+ 2B_{11}\frac{\partial u}{\partial x}\frac{\partial \psi_x}{\partial x} + 2B_{22}\frac{\partial v}{\partial y}\frac{\partial \psi_y}{\partial y} + 2B_{66}\!\left(\frac{\partial u}{\partial y}+\frac{\partial v}{\partial x}\right)\!\left(\frac{\partial \psi_x}{\partial y}+\frac{\partial \psi_y}{\partial x}\right) \\
&+ 2B_{12}\!\left(\frac{\partial u}{\partial x}\frac{\partial \psi_y}{\partial y}+\frac{\partial v}{\partial y}\frac{\partial \psi_x}{\partial x}\right) \\
&+ 2B_{16}\!\left(\frac{\partial u}{\partial x}\!\left[\frac{\partial \psi_x}{\partial y}+\frac{\partial \psi_y}{\partial x}\right]+\frac{\partial \psi_x}{\partial x}\!\left[\frac{\partial u}{\partial y}+\frac{\partial v}{\partial x}\right]\right) \\
&+ 2B_{26}\!\left(\frac{\partial v}{\partial y}\!\left[\frac{\partial \psi_x}{\partial y}+\frac{\partial \psi_y}{\partial x}\right]+\frac{\partial \psi_y}{\partial y}\!\left[\frac{\partial u}{\partial y}+\frac{\partial v}{\partial x}\right]\right) \\
&+ D_{11}\!\left(\frac{\partial \psi_x}{\partial x}\right)^2 + D_{22}\!\left(\frac{\partial \psi_y}{\partial y}\right)^2 + D_{66}\!\left(\frac{\partial \psi_x}{\partial y}+\frac{\partial \psi_y}{\partial x}\right)^2 \\
&+ 2D_{12}\frac{\partial \psi_x}{\partial x}\frac{\partial \psi_y}{\partial y} + 2D_{16}\frac{\partial \psi_x}{\partial x}\!\left(\frac{\partial \psi_x}{\partial y}+\frac{\partial \psi_y}{\partial x}\right) \\
&+ 2D_{26}\frac{\partial \psi_y}{\partial y}\!\left(\frac{\partial \psi_x}{\partial y}+\frac{\partial \psi_y}{\partial x}\right) \\
&+ A_{44}\!\left(\frac{\partial w}{\partial y}+\psi_y\right)^2 + A_{55}\!\left(\frac{\partial w}{\partial x}+\psi_x\right)^2 \\
&+ 2A_{45}\!\left(\psi_x+\frac{\partial w}{\partial x}\right)\!\left(\psi_y+\frac{\partial w}{\partial y}\right)\Bigg]
\end{aligned}
\tag{1}
$$

The laminate stiffness coefficients occurring in eqn (1) are defined as

$$
(A_{ij}, B_{ij}, D_{ij}) = \int_{-h/2}^{h/2} Q_{ij}(1, z, z^2)\,\mathrm{d}z \qquad i,j = 1,2,6
\tag{2}
$$

and

$$
A_{ij} = k_i k_j \int_{-h/2}^{h/2} Q_{ij}\,\mathrm{d}z \qquad i,j = 4,5
\tag{3}
$$

Here, Q_{ij} for $i,j = 1,\,2,\,6$ are transformed plane-stress reduced stiffness coefficients and Q_{ij} for $i,j=4$, 5 are through-thickness shear stiffness coefficients. The parameters $k_i k_j$ are the shear correction factors which are determined in the present work using the method of Whitney.[6]

The potential energy per unit middle surface area of the applied in-plane stresses σ_x^0, σ_y^0 and τ_{xy}^0 is

$$
\begin{aligned}
dV_g = \frac{1}{2}h \Bigg[& \sigma_x^0 \left[\left(\frac{\partial u}{\partial x}\right)^2 + \left(\frac{\partial v}{\partial x}\right)^2 + \left(\frac{\partial w}{\partial x}\right)^2 \right] \\
& + \sigma_y^0 \left[\left(\frac{\partial u}{\partial y}\right)^2 + \left(\frac{\partial v}{\partial y}\right)^2 + \left(\frac{\partial w}{\partial y}\right)^2 \right] \\
& + 2\tau_{xy}^0 \left[\frac{\partial u}{\partial x}\frac{\partial u}{\partial y} + \frac{\partial v}{\partial x}\frac{\partial v}{\partial y} + \frac{\partial w}{\partial x}\frac{\partial w}{\partial y} \right] \\
& + \frac{h^2}{12} \left\{ \sigma_x^0 \left[\left(\frac{\partial \psi_x}{\partial x}\right)^2 + \left(\frac{\partial \psi_y}{\partial x}\right)^2 \right] + \sigma_y^0 \left[\left(\frac{\partial \psi_x}{\partial y}\right)^2 + \left(\frac{\partial \psi_y}{\partial y}\right)^2 \right] \right. \\
& \left. + 2\tau_{xy}^0 \left[\frac{\partial \psi_x}{\partial x}\frac{\partial \psi_x}{\partial y} + \frac{\partial \psi_y}{\partial x}\frac{\partial \psi_y}{\partial y} \right] \right\} \Bigg]
\end{aligned}
\tag{4}
$$

The maximum kinetic energy per unit middle surface area for harmonic motion with circular frequency p (and assuming that material density ρ is uniform through the thickness) is

$$
dT = \frac{1}{2}\rho h p^2 \left(u^2 + v^2 + w^2 + \frac{h^2}{12}[\psi_x^2 + \psi_y^2] \right)
\tag{5}
$$

The above equations related to the use of SDPT can be modified to embrace the use of CPT. In CPT the Kirchhoff normalcy assumption is invoked and the rotations ψ_x and ψ_y are directly related to the deflection w by way of the equations

$$
\psi_x = -\partial w/\partial x \qquad \psi_y = -\partial w/\partial y
\tag{6}
$$

The displacement at a general point in the plate flat now depends upon only three fundamental quantities, namely u, v and w. In CPT the strain energy per unit middle surface area is obtained on substituting eqns (6) into eqn (1). It is noted that then the A_{44}, A_{55} and A_{45} laminate stiffness coefficients vanish and that the strain energy depends upon second spatial derivatives of w as well as on first derivatives of u and v; in contrast, in SDPT only first spatial derivatives of u, v, w, ψ_w and ψ_y are involved, as well as ψ_x and ψ_y themselves. Furthermore, in CPT the potential energy of applied in-plane stresses and the kinetic energy, per unit middle surface area, are as in eqns (4) and (5) but with the terms involving ψ_x and ψ_y removed.

Multi-term Finite Strip Analysis

Plate flats of the type shown in Fig. 1 are modelled with one or more finite strips of width $b \le B$. A typical individual finite strip in SDPT analysis is shown in Fig. 2(a) and its spatial displacement field corresponds to the assumption of each of the give fundamental quantities as a summation of products of longitudinal (x-direction) series terms and crosswise (y-direction) polynomial functions. Thus, for the particular case of both ends of the structure being supported by diaphragms[1]

$$
\begin{Bmatrix} u \\ v \\ w \\ \psi_y \\ \psi_x \end{Bmatrix} = \sum_{i=r1}^{r2}
\begin{bmatrix}
C_i(x) & 0 & 0 & 0 & 0 \\
0 & S_i(x) & 0 & 0 & 0 \\
0 & 0 & S_i(x) & 0 & 0 \\
0 & 0 & 0 & S_i(x) & 0 \\
0 & 0 & 0 & 0 & C_i(x)
\end{bmatrix}
$$
$$
\times
\begin{bmatrix}
\phi(y) & 0 & 0 & 0 & 0 \\
0 & \phi(y) & 0 & 0 & 0 \\
0 & 0 & \phi(y) & 0 & 0 \\
0 & 0 & 0 & \phi(y) & 0 \\
0 & 0 & 0 & 0 & \phi(y)
\end{bmatrix}
\begin{Bmatrix} A_1 \\ A_2 \\ \vdots \\ \vdots \\ A_{5n+5} \end{Bmatrix}_i
\tag{7}
$$

Here, $\phi(y)$ is a row matrix defined as

$$
\phi(y) = \begin{bmatrix} 1 & y & y^2 & \cdots & y^n \end{bmatrix}
\tag{8}
$$

and represents a crosswise polynomial function of order n. The longitudinal functions $C_i(x)$ and $S_i(x)$ are simple cosine and sine functions, defined as

$$
C_i(x) = \cos\frac{i\pi x}{A} \qquad S_i(x) = \sin\frac{i\pi x}{A}
\tag{9}
$$

The quantities i, $r1$ and $r2$ are integer values with $r2 \ge r1$.

The column matrix at the right-hand side of eqn (7) is the list of generalised displacement coefficients corresponding to the ith terms of the longitudinal series and these coefficients can be related in standard fashion to the strip local degrees of freedom. The latter are located at $(n+1)$ reference lines across the strip in the manner shown in Fig. 2(a) for the specific case of cubic crosswise polynomial interpolation: the indicated freedoms are those that apply for each longitudinal series term. Various types of strip are generated by varying the order of crosswise polynomial

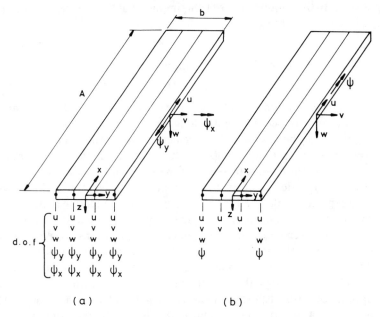

FIG. 2. Individual finite strips for (a) SDPT analysis (cubic interpolation), and (b) CPT analysis.

interpolation, with consequent change in the number of strip reference lines.

In CPT analysis an individual finite strip is shown in Fig. 2(b) and here only one type of strip is considered, corresponding to the assumption of cubic crosswise interpolation of each of the three translational displacements. The spatial strip displacement field, for diaphragm ends, is represented as[1]

$$
\begin{Bmatrix} u \\ v \\ w \end{Bmatrix} = \sum_{i=r1}^{r2} \begin{bmatrix} C_i(x) & 0 & 0 \\ 0 & S_i(x) & 0 \\ 0 & 0 & S_i(x) \end{bmatrix} \begin{bmatrix} \phi(y) & 0 & 0 \\ 0 & \phi(y) & 0 \\ 0 & 0 & \phi(y) \end{bmatrix} \begin{Bmatrix} A_1 \\ A_2 \\ \vdots \\ A_{12} \end{Bmatrix}_i \tag{10}
$$

In the CPT approach C^1-type continuity is required for the w displacement and thus the arrangement of strip degrees of freedom (per longitudinal series term) is as shown in Fig. 2(b), with $\psi = \partial w / \partial y$ ($\equiv -\psi_y$ in the SDPT approach).

In the above the structure, end conditions have been stated to be diaphragms. In the context of SDPT this implies that the specific conditions at $x = 0$ and $x = A$ are

$$v = w = \psi_y = N_x = M_x = 0 \tag{11}$$

whilst in the context of CPT the conditions are

$$v = w = N_x = M_x = 0 \tag{12}$$

where N_x and M_x are the direct force per unit length in the x-direction and the bending moment per unit length along the x-direction, respectively. For both approaches the assumed displacement field satisfies exactly all the stated end conditions for a plate strip with orthotropic material properties. Where the material is anisotropic the force-type end conditions are not satisfied and some degree of over-constraint will apply, though this effect would be expected to be small for plate structures of moderate anisotropy.

Although attention has been restricted to diaphragm end conditions in the description given here and in the numerical applications given below, it is emphasised that the basic analysis is a more general one, encompassing end conditions which are combinations of fixed, free and diaphragm ends. This is achieved through the use of appropriate beam functions in the descriptions of strip displacement fields, in place of the $C_i(x)$ and $S_i(x)$ functions appearing in eqns (7) and (10). The beam functions are the normalised modes of vibration of Timoshenko beams in the context of SDPT analysis, and of Bernoulli–Euler beams in the context of CPT analysis. For single plates, use of the Timoshenko beam functions is described by Dawe et al.,[7−9] and of the Bernoulli–Euler beam functions by Leissa,[10] for example.

Single-term Finite Strip Analysis

The basic assumption here is that the mode shape of buckling or vibration is purely sinusoidal in the longitudinal direction and hence it is only necessary to consider behaviour over one prescribed half-wavelength. This assumption is perfectly correct where the ends of the structure are supported by diaphragms, the plate flats have orthotropic material properties and no applied shear stress is present. Then the analysis simply becomes a special case of the multi-term analysis with $r1 = r2 = 1$ (in eqn (7) or (10)) and $i\pi x/A$ (in eqn (9)) replaced by $\pi x/\lambda$, where λ is the prescribed half-wavelength. However, where material anisotropy and/or applied shear stress are present the nodal lines across the structure become curved and skewed and the displacements must be represented as complex quantities. Realistic predictions of buckling stresses or frequencies of vibration will then be obtained only for long structures where end effects can be ignored.

In SDPT finite strip analysis of long structures the complex spatial strip displacement field is represented as[2]

$$
\begin{Bmatrix} u \\ v \\ w \\ \psi_y \\ \psi_x \end{Bmatrix} = \mathrm{Re} \left(e^{j\xi} \begin{bmatrix} j & 0 & 0 & 0 & 0 \\ 0 & 1 & 0 & 0 & 0 \\ 0 & 0 & 1 & 0 & 0 \\ 0 & 0 & 0 & 1 & 0 \\ 0 & 0 & 0 & 0 & j \end{bmatrix} \cdot \right.
$$

$$
\left. \times \begin{bmatrix} \phi(y) & 0 & 0 & 0 & 0 \\ 0 & \phi(y) & 0 & 0 & 0 \\ 0 & 0 & \phi(y) & 0 & 0 \\ 0 & 0 & 0 & \phi(y) & 0 \\ 0 & 0 & 0 & 0 & \phi(y) \end{bmatrix} \begin{Bmatrix} A_1 \\ A_2 \\ \vdots \\ \vdots \\ A_{5n+5} \end{Bmatrix} \right) \tag{13}
$$

Here, $j = \sqrt{-1}$, $\xi = \pi x / \lambda$, Re( ) denotes the real part of the quantity inside the parentheses, $A_1 \ldots A_{5n+5}$ are generalised coefficients and $\phi(y)$ is defined by eqn (8). In the corresponding CPT analysis the field is

$$
\begin{Bmatrix} u \\ v \\ w \end{Bmatrix} = \mathrm{Re} \left(e^{j\xi} \begin{bmatrix} j & 0 & 0 \\ 0 & 1 & 0 \\ 0 & 0 & 1 \end{bmatrix} \begin{bmatrix} \phi(y) & 0 & 0 \\ 0 & \phi(y) & 0 \\ 0 & 0 & \phi(y) \end{bmatrix} \begin{Bmatrix} A_1 \\ A_2 \\ \vdots \\ A_{12} \end{Bmatrix} \right) \tag{14}
$$

In both cases the degrees of freedom of the single-term finite strips (i.e. a family of strip types for SDPT analysis and a single cubic type for CPT analysis) are the same as those corresponding to one term of the equivalent multi-term strips described earlier (see Fig. 2).

Features of the Analyses

The basic equations and description presented above provide some pertinent details of the basis for the development of finite strip analyses of the buckling and vibration of prismatic plate structures, whether these are of finite length or are 'long'. Of course, there is very considerable further detail and development before the basis is converted into an accurate and efficient capability. This is taken up at length in related works[1,2] and here we wish only to draw attention to some particular features of the analyses.

Important features that have emerged in the description given thus far are as follows.

1. A very broad range of material properties is considered, including bending–stretching coupling and full anisotropy.

2. The applied membrane stress system includes shear stress as well as biaxial direct stresses.

3. In developing the out-of-plane properties of plate flats both the first-order shear deformation plate theory and the classical plate theory are used.

Amongst other significant features not discussed above are the following ones.

4. Multi-level substructuring techniques are used to a very considerable extent and are of great importance since without them the range and complexity of application that could be considered would be drastically reduced. Such techniques are applied both within individual plate flats, through the creation of so-called superstrips, and for the plate structure as a whole. Superstrips are generated by a repetitive, multi-level substructuring procedure so that a single superstrip, with effective freedoms only at its outside edges, incorporates 2^c individual strips where c is zero or a positive integer.

5. Eccentric connections of plate flats, due to the finite thicknesses of such flats meeting at junctions, are taken into account so as to provide an accurate modelling of a structure cross-section.

6. Because of the use of substructuring procedures the final eigenvalue problem is nonlinear and the structure eigenvalues (i.e. buckling stresses or natural frequencies) are determined using an extended Sturm sequence approach. Mode shapes can be found by calculating the relative values of those degrees of freedom which are present in the final eigenvalue problem.

Computer Programs

The theoretical analyses have been converted into four computer programs which are distinguished one from another by whether the shear deformation or classical plate theory is used and whether the analysis is of the multi-term or single-term type.

The two programs for the multi-term analysis of structures of finite length are named BAVAMPAS (Buckling And Vibration Analysis of Multi-term Plate Assemblies using SDPT) and BAVAMPAC (as for BAVAMPAS except that the last letter of BAVAMPAC indicates CPT). The two programs for the single-term, complex-algebra analysis of long structures are named BAVPAS (Buckling And Vibration of Plate Assemblies using SDPT) and BAVPAC (as for BAVPAS but using CPT). It is noted that for diaphragm-supported plate structures of orthotropic material and free of applied shear stress, the BAVAMPAS and BAVPAS programs will yield identical results for a given mode of longitudinal half-wavelength λ, since all arithmetic in BAVPAS is then real: similar remarks apply to BAVAMPAC and BAVPAC. It is further noted that BAVPAC, with superstrip, yields effectively identical results in numerical applications to those obtained using the single-term CPT program VIPASA[11] and that BAVPAC is more versatile in the permitted material properties.

The four programs are written in Fortran 77 and were developed by the authors on a DEC VAX 11/750 computer in the Department of Civil Engineering at the University of Birmingham.

NUMERICAL APPLICATIONS

Only a small section of applications of the four computer programs is described in this section. These applications range from single plates to quite complicated plate structures but do not cover the full scope of the programs. Attention is restricted here to symmetric laminations (i.e. all $B_{ij} = 0$) and to structures with diaphragm-supported ends, and in all the applications the effect of eccentric connections is ignored. (Some further numerical applications are described elsewhere.[2,8,9,12])

It should be noted that where reference is made to superstrip this means here a superstrip comprising 32 individual finite strips (i.e. 2^5 strips) of cubic polynomial order, both in SDPT and CPT analyses, and one superstrip is used per component plate flat or whole plate.

Buckling of Long Anisotropic Plate under Longitudinal Compression

The plate is notionally of infinte length, has simply-supported longitudinal edges and is subjected to a uniform longitudinal compressive stress σ_x^0. It comprises a single layer of material with the properties

$$E_L/E_T = 10 \qquad G_{LT}/E_T = G_{TT}/E_T = 0.25 \quad \text{and} \quad v_{LT} = 0.3$$

The plate width is B and two plate thicknesses of $0.01B$ and $0.1B$ are considered in turn.

Due to the material anisotropy, the nodal lines of the buckled mode are curved and skewed and it is appropriate to seek solutions using the single-term, complex-algebra programs. Results for the SDPT program BAVPAS are given in Table 1, for both plate thicknesses, based on the use of various numbers (N_s) of individual finite strips of various polynomial orders (n) and of a superstrip. For $h/B = 0.01$ the half-wavelength of buckling is

TABLE 1

Buckling of long anisotropic plates under longitudinal compression:
SDPT values of $(\sigma_x^0)_{cr} B^2 h / \pi^2 D_{11}$

h/B	n	$N_s = 1$	$N_s = 2$	$N_s = 3$	$N_s = 4$	Superstrip
0·1	3	3·064 8	2·710 0	2·697 9	2·696 8	2·696 5
	4	2·735 9	2·696 7	2·696 5	2·696 5	
	5	2·698 2	2·696 5	2·696 5	2·696 5	
	6	2·696 7	2·696 5	2·696 5	2·696 5	
0·01	3	7·223 8	5·415 4	4·371 3	3·967 5	3·525 2
	4	5·006 6	3·663 8	3·553 3	3·533 5	
	5	3·741 5	3·529 6	3·525 5	3·525 2	
	6	3·541 1	3·525 2	3·525 2	3·5252	

$\lambda = 10B/11$ whilst for $h/B = 0.1$ it is $\lambda = 10B/13$. It is noted that the CPT buckling factor calculated using BAVPAC, with superstrip, is 3·5352 with $\lambda = 10B/11$ for both plate thicknesses. For the quicker geometry this represents an overestimation—due to the neglect of through-thickness shear deformation effect—of 31% as compared to the SDPT result, together with a different buckled mode.

Buckling of Isotropic and Orthotropic Square Plates under Shear

Here, the buckling under applied shear stress τ_{xy}^0 of two simply-supported square plates, of side length A, is considered using both the SDPT and CPT multi-term programs with superstrip. The first plate is of isotropic material and has thin geometry, with $h/A = 0.01$. The second plate is a five-layer, cross-ply orthotropic laminate with a $0°/90°/0°/90°/0°$ lay-up and with $h/A = 0.1$. The thickness of each of the $0°$ plies is $h/6$ and that of each of the $90°$ plies is $h/4$. The material properties of all layers are identical with

$$E_L/E_T = 30 \qquad G_{LT}/E_T = 0.6 \qquad G_{TT}/E_T = 0.5 \qquad \nu_{LT} = 0.25$$

and the calculated shear correction factors are $k_4^2 = 0.591\,39$ and $k_5^2 = 0.87\,323$.

The manner of convergence of BAVAMPAC and BAVAMPAS results for the buckling factor with respect to the number, r, of longitudinal series terms is shown in Table 2 and good monotonic convergence is demonstrated. For the thin isotropic plate the SDPT solution is about 0·2% below that of the corresponding CPT solution, representing a very small shear deformation effect. For this plate a comparative classical solution[13] is $(\tau_{xy}^0)_{cr} A^2 h / \pi^2 D = 9.34$. The gross differences between the CPT and SDPT

TABLE 2
Buckling of shear-loaded square plates: values of
$(\tau^0_{xy})_{cr} A^2 h/\pi^2 D$

r	Isotropic plate		Orthotropic laminate	
	CPT	SDPT	CPT	SDPT
2	10·388	10·367	4·076	3·679
3	9·378	9·356	3·774	1·693
4	9·366	9·343	3·765	1·665
5	9·334	9·311	3·757	1·648
6	9·332	9·309	3·756	1·644

predictions for the orthotropic laminate give an indication of how important shear deformation effects can be in some situations, though admittedly the considered laminate is of somewhat extreme thickness for an elastic buckling analysis.

Free Vibration of Orthotropic Square Tube

The tube has the cross-section shown in Fig. 3 and is of length $10B$ with diaphragm ends. The individual walls of the tube are of the same five-layer, cross-ply construction as defined for the preceding shear-loaded plate buckling application. Natural frequencies have been calculated using the finite strip method, with superstrip, for the first ten modes of vibration of the tube, based on the use of SDPT and CPT in turn. The nature of the problem is such that results can be generated using either BAVPAS or BAVAMPAS in the context of SDPT, and using either BAVPAC or BAVAMPAC in the context of CPT.

The results obtained are recorded in Fig. 3. Mode shapes are shown in the form of views on the tube cross-section, with the number of longitudinal half-waves indicated, and calculated values of a frequency factor Ω are given. This factor is defined as

$$\Omega = p \frac{A^2}{h} \left[\frac{\rho}{(Q_{11})_t} \right]^{1/2}$$

where

$$(Q_{11})_t = \sum_{l=1}^{nl} \frac{(Q_{11})_t (h_l - h_{l-1})}{h} = 0.517\,745 E_L$$

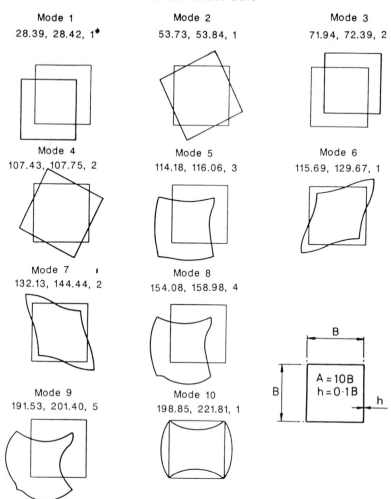

FIG. 3. Mode shapes and frequencies of a square tube. (*The three values given for each mode are the SDPT value of Ω, the CPT value of Ω, and the number of longitudinal half-waves, respectively.)

The influence of through-thickness shear (and to a much lesser extent of rotary inertia) on natural frequency is markedly dependent upon mode shape, being significant for local modes in which the individual walls are highly deformed locally. Thus, for each of modes 6–10 there is a greater than 3% difference between forecasts based on SDPT and on CPT,

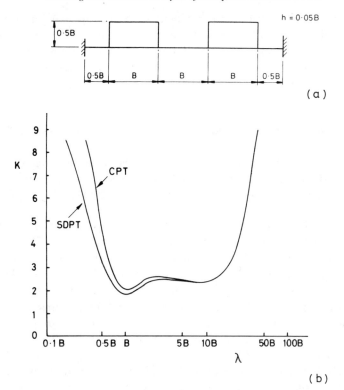

FIG. 4. Buckling under uniform longitudinal compressive stress of a long orthotropic top-hat-stiffened panel. (a) Cross-sectional geometry; (b) K versus λ plot.

including overestimations of CPT forecasts of 9·3, 11·5 and 12·1% for modes 7, 10 and 6, respectively.

Buckling of Orthotropic Long, Top-Hat-Stiffened Panel under Longitudinal Compression

Here again the plate structure considered, whose cross-section is shown in Fig. 4(a), is formed of plate flats of the same five-layer, cross-ply construction as defined earlier. The longitudinal edges of the panel are fully clamped and analysis of the buckling of the orthotropic structure is made at many specific assumed values of half-wavelength λ using the single-term SDPT and CPT programs BAVPAS and BAVPAC in turn (with superstrip). Graphical presentation of results is given in Fig. 4(b), with a buckling factor K plotted against λ on a logarithmic scale: here $K =$

$(\sigma_x^0)_{cr} B^2 h / \pi^2 D_{11}$. It can be seen that short-wavelength buckling is critical, corresponding to a local mode, and that for such local modes there are very substantial differences between SDPT and CPT predictions, due to the effect of through-thickness shear deformation. These differences diminish for coupled modes and become negligible for overall modes at long wavelengths.

Buckling under Combined Loading of an Anisotropic, Square Hat-Stiffened Panel

The subject problem here relates to one of a group of seven stiffened panels considered by Stroud et al. in a NASA report:[14] the particular panel studied here is referred to as Example 5 in the report and full details of it are recorded therein.

The overall geometry of the panel and its loading are shown in Fig. 5(a).The panel has a square planform with side length 762 mm, its ends are diaphragm-supported and its longitudinal edges are simply-supported. The cross-section of the panel consists of six repeating sections of width 127 mm of the type shown in Fig. 5(b). The material is laminated graphite–epoxy composite with plies aligned at $0°$ or $\pm 45°$ to the longitudinal axis. The lamination is symmetric but is such that significant bending–twisting anisotropy is present in the component flats. The loading comprises longitudinal compressive force N_x^0 and shearing force N_{xy}^0 per unit width of panel. The longitudinal force is distributed between the component flats on the assumption of uniform longitudinal pre-buckling strain whereas the shearing force is distributed on the assumption that the hat stiffeners do not twist.

Stroud et al. have presented accurate finite element solutions, using almost 2000 rectangular elements, to the panel buckling problem for six combinations of the ratio N_x^0/N_{xy}^0; their solutions are generated using a computer program given the acronym EAL and provide a basis for comparison for the finite strip method results of the present study. Figure 5(c) shows, by way of an interaction diagram, the close comparison between the EAL buckling predictions and those of the multi-term CPT program BAVAMPAC, with $r = 3$ and $r = 7$ in turn. Results have also been generated using the SDPT program BAVAMPAS but these are indistinguishable from the BAVAMPAC results in a graphical presentation because through-thickness shear effects are insignificant for this application wherein the component flats are very thin. Incidentally the buckling of the other six NASA panels has also been predicted successfully using BAVAMPAC and BAVAMPAS.

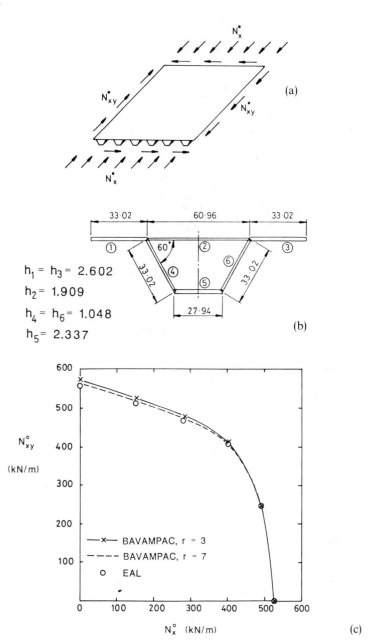

FIG. 5. Buckling under combined loading of an anisotropic square hat-stiffened panel. (a) General geometry and loading; (b) details of repeating structural element (with dimensions in mm); (c) interaction buckling curve.

CONCLUSIONS

Analysis capabilities for the prediction of buckling stresses and natural frequencies of vibration of composite prismatic plate structures have been described. Collectively these capabilities are very broad-ranging, encompassing structures which may be either of finite length or long, and whose properties can be based on either first-order shear deformation plate theory or classical plate theory. Furthermore, the permitted description of material properties is very general, applied shear stress may be present, eccentric connections of plate flats are accommodated and the use of multi-level substructuring techniques allows the efficient solution of complicated problems. Four interrelated computer programs have been developed from the analysis procedures and something of their capability has been demonstrated here in a small number of numerical applications, although the scope of the programs is considerably wider than these applications indicate.

ACKNOWLEDGEMENT

The authors are pleased to acknowledge the financial support of Procurement Executive, Ministry of Defence, during part of the investigation reported here.

REFERENCES

1. DAWE, D. J. and PESHKAM, V., Buckling and vibration of finite-length composite prismatic plate structures with diaphragm ends: Part I, Finite strip formulation. (Submitted.)
2. DAWE, D. J. and CRAIG, T. J., Buckling and vibration of shear deformable prismatic plate structures by a complex finite strip method. *Int. J. of Mech. Sci.*, **30** (1988) 77–99.
3. MINDLIN, R. D., Influence of rotary inertia and shear on flexural motion of isotropic elastic plates. *J. of Appl. Mech.*, **18** (1951) 31–8.
4. YANG, P. C., NORRIS, C. H. and STAVSKY, Y., Elastic wave propagation in heterogeneous plates. *Int. J. of Solids and Structures*, **2** (1966) 665–84.
5. WHITNEY, J. M. and PAGANO, N. J., Shear deformation in heterogeneous anisotropic plates. *Jnl of Appl. Mech.*, **37** (1970) 1031–6.
6. WHITNEY, J. M., Shear correction factors for orthotropic laminates under static load. *Jnl of Appl. Mech.*, **40** (1973) 302–4.
7. DAWE, D. J. and ROUFAEIL, O. L., Rayleigh–Ritz vibration analysis of Mindlin plates. *Jnl of Sound and Vibration*, **69** (1980) 345–59.

8. ROUFAEIL, O. L. and DAWE, D. J., Vibration analysis of rectangular Mindlin plates by the finite strip method. *Computers and Structures*, **12** (1980) 833–42.

9. DAWE, D. J. and CRAIG, T. J., The vibration and stability of symmetrically-laminated composite rectangular plates subjected to in-plane stress. *Composite Structures*, **5** (1986) 281–307.

10. LEISSA, A. W., The free vibration of rectangular plates. *Jnl of Sound and Vibration*, **31** (1973) 257–93.

11. WITTRICK, W. H. and WILLIAMS, F. W., Buckling and vibration of anisotropic or isotropic plate assemblies under combined loadings. *Int. J. of Mech. Sci.*, **16** (1974) 209–39.

12. PESHKAM, V. and DAWE, D. J., Buckling and vibration of finite-length composite prismatic plate structures with diaphragm ends: Part II, Computer programs and buckling applications. (Submitted.)

13. TIMOSHENKO, S. P. and GERE, J., *Theory of Elastic Stability*. 2nd edn. McGraw-Hill, New York, 1971.

14. STROUD, W. J., GREENE, W. H. and ANDERSON, M. S., Buckling loads of stiffened panels subjected to combined longitudinal compression and shear: Results obtained with PASCO, EAL and STAGS computer programs. NASA TP 2215, 1984.

25

Torsional Buckling and Post-buckling of a CFC Geodetic Cylinder

J. S. Sandhu, K. A. Stevens and G. A. O. Davies

Aeronautics Department, Imperial College, London SW7 2AZ, UK

ABSTRACT

A linear buckling and large displacement non-linear finite element analysis have been used to model composite cylindrical geodetic shells. The non-linear analysis is based on the Co-Rotational formulation in which the rigid body motion is removed from the overall motion of the element. The element deformations are then measured in a local convected axes system. The finite element analysis has been compared with experimental results for the cylindrical geodetic shell under torsion. Initial linear buckling results produced errors of 30% when compared with the experimental buckling torque. The non-linear analysis did not greatly reduce this error, but it did predict the correct overall behaviour of the shell and is essential to predict the final failure load. Recent work has shown that the local joint flexibility is the dominant factor in the shell behaviour. When this is incorporated in the finite element models the error in the buckling torque is reduced to about 7%.

INTRODUCTION

The rear fuselage of a helicopter, which supports the tail rotor, is in the form of a conical shell. The behaviour of conical isotropic shells and stiffened shells under a variety of loading is well understood. In particular, the effect of compression or bending loads have been studied by a number of researchers and it is known that the unstiffened conical shell can be as imperfection sensitive as the cylinder. However, a 'new' form of shell is likely to be used in future helicopter fuselages to overcome the impact damage problem. These shells may consist of an open framework in which the slender component bars follow the geodetic lines of the shell. A famous example of the use of geodetic construction was by Barnes Wallis in the design of the Vickers Wellington, a Second World War bomber which exhibited a high level of damage tolerance.

In the contemporary manifestation the geodetics will be in the form of a tape- or filament-wound composite laminate made from carbon fibre epoxy or a thermoplastic carbon composite. The choice of the geodetic form enables the tape to lie flat on the shell surface.

Previous work on the analytic and experimental analysis of geodetic composite shells has been done at Westland Helicopters Ltd.[1] An investigation of various structural configurations for the rear fuselage concluded that a good contender is the geodetic composite shell covered by a thin skin. At Westlands finite element analysis (FEA) models using MSC/ NASTRAN were developed for cylindrical carbon fibre reinforced plastic (CFRP) and glass fibre reinforced plastic (GFRP) geodetic shells. Experimental tests were conducted, and although the predicted buckling mode shape was correct, the predicted buckling load was found to be 40% greater than the experimentally observed buckling torque.

The only other available work on geodetic shells is by Hayashi,[2] who used the buckling theory of orthotropic cylinders to obtain buckling loads under compression, bending and torsion.

This paper will be concerned with the theoretical and experimental behaviour of a geodetic CFRP cylinder loaded in torsion. It represents the first part of a programme of research directed towards the fundamental understanding of the structural characteristics of geodetic cones. A predictive capability for stiffness and strength is required which implies non-linear large displacement post-buckling analysis for these lightweight structures which will be prone to instability. To obtain an initial estimate for the buckling load a linear buckling finite element analysis (LBFEA) has been developed. Experimental investigations have shown that very large post-buckling deformations occur before eventual failure.

LINEAR FINITE ELEMENT ANALYSIS

In order to analyse effectively geodetic shell structures the geometry of the curved shell members must be accurately represented. To do this a four-noded, cubic interpolation, isoparametric curved beam element was incorporated into the finite element code ICB-FINEL.[3]

Previous researchers have presented element formulations based on appropriate beam theory.[4,5] In order to take advantage of the beam theory concept and to represent the orthotropic properties of the geodetic construction, it is essential to develop a local cartesian frame (r, s, t), where r will be normal to the cross-section, while s and t lie along the principal axes of inertia. A method for defining this axes system is given in the Appendix.

The element formulation assumes that the nodes are located at the centroid of the cross-section. The rotations of section normals are separately interpolated so that shear deformation is accounted for.

NON-LINEAR FINITE ELEMENT ANALYSIS

The use of non-linear finite element methods in structural analysis has increased considerably over the last decade. The two common formulations for the non-linear problem are the Total Lagrangian (TL) and Updated Lagrangian (UL) methods.[6] The TL method uses the undeformed initial state of the structure as a reference for all physical quantities, while the UL method refers all physical quantities to the last configuration calculated.

A third approach is the Co-Rotational (CR) formulation in which a local cartesian coordinate system attached to each finite element continuously translates and rotates with the element as the deformations proceed. The reference configuration for strains and stresses is by this procedure continuously changing in accordance with the rigid body motion of the element.

The dominant motion in the large displacement analysis of a space frame is attributable to finite rotations, the elastic strains remaining small. Thus for the finite element model of the geodetic shell the buckling deformation will to a large extent consist of rigid body motion of the elements. If this rigid body motion is eliminated from the total displacements, the deformational part which remains is always a small quantity relative to the local element axes. Thus the CR formulation incorporated with the small deflection beam theory would appear to be an efficient method for large displacement analysis.

The present large displacement analysis[7] is based on the work of Bates,[8] who used two fundamental theorems, Euler's theorem of rigid body rotation and the polar decomposition theorem (PDT) of continuum mechanics to develop a rigorous finite rotation theory.

NON-LINEAR FORMULATION

Euler's theorem asserts that the change in orientation of a rigid body, fixed at any point, can be achieved by a rotation about some axis through this point. So, if a rigid body is rotated into a new configuration about a fixed point, then there is only one line passing through this point which remains invariant during the motion, i.e.

$$\mathbf{e} = R\mathbf{e} \tag{1}$$

where $\mathbf{e}$ is a unit vector directed along the fixed axis of rotation and R is the transformation matrix given by

$$\mathbf{X} = R\mathbf{X}^0 \qquad \mathbf{X}^0 = R^{\mathrm{T}}\mathbf{X} \tag{2}$$

where $\mathbf{X}^0$ is the original and $\mathbf{X}$ the final position vectors of a point, both measured in the fixed space X_i.

Assuming the elements of R are known, it is possible to deduce both the angle and the axis of rotation to which R corresponds.[7,8] An explicit definition of R in terms of generalised rotation coordinates can be obtained by geometrical arguments.[8]

Since rigid body motion makes no contribution to the potential energy of a structure, PDT and Euler's theorem can be applied to obtain directly the deformation itself. The PDT states that any general motion is equivalent to a rigid body motion followed by a pure deformation.

To find the generalised strains at any point on the reference line of the beam the deformation may first be isolated by removing the rigid body motion of this point from the total nodal displacements. This is equivalent to transforming the nodal displacements to a set of convected axes which are attached to and rotate with the point in question.

For the proper modelling of buckling problems the total potential energy should include all terms that are quadratic in the generalised displacements. So when deriving the kinematic relations, the rotation matrix R should be approximated to at least second order. If linear kinematic relations are used, or significant quadratic terms are excluded, then the ability of the model to represent certain forms of buckling may be severely affected.

The transformation from the unconvected state (r^0, s^0, t^0) to the convected state (r, s, t) must be considered in two stages, each involving a Euler rotation of the localised triad. Using the PDT, the total rotation θ can be decomposed into a rigid body rotation $\check{\theta}$ followed by a small transverse shear deformation $\hat{\theta}$, so

$$R(\theta) = R(\hat{\theta})R(\check{\theta}) \tag{3}$$

where the required two-stage transformation can be written

$$T = R^{T}(\hat{\theta})R(\theta)T^{0} = R(\check{\theta})T^{0} \tag{4}$$

where

$$T^{0} = \{\mathbf{r}^{0}, \mathbf{s}^{0}, \mathbf{t}^{0}\} \qquad T = \{\mathbf{r}, \mathbf{s}, \mathbf{t}\} \tag{5}$$

Now the small transverse shear deformation $\hat{\theta}$ can be determined by evaluating the rotation of the normal to the cross-section in the convected state. Hence the transverse shear deformation rotation matrix $R(\hat{\theta})$ can be found and then the rigid body rotation matrix $R(\check{\theta})$ can be calculated using eqn (3). Using eqn (4) the convected state (r, s, t) can be obtained. This leads to the convected nodal displacements associated with the pure deformation.[8] At node i

$$\hat{\mathbf{u}}_{i} = (\mathbf{p}_{i} - \mathbf{p}) - R(\check{\theta})(\mathbf{p}_{i}^{0} - \mathbf{p}^{0}) \tag{6}$$

where the current coordinates $\mathbf{p}_{i}, \mathbf{p}$ and current displacements $\mathbf{u}_{i}, \mathbf{u}$ are known, and then the determination of the initial coordinates $\mathbf{p}_{i}^{0}, \mathbf{p}^{0}$ is straightforward.

The convected nodal rotations are obtained by removing the rigid body rotation $\check{\theta}$ from the total nodal rotations θ_{i}.

To find the tangential stiffness of the element in the current configuration, it is necessary to calculate the internal increment of virtual work associated with the incremental motion. The incremental motion of a point is given by

$$\begin{bmatrix} \dot{u}_{r} \\ \dot{u}_{s} \\ \dot{u}_{t} \end{bmatrix} = \begin{bmatrix} \dot{\bar{u}}_{r} \\ \dot{\bar{u}}_{s} \\ \dot{\bar{u}}_{t} \end{bmatrix} + [\{R(\dot{\theta}_{L}) - I\}R(\hat{\theta}_{L})] \begin{bmatrix} 0 \\ s \\ t \end{bmatrix} \tag{7}$$

where $\dot{\bar{u}}_{r}, \dot{\bar{u}}_{s}, \dot{\bar{u}}_{t}$ are the incremental displacements at $s = t = 0$. As mentioned earlier, Bates[8] has shown that using a first-order approximation for R in eqn (7) leads to an incomplete geometric stiffness matrix which results in considerable errors in critical load level.

Using the second-order approximation for R in eqn (7) and substituting the incremental displacements into the incremental strain field calculated from the Green's strain tensor leads to the explicit definition of the six generalised incremental strains.[7,8] After the application of the principle of virtual work the linear and geometric stiffness matrices of the element can be obtained.

The UL form of the virtual work equilibrium equations can be written as[7,8]

$$\int_{L} D\dot{e}\delta\dot{e}\,dr + \delta\dot{W}_{n1} = \delta W_{ex} - \int_{L} S\delta\dot{e}\,dr \tag{8}$$

where $\dot{e}$ is the linear part of the incremental Green strain tensor, $\delta \dot{W}_{n1}$ is the non-linear increment of virtual work, δW_{ex} is the external virtual work in a neighbouring equilibrium configuration occupied after the application of the next load increment, S is the vector of generalised stresses and D the stress–strain relationship.

The linear and non-linear strain–displacement matrices are introduced via the relations

$$\dot{e} = B_1 \dot{\Delta} \qquad \dot{G} = B_{n1} \dot{\Delta} \tag{9}$$

where the vector $\dot{G}$ lists the nine incremental displacement derivatives that appear in $\delta \dot{W}_{n1}$ and $\dot{\Delta}$ is the vector of generalised displacements.

After the application of the virtual work equations the standard form is obtained as

$$(K_1 + K_{n1}) \dot{\Delta} = R_{ex} - R_{in} \tag{10}$$

where

$$K_1 = \int_L B_1^T D B_1 \, dr \qquad K_{n1} = \int_L B_{n1}^T Q B_{n1} \, dr \qquad R_{in} = \int_L B_1^T S \, dr \tag{11}$$

and R_{ex} is the vector of external loads and Q is the initial stress matrix.

LINEAR BUCKLING ANALYSIS

According to the energy criterion of stability, the critical (limit of bifurcation) state is characterised by the vanishing of the second variation of the total potential energy of the structure.[9] In finite element formulation the corresponding criterion is that the determinant of the total (tangent) stiffness matrix is zero.

The linear eigenvalue problem has been applied extensively for plate and shell buckling problems.[10] Once the linear and geometric stiffness matrices K_1 and K_{n1} respectively have been formed the basic linear eigenvalue problem can be solved to obtain the initial buckling load and buckled mode shape. This is given by

$$K_1 \Delta_E = -\lambda_E K_{n1} \Delta_E \tag{12}$$

where λ_E is the eigenvalue (buckling load factor) and Δ_E is the eigenvector (buckling mode).

IMPLEMENTATION

The proposed linear buckling and CR formulations were implemented into ICB-FINEL[3] on an Apollo 3000. The eigenvalue problem was solved using the Lanczos method.

In order to overcome the difficulties associated with limit points, snap-through and snap-back problems, the Crisfield arc-length method in which the arc length of the incremental displacement vector is kept constant during the equilibrium iteration has been implemented.[10]

The incremental equations of equilibrium are solved using either modified or full Newton–Raphson iteration, convergence being monitored by a displacement criterion (usually set to 0·001).

Many example problems were analysed for the linear and non-linear analysis, which demonstrated the efficiency and accuracy of the proposed element and method of analysis.[7,10]

EXPERIMENTAL ANALYSIS

A cylindrical CFRP high modulus geodetic shell was tested in torsion. Figure 1 shows the cylindrical geodetic shell. There are six clockwise and six anti-clockwise geodetic windings along the shell. Each member cross-section is 2 mm thick and 2·75 mm wide. At the joints the members are interleaved and cross orthogonally. The shell was 600 mm in length, 151 mm internal diameter and was fully built-in at both ends into aluminium end fittings using an epoxy resin.

The specimen was extensively strain gauged and the applied torque and end rotations at the load application point were monitored. Rotations at three other positions along the length were also monitored using three pairs of displacement transducers (LVDTs). In order to investigate shape changes of a particular cell of the structure clip gauges were used. Damage initiation was monitored using acoustic emission equipment.

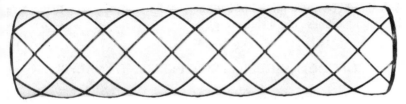

FIG. 1. Cylindrical CFRP geodetic shell.

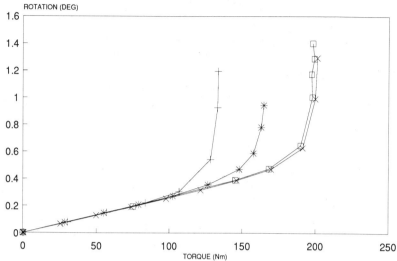

FIG. 2.   Rotation/applied torque at LVDT2. —+—, T1; —*—, T2; —⊟—, T3; —×—, T4.

EXPERIMENTAL RESULTS

Testing of the shell confirmed a phenomenon which had previously been observed on an unmonitored shell. The torsional buckling load was found to increase with successive loading in opposite directions, until a constant buckling torque of about 200 N m was achieved for loading in the clockwise and anti-clockwise directions. 'Buckling torque' is the point when a stationary or near-stationary point is reached on the torque/displacement plot. The increasing buckling torque phenomenon is illustrated in Fig. 2 for the torque/rotation plots at the middle LVDT position. This shows the increasing buckling torque from the initial torque loading T1 to the point at which successive loads (in either direction) produced almost identical response at T3 and T4. The structure is clearly seen to buckle at increasing torque until an almost constant level is reached at about 200 N m.

It should be noted that although the buckling torque is seen to increase, the initial stiffness is constant (Fig. 2).

GEODETIC SHELL FINITE ELEMENT ANALYSIS

The cylindrical geodetic shell finite element model used in the analysis is shown in Fig. 3. One four-noded element is used between each joint, a total

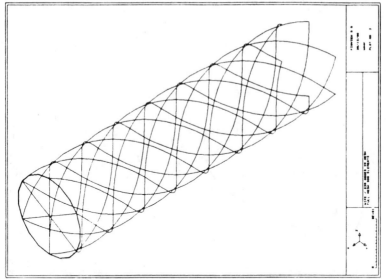

FIG. 3. FEA mesh for cylindrical geodetic shell.

of 511 nodes and 204 elements. The element is curved and twisted between the joints. The shell was assumed to be fully built-in at both ends with a circular end loading frame which was assumed infinitely stiff (Fig. 3).

The LBFEA of the shell produced a buckling torque of 288 N m and the buckling mode shown in Fig. 4.

The mode shape is identical to the test results and that observed at Westlands;[1] an ovalling of the section was seen to travel and rotate along the length of the shell. The error between the experimental and LBFEA torques is about 30%. This compares with an error of about 40% observed by Bowhay and Crannage;[1] it is suspected that their model was constructed using straight elements between the joints (their shell was constructed of closer spaced members than the one in this work).

To investigate the effect of using straight members to model the cylindrical shell, a FEA model with straight elements between the joints was analysed. The mode shape was as previously predicted but the LBFEA buckling torque was 452·5 N m, an error of about 55%. This shows the importance of using curved elements between the joints.

To reconcile the two we decided first of all to look at uncertainties in the basic finite element model. A parametric study was conducted to investigate the effects of boundary conditions, out-of-plane bending stiffness $E_{rr}I_s$, in-plane bending stiffness $E_{rr}I_t$ and initial imperfections.

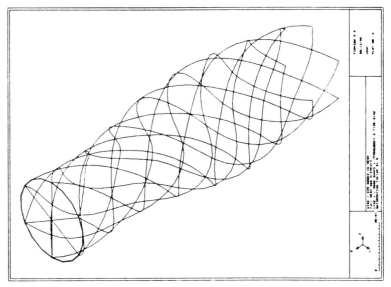

Fig. 4.　Linear buckling FEA mode 1 for cylindrical shell.

PARAMETRIC STUDY

Originally the FEA model was assumed to be fully built-in at both ends. In order to investigate the effects of differing boundary conditions the LBFEA was performed with boundary conditions ranging from fully built-in with no axial freedom to simply supported at both ends with axial freedom. The results showed only about 10% variation in buckling torques between the extreme cases. The torsional stiffness remained virtually unchanged. It is suspected that the post-buckling behaviour would be more sensitive to boundary conditions.

The effects of varying the in-plane and out-of-plane bending stiffnesses on the LBFEA buckling torque and torsional stiffness demonstrated that for large changes in the bending stiffnesses the change in buckling torques were relatively small. As expected the out-of-plane bending stiffness has a greater influence on the buckling torque.

To investigate the effects of initial imperfections on the LBFEA buckling torque of the cylindrical geodetic shell an initial geometrical imperfection was applied to the shell structure. The imperfection was in the form of an 'ovalling' of the section defined by the imperfection parameter I_p, which is a measure of the change from the original circle cross-section to the oval

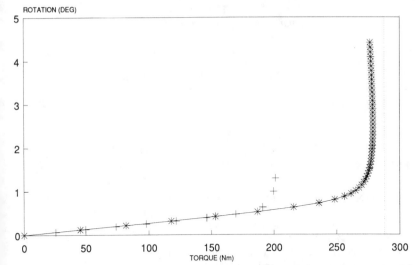

FIG. 5. Rotation/applied torque at LVDT2. —+—, Experiment T4; —*—, NLFEA model; ····, LBFEA buckling torque.

cross-section. Results showed that for relatively small imperfections ($I_p < 5\%$) the LBFEA buckling torque and torsional stiffness variations were small (less than 10%).

NON-LINEAR FINITE ELEMENT ANALYSIS

The non-linearity of the tested shell demonstrates the need for the non-linear FEA (NLFEA) to predict accurately the shell buckling torque and final failure. The NLFEA for the shell is shown in Fig. 5 for the middle LVDT position. A large difference is seen between the experimental and NLFEA results, although the trend appears the same. The NLFEA predicts a buckling torque of about 279 N m. The linear part of the results compare very well.

JOINT IN-PLANE FLEXIBILITY

The geodetic joints are formed by interleaving the unidirectional tapes which cross orthogonally during the winding process.

The latest investigations have concentrated upon a characterisation of

TABLE 1
Torsional buckling torque and stiffness summary

	Model 1		Model 2		Experimental
	LBFEA	NLFEA	LBFEA	NLFEA	
Buckling torque (N m)	288·3	279·6	227·8	214·6	200·0
	(+30%)	(+28%)	(+12%)	(+7%)	
LVDT1 (N m/deg)	861·5	866·8	925·7	937·9	809·0
	(+6%)	(+7%)	(+13%)	(+14%)	
LVDT2 (N m/deg)	375·5	375·0	367·0	366·1	388·9
	(−4%)	(−4%)	(−6%)	(−6%)	
LVDT3 (N m/deg)	265·0	264·9	262·7	262·6	252·9
	(+5%)	(+5%)	(+4%)	(+4%)	

Note $I_p = 0.7\%$ initial imperfection.
LVDT stiffnesses for NLFEA are for first increment.
(%) is error between FEA and experiment.
Model 1 is original model with no joint flexibility.
Model 2 is new refined model with joint flexibility.

these joints which is not represented in the FEA model. This relates to the consequences of a scissoring flexibility between cross beams. Experimental work has been done to determine this flexibility and this will be reported upon in a future paper.

We can report, however, that the introduction of this additional flexibility does have a significant effect on the theoretically determined buckling torque. Table 1 shows that for the NLFEA buckling torque the error is about 7%.

CONCLUSIONS

The element and the analysis was found to be accurate and efficient.

The experimental analysis of the cylindrical CFRP geodetic shell under torsion was found to produce an increasing buckling torque with a constant torsional stiffness for successive load reversals, until a constant repeatable buckling torque of about 200 N m was reached. This is thought to be due to realignment of fibres to produce a better overall geodetic path for the fibres.

Using a curved element the linear buckling FEA was found to produce a significant improvement in predicting the buckling torque over a previous analysis which used a straight element. But the error was still 30% compared with the experimental buckling torque.

Changing the boundary conditions from fully built-in to simply supported was found to affect the LBFEA buckling torque by only about 10%.

The out-of-plane bending stiffness has a greater effect on the overall buckling torque than the in-plane bending stiffness. But a reduction in out-of-plane bending stiffness of over 50% would be required to match the experimental and LBFEA buckling torques.

Small initial geometrical imperfections had no significant effect on the overall LBFEA buckling torque and torsional stiffness.

It appears that the effects of changing boundary conditions, bending stiffnesses and initial geometrical imperfections do not account for the 30% difference between experimental and LBFEA buckling torques.

The non-linear FEA produced little change in the buckling torque, but the predicted trend was the same as the experimental results. The NLFEA is essential to predict final failure.

Recent investigations show the importance of representing joint flexibility in the FEA. The results are then dramatically improved with errors below 10%.

REFERENCES

1. BOWHAY, D. A. and CRANNAGE, M. A., An advanced tailcone structural configuration study with particular reference to geodetic construction. WHL ATN 296, Westland Helicopters Limited, Yeovil, January 1986.
2. HAYASHI, T., Buckling strength of cylindrical geodesic structures. *Proc. Japan/ US Conf.*, ed. K. Kawata and T. Akasaka. Tokyo, 1981.
3. HITCHINGS, D., *FINEL User Manual.* Imperial College, Aeronautics Department, London.
4. JIROUSAK, J., A family of variable section curved beam and thick-shell or membrane-stiffening isoparametric elements. *Int. J. Num. Meth. Engng*, **17** (1981) 171.
5. ZHANG, S. H. and LYONS, L. P. R., A thin-walled box beam finite element for curved bridge analysis. *Comput. Struct.*, **18** (1984) 1035.
6. BATHE, K. J., *Finite Element Procedures in Engineering Analysis.* Prentice–Hall, New Jersey, 1982.
7. SANDHU, J. S., STEVENS, K. A. and DAVIES, G. A. O., *Geodetic Shell Structures.* Imperial College, Aeronautics Department, London, October 1988.
8. BATES, D. N., The mechanics of thin-walled structures with special reference to finite rotations. PhD Thesis, University of London, 1987.

9. THOMPSON, J. M. T. and HUNT, G. W., *A General Theory of Elastic Stability.* John Wiley, London, 1973.
10. SANDHU, J. S., STEVENS, K. A. and DAVIES, G. A. O., *Geodetic Shell Structures.* Imperial College, Aeronautics Department, London, October 1987.
11. LIPSCHUTZ, M. M., *Differential Geometry.* Schaum's outline series, McGraw-Hill, New York, 1969.

APPENDIX: LOCAL REFERENCE FRAME

The local reference frame (r, s, t) can be defined by the three unit vectors $(\mathbf{r}_i, \mathbf{s}_i, \mathbf{t}_i)$ at the node i. The cross-section must be able to 'follow' a surface such that t is always normal to the surface (i.e. $\mathbf{t}$ is the principal normal vector).

If $(\mathbf{i}, \mathbf{j}, \mathbf{k})$ are unit vectors in the global axes direction (X_1, X_2, X_3) respectively, then the position vector of a node i is given by

$$\mathbf{X} = X_1\mathbf{i} + X_2\mathbf{j} + X_3\mathbf{k} \tag{A.1}$$

The tangent vector along the r direction is given by[11]

$$\mathbf{r}_i = J^{-1}\left[\frac{\partial^2 X_1}{\partial \xi^2}\mathbf{i} + \frac{\partial^2 X_2}{\partial \xi^2}\mathbf{j} + \frac{\partial^2 X_3}{\partial \xi^2}\mathbf{k}\right] \tag{A.2}$$

where the Jacobian factor is given by

$$J = \left[\left(\frac{\partial X_1}{\partial \xi}\right)^2 + \left(\frac{\partial X_2}{\partial \xi}\right)^2 + \left(\frac{\partial X_3}{\partial \xi}\right)^2\right]^{1/2} \tag{A.3}$$

The vector $\mathbf{t}_i$ could be defined using an additional reference node at each node position,[4] but an alternative is to use the curvature vector,[11] then

$$\mathbf{t}_i = \frac{\partial^2 \mathbf{X}}{\partial \xi^2} \bigg/ \left|\frac{\partial^2 \mathbf{X}}{\partial \xi^2}\right| \tag{A.4}$$

so

$$\mathbf{t}_i = -R^{-1}\left(\frac{\partial^2 X_1}{\partial \xi^2}\mathbf{i} + \frac{\partial^2 X_2}{\partial \xi^2}\mathbf{j} + \frac{\partial^2 X_3}{\partial \xi^2}\mathbf{k}\right) \tag{A.5}$$

where

$$R = \left[\left(\frac{\partial^2 X_1}{\partial \xi^2}\right)^2 + \left(\frac{\partial^2 X_2}{\partial \xi^2}\right)^2 + \left(\frac{\partial^2 X_3}{\partial \xi^2}\right)^2\right]^{1/2} \tag{A.6}$$

is the curvature vector.

The negative sign in eqn (A.5) is introduced so that the curvature vector which is normally directed towards the origin is directed outwards.[10] The vector $\mathbf{t}_i$ will coincide with the normal to the surface as required.

The vector $\mathbf{s}_i$ can be obtained by vector multiplication:

$$\mathbf{s}_i = -\mathbf{r}_i \otimes \mathbf{t}_i \tag{A.7}$$

26

High-Order Analysis of General Multi-layered Rectangular Plates Subjected to Transverse Loading

J. T. Mottram

Department of Engineering, University of Warwick,
Gibbet Hill Road, Coventry CV4 7AL, UK

ABSTRACT

This paper gives a parametric study on shear deformation of two eight-layered quasi-isotropic square plates using the high-order procedure of Lo et al. The two laminates contain the same ply orientations, one with a symmetrical $(90°, 45°, 0°, -45°)_s$ arrangement, and the other with an antisymmetrical $(90°, 45°, 0°, -45°, 45°, 90°, -45°, 0°)$ arrangement. Results with uniform distributed load and simply-supported edges are provided through normalised central displacement and upper surface stresses for a range of span-to-thicknesses from 4 to 100. For the two lay-up arrangements displacement at all spans differs by less than 1%. Although surface stress results have different magnitudes there is similarity in the characteristic trend as span-to-thickness is reduced. The parametric comparison suggests that shear deformation may be stacking sequence independent.

Furthermore, a number of lamina material property sets are used to demonstrate their dramatic effect on predictions. This shows that the shear behaviour of general laminates is a complex function of lamina properties, and that with test data chosen classical theory becomes inappropriate when span-to-thickness is below 30.

INTRODUCTION

The potential of advanced fibre-reinforced layered composites for structural members,[1] has produced considerable research effort on the

503

bending behaviour of laminates. To begin with Classical Plate Theory, which employs Kirchhoff's hypothesis, that normals remain straight and normal, was successful in determining the bending of thin laminates.[2] It soon became apparent that filamentary materials were susceptible to thickness effects because their ratio of longitudinal to shear moduli is high. Typically, the ratio for a uni-directional high performance fibre/polymer will be between 10 and 30 times that of structural metals.

Reissner[3] and Mindlin[4] developed isotropic transverse shear deformation theories by the late 1950s. These were furthered by a number of workers[5-9] from the late 1960s into high-order theories that coped with orthotropic layered examples. In parallel with these developments, Pagano and co-workers[10-12] produced exact elasticity solutions for several simply-supported laminate problems, by choosing Fourier series approximations for displacement fields. In particular a thorough investigation was performed of the test case, consisting of a rectangular $(A = 3B)$, equal thickness, cross-ply $(0°, 90°, 0°)$ plate; with the lamina material property data set $E_{11}/E_{22} = 25$; $G_{12}/E_{22} = G_{13}/E_{22} = 0.5$, $G_{23}/E_{22} = 0.2$, $v_{12} = v_{13} = v_{23} = 0.5$, and subjected to sinusoidal pressure load. Shear deformation was monitored as the span-to-thickness ratio was varied from 4 to 100. For this problem classical theory becomes ineffective when the ratio is < 40.

The aforementioned analyses are all of a single type where deformation in plate domain is defined by a single set of displacement functions. With the ever-growing refinement to finite element methods during the 1970s and 1980s it is not surprising that a number of orthotropic shear flexible elements have been formulated. Often, to ascertain modelling limitations of the elements Pagano's cross-ply problem was taken as a test case. It has been shown[13] that finite element predictions for central transverse displacement are accurate. But that unless an excessive number of degrees of freedom are involved, stresses are likely, over the span-to-thicknesses taken, to be less accurate than the classical equivalent. To date, there is a limited number of exact solutions, to verify numerical procedures, such as the finite element method, for a wide spectrum of laminate bending problems. Of these, the majority have symmetrical cross-ply layers.[10,11] There exists few analyses for anti-symmetrical[12,14,15] and none for general stacking sequences. This indicates a requirement for an easy-to-use procedure which accurately predicts the elastic shear deformation of general multi-layered plates. This paper discusses such a procedure with particular reference to a parametric study of two eight-layered quasi-isotropic laminates with symmetrical and anti-symmetrical stacking sequences.

METHOD OF ANALYSIS

An easy-to-use shear deformation solution for general multi-layered rectangular plates is available with the pinnacle of high-order theories for plate deformation attributed to Lo *et al.*[16,17] The technique is based on the third order domain displacement functions

$$u = u^0(x, y) + z\Psi_x(x, y) + z^2\xi_x(x, y) + z^3\phi_x(x, y)$$
$$v = v^0(x, y) + z\Psi_y(x, y) + z^2\xi_y(x, y) + z^3\phi_y(x, y) \qquad (1)$$
$$w = w^0(x, y) + z\Psi_z(x, y) + z^2\xi_z(x, y)$$

where u^0, Ψ_x, ξ_x, ϕ_x, etc., are 11 independent generalised coefficients to define deformation.

Eleven pertinent governing equilibrium equations are formulated from the principle of stationary potential energy. Displacements, strains and stresses are made available by substituting Fourier series approximations for displacement functions (as defined by (1)), and load conditions, into the equilibrium equations in terms of generalised coefficients. Edge boundary conditions are satisfied by an appropriate choice of Fourer series. A detailed explanation of the method has been given in the papers of Lo *et al.* and will not be reproduced here.

Presented now are the displacement functions and load form for the situation of uniform pressure load. Consider a rectangular plate, side lengths A and B, thickness h, subjected to a constant pressure on the top surface, $z = h/2$, of the form,[18]

$$q_z = \frac{16q_0}{\pi^2} \sum_{m=1}^{\infty} \sum_{n=1}^{\infty} \frac{1}{mn} \sin\frac{m\pi x}{A} \sin\frac{n\pi y}{B} \qquad (2)$$

q_0 is the magnitude of pressure load. The summation term is zero when m or n are even. For simply supported edges the corresponding Fourier series for an assumed displacement field are

$$u = \sum_{m=1}^{\infty} \sum_{n=1}^{\infty} (K_{7mn} + zK_{1mn} + z^2 K_{8mn} + z^3 K_{2mn}) \cos\frac{m\pi x}{A} \sin\frac{n\pi y}{B}$$

$$v = \sum_{m=1}^{\infty} \sum_{n=1}^{\infty} (K_{9mn} + zK_{3mn} + z^2 K_{10mn} + z^3 K_{4mn}) \sin\frac{m\pi x}{A} \cos\frac{n\pi y}{B} \qquad (3)$$

$$w = \sum_{m=1}^{\infty} \sum_{n=1}^{\infty} (K_{5mn} + zK_{11mn} + z^2 K_{6mn}) \sin\frac{m\pi x}{A} \sin\frac{n\pi y}{B}$$

K_{imn} are a set of K_i, $i = 1$–11 constants for each non-zero summation. To evaluate these constants loading and displacement forms were substituted into the governing equilibrium equations. The resulting expression in matrix notation is

$$\sum_{m=1}^{\infty}\sum_{n=1}^{\infty} [L]\{K\} = \sum_{m=1}^{\infty}\sum_{n=1}^{\infty} \{F\} \tag{4}$$

$[L]$ is a fully populated 11×11 stiffness matrix as reported in Ref. 16 for the case of a sinusoidal load. Elements in $[L]$ are dependent on lamina material properties, stacking sequence, summation terms m and n, and the trigonometry part of eqn (3).

Based on the governing expression (4), a series of FORTRAN programmes have been written to predict shear deformation of multi-layered plates and strips, with one of three load types, namely sinsusoidal, uniform-distributed and patch.

VERIFICATION OF PROCEDURE

Cross-Ply

In order to demonstrate the accuracy of the high-order approach Table 1 gives a comparison with the exact elasticity solution of Pagano's test case, which is outlined in the introduction. For future comparisons the lamina material property set will be referred to as Material I. With this type of problem, eqn (4) is greatly simplified for the following reasons:

(i)   Sinusoidal pressure load is modelled by

$$q_z = q_0 \sin\frac{\pi x}{A}\sin\frac{\pi y}{B};$$

(ii) with $m = n = 1$ only, summations in the displacement functions (3) can be dropped;

(iii) symmetry in stacking about mid-plane enables eqn (4) to be divided into its flexural and in-plane contributions; and

(iv) all trigonometric terms in eqn (4) can be cancelled out because coupling stiffness terms are zero for cross-ply laminates.

Application of these four factors to the test case gives the flexural component of eqn (4) as

$$[C]\{K\} = \{q\} \tag{5}$$

TABLE 1
Three-layer (0°, 90°, 0°) simply-supported rectangular plate (B = 3A) subjected to sinusoidal loading

s	$\bar{\sigma}_x\left(\frac{A}{2},\frac{B}{2},\pm\frac{h}{2}\right)$ High-order	Pagano Flex.	Pagano In-pl.	$\bar{\sigma}_y\left(\frac{A}{2},\frac{B}{2},\pm\frac{h}{6}\right)$ High-order	Pagano Flex.	Pagano In-pl.	$\bar{\omega}\left(\frac{A}{2},\frac{B}{2},0\right)$ High-order	Pagano Flex.	Pagano In-pl.
2	2·13 −1·62	±1·80	+0·228	0·230 −0·268	±0·217	0·025 4	8·17	7·786	0·0
4	1·14 −1·10	±1·09	+0·023	0·109 −0·119	±0·105	0·007	2·82	2·36	
10	0·726 −0·725	±0·712	0·003	0·041 8 −0·043 5	±0·040 4	0·0	0·919	0·867	
20	±0·650	±0·646	0·0	0·029 4 −0·029 9	±0·028 9		0·610	0·596	
50	±0·628	±0·627		±0·025 9	±0·025 6		0·520	0·518	
100	±0·624	±0·624		±0·025 3	±0·025 2		0·508	0·507	
CPT	±0·623			±0·025 2			0·503		

where terms in $[C]$ and $\{q\}$ are reported in Appendix II of Ref. 13, and

$$\{K\} = \{K_1, K_2, K_3, K_4, K_5, K_6\}$$

Non-dimensional quantities presented are defined with respect to data by

$$\bar{\sigma}_x, \bar{\sigma}_y = \frac{1}{q_0 s^4}(\sigma_x, \sigma_y)$$

and

$$\bar{w} = \frac{100 E_{22} w}{q_0 h s^4}$$

(6)

where the span-to-thickness ratio $s = A/h$.

Combined flexural and in-plane contributions must be taken when comparing with exact solutions. Over the span-to-thickness range quoted the difference is $< 5\%$. For $s > 10$ the deviation is below 2% and the in-plane contribution becomes insignificant. As s approaches 100 the two three-dimensional analyses are indistinguishable from classical plate theory.

Symmetrical Angle-Ply

For general laminates, where lamina are now orientated at $\theta°$ to the principal axes, twisting coupling stiffnesses are non-zero and solution to

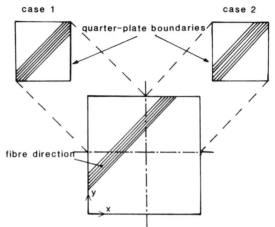

FIG. 1.   Quarter-plate modelling.

eqn (4) is dependent on trigonometric terms. One source of this complication arises from the assumed displacements given by eqn (3). Presence of general lamina induce inherent twisting into the bending response. This means that quarter-plate symmetry, implicit by eqn (5), is no longer strictly correct. Fortunately, it has been shown[13] that by ignoring trigonometric terms and taking the mean of the two quarter-plate models, as illustrated in Fig. 1, acceptable accuracy is attained.

Table 2 presents flexural $\bar{w}$ values for a four-layered symmetrical angle-

TABLE 2

Four-layer (45°, −45°, −45°, 45°) simply-supported square plate subjected to sinusoidal loading

s	$\bar{w}\left(\dfrac{A}{2},\dfrac{B}{2},0\right)$				
	High order		Average	Reddy	$\dfrac{\text{AV-RD}}{\text{AV}} \times 100\%$
	Case 1 $(45°, -45°)_s$	Case 2 $(-45°, 45°)_s$			
5	0·506	1·147	0·827	0·760	8·1
7·5	0·273	0·801	0·537	0·485	9·7
10	0·192	0·679	0·436	0·410	6·0
20	0·114	0·562	0·338	0·315	6·8
50	0·092	0·528	0·310	0·275	11·3

ply $(45°, -45°)_s$, square plate with sinusoidal load. Lamina material properties (Material II), were, $E_{11}/E_{22} = 40$, $G_{12}/E_{22} = G_{13}/E_{22} = 0.6$, $G_{23}/E_{22} = 0.5$, and $v_{12} = v_{13} = v_{23} = 0.25$.

The in-plane contribution, which is small, has been neglected to be compatible with a finite element solution. Comparison is provided with an eight-noded penalty shear finite element developed by Reddy.[19] The values of $\bar{w}$, as predicted with the two quarter-plate cases 1 and 2, are very different. Over the s range reported the average is in the region of 8.0% higher than the equivalent finite element model. This problem is an extreme type as all layers are orientated at $+45°$ or $-45°$. It can be shown[13] that as the proportion of $\pm 45°$ plies is reduced there is a rapid improvement in deviation between quarter-plate models. This is to be expected as the laminate stiffnesses and quarter-plate boundary conditions approach those of a cross-ply example.

Anti-symmetrical Angle-Ply

Preceding test cases and other examples in Ref. 13 illustrate the power of the high-order procedure for laminates with symmetrical stacking sequences. Now the procedure will be applied to a problem with an anti-symmetrical arrangement. For this class of laminate all coupling stiffnesses are definite. Flexural and in-plane deformations are therefore coupled and so the solution involves all eleven generalised coefficients defined by eqn (1).

Table 3 presents central displacement for the four-layered antisymmetrical laminate $(45°, -45°, 45°, -45°)$ which possesses lamina

TABLE 3

Four-layer (45°, −45°, 45°, 45°) simply-supported square plate subjected to sinusoidal loading

s	$\bar{w}\left(\dfrac{A}{2}, \dfrac{B}{2}, 0\right)$				
	High-order		Average	Reddy	$\dfrac{\text{AV-RD}}{\text{RD}} \times 100\%$
	Case 1	Case 2			
5	0.721	0.750	0.736	0.627	14.8
7.5	0.415	0.423	0.419	0.377	10.0
10	0.311	0.315	0.313	0.275	12.1
20	0.213	0.213	0.213	0.191	10.3
50	0.186	0.186	0.186	0.167	10.2

TABLE 4
Eight-layered symmetrical and anti-symmetrical laminates when s equals 100

Lay-up		$\bar{w}_{100}$	$\bar{\sigma}_{x100}\left(+\dfrac{h}{2}\right)$	$\bar{\sigma}_{y100}\left(+\dfrac{h}{2}\right)$
		Material I		
Sym	Case 1	0·393	0·023 8	0·462
	Case 2	0·863	0·052 5	1·043
	Ave	0·628	0·038 2	0·752
Anti	Case 1	0·623	0·037 2	0·454
	Case 2	0·622	0·050 9	0·682
	Ave	0·623	0·044 1	0·568
		Material II		
Sym	Case 1	0·251	0·014 0	0·461
	Case 2	0·576	0·032 2	1·107
	Ave	0·413	0·023 1	0·784
Anti	Case 1	0·409	0·022 5	0·459
	Case 2	0·409	0·031 8	0·706
	Ave	0·409	0·027 2	0·582
		Material III		
Sym	Case 1	0·391	0·022 0	0·459
	Case 2	0·860	0·048 4	1·037
	Ave	0·626	0·035 2	0·748
Anti	Case 1	0·619	0·034 2	0·451
	Case 2	0·619	0·047 7	0·678
	Ave	0·619	0·041 0	0·565
		Material IV		
Sym	Case 1	0·252	0·015 2	0·469
	Case 2	0·578	0·035 0	1·111
	Ave	0·415	0·025 1	0·790
Anti	Case 1	0·411	0·024 5	0·460
	Case 2	0·411	0·033 8	0·709
	Ave	0·411	0·029 2	0·584

properties identical to the symmetrical test case. The values of $\bar{w}$ determined by the two quarter-plate models are very similar, and this contrasts with the symmetrical example. For the range of s compared the average is about 11·5% higher than Reddy's finite element. This is slightly higher than found for the symmetrical case.

PARAMETRIC STUDY OF TWO EIGHT-LAYERED LAMINATES

With the capability of predicting shear response for general laminates the effect of stacking sequence could be examined. To accomplish this task studies were made using two square quasi-isotropic eight-layered laminates, each with two lamina orientated at $0°$, $90°$, $45°$ and $-45°$ to global axes, having stacking sequences which are symmetrical and anti-symmetrical. Lay-up arrangements for case 1 quarter-plate modelling are $(90°, 45°, 0°, -45°)_s$ and $(90°, 45°, 0°, -45°, 45°, 90°, -45°, 0°)$, respectively. The load was a uniformly distributed pressure on the upper surface so numerical results were obtained with eqns (2)–(4). Solution convergence was achieved if all summation terms up to m and n equal 20 were included. Recorded in Table 4 are non-dimensional central displacement and stresses when s is 100 and it is acceptable to classify the plate as thin. The top two blocks of results are for Material I and Material II lamina properties. These are data sets often employed in the verification of numerical analyses concerned with shear deformation. The other two blocks are for Material III and Material IV properties where ratios E_{11}/E_{22} for Material I and Material II data have been interchanged. The four lamina material data sets are given in Table 5. Because the physical structure of advanced filamentary composites can conceivably provide such lamina properties it

TABLE 5
Lamina material properties for parametric study

Material	Symbol	$\dfrac{E_{11}}{E_{22}}$	$\dfrac{G_{12}}{E_{22}} = \dfrac{G_{13}}{E_{22}}$	$\dfrac{G_{23}}{E_{22}}$
I	■	25	0·5	0·2
II	●	40	0·6	0·5
III	□	25	0·6	0·5
IV	○	40	0·5	0·2

In all cases $v_{13} = v_{23} = v_{12} = 0·25$.

does not seem unreasonable to take such a wide variation in material data for the parametric study, even though Materials III and IV do not represent specific composite material. To illustrate quarter-plate models and stacking sequence, results from case 1 and case 2 are presented with their average. Average $\bar{w}_{100}$ values are nearly the same for both laminates. For the symmetrical laminate there is an increase from case 1 to case 2 in excess of 100%. This compares with virtually zero change for the anti-symmetrical. With the choice of material data sets $\bar{w}_{100}$ varies by 52%. Quarter-plate

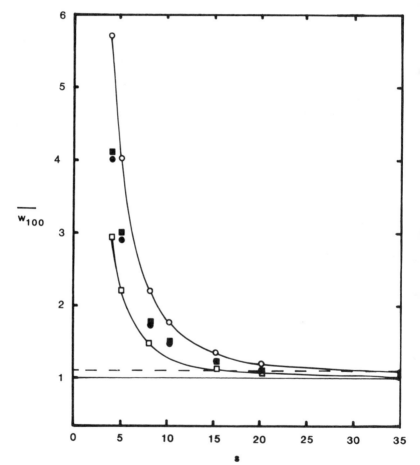

FIG. 2. Normalised central displacement vs span-to-thickness for eight-layered, symmetrical and anti-symmetrical square plates, subjected to uniform distributed load. (■, material I; ●, material II; □, material III; ○, material IV.)

modelling has a profound effect on stresses, $\bar{\sigma}_{x100}$, and $\bar{\sigma}_{y100}$, for both laminates. However, the difference between case 1 and case 2 for the symmetrical is about three times more than for the anti-symmetrical. Changes made to data sets makes little impact on the $\bar{\sigma}_{y100}$ predictions. Whereas the same data sets change $\bar{\sigma}_{x100}$ by over 60%. Absolute maximum and minimum global surface stresses are obtained with the symmetrical stacking sequence.

Plotted in Figs 2–6 are central displacement and surface stresses for the laminates as a function of s. For ease of interpretation data has been

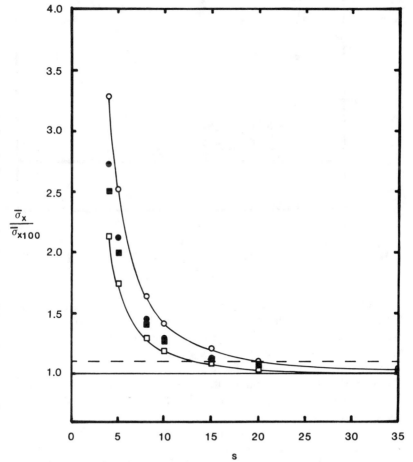

FIG. 3. Normalised central surface x stress vs span-to-thickness for eight-layered, symmetrical and square plate, subjected to uniform distributed load.

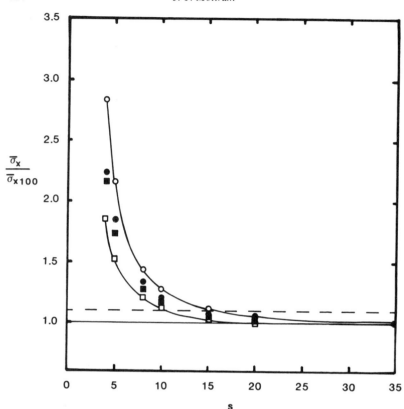

FIG. 4.   Normalised central surface x stress vs span-to-thickness for eight-layered, anti-
symmetrical square plate, subjected to uniform distributed load.

normalised by dividing $\bar{w}$, etc., by corresponding $\bar{w}_{100}$, etc., as given in Table 4. Appearance of the shear effect is not until s is below 35 so plots do not extend past this value of span-to-thickness. To distinguish between Material I–IV results each set has been allocated a symbol. These are given alongside the lamina data in Table 5.

Once normalised, one cannot differentiate between central displacements from the two stacking sequence. Hence Fig. 2 is for both laminates. Stresses are distinct and Figs 3–6 present all the information. Figure 5 giving normalised y-direction stress for the symmetrical example is uncharacteristic. One would have expected to observe a smooth transition from shear dominance to the independent behaviour. It is unclear why this particular set of results does not fit in with the rest.

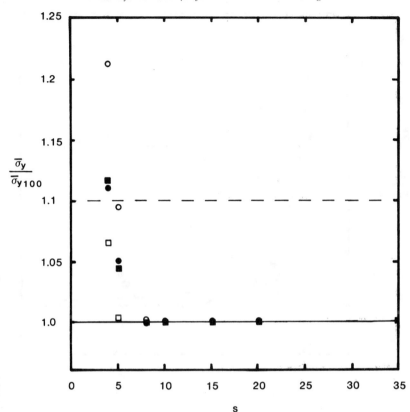

Fig. 5. Normalised central surface *y* stress vs span-to-thickness for eight-layered, symmetrical and square plate, subjected to uniform distributed load.

For all five parameters plotted, curves of relative shear response do not cross. This would not be the situation if absolute non-dimensional values had been presented. There is an enormous change in magnitude of shear dominance, as monitored by results when *s* is 4, not only for material data, but also between the five parameters. From a study of plots it is clear that the material data sets in ascending order of shear deformation are III, I, II, IV. The normalised technique employed to compare results from the parametric study has thrown up an interesting feature. This being the small difference between Material I and II data sets. This suggests that where both sets have been used in the verification of numerical methods[19] no actual modelling difference has been introduced.

In the figures the dashed line indicates the value at which the shear

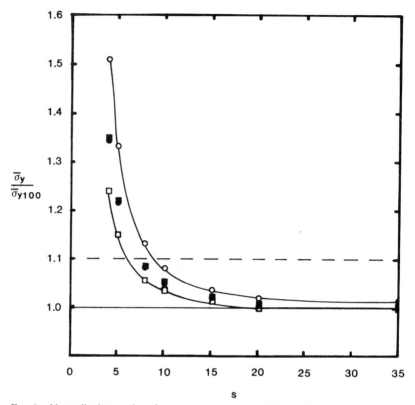

FIG. 6. Normalised central surface *y* stress vs span-to-thickness for eight-layered, anti-symmetrical square plate, subjected to uniform distributed load.

component is 10% of elastic. Where this line intercepts the curves gives minimum *s* at which analytical work could permissibly ignore shear. Taking Material IV curves, Fig. 2 shows that this value of *s* is 30 for central displacement. Figures 3, 4 and 6 show that *s* is somewhat lower for stresses lying between 13 and 20. This is the same observation as made by Pagano studying simple cross-ply problems.[10]

Examination of all plots at $s = 4$ shows that there is no pattern to the spread provided by the four data sets. For *y*-direction stress this spread is in the order of 20%. This increases to 55% for *x*-direction stress, and even higher still, to 96%, for central displacement. It can also be noted that an increase in the ratio E_{11}/E_{22} while maintaining the other ratios constant causes shear deformation to be more dominant.

Finally, Table 6 presents a matrix of normalised central displacement for

TABLE 6
Shear deformation for the eight-layered symmetrical laminate with changes to lamina material properties

$\dfrac{G_{12}}{E_{22}} = \dfrac{G_{13}}{E_{22}}$	0·2	0·35	0·5
	$\dfrac{\bar{w}_4}{\bar{w}_{100}}$		
$\dfrac{G_{23}}{E_{22}}$			
0·2	5·85	4·76	4·10
0·35	4·56	3·93	3·52
0·5	3·83	3·41	3·11

$$\frac{E_{11}}{E_{22}} = 25.$$

the symmetrical example at s equal 4, when ratios $G_{12}/E_{22} = G_{13}/E_{22}$ and G_{23}/E_{22}, are each taken to be 0·2, 0·35 and 0·5. E_{11}/E_{22} was constant at 25. This parametric study provides the following information:

(i) Over the range quoted, changing G_{23}/E_{22}, with constant G_{12}/E_{22}, alters predictions by 30% more than the corresponding change to G_{12}/E_{22}, with constant G_{23}/E_{22}.

(ii) The higher the values of G_{23}/E_{22} or G_{12}/E_{22} ratios, the less the central displacement alters when a change is made to the other ratio.

(iii) Response of the symmetrical laminate due to shear deformation is a complex function of lamina material properties.

CONCLUSIONS

The high-order procedure can be used to investigate shear deformation in general multi-layered orthotropic simply supported plates subjected to transverse loading. Non-dimensional central displacement and surface stress results from a number of test cases are used to verify the procedure.

A parametric study was made with two eight-layered, quasi-isotropic laminates, which possess symmetrical and anti-symmetrical stacking sequences. To obtain reliable results the average of two quarter-plate models must be taken. Quarter-plate modelling is seen to be more critical for the symmetrical arrangement. The overall shear response of the two laminates was not very different. Normalised central displacement was identical over a span-to-thickness, *s*, range of 4–100. Equivalent stresses were of different magnitudes, but had the same characteristic curve.

With 10% shear contribution, the level at which one should not consider applying classical thin plate theory, the parametric study indicates that this is reached when *s* equals 30 for displacement, and 20 for surface stresses.

Four lamina material property data sets highlight the considerable effect of properties on resultant shear deformation. Two of these sets are often employed in numerical methods to establish their shear flexible limits. It is demonstrated here that these data provide practically identical shear contribution over the whole *s* range. It is finally concluded that shear deformation is a complex function of lamina properties and that the minimum span-to-thickness for thin plate theory will vary accordingly.

REFERENCES

1. *Advanced Composites: The Latest Developments*. Proc. 1986 Conf. Engng Soc. Detroit, ASM Conf., 1987.
2. ASHTON, J. E., HALPIN, J. C. and PETIT, P. H., *Primer on Composite Materials: Analysis*. Technomic, Stamford, 1969.
3. REISSNER, E., The effect of transverse shear deformation on the bending of elastic plates. *J. Appl. Mech.* (1985) A-69, A-77.
4. MINDLIN, R. D., Influence of rotary inertia and shear on flexural motion of isotropic elastic plates. *J. Appl. Mech.*, **18** (1951) 31–8.
5. GIRKMAN, K. and BEER, R., Application of Eric Reissners' refined plate theory to orthotropic plates. *Oster. Ingenieur-Archiu*, **12** (1958) 101–10.
6. YANG, D. C., NORRIS, C. H. and STAVSKY, Y., Elastic wave propagation in heterogeneous plates. *Int. J. Solids Struct.*, **2** (1966) 665–84.
7. WHITNEY, J. M., The effect of transverse shear deformation on the bending of laminated plates. *J. Comp. Mater.*, **3** (1969) 534–47.
8. AMBARTSUMYAN, S. A., *Theory of Anisotropic Plates*, eds J. E. Ashton and T. Cheron. Technomic, Stamford, 1970.
9. WHITNEY, J. M. and SUN, C. T., A refined theory for laminated anistropic cylindrical shells. *J. Appl. Mech., Trans. ASME*, Series E, **41**(2) (1974) 471–6.
10. PAGANO, N. J., Exact solution for rectangular bi-directional composites and sandwich plates. *J. Comp. Mater.*, **4** (1970) 20–34.
11. PAGANO, N. J. and HATFIELD, S. J., Elastic behaviour of multi-layered bidirectional composites. *J. Am. Inst. Aeronaut. & Astronaut.*, **10** (1972) 931–3.
12. WHITNEY, J. M. and PAGANO, N. J., Shear deformation in heterogeneous anistropic plates. *J. Appl. Mech.*, **37** (1970) 1031–6.

13. MOTTRAM, J. T., High-order analysis of generally symmetrical laminated plates under transverse loading. *Comp. Struct.* (1989), **12**, in press.
14. WHITNEY, J. M., The effect of boundary conditions on the response of laminated composite plates. *J. Comp. Mater.*, **4** (1970) 192–203.
15. WHITNEY, J. M., Stress analysis of thick laminated composite and sandwich plates. *J. Comp. Mater.*, **6** (1972) 426–40.
16. LO, K. H., CHRISTENSEN, R. H. and WU, E. M., A high-order theory of plate deformation—Part I: Homogeneous plates. ASME paper No. 77, WA/APM-23.
17. LO, K. H., CHRISTENSEN, R. H. and WU, E. M., A high-order theory of plate deformation—Part II: Laminated plates. ASME paper No. 77, WA/APM-24.
18. TIMOSHENKO, S. and WOINOWSKY-KRIEGER, S., *Theory of Plates and Shells*, 2nd edn. Engineering Societies Monographs, McGraw-Hill, New York, 1950.
19. REDDY, J. N., A penalty plate bending element analysis of laminated anisotropic composite plates. *Int. J. Numer. Methods Engng*, **15** (1980) 1187–206.

27

Stress Analysis of a Composite Plate Based on a New Plate Theory

JI-LIANG DOONG, CHIN-PING FUNG and CAINAN LEE

*Department of Mechanical Engineering,
National Central University,
Chung-Li, 32054 Taiwan*

ABSTRACT

In this paper, the governing equations of a simply supported laminated plate based on a new plate theory are presented. Using the seven kinematic variables theory, static behaviour and stress analysis are studied. The present higher-order plate theory satisfies the boundary conditions that the transverse shear stresses vanish on the top and bottom surfaces of the plate. And the benefit of the significant simplification can be seen to compare with the Lo et al. higher-order plate theory. Comparison with exact elasticity solutions shows excellent agreement. Also, present results are compared with the Mindlin plate theory results, Reddy's results and eleven variables higher-order plate theory results based on Lo et al.

NOTATION

a, b, h	Dimensions of plate in x, y, z directions
$A_{ij}, B_{ij}, D_{ij}, E_{ij},$ F_{ij}, G_{ij}, H_{ij}	Laminate stiffness
C_{ij}	Components of the anisotropic stiffness matrix
E_i, G_{ij}	Young's modulus and shear modulus
$\bar{P}_i, \Delta P_i$	The applied and perturbing surface tractions
q, q_o	Normal pressure and its peak value
u_i	Displacement of plate in x, y, z directions
u_x, u_y, w	Displacement of plate ($z = 0$) in x, y, z directions
$\bar{W}$	$10(Wh^3 E_2/q_o a^4)$, $W = W(a/2, b/2)$

v_{ij}	Poisson's ratios
$\xi_x, \xi_y, \xi_z, \phi_x, \phi_y$	Higher-order shear deformation terms
ρ	Density
$\bar{\sigma}_1$	$\sigma_1(a/2, b/2, h/2)h^2/q_o a^2$
$\bar{\sigma}_2$	$\sigma_2(a/2, b/2, h/2)h^2/q_o a^2$
$\bar{\sigma}_4$	$\sigma_4(a/2, 0, 0)h/q_o a$
$\bar{\sigma}_5$	$\sigma_5(0, b/2, 0)h/q_o a$
$\bar{\sigma}_6$	$\sigma_6(0, 0, h/2)h^2/q_o a^2$
ψ_x, ψ_y, ψ_z	Rotations of plate in x, y, z directions

INTRODUCTION

The classical laminated plate theory, in which it is assumed that normals to the midplane before deformation remain straight and normal to the plane after deformation, because of the relatively soft interlaminar shear modulus in high performance composites, becomes very inaccurate for determining gross response, such as plate deflection, and the internal stress distribution. These results are due to the neglect of transverse shear strain in the classical theory. Many different laminated plate theories[1−3] have been proposed which are intended to improve upon the classical laminated plate theory by accounting for the effects of the transverse shear strain in the plate. In Reissner–Mindlin type theory,[1] the displacement field accounts for linear or higher-order variations of midplane displacement through thickness, but the exact solutions[4,5] for composite material structure indicate that Mindlin-type theories do not adequately model the behaviour of highly orthotropic composite structures. Thick laminated plates are basically three-dimensional problems and they are intractable as the number of layers becomes moderately large. To improve the accuracy of Mindlin-type plate theories and retain the advantages of being able to treat three-dimensional problems as two-dimensional analysis, a suitable higher-order plate theory is necessary.

Several high-order theories[6−9] have been developed for the analysis of composite plates. Reddy[8] studied the bending problem using a principle of virtual displacements to derive a higher-order theory in which the displacement field is chosen to satisfy the stress-free boundary conditions. Also, deflection and stress of laminated plates have been studied by using this theory in conjunction with the finite element.[9] A critical evaluation of new plate theories by Bert[10] indicated that the theory of Lo et al.[6] provides an accurate prediction of the non-linear bending stress distribution.

Using the theory developed by Lo et al.[6] a series of studies about homogeneous, laminated and bimodulus plates had been carried out.[11,12] In those studies, very accurate results were obtained. However, Lo et al. higher-order plate theory is complex, and the displacement field of the theory does not satisfy the conditions of zero transverse shear stress on the top and bottom surfaces of the plate (stress-free boundary conditions), so a new plate theory simplified from the Lo et al. theory with seven kinematic variables of a composite plate is developed in the present study. The deflection and stress of laminated plates are solved to compare with three-dimensional elasticity solutions,[13] the Reddy result[8] and previous higher-order theory results.

FORMULATION

Hamilton's principle can be used to derive the non-linear plate equations, which are then perturbed by an incremental deformation and linearised to obtain the desired governing equations. The procedure is the same as that used in Refs 14 and 15. Hamilton's principle may be expressed as

$$\delta \int_{t_0}^{t_1} (U - K - W_e - W_i) \, dt = 0 \tag{1}$$

where U is the strain energy, K the kinetic energy, W_e the work of external forces and W_i the work of internal forces. Taking ϕ as the strain energy per unit initial volume, we use the combination of Trefftz stresses and material (Lagrangian) strains to express it.

$$U = \int_{V_o} \phi \, dV \qquad K = \frac{1}{2} \int_{V_o} \rho \dot{u}_i \dot{u}_i \, dV$$

$$W_e = \int_{S_o} P_i^* u_i \, dS \qquad W_i = \int_{V_o} X_i^* u_i \, dV \tag{2}$$

and

$$\delta U = \int_{V_o} \delta \phi \, dV = \int_{V_o} \sigma_{ij}^* \delta e_{ij} \, dV \tag{3}$$

$$P_s^* = \sigma_{ij}^* n_i (\delta_{js} + u_{s,j}) \qquad \delta_{js} = \begin{Bmatrix} 0 & j \neq s \\ 1 & j = s \end{Bmatrix} \tag{4}$$

where the σ_{ij}^* are Trefftz stress components referred to the material coordinates, the e_{ij} are material strains referred to the material coordinates, the u_i are Cartesian displacement components referred to the spatial frame, the P_i^* are Cartesian components of the surface force per unit initial area, the X_i^* are Cartesian body force components per unit initial volume, ρ is the initial mass density, V_0 the initial volume, and S_0 the initial boundary surface. Substituting eqns (2)–(4) into eqn (1) yields

$$\delta \int_{t_0}^{t_1} \left[\int_{V_0} (\phi - X_i^* u_i - \tfrac{1}{2}\rho \dot{u}_i \dot{u}_i) \, dV - \int_{S_0} P_i^* u_i \, dS \right] dt = 0 \tag{5}$$

Assume that the stresses and applied forces do not vary. Then taking the variation, integrate the kinetic energy term by parts with respect to time. Equation (5) becomes

$$\int_{t_0}^{t_1} \left[\int_{V_0} (\sigma_{ij}^* \delta e_{ij} - X_i^* \, \delta u_i + \rho \ddot{u}_i \, \delta u_i) \, dV - \int_{S_0} P_i^* \, \delta u_i \, dS \right] dt = 0 \tag{6}$$

The Lagrangian strain components are related to the displacement by

$$e_{ij} = \tfrac{1}{2}(u_{i,j} + u_{j,i} + u_{k,i} u_{k,j}) \qquad (i,j,k = 1,2,3) \tag{7}$$

Firstly, assume the same form for the displacements as that of Lo *et al.*,[6] then introduce the condition that the transverse shear stresses, $\sigma_{xz} = \sigma_5$ and $\sigma_{yz} = \sigma_4$, vanish on the top and bottom surfaces of the plate, and these conditions are equivalent to the requirement that the corresponding strains be zero on these surfaces for orthotropic plates or plates laminated of orthotropic layers. We have $\varepsilon_5(x, y, \pm h/2, t) = 0$ and $\varepsilon_4(x, y, \pm h/2, t) = 0$, the displacement field becomes

$$u_1 = u_x + z[\psi_x - z\psi_{z,x}/2 - 4(z/h)^2(w_{,x} + \psi_x + h^2\xi_{z,x}/4)/3]$$
$$u_2 = u_y + z[\psi_y - z\psi_{z,y}/2 - 4(z/h)^2(w_{,y} + \psi_y + h^2\xi_{z,y}/4)/3] \tag{8}$$
$$u_3 = w + z\psi_z + z^2\xi_z$$

Material parameters are defined:

$$(A_{ij}, B_{ij}, D_{ij}, E_{ij}, F_{ij}, G_{ij}, H_{ij}) = \int C_{ij}(1, z, z^2, z^3, z^4, z^5, z^6) \, dz$$
$$(i,j = 1,2,3,4,5,6) \tag{9}$$

And the stress–displacement relations are given by:

$$\bar{\sigma}_{xx} = C_{11}u_{1,x} + C_{12}u_{2,y} + C_{13}u_{3,z} + C_{16}(u_{1,y} + u_{2,x})$$
$$\bar{\sigma}_{yy} = C_{12}u_{1,x} + C_{22}u_{2,y} + C_{23}u_{3,z} + C_{26}(u_{1,y} + u_{2,x})$$
$$\bar{\sigma}_{zz} = C_{13}u_{1,x} + C_{23}u_{2,y} + C_{33}u_{3,z} + C_{36}(u_{1,y} + u_{2,x})$$
$$\bar{\sigma}_{yz} = C_{44}(u_{2,z} + u_{3,y}) + C_{45}(u_{1,z} + u_{3,x}) \tag{10}$$
$$\bar{\sigma}_{xz} = C_{55}(u_{1,z} + u_{3,x}) + C_{45}(u_{2,z} + u_{3,y})$$
$$\bar{\sigma}_{xy} = C_{66}(u_{1,y} + u_{2,x}) + C_{16}u_{1,x} + C_{26}u_{2,y} + C_{36}u_{3,z}$$

where C_{ij} are the components of the anisotropic stiffness matrix used in Ref. 16.

Substitute eqns (7) and (8) into eqn (6). In the volume integral perform the integration with respect to z through the thickness of the plate from $-h/2$ to $h/2$. The last integral over the bounding surface is done as in Ref. 15 and in this study of bending problems we neglect body forces, lateral loads, surface forces and all other initial stress but normal-pressure loading q which is sinusoidally distributed as

$$q = q_o \sin(\pi x/a) \sin(\pi y/b) \tag{11}$$

Finally, perform all necessary partial integrations needed to remove derivatives from the variations of the displacements and collect terms that contain variations of the same displacements to find the governing equations. The displacements are independent and their variations are arbitrary, also the simplified conditions are considered, the governing equations can be obtained and simplified to:

$$Q_{1,x} + Q_{2,y} = 0 \qquad Q_{2,x} + Q_{3,y} = 0$$
$$Q_{4,x} + Q_{5,y} + (4/3h^2)(Q_{18,xx} + 2Q_{19,xy} + Q_{20,yy} - 3Q_{15,x} - 3Q_{16,y}) + q = 0$$
$$Q_{6,x} + Q_{7,y} - Q_4 - (4/3h^2)(Q_{18,x} + Q_{19,y} - 3Q_{15}) = 0$$
$$Q_{7,x} + Q_{8,y} - Q_5 - (4/3h^2)(Q_{19,x} + Q_{20,y} - 3Q_{16}) = 0$$
$$-Q_{11} + (Q_{12,xx} + 2Q_{13,xy} + Q_{14,yy})/2 = 0$$
$$-2Q_{17} + (Q_{18,xx} + 2Q_{19,xy} + Q_{20,yy})/3 = 0$$
$$\tag{12}$$

For the simply supported cross-ply laminated plate, the boundary conditions are, along the x-constant edges:

$$u_y = 0 \quad \psi_y = 0 \quad w = 0 \quad \psi_z = 0 \quad \xi_z = 0 \quad \bar{N}_{xx} + \Delta N_{xx} = Q_1$$
$$\bar{P}_{xx} + \Delta P_{xx} = -4Q_{18}/3h^2 \qquad \bar{M}_{xx} + \Delta \bar{M}_{xx} = Q_6 - 4Q_{18}/3h^2 \tag{13}$$
$$\bar{M}^*_{xx} + \Delta M^*_{xx} = -Q_{12}/2 \qquad \bar{T}_{xx} + \Delta T_{xx} = -Q_{18}/3$$

along the y-constant edges:

$$u_x = 0 \quad \psi_x = 0 \quad w = 0 \quad \psi_z = 0 \quad \xi_z = 0 \quad \bar{N}_{yy} + \Delta N_{yy} = Q_3$$
$$\bar{P}_{yy} + \Delta P_{yy} = -4Q_{20}/3h^2 \qquad \bar{M}_{yy} + \Delta M_{yy} = Q_8 - 4Q_{20}/3h^2 \tag{14}$$
$$\bar{M}^*_{yy} + \Delta M^*_{yy} = -Q_{14}/2 \qquad \bar{T}_{yy} + \Delta T_{yy} = -Q_{20}/3$$

The $Q_1 \sim Q_{20}$ are given in the Appendix. And

$$[(\bar{N}_{ii}, \bar{P}_{ii}, \bar{M}^*_{ii}, \bar{M}_{ii}, \bar{T}_{ii}), (\Delta N_{ii}, \Delta P_{ii}, \Delta M_{ii}, \Delta M^*_{ii}, \Delta T_{ii})]$$

$$= \int [(\bar{P}_i, \Delta P_i)(1, -4z^3/3h^2, z - 4z^3/3h^2, -z^2/2, -z^3/3)] \, dz$$

$$(i = x, y)$$

displacements of the following form satisfy the boundary conditions in eqns (13) and (14).

$$(u_x, \psi_x) = (hU, \ \psi_x)\cos(\pi x/a)\sin(\pi y/b)$$
$$(u_y, \psi_y) = (hV, \psi_y)\sin(\pi x/a)\cos(\pi y/b) \quad (15)$$
$$(w, \psi_z, \xi_z) = (hW, \psi_z, \zeta_z/h)\sin(\pi x/a)\sin(\pi y/b)$$

Equation (10) is obtained by substituting the assumed displacement fields of eqn (15) into the equations of motion (12)

$$[C]\{\Delta\} = \{Q\} \quad (16)$$

where

$$\{\Delta\} = \{U, V, W, \psi_x, \psi_y, \psi_z, \zeta_z\}^\tau,$$
$$\{Q\} = \{0, 0, -q_o, 0, 0, 0, 0\}^\tau$$

The elements of the symmetric coefficient matrix $[C]$ are given in the Appendix.

RESULTS AND DISCUSSION

As a first example, we take the case of a rectangular, cross-ply laminate under sinusoidal load to verify the accuracy of the present higher-order

TABLE 1
Comparison of deflection and transverse shear stresses for rectangular three-layer laminate plate $(0°/90°/0°)$ $(b = 3a,\ E_1 = 25E_2,\ E_3 = E_2,$ $G_{12} = G_{31} = 0.5E_2,\ G_{23} = 0.2E_2,\ v_{12} = v_{13} = v_{23} = 0.25)$

a/h		$\bar{W}$	$\bar{\sigma}_4$	$\bar{\sigma}_5$
2	A	8·17	0·066 8	0·257
	B	8·063 7	0·066 4	0·237 0
	C	8·203 0	0·062 9	0·231 0
	D	7·894 4	0·069 1	0·235 1
4	A	2·82	0·033 4	0·351
	B	2·652 6	0·034 0	0·272 6
	C	2·660 0	0·031 5	0·279 3
	D	2·641 1	0·034 8	0·272 4

A: Pagano's result.[13]
B: present plate theory.
C: 11 variables higher-order plate theory.
D: Reddy's result.[8]

solutions. In Table 1, a comparison of deflection and transverse shear stresses is made among a three-dimensional elasticity solution,[13] the 11 kinematic variables higher-order theory results based on the Lo *et al.* displacement field, Reddy's higher-order theory results[8] and the present plate theory results. As one can see, there is excellent agreement between the present solutions and the three-dimensional elasticity solutions. Also, the present solutions are nearly the same as the 11 variables solutions and the present higher-order plate theory is more accurate than Reddy's refined higher-order plate theory. However, the stress-free boundary conditions are satisfied by the present theory and it is significant simplification to compare it with the 11 variables plate theory. From the above explanation, the present plate theory is more suitable than 11 variables plate theory.

Table 2 shows the variations of deflection and stresses for rectangular laminate from thick plate to thin plate; its material properties are the same as that in Table 1. Again, the present plate theory result coincides with the three-dimensional elasticity solution very well. Also, the present theory is

TABLE 2

Comparison of deflection and stresses for rectangular three-layer laminate plate (0°/90°/0°); the material properties are the same as that given in Table 1

a/h		$\bar{W}$	$\bar{\sigma}_1$	$\bar{\sigma}_2$	$\bar{\sigma}_4$	$\bar{\sigma}_5$	$\bar{\sigma}_6$
4	A	2·82	1·14	0·109	0·0334	0·351	0·0269
	B	2·6526	1·0332	0·1018	0·0340	0·2726	0·0264
	D	2·6411	1·0356	0·1028	0·0348	0·2724	0·0263
	E	2·3626	0·6130	0·0934	0·0257	0·1566	0·0205
10	A	0·919	0·726	0·0418	0·0152	0·420	0·0120
	B	0·8632	0·6923	0·0395	0·0168	0·2859	0·0115
	D	0·8622	0·6924	0·0398	0·0170	0·2859	0·0115
	E	0·8030	0·6214	0·0375	0·0133	0·1578	0·0105
20	A	0·610	0·650	0·0294	0·0119	0·434	0·0093
	B	0·5940	0·6407	0·0287	0·0139	0·2880	0·0091
	D	0·5937	0·6407	0·0289	0·0139	0·2880	0·0091
	E	0·5784	0·6228	0·0283	0·0113	0·1580	0·0088
100	A	0·508	0·624	0·0253	0·0108	0·439	0·0083
	B	0·5070	0·6240	0·0251	0·0129	0·2886	0·0083
	D	0·5070	0·6240	0·0253	0·0129	0·2886	0·0083
	E	0·5064	0·6233	0·0253	0·0106	0·1581	0·0083
	F	0·503	0·623	0·0252			0·0083

E: first-order plate theory.
F: classical plate theory.

TABLE 3

Values of deflection and stresses with the various aspect ratio a/b ($E_1 = 25E_3$, $a/h = 10$, $h_i = h/3$, $E_2 = E_3$)

a/b		$\bar{W}$	$\bar{\sigma}_1$	$\bar{\sigma}_2$	$\bar{\sigma}_4$	$\bar{\sigma}_5$	$\bar{\sigma}_6$
0·3	B	0·865 6	0·694 4	0·324	0·014 6	0·286 4	0·010 4
	C	0·872 4	0·714 3	0·032 5	0·013 6	0·307 2	0·010 5
	E	0·805 2	0·623 2	0·030 9	0·011 6	0·158 1	0·009 5
0·9	B	0·750 9	0·598 6	0·232 5	0·085 7	0·255 6	0·026 4
	C	0·756 4	0·615 9	0·234 4	0·084 2	0·274 1	0·026 8
	E	0·702 3	0·539 7	0·218 2	0·063 6	0·142 0	0·024 0
1·5	B	0·485 8	0·382 9	0·382 0	0·183 1	0·173 8	0·027 1
	C	0·488 4	0·393 8	0·384 2	0·183 2	0·185 9	0·027 5
	E	0·463 3	0·351 3	0·367 3	0·134 7	0·098 5	0·024 8
2·1	B	0·257 0	0·199 6	0·351 4	0·218 8	0·097 7	0·018 9
	C	0·257 8	0·205 3	0·352 3	0·219 8	0·104 3	0·019 1
	E	0·250 4	0·186 3	0·348 2	0·102 7	0·056 6	0·013 7
2·7	B	0·136 1	0·104 0	0·266 8	0·208 4	0·055 1	0·011 9
	C	0·136 4	0·107 1	0·267 0	0·209 3	0·058 7	0·012 1
	E	0·134 5	0·097 8	0·271 5	0·156 4	0·032 4	0·010 8
3·0	B	0·102 0	0·077 3	0·228 5	0·197 5	0·042 6	0·009 6
	C	0·102 2	0·079 7	0·228 6	0·198 3	0·045 3	0·009 7
	E	0·101 3	0·072 8	0·235 4	0·148 8	0·025 2	0·008 6

TABLE 4

Comparison of deflection and stresses with the various value of Young's modulus in z-direction, E_3, for three-ply laminated plate ($E_1/E_2 = 40$, $G_{23} = 0.5E_2$, $G_{12} = G_{31} = 0.6E_2$, $v_{12} = v_{13} = 0.25$, $v_{23} = 0.01$, $a/h = 4$, $b/a = 3$)

E_3/E_2		$\bar{W}$	$\bar{\sigma}_1$	$\bar{\sigma}_2$	$\bar{\sigma}_4$	$\bar{\sigma}_5$	$\bar{\sigma}_6$
0·01	B	1·982 9	0·656 7	0·125 1	0·023 1	0·428 7	0·016 3
0·02	B	1·811 3	0·828 3	0·114 5	0·022 3	0·421 0	0·017 8
0·10	B	1·655 9	0·983 6	0·104 9	0·021 6	0·414 4	0·019 3
0·20	B	1·635 1	1·004 4	0·103 6	0·021 5	0·413|0	0·019 4
1·0	B	1·618 3	1·021 7	0·102 5	0·021 4	0·412 2	0·019 6
10·0	B	1·613 6	1·031 2	0·100 7	0·021 3	0·411 8	0·019 6
50·0	B	1·609 1	1·059 4	0·093 3	0·021 0	0·410 6	0·019 6
200·0	B	1·586 3	1·211 8	0·052 8	0·019 2	0·404 0	0·019 4
	E	1·633 0	0·615 9	0·107 9	0·014 9	0·277 0	0·017 4

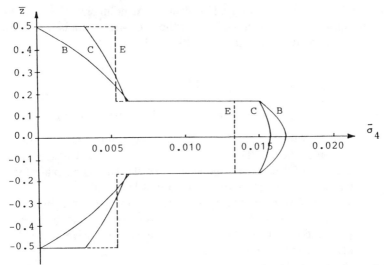

FIG. 1. Variations of transverse shear stress $\bar{\sigma}_4$ for rectangular three-layer laminate plate $(a/h = 10, b/a = 3, E_1/E_2 = 25, E_3 = E_2, h_2/h_1 = h_2/h_3 = 1 \cdot 0)$.

close to Reddy's result in stress analysis, especially in $\bar{\sigma}_4$ and $\bar{\sigma}_5$, and they can significantly be distinguished only in thicker plates. The stress-free boundary condition which is satisfied by the present theory and Reddy's may explain this phenomenon. The effect of aspect ratio on deflection and stresses is given in Table 3, where $a/h = 10$ and $h_i = h/3$. It is observed that the larger the aspect ratio a/b, the lower the deflection, $\bar{\sigma}_1$ and $\bar{\sigma}_5$, and the aspect ratio hardly affects the $\bar{\sigma}_6$. In Table 4, we consider the influence of Young's modulus in the z-direction, E_3, on the deflection and stresses of laminated plates. As it can be seen, E_3 really affects these properties. However, the effect of E_3 cannot be found in Mindlin plate theory and Reddy's refined higher-order plate theory. From comparison of Tables 1–4, it can be concluded that the accuracy of the present theory is enough and better than Reddy's, also, the effect of E_3 is considered and its simplicity makes it suitable for the study of plate problems.

In the present plate theory, the displacement field is obtained by modifying that of Lo *et al.*'s theory to satisfy the stress-free boundary condition, so we are interested in the variation of transverse shear stresses and their boundary conditions. Figure 1 illustrates the condition of transverse shear stress $\bar{\sigma}_4$, and it is observed that present theory coincides with the 11 variables plate theory near the mid-plane of the plate, but the closer it is to top and bottom surfaces of the plate, the more distant the present theory is from the other. However, the Mindlin plate theory and the 11 variables plate theory do not satisfy the stress-free boundary condition and near the plate surface their results are doubtful.

CONCLUSIONS

The preliminary results presented here indicate the following:

1. The 11 variables higher-order theory result based on the displacement field of the Lo *et al.* theory is complex and also its results are doubtful near the free surface of the laminated plate for the stress-free boundary condition cannot be satisfied.
2. The present plate theory coincides with the three-dimensional elasticity solution[13] very well and it is more accurate than Reddy's plate theory.[8] The influence of E_3 which is exclusive in Reddy's is also considered in this study.

ACKNOWLEDGEMENT

The authors thank the National Science Council for providing support for this work through grant NSC75-0401-E008-07.

REFERENCES

1. MINDLIN, R. D., Influence of rotary inertia and shear on flexural motions of isotropic, elastic plates. *ASME J. Appl. Mech.*, **18** (1951) 316–23.
2. WHITNEY, J. M., Stress analysis of thick laminated composite and sandwich plates. *J. Compos. Mater.*, **6** (1972) 426–40.
3. NOOR, A. K., Stability of multilayered composite plates. *Fiber Sci. Technol.*, **8** (1975) 81–9.
4. DONG, S. B. and NELSON, R. B., On natural vibration and waves in laminated orthotropic plates. *J. Appl. Mech.*, **39** (1972) 739–45.
5. KULKARNI, S. V. and PAGANO, N. J., Dynamic characteristics of composite laminates. *J. Sound Vib.*, **23** (1972) 127–43.
6. LO, K. H., CHRISTENSEN, R. M. and WU, E. M., A high-order theory of plate deformation, part 2: laminated plates. *J. Appl. Mech.*, **44** (1977) 669–76.
7. BHIMARADDI, A. and STEVEN, L. K., A higher order theory for free vibration of orthotropic, homogeneous, and laminated rectangular plates. *J. Appl. Mech.*, **51** (1984) 195–8.
8. REDDY, J. N., A simple higher-order theory for laminated composite plates. *J. Appl. Mech.*, **51** (1984) 745–52.

9. PHAN, N. D. and REDDY, J. N., Analysis of laminated composite plates using a high-order shear deformation theory. *Int. J. Numer. Methods Engng*, **21** (1985) 2201–20.
10. BERT, C. W., A critical evaluation of new plate theories applied to laminated composites. *Compos. Struct.*, **2** (1984) 329–47.
11. DOONG, J. L., CHEN, T. J. and CHEN, L. W., Vibration and stability of an initially stressed laminated plate based on a higher-order deformation theory. *Compos. Struct.*, **7** (1987) 285–309.
12. FUNG, C. P. and DOONG, J. L., Bending of a bimodulus laminated plate based on a higher-order shear deformation theory. *Compos. Struct.*, **10** (1988) 121–44.
13. PAGANO, N. J., Exact solution for rectangular bidirectional composites and sandwich plates. *J. Compos. Mater.*, **4** (1970) 20–34.
14. BRNNELLE, E. J. and ROBERTSON, S. R., Initially stressed Mindlin plates. *AIAA J.*, **12** (1974) 1036–45.
15. DOONG, J. L. and LEE, C., Vibration and stability of laminated plates based on a new plate theory. *J. CSME*, **9** (1988).
16. JONES, R. M., *Mechanics of Composite Materials*. McGraw-Hill, New York, 1975.

APPENDIX

$$
\begin{bmatrix} [L_1] \\ [L_2] \\ Q_{11} \\ [L_3] \\ Q_{17} \\ [L_4] \end{bmatrix} = \begin{bmatrix} [A_1] & [B_1] & [A_2] & [D_1] & 2[B_2] & [E_1] \\ [B_1] & [D_1] & [B_2] & [E_1] & 2[D_2] & [F_1] \\ [A_3] & [B_3] & A_{33} & [D_3] & 2B_{33} & [E_3] \\ [D_1] & [E_1] & [D_2] & [F_1] & 2[E_2] & [G_1] \\ [B_3] & [D_3] & B_{33} & [E_3] & 2D_{33} & [F_3] \\ [E_1] & [F_1] & [E_2] & [G_1] & 2[F_2] & [H_1] \end{bmatrix} \begin{bmatrix} [T_1] \\ [T_2] \\ [\psi_z] \\ [T_3] \\ [\xi_z] \\ [T_4] \end{bmatrix}
$$

$$
\begin{bmatrix} [L_5] \\ [L_6] \end{bmatrix} = \begin{bmatrix} [A_4] & [A_4] & [B_4] & 2[B_4] & [D_4] & 3[D_4] \\ [D_4] & [D_4] & [E_4] & 2[E_4] & [F_4] & 3[F_4] \end{bmatrix} [L_7]
$$

$$
[L_7] = [[T_5] \quad [T_6] \quad [T_7] \quad [T_8] \quad [T_9] \quad [T_{10}]]^\tau
$$

$$
[L_1] = [Q_1 \quad Q_2 \quad Q_3]^\tau \qquad [L_2] = [Q_6 \quad Q_7 \quad Q_8]^\tau
$$

$$
[L_3] = [Q_{12} \quad Q_{13} \quad Q_{14}]^\tau \qquad [L_4] = [Q_{18} \quad Q_{19} \quad Q_{20}]^\tau
$$

$$
[L_5] = [Q_4 \quad Q_5]^\tau \qquad [L_6] = [Q_{15} \quad Q_{16}]^\tau
$$

$$
[\Gamma_1] = \begin{bmatrix} \Gamma_{11} & \Gamma_{16} & \Gamma_{16} & \Gamma_{12} \\ \Gamma_{16} & \Gamma_{66} & \Gamma_{66} & \Gamma_{26} \\ \Gamma_{12} & \Gamma_{26} & \Gamma_{26} & \Gamma_{22} \end{bmatrix} \qquad (\Gamma = A, B, D, E, F, G, H)
$$

$$
[\Gamma_2] = [\Gamma_{13} \quad \Gamma_{36} \quad \Gamma_{23}]^\tau \qquad (\Gamma = A, B, D, E, F)
$$

$$
[\Gamma_3] = [\Gamma_{13} \quad \Gamma_{36} \quad \Gamma_{36} \quad \Gamma_{23}] \qquad (\Gamma = A, B, D, E, F)
$$

$$[\Gamma_4] = \begin{bmatrix} \Gamma_{55} & \Gamma_{45} \\ \Gamma_{45} & \Gamma_{55} \end{bmatrix} \qquad (\Gamma = A, B, D, E, F)$$

$$[T_1] = [u_{,x} \quad u_{x,y} \quad u_{y,x} \quad u_{y,y}]^{\tau}$$

$$[T_2] = [\psi_{x,x} \quad \psi_{x,y} \quad \psi_{y,x} \quad \psi_{y,y}]^{\tau}$$

$$[T_3] = -\tfrac{1}{2}[\psi_{z,xx} \quad \psi_{z,xy} \quad \psi_{z,yx} \quad \psi_{z,yy}]^{\tau}$$

$$[T_4] = -\frac{4}{3h^2} \begin{bmatrix} w_{,xx} + \psi_{x,x} + h^2\xi_{z,xx}/4 \\ w_{,xy} + \psi_{x,y} + h^2\xi_{z,xy}/4 \\ w_{,xy} + \psi_{y,x} + h^2\xi_{z,xy}/4 \\ w_{,yy} + \psi_{y,y} + h^2\xi_{z,yy}/4 \end{bmatrix}$$

$$[T_5] = [w_{,x} \quad w_{,y}]^{\tau} \qquad [T_6] = [\psi_x \quad \psi_y]^{\tau}$$

$$[T_7] = [\psi_{z,x} \quad \psi_{z,y}]^{\tau} \qquad [T_8] = -[T_7]/2$$

$$[T_9] = [\xi_{z,x} \quad \xi_{z,y}]^{\tau}$$

$$[T_{10}] = {}^{-}4[w_{,x} + \psi_x + h^2\xi_{z,x}/4 \quad w_{,y} + \psi_y + h^2\xi_{z,y}/4]/3h^2$$

Coefficents of matrix $[C_{rs}]$ in eqn (16) $(\alpha = m\pi/a, \ \beta = m\pi/b, \ \gamma = 4/3h^2)$

$$C_{1,1} = -A_{11}\alpha^2 - A_{66}\beta^2$$

$$C_{1,2} = -(A_{12} + A_{66})\alpha\beta$$

$$C_{1,3} = \gamma\alpha(E_{11}\alpha^2 + E_{12}\beta^2 + 2E_{66}\beta^2)$$

$$C_{1,4} = [(\gamma E_{11} - B_{11})\alpha^2 + (\gamma E_{66} - B_{66})\beta^2]/h$$

$$C_{1,5} = (\gamma E_{12} + \gamma E_{66} - B_{12} - B_{66})\alpha\beta/h$$

$$C_{1,6} = \{A_{13}\alpha + \alpha/2[D_{11}\alpha^2 + (D_{12} + 2D_{66})\beta^2]\}/h$$

$$C_{1,7} = \alpha[E_{11}\alpha^2 + 6B_{13} + (E_{12} + 2E_{66})\beta^2]/3h^2$$

$$C_{2,2} = -A_{22}\beta^2 - A_{66}\alpha^2$$

$$C_{2,4} = C_{1,5}$$

$$C_{2,5} = [(\gamma E_{66} - B_{66})\alpha^2 + (\gamma E_{22} - B_{22})\beta^2]/h$$

$$C_{2,6} = \beta[2A_{23} + (D_{12} + 2D_{66})\alpha^2 + D_{22}\beta^2]/2h$$

$$C_{2,7} = \beta[(E_{12} + 2E_{66})\alpha^2 + E_{22}\beta^2 + 6B_{23}]/3h^2$$

$$S_1 = 6\gamma D_{55} - A_{55} - 9\gamma^2 F_{55}$$

$$S_2 = 6\gamma D_{44} - A_{44} - 9\gamma^2 F_{44}$$

$$S_3 = F_{12} - \gamma H_{12} + 2F_{66} - 2\gamma H_{66}$$

$$S_4 = E_{12} + 2E_{66} - \gamma G_{12} - 2\gamma G_{66}$$

$$C_{3,3} = S_1\alpha^2 + S_2\beta^2 - \gamma^2[H_{11}\alpha^4 + H_{22}\beta^4 + (2H_{12} + 4H_{66})\alpha^2\beta^2]$$

$$C_{3,4} = \alpha[S_1 + S_3\gamma\beta^2 + (F_{12} - \gamma H_{11})\gamma\alpha^2]/h$$

$$C_{3,5} = \beta[S_2 + S_3\gamma\alpha^2 + (F_{22} - \gamma H_{22})\gamma\beta^2]/h$$

$$C_{3,6} = \gamma[-G_{11}\alpha^4/2 - (2G_{66} + G_{12})\alpha^2\beta^2 - E_{13}\alpha^2 - G_{22}\beta^4/2 - E_{23}\beta^2]/h$$

$$C_{3,7} = \gamma[-(2H_{12} + 4H_{66})\alpha^2\beta^2 - H_{11}\alpha^4 - H_{22}\beta^4 - 6F_{13}\alpha^2 - 6F_{23}\beta]/3h^2$$

$$C_{4,4} = [(2\gamma F_{11} - D_{11} - \gamma^2 H_{11})\alpha^2 + S_1 + (2\gamma F_{66} - D_{66} - \gamma^2 H_{66})\beta^2]/h^2$$

$$C_{4,5} = [(2\gamma F_{12} - D_{12} + 2\gamma F_{66} - D_{66} - \gamma^2 H_{12} - \gamma^2 H_{66})\alpha\beta/h^2$$

$$C_{4,6} = \alpha[(E_{11} - \gamma G_{11})\alpha^2 + S_4\beta^2 + 2(B_{13} - \gamma E_{13})]/2h^2$$

$$C_{4,7} = \alpha[(F_{11} - \gamma H_{11})\alpha^2 + S_3\beta^2 + (6D_{13} - 6\gamma F_{13})]/3h^3$$

$$C_{5,5} = [(2\gamma F_{66} - D_{66} - \gamma^2 H_{66})\alpha^2 + S_2 + (2\gamma FF_{22} - D_{22} - \gamma^2 H_{22})\beta^2]/h^2$$

$$C_{5,6} = \beta[S_4\alpha^2 + (2B_{23} - 2\gamma E_{23}) + (E_{22} - \gamma G_{22})\beta^2]/2h^2$$

$$C_{5,7} = \beta[S_3\alpha^2 + (F_{22} - \gamma H_{22})\beta^2 + (6D_{23} - 6\gamma F_{23})]/3h^3$$

$$C_{6,6} = -[(2F_{12} + 4F_{66})\alpha^2\beta^2 + 4D_{13}\alpha^2 + 4D_{23}\beta^2 \\ + F_{11}\alpha^4 + F_{22}\beta^4 + 4A_{33}]/4h^2$$

$$C_{6,7} = -[(2G_{12} + 4G_{66})\alpha^2\beta^2 + 8E_{13}\alpha^2 + 8E_{23}\beta^2 \\ + G_{11}\alpha^4 + 12B_{33} + G_{22}\beta^4]/6h^3$$

$$C_{7,7} = -[(2H_{12} + 4H_{66})\alpha^2\beta^2 + 12F_{13}\alpha^2 \\ + 12F_{23}\beta^2 + H_{11}\alpha^4 + H_{22}\beta^4 + 36D_{33}]/9h^4$$

28

Stress Concentration of the Composite Material Used by the X-29A Forward-Swept Wing Aircraft under Various Temperature Levels

HSIEN-YANG YEH

Mechanical Engineering Department, California State University, Long Beach, California 90840, USA

and

HSIEN-LIANG YEH

Civil Engineering Department, University of Texas, Arlington, Texas 76019, USA

ABSTRACT

The theory of anisotropic elasticity was used to evaluate the anisotropic stress concentration factors of a composite laminated plate containing a small circular hole. The results predicted by the constant strain approach matched the testing data very well at room temperature (70°F) and were about 10–15% more conservative than the experimental data at elevated temperature (200°F). At low temperature (−60°F) the results predicted by the mixture rule approach provided good correlation with the experimental data. Furthermore, the experimental data showed the stress concentration decreased as the temperature increased. A simple random statistical study indicated that a fairly isotropic laminated plate is reached if the number of lamina with arbitrary orientation is >40.

INTRODUCTION

The X-29A advanced technology demonstrator is sponsored by the Defense Advanced Research Projects Agency with support from NASA and the Air Force.

FIG. 1.   X-29A airplane.

The X-29A features a unique forward-swept wing (Fig. 1), made of composite materials which offers weight reduction of as much as 20% in comparison with convention aft-swept wings.

A forward-swept wing is prone to structural divergence, because as dynamic pressure increases, forces tend to bend the leading edge up. If a divergent speed were reached, a cycle of leading edge bending, increased local angle of attack and greater wing loading could grow to cause a structural failure. The wing's high rigidity prevents divergence from occurring within the X-29's flight envelope.

Because of their importance in aircraft design applications, laminated, continuous-fiber reinforced-resin matrix composites containing through cutouts have been the subject of considerable study (Refs 1–6). In this paper, anisotropic plate theory was used to calculate the anisotropic stress concentration factors (SCF) for the X-29A composite plate containing a circular hole.

ANALYSIS

Let axes 1, 2 be the principal coordinate axes of the laminated plate, and let axes L, T be the principal material axes of the single composite ply shown in Fig. 2.

For an anisotropic plate containing a circular hole subjected to remote uniaxial tensile stress σ_∞, acting at an angle ϕ with respect to the principal

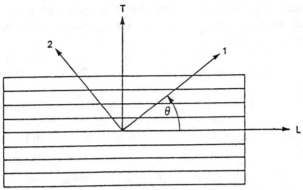

FIG. 2. Rotation from material axis L to laminated plate axis 1.

elastic axis 1 of the plate (Fig. 3), the tangential stress, σ_α and tangential stress concentration factor, $K \equiv \sigma_\alpha/\sigma_\infty$ along the circular hole boundary may be expressed as[3]

$$K \equiv \frac{\sigma_\alpha}{\sigma_\infty} = \frac{E_\alpha}{E_1} \{ [-\cos^2 \phi + (k+n)\sin^2 \phi]k\cos^2 \alpha$$
$$+ [(1+n)\cos^2 \phi - k\sin^2 \phi]\sin^2 \alpha$$
$$- n(1+k+n)\sin \phi \cos \phi \sin \alpha \cos \alpha \}$$

where E_α is the modulus of elasticity in the α direction (Fig. 3).

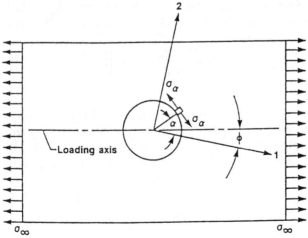

FIG. 3. Tension at an angle to a principal elastic axis 1 of an anisotropic plate with a circular hole.

To evaluate the modulus of elasticity of a laminated plate, both the mixture rule approach and the constant strain approach could be used. In the mixture rule approach, the transformed ply-elastic constants $\{\bar{\bar{E}}_1, \bar{\bar{E}}_2, \text{etc.}\}$ with respect to the $\{1, 2\}$ system can be related to the material constants $\{E_L, E_T, \text{etc.}\}$ with respect to the $\{L, T\}$ system through the transformation equations.[3]

If the composite plate is made of N single plies with different fiber orientations, then by using the mixture rule, the engineering elastic constants $\{\bar{E}_1, \bar{E}_2, \bar{G}_{12}, \bar{v}_{12}, \bar{v}_{21}\}$ for the composite plate can be written as[1]

$$\bar{E}_1 = \frac{1}{N} \sum_{j=1}^{N} \bar{\bar{E}}_1(\Theta_j) \qquad \bar{E}_2 = \frac{1}{N} \sum_{j=1}^{N} \bar{\bar{E}}_2(\Theta_j) \qquad \text{etc.}$$

In the constant strain approach, it is assumed that the strain remains constant across the laminate thickness and the inplane stress–strain relation for a laminate is used and it is actually the stress resultant versus inplane strain relation.[5,6]

The calculation of the effective engineering elastic constants $\{\bar{E}_1, \bar{E}_2, \bar{G}_{12}, \bar{v}_{12}\}$ is performed by relating the compliance components to the inplane engineering constants under uniaxial tension along the 1-axis.[1]

On top of these two approaches, in evaluating the modulus of elasticity of a laminated plate, one interesting question is how many plies with arbitrary orientation are required to make a laminated plate isotropic? To answer this question, a simple random statistical approach could be used. Starting from the constitutive equation for a generally orthotropic lamina:[7]

$$\begin{Bmatrix} \sigma_x \\ \sigma_y \\ \tau_{xy} \end{Bmatrix} = \begin{bmatrix} \bar{Q}_{11} & \bar{Q}_{12} & \bar{Q}_{16} \\ \bar{Q}_{12} & \bar{Q}_{22} & \bar{Q}_{26} \\ \bar{Q}_{16} & \bar{Q}_{26} & \bar{Q}_{66} \end{bmatrix} \begin{Bmatrix} \varepsilon_x \\ \varepsilon_y \\ \gamma_{xy} \end{Bmatrix}$$

Using uniformly distributed random real numbers between zero and one to generate the arbitrary orientation angles for different plies, collect the random sample, evaluate the sample mean, sample standard deviation and compare the normalized stiffness matrix with that in the isotropic case:

$$[\bar{Q}]_{\text{normalized isotropic}} = \begin{bmatrix} 1 & v & 0 \\ v & 1 & 0 \\ 0 & 0 & \dfrac{1-v}{2} \end{bmatrix}$$

finally, tabulate the results.

RESULTS

The X-29A forward-swept wing composite plate is made up of 40 plies with a total thickness of 0.56 cm (0.22 in). The stacking sequence and the ply-engineering elastic constants are given by

$$[\pm 45|0_4| \pm 45|90_2| \pm 45|0_4| \pm 45|0_2]_s$$

$-60°F$	$70°F$	$200°F$
$E_L = 19.23 \times 10^6$ psi	$E_L = 18.76 \times 10^6$ psi	$E_L = 18.29 \times 10^6$ psi
$E_T = 1.71 \times 10^6$ psi	$E_T = 1.57 \times 10^6$ psi	$E_T = 1.44 \times 10^6$ psi
$G_{LT} = 0.858 \times 10^6$ psi	$G_{LT} = 0.82 \times 10^6$ psi	$G_{LT} = 0.785 \times 10^6$ psi
$v_{LT} = 0.312$	$v_{LT} = 0.312$	$v_{LT} = 0.312$

The tangential stresses σ_α around a circular hole in a laminated X-29A composite plate were calculated for three loading cases: $\phi = 0$ (loading in axis-1 direction), $\phi = \pi/4$, and $\phi = \pi/2$ (loading in axis-2 direction). The results obtained from the constant strain approach and the mixture rule approach at three different temperature levels ($-60°F$, $70°F$, and $200°F$) are shown in Table 1.

Figures 4 and 5 indicate that the locations of the maximum tangential stress points could shift by changing the fiber orientation with respect to the loading axis. Figure 6 shows that at elevated temperature ($200°F$), the maximum stress concentration factor K reached the peak value of 3 at four locations ($\alpha = 120°, 300°$ by the constant strain approach and $\alpha = 105°$ and $285°$ by the mixture rule approach).

TABLE 1
Stress concentration factor correlation

Data for phi $= 0°$ Alpha $= 90°$

Load (lbs)	$-60°F$			$70°F$			$200°F$		
	SCF expmt	SCF mix rule	SCF const strain	SCF expmt	SCF mix rule	SCF const strain	SCF expmt	SCF mix rule	SCF const strain
600	4.09	4.53	3.71	3.53	4.58	3.71	3.00	4.64	3.72
1 000	4.36	4.53	3.71	3.71	4.58	3.71	3.22	4.64	3.72
2 000	4.36	4.53	3.71	3.98	4.58	3.71	3.29	4.64	3.72
3 000	4.36	4.53	3.71	3.88	4.58	3.71	3.39	4.64	3.72
4 000	4.36	4.53	3.71	3.84	4.58	3.71	3.38	4.64	3.72

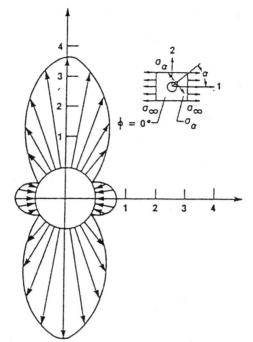

FIG. 4.   SCF of X-29 laminated plate (constant strain).

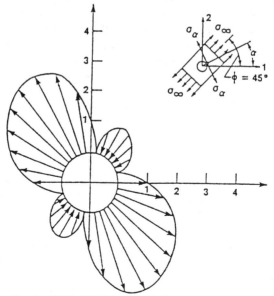

FIG. 5.   SCF of X-29 laminated plate (constant strain).

SCF CORRELATION FOR PHI= 45°, 200° F

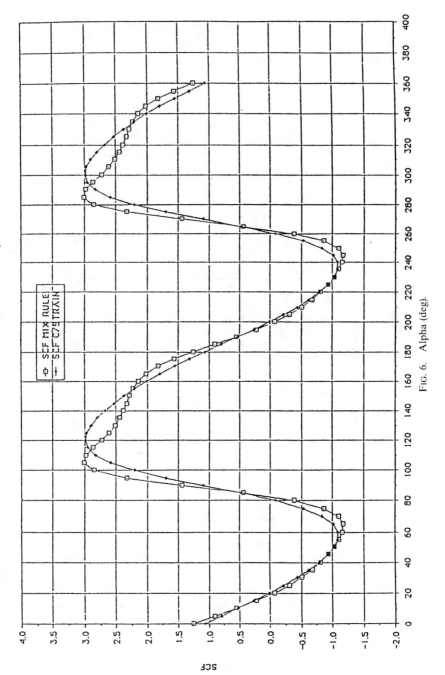

FIG. 6. Alpha (deg).

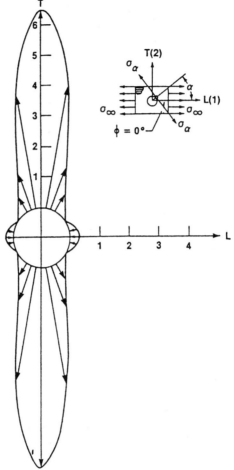

FIG. 7.   SCF of X-29 single ply.

For comparison purposes, similar calculations were made for a simple ply of the X-29A composite. Figures 7, 8 and 9 show that the maximum stress concentration factor K reaches the peak values of 6·4, 4·0 and $-3·5$ respectively.

The random statistical results from evaluation of the modulus of elasticity for a laminated plate with arbitrary ply orientation are listed in Table 2. These results indicate that a laminated plate made by the specific graphite/epoxy material ($E_L = 18·76 \times 10^6$ psi, $E_T = 1·57 \times 10^6$ psi, $G_{LT} = 0·82 \times 10^6$ psi, $\nu_{LT} = 0·312$) will provide a fairly isotropic characteristic if the

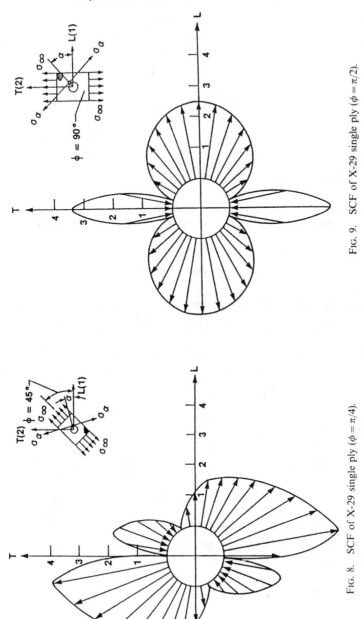

FIG. 9.   SCF of X-29 single ply ($\phi = \pi/2$).

FIG. 8.   SCF of X-29 single ply ($\phi = \pi/4$).

TABLE 2

N	10	20	30	40	50	60	70	80
$\dfrac{Q}{\bar{Q}_{11}}$	1	1	1	1	1	1	1	1
$\bar{Q}_{22}$	1·16	1·11	1·10	1·09	1·09	1·08	1·08	1·08
$\bar{Q}_{12}$	0·34	0·34	0·33	0·33	0·33	0·33	0·33	0·33
$\bar{Q}_{16}$	0·01	0·01	0·01	0·01	0·01	0·01	0·01	0·01
$\bar{Q}_{26}$	−0·02	−0·02	−0·02	−0·01	−0·01	−0·01	−0·01	−0·01
$\bar{Q}_{66}$	0·39	0·38	0·37	0·37	0·37	0·37	0·37	0·37

N = number of plies with arbitrary orientation.

number of lamina with random orientation is > 40. A further discussion of this random statistical study will be presented in another paper.[8]

CONCLUSION

The theory of anisotropic elasticity was used to evaluate the anisotropic stress concentration factors for laminated OBX-29A (forward-swept wing) research aircraft composite plates of three different temperature levels.

It is well known that the usual isotropic material stress concentration factor is three. However, the analysis showed that the anisotropic stress concentration factor could be greater than three for composite materials, and the locations of the maximum tangential stress points could shift by the change of fiber orientation with respect to the loading axis.

Both the mixture rule approach and the constant strain approach were used to calculate stress concentration factors at room temperature. The results obtained by the mixture rule approach were about 20% different from the experimental data. However, the results predicted by the constant strain approach matched the test data very well. This showed the importance of the inplane shear effect on the evaluation of stress concentration factors for the laminated X-29A composite plate. A further investigation on the inplane shear effect will need a three-dimensional model from anisotropic elasticity plus the interlaminar stress analysis.

At low temperature (− 60°F), the results predicted by the mixture rule approach provided good correlation with the experimental data. At elevated temperature (200°F), the results calculated from the constant strain approach were about 10–15% more conservative than the experimental data. These indicated both the advantages and the limitations of different analytical models in predicting stress concentration factors at various temperature levels. Furthermore, the experimental data showed the stress concentration factors decreased as the temperature increased.

It was found that through the lamination process the stress concentration factor could be reduced drastically, and therefore the structural performance could be improved. The next logical step in the study of the anisotropic stress concentration problem would be to know the optimum lamination process and to obtain the minimum stress concentration factors of laminate plates. A simple random statistical study indicated that for the specific graphite/epoxy used by the X-29A forward-swept wing aircraft, a fairly isotropic laminated plate is reached if the number of lamina with arbitrary orientation is > 40.

ACKNOWLDEGEMENTS

This work was supported by the NASA-ASEE Summer Faculty Fellowship in 1988. The authors are greatly indebted to Donald C. Bacon, Jr, assistant chief of research engineering division, V. Michael DeAngelis, branch chief of aerostructures, Don Black, deputy branch chief of aerostructures, Lawrence F. Reardon, group leader of structural test operation and Dr William L. Ko, aerospace engineer for invaluable advice, experimental data collection and helpful discussions.

REFERENCES

1. KO, W. L., Stress concentration around a small circular hole in the HiMAT composite plate. NASA TM-86038, December 1985.
2. SAVIN, G. N., *Stress distribution around holes*. NASA TT F-607, 1970.
3. LEKHNITSKII, S. G., *Anisotropic Plates*. Gordon and Breach, New York, 1968.
4. NUISMER, R. J. and WHITNEY, J. M., Uniaxial failure of composite laminates containing stress concentrations. *Fracture Mechanics of Composites*. ASTM ATP 593, 1975, pp. 117–42.
5. YEH, H. Y., *Stress concentration around circular hole in a composite material specimen representative of the X-29A forward-swept wing aircraft*. NASA CR-179435, Aug. 1988.
6. YEH, H. Y., *Temperature effect on stress concentration around circular hole in a composite material specimen representative of X-29A forward-swept wing aircraft*. NASA CR-179439, Aug. 1988.
7. TSAI, S. W. and HAHN, H. T., *Introduction to Composite Materials*. Technomic Publishing Company, Inc., Westport, CT, 1980.
8. YEH, H. Y. and YEH, H. L., A simple random statistical study about mechanical properties of composite laminated plates (submitted).

29

Sensitivity Studies in the Outgassing of Spacecraft Composites

ROBERT D. KARAM

Fairchild Space Company, Germantown, Maryland 10874-1181, USA

ABSTRACT

A systematic approach is described for analyzing outgassing in spacecraft composites. Simplifications that apply to practical cases are listed then used to derive tractable mathematical expressions. The boundary condition in the model is related to temperature and surface treatment.

Outgassing of a platform subjected to sinusoidal heating is discussed in a numerical example. Sensitivity to small temperature fluctuations does not appear significant, but variations in mean temperature, energy levels, and surface parameters produce very different estimates of outgassing rates. It is recommended that thermal/diffusion tests be performed on samples fabricated simultaneously with flight hardware.

INTRODUCTION

Considerable publicity has been given in conferences and in the literature to contamination caused by the outgassing of spacecraft composites.[1] But contamination is just one of several problems associated with the process. For example, there are changes that take place in the physical character of the outgassing structure itself which can be detrimental to its own temperature control and dimensional stability.[2] There is also the problem of electrostatic charge build-up and discharge which accompany floating particles in space and which can disrupt the function of nearby electronics.[3] Finally, escaping particles from a spacecraft surface induce a momentum which must be considered in the design of the attitude control system. An appreciation for this is found in the requirement for the Ocean Topography Experiment (TOPEX) satellite, that the spacecraft position be predictable to an accuracy of a few centimeters in a 1330-km orbit. It is estimated that induced accelerations of the order of 10^{-10} m/s^2 are sufficiently large to be included in precision orbit determination (POD).[4]

In a previous study[5] a thermal diffusion model was developed to examine impurity profiles of composite panels exposed to orbital heating. An important finding was that the calculated value of residual impurity was strongly dependent on temperature and diffusion constants. Small variations in these parameters translated into very different estimates in the time it takes for the outgassing of an assigned amount of impurity. The results emphasized the need for accurate determination of temperature and material constants.

The investigation is extended in this paper to explore whether similar sensitivities are displayed when the boundary conditions are varied. Unlike temperature, which is usually predictable and can be monitored, the actual physical state of outgassed particles at a surface may not be readily known, especially if there is particle exchange with adjacent surfaces. It was assumed in Ref. 4 that concentration vanishes at the surface, in conformity with the notion that particles emerging into space vacuum escape almost instantly. But it appears that this is a special case of a general situation in which surface treatment and temperature dictate the status. At sufficiently low surface temperatures, outgassing rates may be so slow as to be justifiably ignored. At very high temperatures, the assumption of instant escape may become fairly accurate. In the general case, emission is found not to continue indefinitely but to reach an equilibrium state in which the surface condition balances the body concentration activity.

These considerations are outlined and examined with respect to heating in a space environment. Realistic assumptions are made in order to simplify the analysis and derive expressions amenable to numerical computations. It is shown that the approach yields a useful scheme for assessing times for the release of contaminants.

An example is given to illustrate application of the theory and to highlight sensitivity to boundary parameters. It is concluded that accurate determination of outgassing rates in composites requires meticulous ground testing as well as temperature monitoring in orbit.

GENERAL REVIEW

Simplifications and Assumptions

It is difficult to formulate models which depict exactly the thermal and mass diffusion in composites. Part of the complexity lies in nonuniform interaction of fibers with the matrix, which is the result of incomplete mixing and curing. Hence there are uncertainties in almost any representation of the paths for heat and particle transfer. The usual approach is to assume localized uniformity when writing the field equations, then to ascribe a second-order tensor character to the transport properties, which are measured by testing samples obtained from the same lot as flight hardware. The errors due to this idealization will be somewhat dampened by averaging data from an increasing number of samples.

Other simplifications are listed below. The ultimate justification for all the assumptions is that they do provide a workable tool for assessing trends of actual behavior.

1. Outgassing is due solely to desorption, which occurs when absorbed gases and other impurities become volatile in space vacuum. Fragmentation from loose remnants and molecular chain break-off are assumed to have been completely exhausted in preflight bake-outs. Also there are no chemically active components.
2. Average unchanging values of density, specific heat, and the components of thermal conductivity can be used. This does not appear to be contradicted by test or flight data when the temperature is within the range usually produced by thermal control.[2] It can also be stated that any contribution to the transport properties from volatiles may be ignored due to their inherently low values.
3. The coefficient of diffusivity depends only on temperature. The evidence is overwhelmingly in support of this.[6] Dependence on porosity and other microstructural characteristics will be assumed already contained in available data on the coefficients.
4. Effects due to pressure and shear flow through the pores can be eliminated in view of the very low and constant pressure in a space environment.
5. Mass diffusion is many orders of magnitude slower than heat transfer. Hence the equations of conservation of mass and energy can be solved separately and connected only through the dependence of the diffusion coefficients on temperature.

It is further noted that many spacecraft structures are considered 'thin' in the sense that a one-dimensional analysis is adequate to describe temperature and concentration distributions. This is particularly the case with composite panels in which only a few lay-ups are needed to meet structural and thermal requirements. Such panels generally have all edges and inactive surfaces sealed and insulated, and mass and heat transfer take

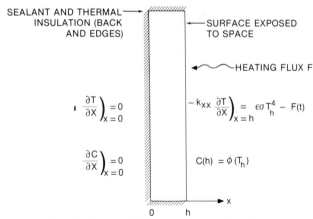

Fig. 1. Thermal-diffusion mathematical model.

place across a large control surface which is exposed to the environment. Heating and outgassing are uniform throughout the control surface except possibly in the vicinity of the edges. The exceptions can be treated separately as edge effects.

Figure 1 is a schematic of the thermal-diffusion model used in the analysis. Extension to cylindrical and spherical coordinates to represent circular booms and antennas requires coordinate transformations but should follow the same arguments presented for the Cartesian system.

The Temperature Distribution

For a composite body with uniform microstructure and unchanging properties, the energy equation is[7]

$$\rho C_p \frac{\partial T}{\partial t} = -\text{div}\,\mathbf{q} \tag{1}$$

where ρ is density, C_p is specific heat, T is temperature, and t is time. In the absence of effects due to internal gas flow, $\mathbf{q}$ is conduction heat flux given by Fourier's law:

$$q_i = -k_{ij}T_{,j} \tag{2}$$

with $k_{ij}(=k_{ji})$ being the components of conductivity. Using the concept of a thin panel, all temperature gradients (but not heat flux) vanish except along the normal to the control surface. Equation (1) becomes

$$\frac{\partial^2 T}{\partial x^2} = \frac{\rho C_p}{k_{xx}} \frac{\partial T}{\partial t} \tag{3}$$

The initial condition is known in accordance with a given timeline and the boundary conditions are (Fig. 1)

$$\left.\frac{\partial T}{\partial x}\right)_{x=0} = 0 \qquad -k_{xx}\left.\frac{\partial T}{\partial x}\right)_{x=h} = \varepsilon\sigma T_h^4 - F(t) \qquad (4)$$

$F(t)$ is absorbed energy from external heating sources (solar, planetary and adjacent emitters) and $\varepsilon\sigma T_h^4$, where ε is emittance and $\sigma = 5\cdot668 \times 10^{-8}$ W/m^2 K^4, is radiated energy at the surface, $x = h$.

Many methods are available for the numerical solution of eqn (3) with boundary conditions (4). An analytical expression is obtained by using the linearization:[8]

$$T_h^4 \approx T_m^4 + 4T_m^3(T_h - T_m) \qquad (5)$$

where T_m is selected with a view to minimizing the errors due to approximation. The method of the Laplace transform will then give

$$T = T_i + 2\int_0^t \sum_{n=1}^{\infty} \frac{\alpha\beta_n^2 \cos\beta_n\xi\, e^{-\alpha\beta_n^2/h^2}}{[h\beta_n^2 + h^2A(1 + hA)]\cos\beta_n}[P(t-\tau) - AT_i]\,d\tau \qquad (6)$$

where

T_i = initial absolute temperature
$P = [3\varepsilon\sigma T_m^4 + F(t)]/k_{xx}$
$A = 4\varepsilon\sigma T_m^3/k_{xx}$
$\alpha = k_{xx}/\rho C_p$
β_n = the roots of the equation $\beta\tan\beta = hA$
$\xi = x/h$

When the flux is simple harmonic, an all-time positive temperature is obtained if

$$F(\omega t) = Q_o + F_o\cos\omega t \qquad (7)$$

with Q_o a constant such that $Q_o > F_o > 0$. The frequency is ω and F_o is amplitude. The temperature is obtained by assuming a sinusoidal solution to eqn (3):

$$T = T_m\left[1 + \frac{F_o}{4Q_o}\,\text{Re}\,\frac{\cosh\beta\xi\, e^{i\omega t}}{\cosh\beta + (k_{xx}T_m/4hQ_o)\beta\sinh\beta}\right] \qquad (8)$$

Here Re stands for 'the real part of', $i = \sqrt{-1}$, and $\beta = (h^2 i\omega/\alpha)^{1/2}$. A convenient selection for T_m is based on the surface temperature at which the absorbed flux is radiated to space during one cycle; that is,

$$\varepsilon\sigma T_m^4 = \frac{1}{2\pi} \int_0^{2\pi} (Q_o + F_o \cos\omega t)\,d\omega t = Q_o \tag{9}$$

The average temperature across the thickness is found by integrating eqn (8) from $\xi = 0$ to $1\cdot0$:

$$T_{av} = T_m\left[1 + \frac{F_o}{4Q_o}\,\mathrm{Re}\,\frac{\sinh\beta\,e^{i\omega t}}{\beta\cosh\beta + (k_{xx}T_m/4hQ_o)\beta^2\sinh\beta}\right] \tag{10}$$

The Concentration Profile

If mass distribution is negligibly slow compared to heat flow, then concentration of matter can be treated separately by the equation

$$\frac{\partial c}{\partial t} = -\mathrm{div}\,\mathbf{J} \tag{11}$$

in which $\mathbf{J}$ is mass flux, t is time, and c is mass of impurity per unit volume of uncontaminated material, divided by its value at time zero. The advantage of this definition for c is that it permits prediction of outgassing as a percentage of original content. Hence the initial condition is that concentration is 100%.

Fick's law applies when there is no fragmentation or chemical activity:

$$J_i = -D_{ij}c_{,j} \tag{12}$$

where $D_{ij}\,(=D_{ji})$ are the components of the diffusivity tensor associated with composites. Utilizing the concept of uniform outgassing in a thin panel,

$$\frac{\partial c}{\partial y} = \frac{\partial c}{\partial z} = 0 \tag{13}$$

and

$$\frac{\partial^2 c}{\partial x^2} = \frac{1}{D_{xx}}\frac{\partial c}{\partial t} \tag{14}$$

The diffusion coefficient is a function of temperature which, in turn, depends on position x and time t. Hence the variation in terms of finite changes is

$$\Delta D_{xx} = \frac{dD_{xx}}{dT}\left(\frac{\partial T}{\partial x}\Delta x + \frac{\partial T}{\partial t}\Delta t\right) \tag{15}$$

For thin structures in orbital flight the temperature slope across the thickness is generally small compared to temperature change with time.

Typical examples include stacked laminates forming antenna dishes and solar cell panels. The temperature in these structures can vary by as much as $\pm 100°C$ as the spacecraft travels between shadow and sunlight, whereas the difference across the thickness at any one time remains just a few degrees. Hence D_{xx} may be considered to be nearly a function of time alone. The accuracy of this assumption can be optimized by referring to the average temperature:

$$T_{av}(t) = \frac{1}{h} \int_0^h T(x, t) \, dx \tag{16}$$

Equation (14) can now be written

$$\frac{\partial^2 c}{\partial x^2} = \frac{1}{D_{xx}(t)} \frac{\partial c}{\partial t} \tag{17}$$

with initial and boundary conditions

$$c(x, 0) = 1.0 \qquad \left(\frac{\partial c}{\partial x}\right)_{x=0} = 0 \qquad c(h) = \phi(T_h) \tag{18}$$

The initial condition corresponds to the definition of c and occurs at the selected initial temperature. The boundary condition at $x = 0$ signifies the presence of a perfect sealant, and the condition at $x = h$ indicates outgassing activity as a function of surface temperature. The function must also include the constants which characterize the influence of the surface treatment.

Introducing

$$\theta = \int_0^t D_{xx}(\tau) \, d\tau \tag{19}$$

eqn (4) becomes

$$\frac{\partial^2 c}{\partial x^2} = \frac{\partial c}{\partial \theta} \tag{20}$$

with the same conditions as in eqn (18), except that time is now replaced by θ in accordance with eqn (19). The solution is (Ref. 7, p. 104)

$$c = \frac{2}{h} \sum_{n=0}^{\infty} \exp\left(-\left[(2n+1)\frac{\pi}{2h}\right]^2 \theta\right) \cos(2n+1)\frac{\pi}{2}\xi$$

$$\times \left\{ \frac{(-1)^n \pi (2n+1)}{2h} \int_0^\theta \exp\left(\left[(2n+1)\frac{\pi}{2h}\right]^2 \tau\right) \psi(\tau) \, d\tau + \frac{(-1)^n}{(2n+1)(\pi/2h)} \right\} \tag{21}$$

where ψ is the image of $\phi(T_h)$ in the time domain. Total mass per unit volume at any instant is found by integrating from $x = 0$ to h and dividing by h:

$$c_{total} = \frac{2}{h^2} \sum_{n=0}^{\infty} e^{-f_n\theta} \left\{ \int_0^\theta e^{f_n\theta} \psi(\tau) \, d\tau + \frac{1}{f_n} \right\} \quad f_n = \left[(2n+1) \frac{\pi}{2h} \right]^2 \quad (22)$$

SENSITIVITY STUDIES

Method of Analysis

Based on the previous discussion, a procedure is developed for evaluating outgassing in composite panels:

1. Obtain the temperature history by solving the energy balance equation independently of contribution by mass flow. The boundary conditions for the heat transfer problem are well established in terms of the physical configuration and external heating.
2. Collect applicable diffusion data. Tests must be performed to establish properties that are not known. It is necessary to determine the effects of temperature and surface treatment on diffusion coefficients and the activity of particle emission at the surface.
3. Replace the instantaneous temperature distribution by an average value taken over the thickness, and transform the diffusion coefficient into a function only of time by inverting the equation for average temperature as a function of time.
4. Estimate residual impurity by using eqn (22) or equivalent. The results can be in terms of the time it takes to expel an assigned percentage of original impurity.

The following example illustrates application to a situation reminiscent of a structure in a near earth orbit. The solution of the simple harmonic problem provides insight into behavior under more complex heating by noting trends and sensitivity to deviations from nominal parameters. The thermophysical properties used in the calculations are from sources[9] which generally emphasize specific samples rather than universal values.

An Example—Sinusoidal Heating

Consider the following data for the panel shown in Fig. 1:

Flux amplitude, $F_0 = 135 \, \text{W/m}^2$ (one tenth a solar constant)
Flux frequency, $\omega = 2\pi/1 \cdot 5/h$ (near earth orbit)

Constant heat input, $Q_o = 200 \, \text{W/m}^2$ (earth heating)
Surface emittance, $\varepsilon = 0.80$ (white coating)
Panel thickness, $h = 0.02 \, \text{m}$
Normal conductivity, $k_{xx} = 1.38 \, \text{W/mK}$ (stacked GFRP laminate)
Density, $\rho = 1605 \, \text{kg/m}^3$ (GFRP)
Specific heat, $C_p = 864 \, \text{J/kg K}$ (GFRP)

$$\beta = (h^2 i\omega/\alpha)^{1/2} = 0.6839 \qquad \alpha = k_{xx}/\rho C_p = 0.995 \times 10^{-6} \, \text{m}^2/\text{s}$$

From eqn (9)

$$T_m = (Q_o/\varepsilon\sigma)^{1/4} = 257.7 \, \text{K}$$

From eqn (8) with $\xi = 1$

$$T_h = 257.7[1 + 0.01619\cos(\omega t - 1.3221)] \tag{23}$$

From eqn (10)

$$T_{av} = 257.7[1 + 0.01592\cos(\omega t - 1.4758)] \tag{24}$$

The difference in the peaks between the two definitions is less than $0.1 \, \text{K}$.

Suppose next that there exist test data which indicate that the diffusion coefficient has the form

$$D_{xx} = a_{xx} e^{-E/T} \tag{25}$$

where, for this particular construction, $a_{xx} = 5.0 \times 10^{-4} \, \text{m}^2/\text{s}$ and E is $4500 \, \text{K}$. This Arrhenius form is regularly observed in measurements of diffusion constants and has a theoretical basis when E is interpreted as energy of activation.[6,10] It should be noted here that for the temperature range under consideration $\alpha/D_{xx} \gg 1.0$.

Tests also provide information on particle activity at the surface as it relates to temperature and surface treatment. The general form is

$$c(h) = 1 - e^{-\gamma/T_h} \tag{26}$$

But the factor γ dominates. Thus, with an excellent sealant, γ is quite large and $c(h)$ approaches 1.0, which indicates that no material escapes the surface. Roughing and perforation can make γ very small and $c(h)$ approaches zero, which is the case of immediate escape once a particle reaches the surface.

Three ranges of γ/T_h will be assessed with variations consistent with the temperature profile of eqn (23). A surface activity factor κ is introduced as follows:

$$\kappa = \gamma/T_h = a/(1 \pm 0.016\,19) \tag{27}$$

where κ can be 0.01, 0.10 or 1.0.

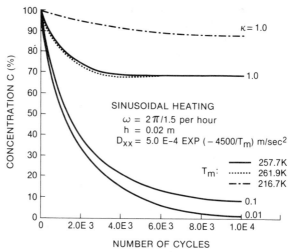

FIG. 2.   Concentration profiles—sinusoidal heating.

Using eqn (19) with the average temperature substituted into eqn (25),

$$\theta = 0.4297 \int_0^{\omega t} \exp\left\{-17.46/[1 + 0.015\,92\cos(\tau - 1.4758)]\right\}\,d\tau \quad (28)$$

The lower limit, $\omega t = 0$, indicates the time at which count begins with $c = 1$.
The integration is performed numerically. Some values are given below as
obtained with a Hewlett–Packard 28S hand calculator:

ωt:	0	$500 \times 2\pi$	$1000 \times 2\pi$	$1500 \times 2\pi$	$2000 \times 2\pi$
$\theta\,(\mathrm{m}^2)$:	0	3.496×10^{-5}	7.240×10^{-5}	10.489×10^{-5}	14.480×10^{-5}

Residual concentration in terms of number of cycles is calculated from eqn
(22) with $\psi(\tau) = 1 - e^{-\kappa}$:

$$c_{\text{total}} = \frac{2}{h^2} \sum_{n=0}^{\infty} \frac{1 - e^{-\kappa}(1 - e^{-f_n\theta})}{f_n} \qquad f_n = \left[(2n + 1)\frac{\pi}{2h}\right]^2 \quad (29)$$

Figure 2 shows the decline of concentration with time for the three ranges
of κ discussed above. Sensitivity to temperature fluctuations as it affects κ is
also indicated. Note that each cycle represents $1.5\,\mathrm{h}$ so that 1500 cycles
constitute 3 months' duration in orbit.

Sensitivity to the activation energy in the diffusion coefficient is
illustrated in Fig. 3. The asymptotic behavior with increasing time makes it

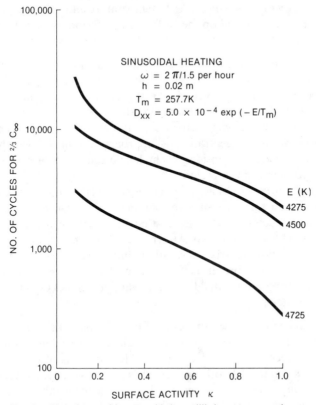

FIG. 3. Time for reaching two-thirds equilibrium concentration.

difficult to make accurate estimates on how long it takes for the release of a large percentage of original content (say 90%), but the reported results for reaching two-thirds equilibrium concentration are representative of the general pattern.

SUMMARY AND DISCUSSION

Some of the problems caused by the outgassing of spacecraft composites are contamination, changes in material properties, electrostatic charge build-up, and induced dynamic forces. Solutions require the identification of the factors which influence diffusion and emission.

A simplified but valid mathematical model was developed to determine the sensitivity of outgassing rates to temperature and diffusion parameters interior to the body and on the surface. Application was illustrated by an example on sinusoidal heating. The following were concluded:

1. Minor temperature fluctuations have a small and predictable effect on concentration history. The influence becomes apparent when the mean temperature is varied. This suggests that outgassing rates can be limited by appropriate thermal control.
2. The surface activity factor dominates emission rates. As expected, equilibrium concentration is approached more slowly for lower values of κ, but a considerably higher percentage of original content is emitted in a given period.
3. Small variations in activation energy translate into large changes in the outgassing rate. Since the parameter E is closely connected to the microstructure, it is expected that the value for a composite body would have to have a large tolerance to account for nonuniformity in construction, matrix distribution, and imperfections in mixing and curing. Therefore it is imperative that tests be performed for measuring E on actual flight hardware or samples closely simulating flight items.

Vacuum chambers used for testing full-scale spacecraft will not provide sufficient surrounding space to simulate the freedom of emission in actual flight. It is recommended that disassembled portions of the structure be evaluated separately. Another approach is to assess the spacecraft behavior by examining data from a large number of specimens fabricated from the same ply lot used for the structure. This method is far less costly since it can be performed in bell jars equipped with contamination sensors and residual product analyzers.

Monitoring flight temperatures is also essential. Some thermistors should be placed in regions which are prone to the effects of outgassing, and others, which serve as calibration units, in areas where outgassing has its least effect. The orbital behavior of the structure can then be assessed by using the measured temperature with test data in an analysis along the lines discussed in this work.

REFERENCES

1. GLASSFORD, P. (Ed.), Optical system contamination: effects, measurement, control. *Proc. SPIE*, Vol. 777, 1987.

2. KARAM, R. D., Thermal engineering of spacecraft composite structures. In *Composite Structures—3*, ed. I. H. Marshall. Elsevier Applied Science, London, 1985.

3. HALL, D. F., Flight experiment to measure contamination enhancement by spacecraft charging. *Proc. SPIE*, Vol. 216, 1980.

4. The TOPEX Satellite System Performance Requirements, JPL Doc. 633-501, 1988.

5. KARAM, R. D., Outgassing of spacecraft composites. In *Composite Structures—4*, ed. I. H. Marshall. Elsevier Applied Science, London, 1987.

6. BARBER, R. M., *Diffusion in and through Solids*. The Macmillan Company, Cambridge, 1941.

7. CARSLAW, H. S. and JAEGER, J. C., *Conduction of Heat in Solids*, 2nd edn. Oxford University Press, London, 1959.

8. KARAM, R. D., Optimum solution of linearized radiation equations. In *Numerical Methods in Thermal Problems*, ed. R. W. Lewis and K. Morgan. Redwood Burn Limited, Swansea, 1979.

9. *Final Design Assessment of ATS Reflector Support Truss*, prepared for Fairchild Industries under Contract SC 72-6, Hercules Inc., Magna, Utah, Sept. 1971.

10. LOOS, A. C. and SPRINGER, G. S., Moisture absorption of graphite–epoxy composites immersed in liquids and in humid air. *J. Comp. Mater.*, **13** (April 1979) 131–47.

30

Vibrational Analysis of Large Antennae for Space Applications

L. HOLLAWAY, M. O'NEILL and M. GUNN

Department of Civil Engineering,
University of Surrey, Guildford, Surrey, GU2 5XH, UK

ABSTRACT

The development of satellites for land mobile communication systems is likely to accelerate in the near future. The demand for both terrestrial and space services is growing rapidly and will require reflectors of over 50 m in diameter and space platforms of 300 m length and width incorporating clusters of reflectors mounted on them. These structural systems will be situated at geostationary orbit, thus providing a wide coverage area of the earth's surface. It is likely that this type of structure will be manufactured as a skeletal system and that the material of construction will be a fibre/polymer composite. The system will be subjected to dynamic forces during translation to and manoeuvring on station.

This paper discusses experimental and analytical techniques which have been used to obtain modal analyses of units of the structural system; the former involves using an impact hammer to excite and a roving accelerometer to detect resonances in the free/free structure. The latter uses a standard computer package to obtain the various resonant frequencies and corresponding modes for the structural system in a free/free condition. The results have shown that for structures based on the unit tetrahedral shape, satisfactory agreement does exist between the two techniques.

The analytical approach for the large size structures is to replace part of the skeletal system by an equivalent plate. This reduces the storage requirement of the computation; 5% errors are introduced by using the plate analogy approach.

1 INTRODUCTION

With the ever expanding requirements of communication services both in developed and isolated regions, the necessity for more sophisticated and capable satellite reflectors becomes critical. Indeed, existing provision of regular services to isolated areas might not be economically feasible without recourse to current satellite technology. Under these circumstances it is necessary to examine and propose suitable configurations for possible use in communications networks in the light of both continuing technological advances and changing communications requirements.

Placing such a large platform supporting many reflector dishes at geostationary orbit (GEO) has many advantages. A single Large Space Structure (LSS) at GEO to support the reflectors would optimise the usage of the available volume of space which may be restricted over developed regions of the earth. Whereas single reflectors require their own power supply units, control electronics and pointing systems, these ancillary services could be shared among many reflectors on the platform, with additional levels of redundancy being provided as required.

During their lifetime the platforms used could be tailored to meet changing demands with dishes being added or replaced as they become obsolete.

Use of a clustered configuration would facilitate routine servicing and additionally the operating life of dishes would be enhanced if attached to a platform whilst in collision with other objects.

A clustered configuration of a set of large parabolic dishes mounted on a supporting tetrahedral platform, and modelled in this investigation, has been designed to cater for the requirements of terrestrial communication, remote sensing and data relay applications; flexibility of use and operational economy are the key built-in features.

Current requirements would indicate that the platform is used to support numerous reflectors varying in size up to 15 m. However, the present study investigates the modal behaviour of a platform when supporting a few very large diameter dishes, from the point of view of both the large inertial influence of the dishes and the provision for possible long term future requirements.

Figure 1 shows an axonemetric view of the LSS modelled for investigation. The dishes have been modelled as continuum systems in order to reduce the storage requirements and running time during the modal analyses of the platform with the mounted reflectors in position.

In the following sections an analytical and an experimental technique

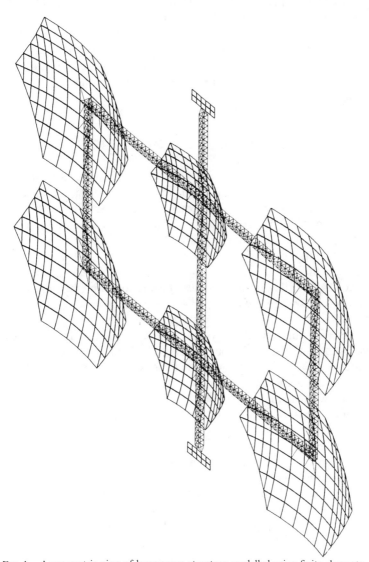

FIG. 1.   Axonometric view of large space structure modelled using finite elements.

will first be used to investigate and to compare the fundamental and higher frequencies of a part of the platform to verify the modelling of the analytical structure. A slab analogy technique for the modal analyses of large reflector dishes has been suggested and the method has been compared with the skeletal system it is replacing. Finally a modal analysis has been performed on the large space platform structure as a skeletal system and the large dishes mounted on the structure have been modelled as continuum systems.

2 THEORETICAL ANALYSIS

Computer analysis of the dynamic behaviour of space structures when performed in the time domain can involve the generation and repeated solution of very large systems of equations, the output often being very difficult to interpret. However, due to the dual nature of time and frequency domains, the structure may equally well be described in terms of its constituent mode shapes, resonant frequencies and corresponding damping ratios.

Once characterised in this manner, the structure may be subjected to a variety of load spectra and its response determined in the frequency domain. If necessary this may then be transformed into time domain data using a suitable algorithm. Whereas a time domain dynamic analysis usually involves the entire structure being analysed for many time steps, in many instances only a few selected modes may be needed to adequately define its behaviour in the modal domain. Load spectra can then be applied to the structure, with significantly reduced solution times.

A modal analysis involves the experimental determination of selected mode shapes, resonant frequencies and associated damping ratios of a structure whose behaviour can then be examined under the application of a variety of load spectra. However, similar information may be obtained by performing an eigenvalue extraction on a model of the structure using the finite element method which yields the undamped natural frequencies (from the eigenvalues) and the mode shapes (eigenvectors).

The ABAQUS finite element package (Hibbit, Karlsson and Sorensen Inc., Providence, Rhode Island, USA) was selected to analyse the cluster configuration. The package has no built-in limits on problem size with use being made of secondary storage automatically as the analysis requires. Consequently it is suited to the extraction of eigenvalues from very large models.

The finite element computer package LUSAS (Finite Element Analyses Ltd, Forge House, Kingston-upon-Thames, UK), has been used for problems not requiring large amounts of storage.

The subspace iteration method, which is presented in Refs 1–3, is the basic algorithm used in both the ABAQUS and LUSAS programs. The procedure is a simultaneous inverse power iteration in which a small set of base vectors is created. A 'subspace' is defined and is transformed by iteration into the space containing the lowest few eigenvectors of the overall system at which point these lowest eigenpairs and hence natural frequencies of vibration are readily available.

ABAQUS then uses the Householder and QR algorithms to extract the required eigenvalues whereas the LUSAS program employs a sparse matric frontal solution technique for this purpose.

3 EXPERIMENTAL INVESTIGATION OF THE ARM OF THE PLATFORM

Practical considerations preclude the making of a complete model of the space platform with a cluster of reflectors, consequently it is necessary to rely exclusively on the results of the finite element analysis for information about its modal behaviour.

It is therefore a prerequisite that the level of accuracy that it is possible to achieve with the finite element analysis is established so that a measure of confidence can be placed in its validity.

A scaled Perspex model of a representative section of an arm of the supporting platform was made. This was supported in a simulated free–free condition and subjected to a series of base load excitation spectra by means of an instrumented impact hammer. The lower modes of resonance were detected using accelerometers strategically placed on the structure. The accelerance frequency response functions (FRF) obtained from these tests were then compared with the natural frequencies and mode shapes from a free–free eigenvalue extraction of a finite element model using the LUSAS package.

3.1 ACQUIRE Package

The experimental work was performed using a DL1200 (Data Laboratories Ltd, 28 Wates Way, Mitcham CR4 4HR, UK) recorder controlled by the ACQUIRE digital signal processing software mounted on an HP series 200 computer. A lack of curve-fitting facilities in the

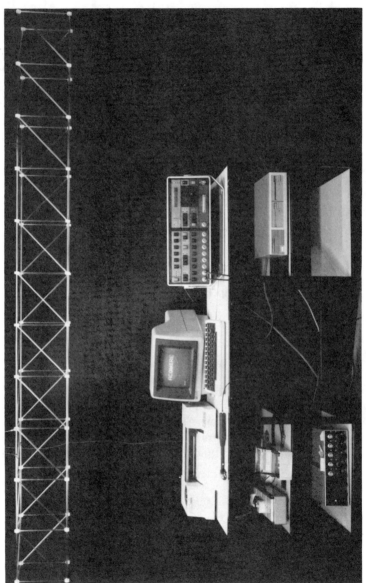

Fig. 2.   Experimental structure consisting of 11 bays.

software required the use of eigenextractions of a finite element model to generate the required mode shapes and frequencies of the structure under test. The existence of these modes and frequencies in the physical structure was then verified by means of the calculation of the FRFs using the fast Fourier transform (FFT) algorithm in the ACQUIRE package.

3.2 DL 1200 Recorder

Time domain force and acceleration signals from the Perspex model were digitally sampled and stored in block sizes of 4096 points using a DL1200 multi-channel recorder prior to processing using ACQUIRE, the sampling intervals used being selected to prevent signal aliasing. A rubber tip was used on the hammer as this was found to excite the required range of frequencies in the structure.

3.3 The Experimental Perspex Model

The experimental structure consisted of 11 bays, each bay forming a cube of side 0·2 m as shown in Fig. 2. The members of the model were manufactured from Perspex and were of tubular cross-section with an external diameter of 6 mm and a wall thickness of 1 mm. The model was braced on an unopposed configuration resulting in a minimum number of joint types.

The effects of the accelerometer on the behaviour of the structure were ignored as the local structural mass of the former was much less than that of the latter for the various modes examined.

4 THE ANALYTICAL AND EXPERIMENTAL ANALYSIS OF THE STRUCTURE

4.1 Analytical Model

The finite element model of the arm was constructed using 6 beam elements per member and its free–free eigenvalues were extracted using the LUSAS program. This mesh density was found to provide acceptable results for the lower global modes and adequately modelled the development of the local modes in the experimental structure whilst remaining within the storage limits of the package. Great care was exercised in accurately modelling the distribution of mass within the structure as this has been found to have great influence upon the accuracy of the results.

Modes 1–6 constitute the 6 rigid body modes for this free/free structure. Modes 7, 8 and 11 are seen to involve the development of shearing across the section of the arms and could be hindered by the provision of suitable cross-bracing. Bending of the arms is developed in modes 9 and 10 with torsion being represented in mode 12.

The results also indicate the instance of numerous local modes starting at mode 13 and appearing between 100 Hz and 120 Hz. These local modes are characterised by the absence of global translations of the model joints; the modes are developed through selective rotations of the joints with consequent bending of the attached beams. Again these modes would be hindered by the provision of suitable cross-bracing members.

4.2 Experimental Model

Using the mode shapes extracted from the analytical analyses, a set of support points was chosen which allowed the free development of all the lower free–free modes of the structure to take place. The node points of the first global bending modes were selected and the structure was suspended by means of light elasticated supports. Particular attention was paid to minimising the intrusion of the supports on modes 7–12; this was due to their importance in the behaviour of the structure.

4.2.1 Excitation and response detection locations

Two regimes of testing were carried out, the first concentrating upon determining the frequencies of the global bending and torsion modes developed and the second focusing upon confirming the existence of the local modes in the structure. To accomplish this the input and output regions were located at member joints of the structure for the first case and on the members of the structure for the latter.

4.3 Results

Table 1 gives the global frequencies for the arm obtained both analytically and experimentally. The experimental results have been collected from a variety of spectra which were taken to ensure the detection of all the lower modes.

The absence of resonances in the region of 100 Hz is evident; the tests failing to detect these resonances because of the co-location of the input and output detection positions with node points for these modes. The discrepancies between the two investigations could be attributed to the effects of material damping in the structure which was not modelled on the finite element analysis; consequently the experimental results should be lower than the corresponding analytical values.

As the impact locations moved away from the joints a significantly reduced frequency range was excited due to the increased flexibility of the

TABLE 1

Global modes detected experimentally and analytically for representative
Perspex section of arm

Mode	Experimental frequencies (Hz)	Analytical frequencies (Hz)	% Difference	Description of mode
1–6	0·0	0·0		RBM
7	46·6	46·6	−0·2	Shearing
8	48·6	48·4	−0·4	Shearing
9	62·3	62·0	−0·5	Bending
10	62·3	62·0	−0·5	Bending
11	70·1	71·4	1·9	Shearing
12	79·6	79·9	0·4	Torsion
13–20	100+	100–120		Local bending

input locations. Although frequencies up to 150 Hz were excited the signal power was low and hence it was not possible to obtain frequency response functions as previously. However, the output FFT of the accelerometer response (Fig. 3) showed an additional broad spread of resonances near 100 Hz; these are in agreement with the development of local modes as predicted by the LUSAS package.

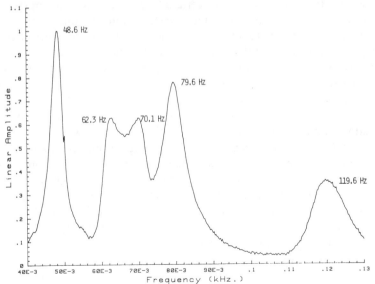

FIG. 3. Typical FFT of accelerometer response from Perspex model.

4.4 Discussion

Good agreement exists between the undamped natural frequencies predicted by the eigenvalue extractions on a finite element model of the representative section of the arm and resonant frequencies of the arm determined experimentally.

The free–free conditions have been shown to be well modelled by the use of light elasticated supports located at the first global bending mode nodal points for the structure.

The finite element analysis has been shown to provide a very good estimate of the free–free natural frequency of the Perspex arm and a high level of confidence can be placed upon its predictions of the modal behaviour of larger free–free structures.

5 ANALYSES BY THE SLAB ANALOGY TECHNIQUE

The analyses of large symmetrical skeletal space structures often requires the use of very powerful computers and efficient programs to cope with the very large sparsely populated matrices generated as part of the description of the structure. The analyses of structures by this means, however, does not completely take into account their regularity and gross behavioural tendencies. It can be expected that certain very large space structures would act essentially as slabs or plates due to their overall structural dimensions and repetitive structural configuration. Therefore, as a preliminary analysis for the determination of the dynamic characteristics of a tetrahedral large space truss structure, simple expressions for stiffness and strengths can be formulated and a slab or plate analogy can be used in place of the actual model to obtain the required frequencies and modes.

In these analogies the projected frame, either single or double layer, is replaced by a plate of identical external dimensions and whose thickness, elastic constants and mass density are calculated using equations derived and based upon the layout of the regular grid. These equations rely on the assumption that the bending stiffness of the frame is provided wholly by the members forming the faces of the frame, with no contributions being made by the bracing members. The structure is pin-jointed. Shear deflection effects are ignored in these analyses. The slab analogy has been developed to take account of the dynamic behaviour of plates by the incorporation of mass density terms into the analogies.[4]

A slab analogy for a 50 m diameter hexagonal plate formed using a three way double layer grid configuration was prepared and modelled using finite

TABLE 2
Comparison of skeletal and slab analogy results for 50 m three way double layer hexagonal grid system

Mode no.	ABAQUS skeletal analysis (Hz)	LUSAS slab analogy analysis (Hz)	Description of mode	% Difference
1–6	0·0	0·0	RBM	
7	5·19	5·21	Bending	0·4
8	5·19	5·27	Bending[a]	1·5
9	8·90	8·92	Bending	0·22
10	10·18	11·08	Bending and Torsion +	8·8
11	12·84	12·82	Bending[a]	0·15
12	18·02	19·05	Bending	5·7

[a] Anticlastic bending 'flutter'.

elements with varying mesh densities and the results compared with those obtained from an analysis of the original space frame.

The LUSAS program was used to perform an eigenvalue extraction under free–free boundary conditions to obtain the fundamental and higher frequencies and corresponding modes for the slab analogies. The original skeletal structure containing 681 pinned joints and 2670 members was similarly analysed using ABAQUS.

A comparison of both sets of results is given in Table 2 and the relevant mode shapes on the first three modes are shown in Fig. 4.

The above slab analogy used a total of 110 shell elements, both three-node and four-node shell elements being used in the model. This resulted in an uneven distribution of nodal degrees of freedom in the model, an aspect that is reflected in the separation of the modes 7 and 8 in the analogy results.

It will be seen that although good agreement is obtained between the slab analogy and original structures for modes 7, 8, 9, 11 and 12, the analogy's prediction for mode 10 is not so good. A variety of meshes were generated to assess their effects on the analogy and the most satisfactory was obtained using the mesh density employed above with the results for more densely meshed models tending to give frequencies further from the exact solution. It is suggested that this is due to the fact that the actual structural behaviour of the discrete (skeletal) model departs from the structural behaviour of a thin plate for this combined bending and torsional mode of deformation.

In this and the preceding sections it has been shown that it is possible to accurately predict the existence of global and local modes in representative

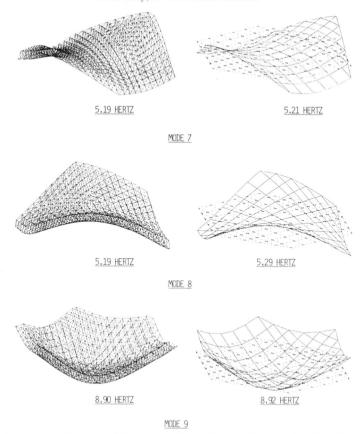

5.19 HERTZ   5.21 HERTZ

MODE 7

5.19 HERTZ   5.29 HERTZ

MODE 8

8.90 HERTZ   8.92 HERTZ

MODE 9

FIG. 4. Axonometric views of skeletal and slab analogy mode shapes for 50 m. Three way double layer hexagonal grid system.

free–free structures by the use of eigenextractions of their finite element models. It has also been shown that reasonable predictions of the modal behaviour of three way double layer grid based space frames, in the form of hexagonal dishes can be obtained through the use of a slab analogy for these systems.

6 THE SPACE FRAME AND CLUSTER

The supporting rectangular platform is fabricated using existing proven tetrahedral space frame technology and measures $150 \times 77\cdot5 \times 2\cdot5$ m. Two

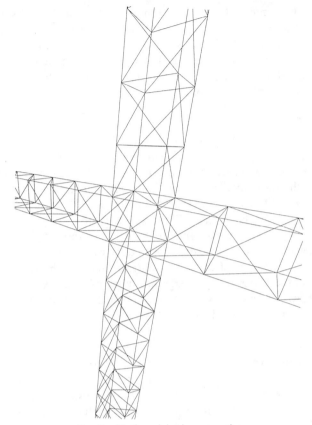

FIG. 5. Platform joint in perspective.

banks of 15 × 5 m solar arrays are supported by arms provided from the platform, their lengths being dictated by the requirement of an unobstructed line of site with the sun. The reflectors and the space frame can be unfurled separately and the component parts would thus be connected at GEO by extravehicular activity (EVA). Taking into account the storage and deployment requirements of the platform section and the need to keep the number of joint types to a minimum, the bracing configuration as successfully used in the 11 bay Perspex model was adopted for use in the frame. An intersection of four arms of the platform is shown in Fig. 5.

Carbon fibre reinforced polyethersulphone has been chosen as the material for use throughout the cluster. This thermoplastic composite has been suitably reinforced to incur a minimum of thermal expansion over a wide temperature range. Representative material properties were obtained from both static and modal tests on tubular sections of the composite. A feature of the composite is its very low material damping characteristics; whereas this allows accurate finite element modelling, it should be borne in mind when designing additional damping mechanisms for the LSS.

7 MODELLING OF CLUSTER

The slab analogy was used to significantly reduce the cost of modelling the dishes. Preliminary eigenextractions from individual dishes under a variety of support conditions indicated that a fundamental frequency of between 1 and 3 Hz for a 75 m flat dish could be expected. These frequencies being at least an order of magnitude greater than those expected for the overall cluster, it was decided that the dishes would most economically be modelled using the slab analogy. This would represent the distribution of stiffness and mass of the dishes without a significant loss of accuracy being incurred. Accordingly a mesh density similar to that used in the slab analogy study discussed earlier, was used to model the 6 dishes.

7.1 Platform
Bar elements were used to model the platform members rather than beam elements to enable a reduction from 6 to 3 degrees of freedom per node in the platform. In all there were 896 nodes and 2876 elements in the platform model, each element having identical material and cross-sectional properties.

7.2 Solar Arrays
The solar arrays were modelled using a set of 8-node shell elements, the latter being used to hinder the development of inextensional modes found to be associated with the use of isolated 4-node shell elements in these ABAQUS analyses.

7.3 Results for the Platform and Cluster of Reflectors
The first six natural frequencies of the cluster are given in Table 3 with the first three mode shapes shown in Fig. 6.

As expected, the overall structural frequencies are much lower than those of the dishes. For this case, a second analysis was performed which replaced the bar elements on the platform with beam elements. Marginally higher

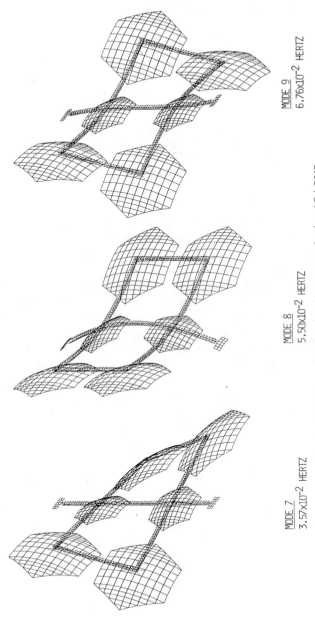

MODE 7
3.57x10^{-2} HERTZ

MODE 8
5.50x10^{-2} HERTZ

MODE 9
6.76x10^{-2} HERTZ

FIG. 6.   First three vibrational modes extracted using ABAQUS.

TABLE 3

Natural frequencies for platform with cluster of reflectors ($\times 10^{-2} Hz$)

Mode no.	Focal lengths of large dishes				
	50 m	70 m	Infinity (flat dish)	Point masses	Rim stiffened cluster
7	3·57	3·61	3·65	3·81	4·13
8	5·50	5·55	5·65	5·38	7·28
9	6·76	6·82	6·90	7·13	8·99
10	10·05	10·07	10·18	9·14	14·68
11	10·20	10·22	11·01	9·14	16·53
12	12·34	12·50	12·72	12·29	17·96

frequencies were obtained, reflecting the increased stiffness of the beam elements, but the values did not differ from the former by more than 0·1%, indicating that axial deformations were predominant in the development of these mode shapes.

A series of parameter studies were performed on the cluster to assess its sensitivity to variations in modelling of the dishes. These included changing the focal length of the dishes from 50 m to 70 m for the larger dishes and from 30 m to 50 m for the small ones. The reflectors were then replaced with flat plates of the same diameters. Table 3 also gives the natural frequencies for these two studies. It is seen that as the dish centres of mass move towards the platform a relatively minor change in global frequencies result. This indicates that the use of flat dishes to support outstands, which develop the required parabolic surface, may be taken as a viable option from a modal analysis point of view. It also indicates that flat dishes may be used as first approximations to replace parabolic reflectors in these analyses. The number of elements required in the finite element configuration for flat dishes is significantly less than those for the corresponding parabolic dishes.

The original reflectors were then modelled by equivalent point masses located at the centres of mass of the dishes; the resulting platform frequencies are given in Table 3. By comparing the point mass results with those for the 50 m focal length dish it can be seen that acceptable approximate results may be obtained from the former solution; certain modes are overestimated whereas others are underestimated, depending upon the participation of dish inertia on each mode. Similar behaviour may be observed when using point masses to replace the flat dishes. In both investigations, however, a significant improvement is obtained in the estimates of the modal behaviour of the systems by modelling the reflectors as flat plates as opposed to point masses.

A further parameter study on the cluster modes was undertaken to determine the effect of varying the displacement support conditions. Additional supports were provided between the platform and rims of each dish. Table 3 also shows the first six natural frequencies.

The results show that by making full use of the available dish stiffness to increase the stiffness of the supporting platform, a significant improvement in the modal behaviour of the cluster results. This is the most effective step discovered in attempts to optimise the above structure and is seen to be especially applicable to clusters containing large dishes.

8 OBSERVATIONS

The current investigation has shown that the structural modelling of equivalent experimental and analytical systems have compared well, thus providing confidence in the finite element eigenextraction technique for the analysis of large free–free skeletal structures.

The use of the slab analogy for the analysis of large space structures has been shown to be satisfactory with a few percent error over the more rigorous skeletal structure approach. This approximate method was employed to model the parabolic dish cluster situated on the skeletal platform with bar elements being used to model the platform behaviour accurately.

It has also been shown that the provision of supports as near as possible to the rims of the large dishes significantly enhances the modal behaviour of the entire cluster even though the lower modes are developed exclusively through the platform arms; supporting the dishes near to their centre does not take full advantage of their inherent stiffness and requires, in addition, more substantial stiffener supports and pointing systems to be designed.

Further optimisation efforts would most likely be concentrated on increasing the fundamental frequency of the cluster by providing additional torsional stiffeners in particular regions of the platform.

The nodal points for the lower modes yield suitable positions for the placement of orbital transfer vehicle (OTV) boosters, as the thrust applied to these points would induce minimum vibrations during orbit transfer and on station positioning manoeuvres.

ACKNOWLEDGEMENT

The authors would like to acknowledge the financial support of the SERC/MOD during this project.

REFERENCES

1. RAMASWAMI, S., Towards optimal solution techniques for large eigenproblems on structural mechanics. PhD Thesis, MIT, 1979.
2. NEWMAN, M. and PIPANO, A., Fast model extractions in Nastran via the FEER, Computer Program, NASA, TM X-2893, 1979.
3. BATHE, K. JAN and WILSON, E. L., Large eigenvalue problems in dynamic analysis. *Proc. ASCE EMG*, **98** (1972).
4. HOLLAWAY, L., FARHAN, A. and GUNN, M. J., A continuum vibrational analysis for a tetrahedral space platform. *Int. J. Space Struct.*, **3**(2) (1988) 104–17.

31

The Study of the Behaviour of a Composite Multi-bolt Joint

L. YANG and L. Y. YE

Aircraft Engineering Department, Northwestern Polytechnical University, Xian, Shaanxi Province, People's Republic of China

ABSTRACT

For a multi-bolt joint the main factor influencing its behavior is the distribution of bolt loadings. These joints are divided into two types: regular arrangements and irregular ones. In this paper the method presented can be used to compute bolt loadings, which include their magnitudes and directions, for arbitrary arrangement in joints. The radial forces on the contact boundary can be solved by the displacement coordination equation. Their resultant force is bolt loading. A simplified method can be utilized to compute a regular arrangement joint. It may take less computer running time. The computation results are quite coincident with the test results. It is shown that this method is very efficient.

1 INTRODUCTION

Bolted joints can transfer a large loading and can easily be dismantled, so they are extensively used in composite structures. For structural connection there are often numbers of bolts and it is simply called a multi-bolt joint.

According to geometric configuration the joints may be divided into two types: one is a regular arrangement multi-bolt joint and another is an irregular one. In order to learn about the behavior of a bolted joint in composite structures, a number of theoretical analyses and experimental studies have been done. Material parameters, fastener parameters and design parameters, etc., every factor affecting the joint behavior, has to be widely approached.[1-5] However, these works are mainly concerned with a single bolt joint. For a multi-bolt joint its behavior is affected by many factors besides the above-mentioned parameters, such as pitch, row distance and side distance; that is geometry parameters and interaction between the two adjacent bolts also affect a multi-bolt joint's behavior. Therefore research on this subject is more complex and difficult. On the other hand, the influence of each factor on the joint strength cannot be obtained by experiment as with a single bolt joint because of the high cost. A few articles have been published in this area.[6-8]

In order to increase joint strength, the laminate of connection range usually includes ±45° plies, 40% at least; width and end distance should be greater than a certain minimum value and clamping force supplied by tightening the bolt should exceed a certain value, etc. These requirements are obtained by study of a single bolt joint, and they are also suitable for multi-bolt joints.

For a multi-bolt joint the main problem is how to determine the loading magnitude and direction of each bolt or, more simply speaking, bolt loading under applied tensile load. Therefore, based on past study, in this paper we focus attention on determining the bolt loading for the above-mentioned two types of multi-bolt joints.

2 ANALYSIS METHOD OF BOLT LOADING FOR A MULTI-BOLT JOINT OF IRREGULAR ARRANGEMENT

Owing to the restriction of structural arrangement, there often are bolt groups of irregular arrangement, or arbitrary arrangement, in the structure. Because of the anisotropic nature of the material the stiffness of a composite laminate is different in different directions. For an irregular arrangement joint under tensile load the supporting stiffness provided by the joining plates is different even in the case of all the bolts with the same diameters. From the test results from a single bolt joint it is shown that the bearing strength will vary with load direction.[9,10] Therefore, when the applied load does not act through a symmetric axial line of the joint or it is an irregular arrangement joint, the analysis method suitable for the metallic joints cannot be simply adapted to determine the bolt loading in composite multi-bolt joints.

For research on bolt loading a joint with three bolts in arbitrary arrangement is taken as a typical example.

In the composite bolted joint, in general, the bolts are made of titanium alloy or steel and the joining plates are both composite laminates, or one is a laminate and the other is a metallic component. Under tension the resultant force of radial action forces between bolt and hole represents the unknown bolt loading.

2.1 Basic Assumptions

The usual assumptions for a single bolt joint[11,12] are adopted in this paper: (1) The bolts are rigid. (2) The influence of clearance and friction between bolt and hole on bolt loading is ignored. (3) Under tension the bolts are not deflected. For the same bolt the distribution laws of radial forces on the two holes of the upper and lower plates are the same. The radial forces are uniform along the plate thickness. (4) The relation between stress and strain is linear.

For multi-bolt joints there are the following approximate relations besides the above assumptions. The hole's radial forces and displacement of upper and lower plates are connected to each other by these relations: (a) The radial displacement of notes are equal and opposite. These notes are located at the two ends of the hole diameter at the same hole of the upper and lower plates respectively. (b) The radial resultant forces are equal and opposite. This resultant force is one of all the radial forces at the same hole of each plate. If the stiffnesses of the two plates are different, this assumption means the mean value of the two resultants of the upper and lower plates.

2.2 Analysis Model

The connected system (Fig. 1) is idealized. One of the two connected plates is taken as a basic plate and called the Ω_0 body. The other is taken as the joining plate and called the Ω_1 body.

At first the boundary condition with zero radial displacement along the boundary in contact with the bolt is assumed; then, instead of the above-mentioned assumption along the contact boundary, the radial concentrative forces on the finite points of the hole edge will be adopted.

For a Ω_0 or Ω_1 body there is a global coordinate system. The directions of the two coordinates are opposite. But node forces are radial, so that a mixture coordinate system is used at the nodes of the hole edge.

In Fig. 1 the 3rd hole is supposed to be fixed and the radial rigid bars are taken as supports. At the edges of the 1st and 2nd holes instead of radial support bars the unknown radial forces R_{ij} are adopted. From the above-mentioned assumption (b) there is

$$\{R_{ij}^1\} = -\{R_{ij}^0\} \tag{1}$$

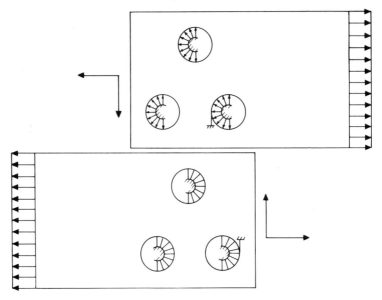

FIG. 1.   The idealization of the connected system.

where subscript i represents hole number and j is node number of this hole edge; superscripts 1 and 0 represent Ω_1 and Ω_0 respectively. The model is shown in Fig. 2.

For the fixed hole the original support stiffness to the bolt is actually changed. But its influence is only limited to local range in consideration of the Saint Venant theory; the bolt loading distribution may be considered as invariable. Any one of the holes can be chosen as fixed and the difference is small. It is confirmed by computation.

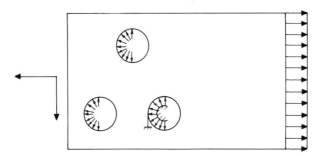

FIG. 2.   The model of a multi-bolt joint with arbitrary arrangement.[1]

2.3 Computing Method and Steps

(a) Equations of displacement coordinates set

From the above-mentioned assumption (a) in this example one obtains

$$
\begin{aligned}
\delta_{11}^1 &= -\delta_{11}^0, \ \delta_{12}^1 = -\delta_{12}^0, \ldots, \ \delta_{1n_1}^1 = -\delta_{1n_1}^0 \\
\delta_{21}^1 &= -\delta_{21}^0, \ \delta_{22}^1 = -\delta_{22}^0, \ldots, \ \delta_{2n_2}^1 = -\delta_{2n_2}^0 \\
\delta_{31}^1 &= -\delta_{31}^0, \ \delta_{32}^1 = -\delta_{32}^0, \ldots, \ \delta_{3n_3}^1 = -\delta_{3n_3}^0
\end{aligned}
\tag{2}
$$

where n_1, n_2 and n_3 are number of notes at 1st, 2nd and 3rd hole edges respectively.

Formula (2) is written as a matrix expression:

$$
\{\delta_{ij}^1\} = -\{\delta_{ij}^0\} \qquad (i = 1, 2, \ldots, i_n; j = 1, 2, \ldots, n_i)
\tag{3}
$$

where i_n is total number of holes; here it equals 3.

(b) Displacement expression of nodes of hole edge

For Ω_1 there is unit loading applied on node 1 of the hole k and $\{\delta_{ij}^1\}_{kl}$ is used to express the radial displacement of node i of hole j. Similarly, unit loading is applied on node 1 of hole k and $\{\delta_{ij}^0\}_{kl}$ is the radial displacement of Ω_0 at the same node of the hole. Under tensile load P, $\{\delta_{ij}^1\}_p$ and $\{\delta_{ij}^0\}_p$ are the radial displacements of Ω_1 and Ω_0 at opposite nodes of the same hole respectively.

Because the relations between stress, displacement and load P are linear the radial displacement of nodes of Ω_1 can be expressed as

$$
\{\delta_{ij}^1\} = \sum_{k=1}^{i_n} \sum_{l=1}^{n_k} \{\delta_{ij}^1\}_{kl} * R_{kl} + \{\delta_{ij}^1\}_p \qquad (i = 1, 2, \ldots, i_n; j = 1, 2, \ldots, n_i)
\tag{4}
$$

where R_{kl} is unknown radial forces at node 1 of hole k, and n_k and n_i are numbers of nodes at the contact boundary of holes k and i respectively.

Similarly, for Ω_0 one obtains

$$
\{\delta_{ij}^0\} = \sum_{k=1}^{i_n} \sum_{l=1}^{n_k} \{\delta_{ij}^0\}_{kl} * R_{kl} + \{\delta_{ij}^0\}_p
\tag{5}
$$

Here the radial force R_{kl} of Ω_1 is taken as the base, and for Ω_0 its direction is opposite.

(c) Solving unknown radial forces

Now Ω_1 is still taken as the base to discuss how to solve the unknown radial forces.

In this example $R_{11}^1, R_{12}^1, \ldots, R_{1n_1}^1, R_{21}^1, R_{22}^1, \ldots, R_{2n_2}^1$ are the radial forces of all nodes at the $(i_n - 1)$ hole edge. Their total number is $(n_1 + n_2)$ and they are independent from each other. $R_{31}^1, R_{32}^1, \ldots, R_{3n_3}^1$ are the radial reaction forces of nodes at a fixed hole. As this hole is fixed at nodes of contact boundary along the radial direction, the displacement equation formula (3) at these nodes will automatically be satisfied. Therefore $(n_1 + n_2)$ independent equations can be obtained by formula (3):

$$\{\delta_{ij}^1\} = -\{\delta_{ij}^0\} \qquad (i = 1, \ldots, i_n - 1; j = 1, \ldots, n_i) \qquad (6)$$

In this example $(i_n - 1)$ is equal to 2.

Substituting formulae (4) and (5) into (6) one can obtain

$$\sum_{k=1}^{i_n - 1} \sum_{l=1}^{n_k} (\{\delta_{ij}^1\}_{kl} + \{\delta_{ij}^0\}_{kl}) * R_{kl} = -(\{\delta_{ij}^1\}_p + \{\delta_{ij}^0\}_p) \qquad (7)$$

$$(i = 1, \ldots, i_n - 1; j = 1, \ldots, n_i)$$

Formula (7) can be written as the following matrix expression:

$$[A]\{R\} = [B] \qquad (8)$$

where $\{R\}$ are unknown radial forces. In this example

$$\{R\} = \{R_{11}, R_{12}, \ldots, R_{1n_1}, R_{21}, R_{22}, \ldots, R_{2n_2}\}^T$$

After solving formula (8), the unknown radial forces of all nodes at the $(i_n - 1)$ hole edge are obtained. And then the finite element technique is utilized to solve radial reaction forces of nodes at the fixed hole.

(d) Determining the range of contact boundary

For a multi-bolt joint when the arrangement of bolts is regular and symmetric, and the tension passes through the axis of symmetry, thus determining the area of contact is easy (this point will be discussed as follows). But when the bolt arrangement is arbitrary or the load does not act through the symmetric axial line of the joint, the contact range of every hole is unknown. Under these conditions the following procedure can be utilized to determine this area.

At the beginning the radial rigid bars along all circular nodes of the fixed hole are adopted. Similarly, the unknown radial forces are applied on all circular nodes of the other $(i_n - 1)$ holes (Fig. 3).

Under tensile load it is expected that all radial forces and bar reaction forces are always in compression at the nodes of the contact area. If these forces on some nodes are tensile, then these forces or bars are deleted, time

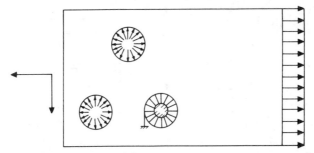

FIG. 3. The model of a multi-bolt joint with arbitrary arrangement.[2]

and again, until all forces are compressive. In this situation the contact boundary is obtained.

For every hole the radial resultant force represents the magnitude and direction of the bolt loading.

2.4 Illustrative Examples

In these examples the property data of the material are:

$$E_1 = 128 \cdot 3800 \times 10^3 \, \text{MPa} \qquad E_2 = 7 \cdot 9478 \times 10^3 \, \text{MPa}$$
$$G_{12} = 3 \cdot 586800 \times 10^3 \, \text{MPa} \qquad v_{12} = 0 \cdot 34$$
$$X_t = 11 \cdot 48364 \times 10^2 \, \text{MPa} \qquad X_c = 8 \cdot 42110 \times 10^2 \, \text{MPa}$$
$$Y_t = 0 \cdot 231280 \times 10^2 \, \text{MPa} \qquad Y_c = 1 \cdot 35926 \times 10^2 \, \text{MPa}$$
$$S = 0 \cdot 609560 \times 10^2 \, \text{MPa}$$

Example 1

A three bolt joint is shown in Fig. 4, where Ω_0 is the steel plate and its thickness t_0 equals 10 mm, and Ω_1 is a $\pm 45°$ laminate and its thickness t_1 equals 2·91 mm. Bolt diameter d equals 5 mm.

The percentage of bolt loading of every bolt is shown in Table 1.

TABLE 1
The results of computation and test

Bolt number	Computation value	Test value	Difference[a]
1	21·2	14·0	
2	39·4	43·0	8·4
3	39·4	43·0	8·4

[a] (Test value − computation value)/Test value.

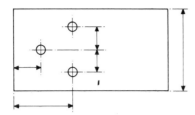

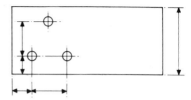

FIG. 4. Three bolt joint in Example 1.   FIG. 5. Three bolt joint in Example 2.

Example 2

A three bolt joint is shown in Fig. 5. There are three joints and their construction is as in Table 2. The tensile load equals 980 N.

For each joint the components of the X and Y directions of the bolt loading are given in Table 3.

From current references collected there are no theoretical or experimental data to be compared with the above results.

TABLE 2
The construction of a three bolt joint

Joint number	Ω_0	Ω_1		
		$0°$	$90°$	$\pm 45°$
A	A1 plate	0·20	0·20	0·60
B	A1 plate	0·40		0·60
C	The same as Ω_1	0·40		0·60

The property data of the A1 plate: $E = 70·56 \times 10^3$ MPa; $v = 0·33$.

TABLE 3
The computation results of three joints

Joint number	Bolt number	X component (N)	Y component (N)
A	1	477·664	11·466
	2	241·374	48·706
	3	290·962	37·240
B	1	458·444	20·678
	2	252·448	42·140
	3	269·108	21·462
C	1	463·834	0·294
	2	258·230	26·754
	3	257·936	26·460

From the results of the examples it is shown that the irregular arrangement is unfavorable and its distribution of bolt loading is more unreasonable.

3 ANALYSIS METHOD OF BOLT LOADING FOR THE MULTI-BOLT JOINT WITH REGULAR ARRANGEMENT

The regular arrangement is, in fact, a kind of special condition of the arbitrary arrangement. In this case the above-mentioned method can be utilized to analyse and compute bolt loading. However, this kind of joint is symmetric and the tensile load usually acts through the symmetric axial line of the joint. Under tension two joining plates cause only a displacement along the load or X axis direction. For every hole the range of contact boundary is symmetric to the X axis. Therefore at the same hole of two connected plates the displacements of each pair of nodes are along the X direction and are equal and opposite. These two nodes are located at the contact boundary on the X axis.

Because it is necessary to satisfy the displacement coordination between each pair of nodes of the same hole the computation is simplified. A joint with three bolts in a row is taken as an example to explain the analysis method of this kind of joint. The model is shown in Fig. 6 (only Ω_1).

In Fig. 6 the 3rd hole is taken as a fixed hole but it is only necessary to fix these nodes on a half-circular boundary of contact. At the 1st and 2nd holes the distributive bolt loadings are applied on the contact boundary of every hole. The cosine distribution law is assumed and their resultant force equal to unity.

The method of solving bolt loading is similar to the above-mentioned one, but unknown bolt loading itself is a concentrated force and its direction is along the X axis.

After solving the bolt loading of the 1st and 2nd holes, the equilibrium condition is utilized to solve the radial reaction resultant force of a fixed hole.

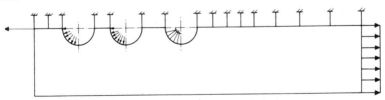

FIG. 6.   The model of a joint with three bolts in a row.

TABLE 4
Laminate parameter

Joint number	0° (%)	90° (%)	±45° (%)
A	40	20	60
B	60		40
C	60	20	20

For a laminate with large anisotropic nature the difference between actual and cosine distribution is considered, so only at the beginning is cosine distribution assumed. Next the law of the radial reaction force of a fixed hole is taken as applying to other holes. This is repeated until the difference between two iterative results is less than a certain small value.

Illustrative Example

A joint with three bolts in a row is shown in Fig. 7. Ω_0 is the A1 plate and its thickness t_0 equals 8 mm; Ω_1 is a composite laminate and its thickness t_1 equals 2·5 mm. Bolt diameter d equals 5 mm.

There are three laminated plates connected to the A1 plate respectively. For the ratios of each ply group see Table 4.

From this example it is shown that the distribution of bolt loading in a row is non-uniform. The difference of longitudinal stiffness $E_x t$ of two joining plates may be taken as the index. If their stiffnesses are different, the bolt loading nearest to the free end of the plate with the greater stiffness is

TABLE 5
The results compared with the test data

Joint number	Bolt number	Computation results (%)	Test data (%)	Difference (%)
A	1	22·35	20·25	
	2	22·40	30·65	
	3	55·25	49·10	12·5
B	1	25·46	24·90	
	2	23·76	23·85	
	3	50·78	51·25	0·9
C	1	25·93	27·60	
	2	24·16	23·90	
	3	49·92	48·50	0·8

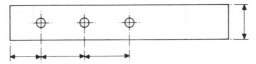

FIG. 7. A joint with three bolts in a row in the example.

greatest. If their stiffnesses more approximate this, non-uniform degree will be improved.

For the results of computation and test see Table 5.

4 CONCLUSION

1. For multi-bolt joints how to determine the bolt loading is an important problem. The bolt loading means its magnitude and direction. In this paper the presented method can be used to compute the bolt loading of the arbitrary arrangement joint. The results of illustrative examples are quite coincident with test data. It is shown that this method is very efficient. Of course, it is also necessary to verify by more experiments.
2. The irregular arrangement is unfavorable. Multi-bolt joints should be as regular and symmetric as possible.
3. The regular arrangement joint is a special condition of the arbitrary arrangement but it is usually used in actual structures. When applied loads act through the symmetric axial line of the joint, the simplified method may be used. The computation results are quite coincident with experimental data. This simplified method may take less computer running time.
4. For the joint with multi-bolts in a row, when the longitudinal stiffness of two joining plates are quite approximate, a non-uniform degree of bolt loading will be improved.

REFERENCES

1. GODWIN, E. W. and MATTHEWS, F. L., A review of the strength of joints in fiber-reinforced plastics. Part 1. *Composites* (July 1980) 155.
2. SONI, S. R., Failure analysis of composite laminates with a fastener hole. *Joining of Composite Materials*. ASTM STP 749, 1981, p. 145.
3. CHANG, F. K., SCOTT, R. A. and SPKINGER, G. S., Strength of mechanically fastened composite joints. *J. Comp. Mater.*, **16** (1982) 470.
4. COLLINGS, T. A., On the bearing strength of CFRP laminates. *Composites* (1982) 241.

5. CHANG, F. K. and SCOTT, R. A., Failure of composite laminates containing pin-loaded holes—method of solution. *J. Comp. Mater.*, **18** (1984) 255.
6. YEN, S. W., Multirow joint fastener load investigation. *Comput. Struct.*, **9**(5) (1978) 483.
7. AGARWAL, B. L., Behavior of multifastener bolted joints in composite materials. *AIAA-80-0307*, 1980.
8. WONG, C. M. S. and MATTHEWS, F. L., A finite element analysis of single and two-hole bolted joints in fiber reinforced plastic. *J. Comp. Mater.*, **16** (1982) 481.
9. MATTHEWS, F. L., The variation of bearing strength with load direction. Symposium: Jointing in Fiber Reinforced Plastic, 1978, p. 126.
10. TSAI, S. W., Bolted joints. *Composite Design* (1981) 1–18.
11. SONI, S. R., Stress and strength analysis of bolted joints in composite laminates. *Comp. Struct.* (1981) 50.
12. YANG, L. and LI, X. H., Contact stress analysis of mechanical fastened joint in composite. Sixth Int. Conf. on Composite Materials, London, 1987.

32

Damage Detection and Damage Mechanisms Analyses in CFRP Laminate Bolted Joints

LIU DA, LIU YI-BING and YING BING-ZHANG

Northwestern Polytechnical University, Xian,
Shaanxi Province, People's Republic of China

ABSTRACT

This paper discusses bearing damage in CFRP laminates around loaded holes under static and fatigue loads. It studies damage initiation and propagation by the 'deply method', 'X-ray radiographs' and 'scanning electron microscope method'. The testing program also was performed to study the effect of temperature and moisture on bolted static bearing strength and its fatigue life. This paper shows delaminations, fiber fractures, splits and interfacial debondings, etc., in CFRP internal laminates. The damage features were recorded by charts and photographs. The test results showed that fiber bundle fracture initiated on the internal laminates and progressed toward the exterior laminates with increasing load. The results also showed bearing areas around loaded holes were related to the fiber orientations of each laminate. The damage specimens were detected and analysed to study damage mechanism.

INTRODUCTION

The application of composite materials in engineering is becoming more and more extensive. Mechanical fastening is one of the common joining methods for laminates. But this connected form will severely lower laminate static strength and its fatigue life.

Bearing failure is a process of damage initiation and gradual loss of stability in CFRP laminates. A major difference between metals and composite materials is that fatigue damage in composite laminates occurs

591

as a result of the accumulation of many cracks, while in metals by the propagation of one dominant crack.[1] In a general way, the bearing damage in composite laminates include matrix cracks, delaminations, fiber fractures and interfacial debondings, etc. The damage propagation in laminates of different fiber orientations are different. Methods for studying damage mechanisms, include acoustic emission, ultrasonic techniques, TBE X-ray radiographs, SEM, stiffness measure[2] and deply methods, etc. Experimental results are listed and some curves or micro-photographs are shown in this paper. These tests showed that fatigue life decreased due to hot-wet environment. Static and fatigue of composite bolted joints were dominated by the growth and propagation of typical damage. Fatigue of composite materials is still an intriguing subject, despite the relatively large amount of experimental research that has been done on it.

In fact composite fatigue is so different from the better known metal fatigue that it requires an entirely new understanding, a different approach.

EXPERIMENTAL PROCEDURE

1 Specimen and Fixture

All of the specimens were cut from $(\pm45^\circ/0_2/\pm45^\circ/0_2/\pm45^\circ/0_2/\overline{90^\circ})_s$ graphite/epoxy (T300/648) laminate. There was a total of 25 plies. The hole in the specimen was drilled to a diameter $5 + 0.1$ mm.

The specimen geometry is described in Fig. 1. The washers between the clevis and the specimen distributed the clampup torque 40 kg cm over an area around the bolt hole.

2 Experiment

Four testing environmental conditions were selected, they were room-temperature and dry (RD) or wet (RW), elevated temperature (82°C) and dry (TD) or wet (TW).

The hole elongation δ was measured using a linear variable displacement

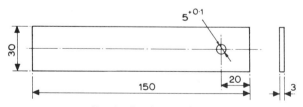

FIG. 1. Specimen (units in mm).

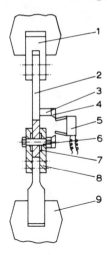

FIG. 2. Bolt-loading apparatus (side view): 1, upper adaptor; 2, specimen; 3, slip; 4, knife-edge; 5, displacement transducer; 6, bolt and nut; 7, washer; 8, Fixture plate; 9, lower adaptor.

transducer installed between the bolt head and the specimen as shown in Fig. 2. For comparison purposes, the oil-mug stroke Δl was also monitored during testing, it was measured by an internal system LVDT with machine MTS 810. An X-Y recorder drew the P–Δl curve and P–δ curve. Figures 3 and 4 give the P–Δl and P–δ curves in static testing and corresponding hysteresis in fatigue testing, separately. Although the displacement magnitude was quite different for those two curves, the curve form was similar in each case. Consequently it was possible to measure the stroke increment instead of measuring the hole permanent elongation directly.

Johnson and Matthews suggested a way to define failure load, i.e. taking

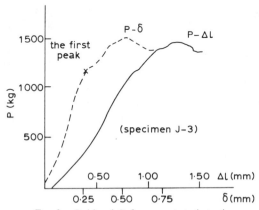

FIG. 3. P–Δl and P–δ curves at static testing.

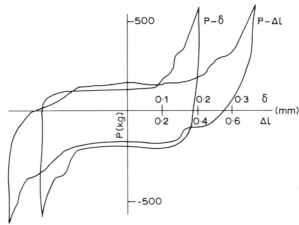

Fig. 4.   Hysteresis P–Δl and P–δ curves in fatigue testing.

the first peak load in the load–extension curve as the static bearing failure load, when the first cracking sound was heard in the specimen.[3] With the RD conditions, the failure load P_d on the average is 1170 kg, corresponding to a deformation δ_d of hole diameter with this load approximating 0·28 mm. Table 1 lists the experimental data from static testing. Since the testing data

TABLE 1
Results of static test

Cond.	$\bar{\delta}_k$ (kg/mm^2)	$\bar{S}_k$	$c_k\%$ (s_k/δ_k)	$\bar{\delta}_b$ (kg/mm^2)	$\bar{S}_b$	$c_b\%$ (S_b/δ_b)	$\bar{\delta}_k/\bar{\delta}_b$	Number
RD (J)	66·78	2·41	3·61	83·97	2·28	2·71	0·80	4
RD (F$_1$)	71·37	5·93	8·28	91·38	4·17	4·56	0·78	4
RD (F$_2$)	65·10	5·52	8·48	85·56	3·19	3·73	0·76	4
RD (D$_2$)	66·02	2·76	4·18	84·60	3·66	4·33	0·78	4
RW (J)	68·97	4·46	6·40	83·30	2·37	2·84	0·83	4
TD (J)	43·68	8·20	18·8	72·20	2·07	2·87	0·60	5
TW (J)				68·82	5·77	8·40		4
TW (F$_1$)	42·75	5·76	13·5	69·75	2·89	4·10	0·61	4
TW (F$_2$)	42·47	2·15	5·05	70·10	4·14	5·90	0·61	4
TW (D)				73·70	1·47	1·90		4

Note:
  $\bar{\delta}_k$, mean damage stress.
  $\bar{S}_k$, root-mean error corresponding to $\bar{\delta}_k$.
  $\bar{\delta}_b$, mean ultimate stress.
  $\bar{S}_b$, root-mean error corresponding to $\bar{\delta}_b$.

TABLE 2
Results of fatigue test

Cond.	P_d (kg)	Δl (mm)	$\bar{N} \times 10^4$	lg/N	$\bar{S}$lg/N	Number
RD (J)	585	0·28	4·01	4·60	0·11	4
RD (F_1)	585	0·28	4·08	3·56	0·25	4
RD (F_2)	585	0·28	5·83	3·98	0·22	4
RD (D)	585	0·28	5·87	3·73	0·20	4
RW (J)	585	0·28	6·45	3·70	0·37	4
TD (J)	440	0·28	0·24	2·36	0·18	4
TW (J)	440	0·28	0·37	2·57	0·17	5
TW (F_1)	440	0·28	0·30	2·45	0·17	6
TW (F_2)	440	0·28	0·29	2·38	0·28	6
TW (D)	440	0·28	0·44	2·55	0·33	6

were more scattered, therefore from the statistical point of view, they must be treated.

Fatigue load amplitude P_f was selected as half of the mean static failure load, i.e., $P_f = \frac{1}{2}P_d = 585$ kg, and the predicted life N was obtained as the constant permanent elongation δ_d of hole attained. In the most severely loaded case, the tension–compression constant amplitude cyclic test ($R = -1$) was selected for all fatigue tests in our study. The testing frequency was 5–8 Hz.

Table 2 shows the results obtained from fatigue testing and Fig. 5 shows typical δ/δ_0–N curves for the J-group in the RW condition. In Fig. 5, it is seen that the ratio between the deformation δ of the hole diameter and the

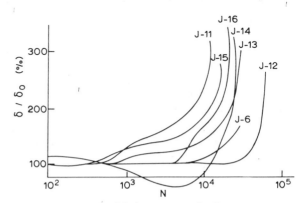

FIG. 5. $\delta/\delta_0/N$ test curves for J-group.

initial deformation δ_i was 200%, it showed fatigue failure in the specimen, and the permanent elongation δ_d of hole was equal to be 0·28 mm.

3 Damage Detection and Damage Mechanism Analysis Around the Loaded Hole

(1) Examination by the 'deply method'[5]

This method permits the individual laminates to be separated for damage examination without destroying the fiber fracture feature. The process for deplying testing is shown in Fig. 6. First, a section of specimen containing the bolted hole was removed to detect each laminate surface, then this section was heated up to 465°C and this temperature kept constant for an hour. Next, the deplied laminate of each specimen was mounted on a work sheet to aid identification and to facilitate handling for microscopic examination and photographing.[6] As the laminates of each specimen were examined, the observed damage details were recorded with sketches and photographs. The sketches of fatigue failures on deplied internal laminates are shown in Fig. 7. It is shown that in fatigue testing the fiber fractures and breakages were severe in internal laminates, while the delamination areas had larger range than the exterior laminate. It was found that fiber fractures initiated on the internal laminates and progressed toward the exterior laminates with increasing load. The typical photomicrographs of internal laminates in static testing provided the damage details adjacent to the bolted hole (Fig. 8). Figure 9 provided a clearer boundary of delamination range. The deplied testing showed the delamination damage between laminates 0° and 0°, or 0° and 90°, occurred with more difficulty, but between laminates ±45° and 0° more easily.

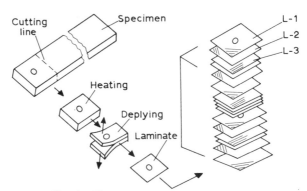

FIG. 6.   The process steps in deplied testing.

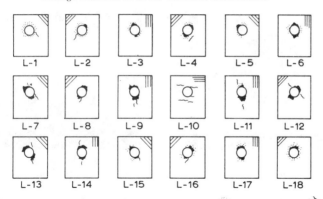

FIG. 7.   Sketches of fatigue failures in internal laminates. ⚬, delamination; ⚬ fiber fracture; ⚬, bearing damage.

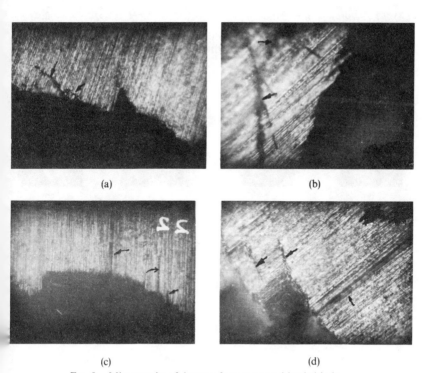

FIG. 8.   Micrographs of damage feature around loaded hole.

FIG. 9. (a) SEM print of delamination area.

FIG. 9. (b) Edge of delamination area SEM print.

The bearing damage areas around the loaded hole in static testing were related to the fiber orientation of each laminate, the defacing of each laminate by the deplied method is shown in Fig. 10, and corresponding sketches are shown in Fig. 11.

(2) TBE radiographs[7]

TBE radiographs were often applied in non-destructive exploration. This method generally obtained additional patterns along the thickness of the specimen.

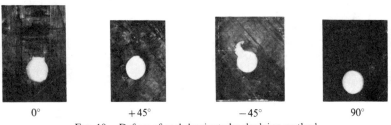

0°   +45°   −45°   90°

FIG. 10.   Deface of each laminate by deplying method.

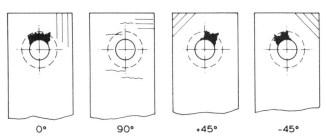

0°   90°   +45°   -45°

FIG. 11.   Sketches of bearing damage area in laminates of different fiber orientation.

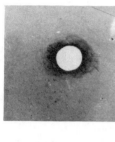

FIG. 12. Outside feature of specimen in static testing.

FIG. 13. X-ray radiograph under static load.

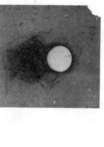

FIG. 14. X-ray radiograph under fatigue load.

FIG. 15. SEM print of bearing half side of hole.

FIG. 16. SEM print of interfaces in composite specimen.

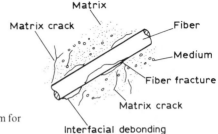

FIG. 17.   The chart of damage mechanism for
            interfacial debonding.

Figure 12 shows the outside feature of the specimen after static testing. This damaged specimen was radiographed in order to detect internal damage. Figure 13 shows that the damage area with black color was located at the bearing half side. In the radiograph of Fig. 14 the damage area of the specimen was located around the hole mainly along the loading direction under the bearing fatigue load.

(3) SEM analysis[8]

The analysis by scanning electron microscope demonstrated that matrix microcracks occurred because of high stress in the bearing areas. These microcracks joined each other through weakness and a main crack was formed (Fig. 15). Figure 16 shows the interfaces of a composite specimen at the RW condition were destroyed. Interfacial debondings enormously lowered the strength of CFRP laminates. The mechanism of fatigue crack propagation in conditioned material is completely different from that in air or in water without conditioning. Because the temperature rose or the medium molecules corroded to matrix, that combined key between fiber and matrix was destroyed. On the other hand stress was transmitted with increasing difficulty. As medium molecules continuously corroded fibers, fiber fracture will occur (Fig. 17).

CONCLUSIONS

1. A loaded hole of CFRP laminates was destroyed because high stress in the bearing area engendered many matrix microcracks. The microcrack propagation resulted in delamination, fiber fracture, etc. This destruction belonged to bearing failure mode. At TD or TW conditions fatigue life under a certain load would be greatly reduced, while bearing strength value was only 72 kg/mm².

2. The scanning electron micrographs showed that interfacial debondings were serious in the moisture-condition, on the other hand, matrix fractures were serious in the air-condition.
3. The three methods, i.e. 'deply method', 'TBE X-ray radiographs' and 'SEM method', respectively have advantages and disadvantages in damage detection for CFRP laminates. The 'deply method' will examine damage features in internal laminates. TBE radiographs generally may obtain the additional photos along the thickness of specimen. The SEM method may be used for the analysis of the damage mechanism.
4. In order to detect cumulative damage, the experiments must be carried out under different load levels.

ACKNOWLEDGEMENTS

The authors wish to express their sincere thanks to Professor Zhang Kai-Da, Mr Jiao Gui-Qing and Guo Wei-Guo for their assistance in these experiments.

REFERENCES

1. HASHIN, Z., Cumulative damage theory for composite materials. *Comp. Sci. and Technol.*, **23** (1985) 1–19.
2. OGIN, S. L., *et al.*, Matrix cracking and stiffness reduction during the fatigue of a $(0/90)_s$ CFRP laminate. *Comp. Sci. and Technol.*, **22** (1985) 21–31.
3. JOHNSON, M. and MATTHEWS, P. L., Determination of safety factors for use when designing bolted joints in GRP. *Composites* (April 1979).
4. YING BING-ZHANG, An investigation of experimental methods for the determination of bearing strength of CFRP laminates. *Proc. 4th Int. Conf. on Composite Structures.*
5. FREEMAN, S. M., Characterization of lamina and interlaminar damage in graphite/epoxy composites by the deply technique. ASTM STP 787, 1982, pp. 50–62.
6. FREEMAN, S. M. and BAILEY, C. D., Detection and verification of internal fiber fracture around loaded hole in graphite/epoxy composites. *11th National SAMPE Technical Conference*, 1979.
7. SENDECKY, G. P. *et al.*, Damage documentation in composites by stereo radiography, ASTM STP 775, 1982, pp. 16–26.
8. THEOCARIS, P. S. *et al.*, Crack propagation in fibrous composite mat materials studied by SEM, *Jnl Comp. Mater.*, **15** (March 1981) 133–41.

33

Effects of Biaxial Loading in Pin Jointed CFRP Plates

W. S. Arnold, I. H. Marshall and J. Wood

Department of Mechanical and Production Engineering,
Paisley College of Technology, Paisley, Scotland, UK

ABSTRACT

Pin jointed isotropic and orthotropic CFRP plates subjected to biaxial loading conditions have been examined using two-dimensional finite element methods. Three distinct modes of uniaxial loading, tension, shear and compression, were initially compared. By introducing both transverse tensile and transverse compressive load components the effects of biaxial loading were then assessed for contrasting load arrangements. Both the plate elasticity and the magnitude of the biaxial load ratio β are shown to exhibit pronounced effects on the final stress state. Under certain load combinations the maximum radial and hoop stresses at the hole boundary are improved. However, this has been shown to be a transitional characteristic and dependent on the magnitude of β. At high absolute values of β the transverse loads dominate the stress state and are considered detrimental to joint strength.

1 INTRODUCTION

Laminated composites are now extensively employed as light weight, high strength alternatives to conventional metal alloys in aircraft and automotive structures. As these components are often joined by mechanical fasteners there has been a number of studies[1-11] to determine the stress distribution around holes in pin loaded composite plates. While most of these investigations have been based on two-dimensional plane

603

stress conditions they have identified some of the fundamental parameters influencing joint strength. Both elasticity methods[1-5] and finite element techniques[6-11] have been used successfully to study the effects of anisotropy, friction, pin stiffness and pin/hole clearance. Of particular interest has been the determination of the contact and hoop stresses at the hole periphery, since these are considered to initiate bearing and tensile failures respectively.

Bickley's early assumption[12] that the contact stresses varied as a cosine distribution when a rivet was pulled sideways in an infinitely wide isotropic plate has been shown to be invalid for orthotropic finite width composite plates. For example, De Jong,[2,3] who investigated the effects of pin load direction, demonstrated that the radial stress distribution alters significantly as the degree of orthotropy changes and that the maximum contact stress tends to occur in line with the principal material direction. Using a three-dimensional approach,[14] the present authors demonstrated that the contact stress distribution is dependent on the nature of the clamping load in bolted joints. Crews and Naik[9] have shown that under combined bearing and bypass loading the contact stresses are clearly dependent on the bypass ratio. In Ref. 9 a finite element procedure was used in which a rigid frictionless pin was simulated by restraining nodes at the hole edge to lie on a circular arc. The contact arc was found by an iterative technique which involved releasing restrained nodes on the hole boundary, which were found to have tensile radial stresses. In an alternative and more convenient approach Eriksson[8] incorporated a contact subroutine into a finite element analysis to study pin elasticity, friction and clearance.

A common aspect of these investigations is that only uniaxial loadings have been considered. Although in most cases joints are primarily designed for uniaxial load conditions, little work has been published on the performance of joints under biaxial loads. Major test experience has indicated the lack of adequate knowledge on the behaviour of joints under combined loads.[13]

Accordingly, the aim of this paper is to evaluate the effects of introducing a state of biaxial loading in both isotropic and orthotropic pin loaded joints.

2 PROCEDURE

The general purpose finite element program PAFEC was used in this study. Meshes were constructed using the orthotropic version of the four- and

TABLE 1

	E_a (GPa)	E_t (GPa)	G_{at} (GPa)	v_{at}
Mild steel	209	209	80·4	0·3
Plate, 0-degree	145·9	10·8	6·4	0·38
Plate, 90-degree	10·8	145·9	6·4	0·028
Plate, $(0°/\pm45°/90°)_s$	57·5	57·5	21·9	0·31

three-noded constant strain elements which have two translational degrees of freedom at each node. A square plate was considered with a plate width to hole diameter ratio of 5. As the pin plate contact arc was dependent on the loading arrangement a specialised contact subroutine within PAFEC was used. Frictionless 'gap elements' were generated between nodes on the hole periphery and adjacent nodes on the bolt surface. These gaps were initially coincident, thereby simulating a frictionless perfect fit bolt. As the problem was solved with gap elements only one run was required for each load configuration.

The bolt material properties were taken to be those of mild steel, while the plate properties were taken for a typical CFRP laminate, as shown in Table 1. In all cases the plate was modelled as a homogeneous orthotropic material. The properties for the quasi-isotropic plate were determined from classical lamination theory.

A schematic representation of the problem is shown in Fig. 1 and a typical displaced shape mesh plot is shown in Fig. 2. With 16 elements subtending each 90-degree arc, the contact arcs were resolved to within 5·625 degrees. As the loading arrangement and plate geometry were symmetrical about the y axis only one half models were required. Loads were transferred from the plate to the bolt by fully restraining the bolt on its central axis, thereby preventing rigid body translation. The mesh performance was verified with previous work by the authors.[11]

In the case where loads are applied uniaxially, with no bypass component, the problem of a perfect fit frictionless pin is linear with respect to the applied load (i.e. the contact area does not vary with applied load[9]). Clearly, with the introduction of transverse loads the contact arc is dependent on the magnitudes and directions of both the axial and transverse load components. For convenience, therefore, a biaxial load ratio β has been introduced to quantify the loading arrangements under consideration. With reference to Fig. 1 the ratio β is given by

$$\beta = F_t/2F_a \tag{1}$$

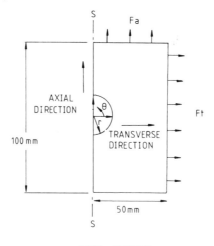

JOINT GEOMETRY

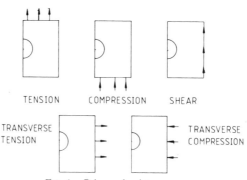

FIG. 1.   Primary load types.

where F_t and F_a are the transverse and axial loads respectively. Hence, for a given β, the contact arc is constant irrespective of the load magnitudes. In addition, where stress results are presented, they have been non-dimensionalised with respect to the bearing stress σ_b given by

$$\sigma_b = 2F_a/(D * t) \qquad (2)$$

where D is the hole diameter and t the plate thickness, which was taken as unity. Therefore, for a given load ratio, the non-dimensionalised stresses should also be invariant with absolute load levels.

The convention adopted here was to assume that all values of F_a were

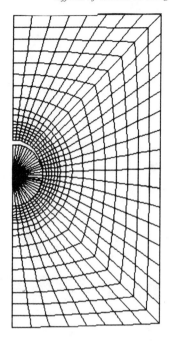

FIG. 2. Displaced shape plot: combined
shear and transverse compression.

positive while the sign of the transverse load was dictated by the nature of
the load (i.e. positive β values correspond to transverse tension while
negative values relate to transverse compression).

Unless otherwise stated, loads were introduced by imposing displacements at the desired node points. Force magnitudes were then evaluated by
summing reactions at these positions.

3 RESULTS

3.1 Uniaxial Loading

Three contrasting modes of uniaxial loading were considered, namely
tension, shear and compression, as shown in Fig. 1. Contact and hoop
stresses at the hole boundary for these configurations in a quasi-isotropic
laminate are compared in Figs 3 and 4 respectively. While there is little
difference in the contact stresses between the tensile and shear configurations, the bearing stresses in the compression load case are significantly
higher. This trend is reversed with reference to the hoop stresses, where they
are at a maximum in the tensile configuration. Interestingly, the location of

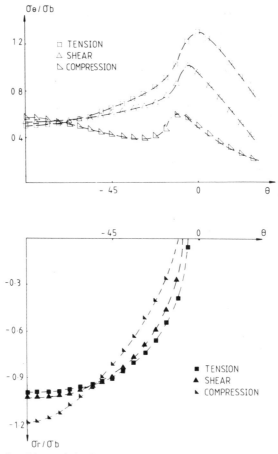

FIG. 3.   Quasi-isotropic laminate: contact and hoop stresses at hole boundary.

the peak hoop stress corresponds to the nodal position immediately following the pin/plate separation point.

One aspect not highlighted on these figures is the effective axial stiffness variations (i.e. the force F_a divided by the axial displacement) between the models. Although in all three configurations the nodes are displaced by a uniform value, the total force required in each case was patently different. It was found that the compression configuration gave the highest axial stiffness, which was approximately 1·4 times that of the tension and shear models.

The effects of material orthotropy were assessed by considering both 0-

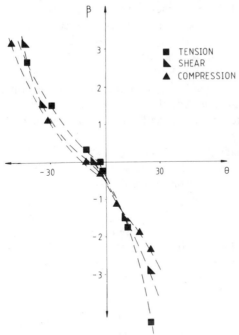

FIG. 4. Quasi-isotropic laminate: variations in contact angle with β.

and 90-degree laminates under similar load conditions. Both the locations and magnitudes of the maximum stress concentration factors, SCFs, are summarised in Table 2. Consistent with the isotropic plate, the contact stresses were highest in the compression model for the 0-degree laminate, while the hoop stresses were highest in the tension configuration. Apart

TABLE 2

		$\sigma_{r\max}/\sigma_{\rm b}$	$\theta_{r\max}$	$\sigma_{h\max}/\sigma_{\rm b}$	$\theta_{h\max}$
Isotropic	Tension	−1·02	−90	1·3	0
	Shear	−1·04	−90	1·05	−5·625
	Compression	−1·18	−90	0·6	−11·25
0-degree	Tension	−1·49	−90	3·6	0
	Shear	−1·51	−90	2·5	0
	Compression	−1·92	−90	0·6	0
90-degree	Tension	−1·42	−22·5	2·35	−90
	Shear	−1·3	−22·5	2·25	−90
	Compression	−1·125	−33·75	2·3	−90

from the expected general increase in SCFs, it was particularly noticeable that the maximum hoop stress in the compression 0-degree model showed no change when compared to the isotropic configuration.

In contrast, results for the 90-degree laminate displayed several distinctive characteristics. For example, the maximum contact stresses were found in the tensile load case and not in the compression configuration, as in the other laminates. Additionally, the peak hoop stresses were of similar magnitudes irrespective of the loading arrangement.

3.2 Biaxial Loading

Clearly, the method of stress superposition which is commonly used to simplify many engineering problems may not be employed by summing fundamental stress components to reach the final stress state since the contact arcs for each of the primary load types are initially different. Consequently, separate runs were required for each loading arrangement considered so that effectively multiple loads were applied simultaneously.

The effects of introducing transverse loads are summarised in Figs 4–7,

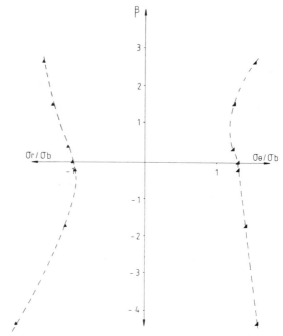

FIG. 5.   Quasi-isotropic laminate: axial tension variations of maximum stress with β.

where the variations in stresses at the hole boundary and their associated contact arcs are plotted with the biaxial load ratio β for a quasi-isotropic laminate. In all three configurations the introduction of transverse tensile loads (positive β) reduces the contact arc while it is increased under transverse compression (negative β), which from a simple mechanics viewpoint appears intuitively correct. Particularly notable are the results for the compression configuration, where, for positive values of β, the contact arc is consistently the smaller whereas for low negative values of β the trend is reversed and the contact arc is the greater of the three configurations.

Whereas this behaviour appears correct under transverse tension, the observed change induced by transverse compression is less obvious. By considering the three primary load types, only in the axial compression model is the upper half of the fastener hole (0–90 degrees) effectively stress free. It therefore appears that for low negative β values stresses in this area are dominated by the transverse compressive stresses, which tend to increase the contact arc and thereby reduce the radial stresses. In contrast, the axial tensile stresses existing in this area in the other configurations counteract this effect.

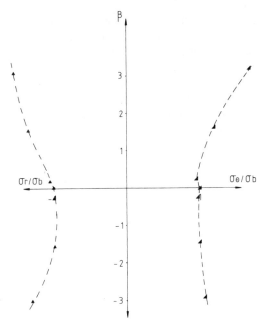

FIG. 6. Quasi-isotropic laminate: axial shear variations of maximum stress with β.

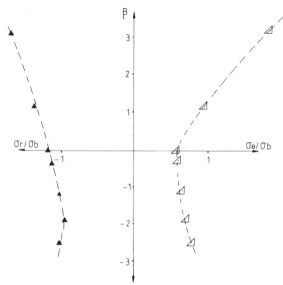

FIG. 7.   Quasi-isotropic laminate: axial compression variations of maximum stress with β.

3.2.1 Transverse compression

In all three configurations the peak contact stresses are initially reduced by the introduction of transverse compressive stresses. Basically this feature is consistent with the observed increased contact arc, which effectively reduces the mean stress levels since the axial load is distributed over a larger area. At lower β values the maximum contact stresses exhibit significant increases in all three configurations. This characteristic is due to the high transverse load component dominating the stress state and is further typified by the migration of the position of maximum stress away from the -90-degree position, as β becomes more negative. In contrast to this, the hoop stresses display no reversal characteristics and are less sensitive under transverse compression, with the maximum values occurring within 5 degrees of the pin/plate separation point.

3.2.2 Transverse tension

Essentially, the peak contact stresses continually increase with β, which is consistent with the associated reduction in the pin/plate contact arc. In all cases the maximum radial stresses were located at -90 degrees. The position of the maximum hoop stresses, however, were dependent on the magnitude of β. At intermediate values, e.g. $\beta = 1$, the peak occurred at -90 degrees while at higher values, e.g. $\beta = 3$, the position shifted to 90 degrees, which is clearly symptomatic of high absolute values of β, where in general the transverse load components govern the resulting stress state. As with the contact stresses under transverse compression, the maximum hoop stresses for positive β exhibit similar stress reversal behaviour in both the tensile and shear configurations. The peak hoop stresses are initially reduced as a result of the compressive stresses generated by the 0-degree position by transverse tensile loads. This behaviour is not observed in the compressive configuration since, for small positive values of β, the position of the peak hoop stress moves to the -90-degree position and hence the improvements at the 0-degree position are not represented in Fig. 7. For all configurations the increased magnitudes at high β values are due to the high tensile hoop stresses generated at the 90-degree position.

Similar analyses conducted on the 0- and 90-degree laminates gave trends close to those of the quasi-isotropic plate. Apart from the higher SCFs, the position of the maximum stresses were less prone to migrate with the introduction of transverse loads.

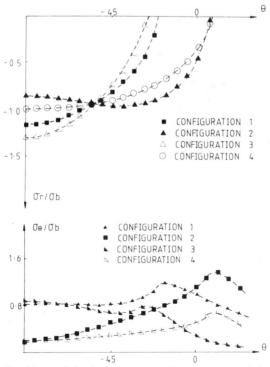

FIG. 8. Quasi-isotropic laminate: contact and hoop stresses at hole boundary.

3.3 Biaxial Loading under Equivalent Load Conditions

Comparisons were made between the following combined load configurations with β values equal to unity:

1. axial tension/transverse tension;
2. axial tension/transverse compression;
3. axial compression/transverse tension;
4. axial compression/transverse compression.

For convenience, loads in these models were applied as stresses to the edges of the plate in preference to displacement constraints which were used in the previous models. This procedure was adopted since load magnitudes

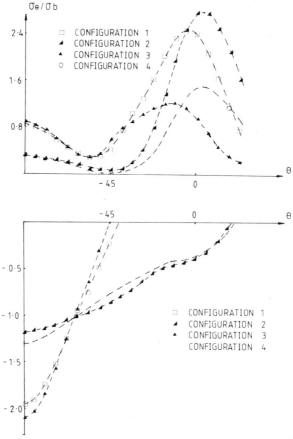

FIG. 9.   0-degree laminate: contact and hoop stresses at the hole boundary.

were more readily controlled. For comparison, configuration 1 was modelled using both the stress and displacement methods and gave stresses at the hole boundary to within 1% of each other. The contact and hoop stresses in the quasi-isotropic laminate are shown in Fig. 8. Clearly the highest contact stresses occur under combined axial compression and transverse tension, while the maximum hoop stresses occur in configuration 2.

Comparable results, which are shown in Fig. 9, for the 0-degree laminate highlight the sensitivity of the contact stresses to the nature of the transverse load. In both cases transverse tension increases the contact stresses while transverse compression reduces them. These results indicate that for an orthotropic laminate, with its principal material direction

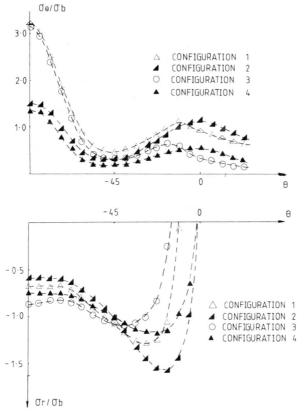

FIG. 10.   90-degree laminate: contact and hoop stresses at the hole boundary.

coincident with the axial direction, the nature of the transverse loading is of more importance than the mode of axial loading. In contrast to the quasi-isotropic laminate, the hoop stresses are reduced in configurations 1 and 2 whereas in configurations 3 and 4 they are increased when compared to the pure axial SCFs.

Results for the 90-degree laminate are shown in Fig. 10. When compared to pure axial loading the effect of introducing transverse loads is dependent on the nature of the loading. The maximum contact stresses are increased in transverse compression and are reduced in transverse tension. The peak hoop stresses are dominated by transverse tensile loads which significantly increase the SCF. Conversely, these SCFs are reduced well below those found under pure axial loading when subjected to transverse compression. These results contrast those of the quasi-isotropic and 0-degree laminates, where the axial load component dominated hoop stresses.

4 FURTHER COMMENTS

No attempt has been made in this paper to predict failure strengths and considerable caution should be taken in the interpretation of these results on a strength basis. Most data have been presented in terms of the maximum stress concentration factors which, in the case of the 0- and 90-degree laminates, are close to or coincident with the principal material directions. Clearly, in cases where matrix cracking dominates the failure process it is the stresses transverse to the fibre direction which are of importance. Obviously, the favoured approach is to incorporate an interactive failure criteria or to adopt the characteristic distance approach as used in Ref. 11 to predict joint strength.

5 CONCLUDING REMARKS

The effects of biaxial loading in three pin loaded laminates have been assessed using finite element techniques. It has been shown that under certain load conditions the introduction of transverse loads can improve the stress state local to a fastener hole. This, however, has been shown to be a transitional characteristic and dependent on the magnitude of the biaxial load ratio β. Furthermore, stresses arising from biaxial loading have been shown to be strongly influenced by the laminates' elastic properties. Irrespective of the load configuration, for high absolute values of β, transverse loads dominate the final stress state and are considered detrimental to joint strength.

ACKNOWLEDGEMENT

This work was carried out with the support of the Procurement Executive, Ministry of Defence.

REFERENCES

1. OPLINGER, D. W. and GANDHI, K. R., Stresses in mechanically fastened orthotropic laminates. *2nd Conf. Fib. Comp. in Flight Vehicle Design*, Ohio, May 1974, pp. 813–41.
2. DE JONG, T., Stresses around pin loaded holes in elastically orthotropic or isotropic plates. *J. Comp. Mater.*, **11** (1977) 313–31.
3. DE JONG, T. and VUIL, H. A., Stresses around pin loaded holes in elastically orthotropic plates with arbitrary load direction. Report LR-333, Department of Aerospace Engineering, Delft University of Technology, Sept. 1981.
4. GARBO, S. P. and OGONOWSKI, J. M., Effects of variances and manufacturing tolerances on the design strength and the life of mechanically fastened composite joints. *AFWAL-TR-81-3401*, Vols 1–3, April 1981.
5. HYER, M. W. and KLANG, E. C., Contact stresses in pin loaded orthotropic plates, Virginia Polytechnic and State University, Report VPI-E-84-14, April 1984.
6. CONTI, P., Influence of geometric parameters on the stress distribution around a pin loaded hole in a composite laminate. *Comp. Sci. Technol.*, **25** (1986) 83–101.
7. WONG, C. S. M. and MATTHEWS, F. L., A finite element analysis of single- and two-hole bolted joints in fibre reinforced plastic. *J. Comp. Mater.*, **15** (1981) 481.
8. ERIKSSON, L. I., Contact stresses in bolted joints of composite laminates. *Comp. Struct.*, **6** (1986) 57–75.
9. CREWS, J. H. JR and NAIK, R. A., Combined bearing and bypass loading in a graphite epoxy laminate. *Comp. Struct.*, **6** (1986) 21–40.
10. CHANG, F. K., Design of composite laminates containing pin loaded holes. *J. Comp. Mater.*, (1984) 279.
11. MARSHALL, I. H., ARNOLD, W. S. and WOOD, J., Observations on bolted connections in composite structures. *Proc. Pan American Congress of Applied Mechanics*, Rio de Janeiro, Brazil, Jan. 1989, pp. 43–6.
12. BICKLEY, W. G., The distribution of stress around a circular hole in a plate. *Phil. Trans. R. Soc., London*, **227A** (1928) 383–415.
13. Private communication from R. F. Mousley, formerly of the Royal Aircraft Establishment, Farnborough.
14. ARNOLD, W. S., MARSHALL, I. H. and WOOD, J., A finite element assessment of clamping effects in composite bolted joints. To be published in *Proc. Inst. Mech. Engrs*, Part G, *J. Aerospace Engng*, 1st edn.

34

Structural Performance of GRP Pipes with Mixed Wall Construction

ZHOU RUN-YU, ZHOU SHI-GANG, XUE YUAN-DE
and QIN RUI-YAO

Tongji University, Shanghai, People's Republic of China

ABSTRACT

In this paper, a new kind of GRP pipe with a polymer concrete core, named a GRPPC pipe, is designed and the manufacturing process is concisely described as well. Based on various contrasting analytical methods for the buried pipes, we adopt one of them for GRPPC pipes. The stresses of the joints are analysed with FEM. Two types of tests are carried out. The properties of the composed material are measured for pipe design. The structural performance is tested and the results prove the reliability of the adopted method of design. The cost of the new pipe is reasonable and acceptable.

1 INTRODUCTION

Glass fibre reinforced plastic pipes have many excellent features, such as high strength, stiffness and light weight, corrosion resistance, low water-flow resistance, convenient installation, etc. Therefore, the use of GRP pipes has been spread worldwide in the sewage and drinking water supply pipelines, crude oil delivery as well as municipal thermal delivery, etc.

GRP pipes can be classified into two groups according to the different pipe wall build-up. One is the pure GRP wall pipe wound with filament, produced by Vetroresina and Owens-Corning Corp. Another is the sandwich pipe, such as the centrifugally cast glass reinforced plastic pipe, manufactured by Johnston Corp. We contrast the mechanical character-istics and the costs of the above mentioned pipes with each other. For the filament winding pipe, due to the low modulus of glass fibre, usually the pipe wall is thick enough to increase the pipe stiffness under service conditions. Thus, its cost is high. For centrifugally cast GRP pipe, the fibre is chopped and its strength isn't fully utilized, so its cost goes up.

To develop the large diameter pipe for drinking water supply pipelines, a new kind of GRP pipe with a polymer concrete core, named GRPPC pipes, has been designed and manufactured in our laboratory. The GRP pipe with mixed wall construction can provide a range of required strengths and stiffnesses, and its cost is reasonable and acceptable.

2 PIPE WALL BUILD-UP AND MANUFACTURING PROCESS

Figure 1 shows a typical pipe wall build-up of GRPPC pipe. The main differences between GRPPC pipes and other GRP pipes are the following:

(a)   The outside and inside reinforced layers of GRPPC pipes are symmetric laminates composed of continuous fibre or woven fabrics reinforced polymer resin. The quantities of hoop fibre and axial fibre are determined after analysing the stresses, deflection and buckling of the pipe under service conditions. The high strength of continuous fibre is adequately used.

(b)   The core layer is polymer concrete, which bears the shear stress that results from applied loads on the pipe. A certain adhesive strength between the core layer and GRP layers is required to ensure the entirety of pipe with mixed wall construction. Resin may be saved by regulating the grade of sand.

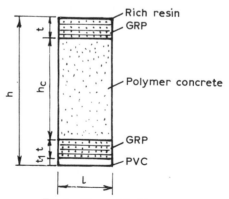

Fɪɢ. 1.   Pipe wall build-up.

(c) The inside covering layer and barrier layer of the pipe consists of PVC plate.

(d) The outside protector of the pipe is a rich resin layer, for a smooth surface and to prevent the fibre from exposure.

The manufacturing method for GRPPC pipe is the following: on the rotational mandrel of chosen diameter, at first we weld PVC plate and handle its surface. Secondly according to the design, we wind fibre or lay woven fabrics to form the internal GRP laminate. And we form a core layer of polymer concrete. Then we form the external GRP laminate similar to the internal GRP laminate. Lastly we put a rich resin layer on the outside surface of the pipe. There are several methods for the manufacture of the polymer concrete core. One of them is to cast polymer concrete into a closed region which is located at the internal GRP laminate and the external GRP laminate.

3 DESIGN METHOD

In this section, we will discuss the design method for the buried pipes.

Generally, the pipes are classified into two types. One is rigid pipe, the ratio $\delta y/D$ of vertical deflection to the pipe diameter is less than 2·5%. The other is flexible, the ratio $\delta y/D$ is 2·5–5·0%.

In pipe design, the stresses must be analysed and the vertical deflection must be calculated and the stability should be checked.

3.1 Stress Analysis

For the buried pipe, the stress analysis is complicated. The variation of the pipe stiffness and the backfill soil compaction leads to an obvious difference of stresses. In analysing the stresses, internal pressure, soil load, traffic load and other live load, vacuum, thermal variation and the stalling force must be considered.

Figure 2(a) shows a simple model for analysing pipe stresses. It is fitted to rigid pipe, and the horizontal passive resistance of the soil is ignored. Figure 2(b) shows another model and two kinds of horizontal resistance are calculated. The load q_L is caused by traffic load and other live load and the load q_s is caused by soil above the pipe. The load form of the third model, suggested by Yemeiliyanof, is analogous to the first model, shown in Fig. 2(a). Considering the interaction between the pipe wall and the soil around the pipe, he calculated the internal forces of the pipe with the method of

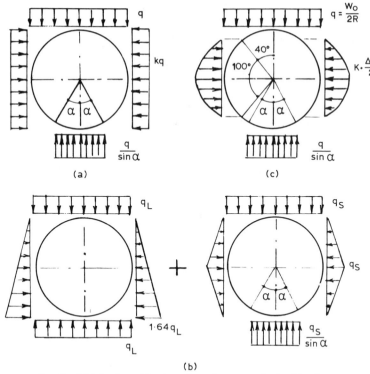

FIG. 2.   Distribution of loads and reaction around a flexible pipe.

elastic theory. Figure 2(c) shows Spangler's model. His formula for the moment is:

$$M = \frac{D_1 K_2 W_0 R}{1 + 0.061 \dfrac{E_0 R^3}{EI}} \tag{1}$$

where

D_1 = deflection lag factor,

K_2 = bedding constant for moment,

W_0 = load on pipe per unit length,

R = mean radius of pipe,

EI = stiffness of pipe wall per unit length,

E = modulus of backfill soil reaction.

Figure 3 shows four curves corresponding to the four models mentioned above in a given working condition. Observing the curves for model 1 and model 2, we find they are constant to $S(= EI/D^3)$. It is inconsistent with the

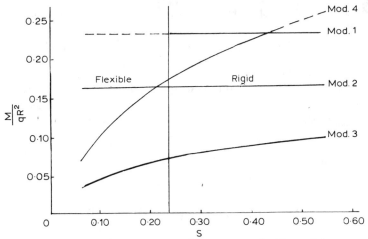

Fig. 3. The relation between bending moment and stiffness of pipe.

reality of pipe. The internal force calculated with model 3 is the lowest. As the ratio of the vertical deflection to the pipe diameter is less than 2·5%, S is about less than 0·25.

According to the results of the experiment, model 4, which is according to Spangler's formula, is usually used for flexible pipe and model 1 is used for the rigid pipe.

The stress of the pipe wall is determined by:

$$\sigma = pR/t + My/I \qquad (2)$$

where

p = internal pressure,
I = moment of inertia of the pipe wall per unit length.

3.2 The Deflection Calculation

The vertical deflection can be calculated using the formulas of the previous models. The formula for model 1 is:

$$\delta y = f(\alpha) \frac{qR^4}{EI} \qquad (3)$$

where $f(\alpha)$ = bedding factor for deflection.

Spangler's formula is

$$\delta y = \frac{D_1 K_1 W_0 R^3}{EI + 0·061 E_0 R^3} \qquad (4)$$

where K_1 = bedding constant.

The allowable deflection (δ) is chosen for various working conditions generally provided by the user. Usually (δ)/D for the buried pipe is less than 5·0%. If the pipe crosses a road (δ)/D should be less than 2·5%.

3.3 The Stability Test

Stability is also an important factor in pipe design. Due to the external pressure, buckling could occur. There are many different formulas for the buckling pressure of a buried pipe. A simple formula is:

$$q_{cr} = 3EI/R^3 \qquad (5)$$

Jan Allgood suggested a formula:

$$q_{cr} = \bar{c}M_{S}EI/D^3 \qquad (6)$$

where

$\bar{c}$ = special factor for buckling equation related to Poisson's ratio for soil,

M_S = type of soil modulus.

Equation (6) is the most useful for calculating the buckling pressure of a buried pipe.

The external pressure q is described by

$$q = q_S + q_L + q_V \qquad (7)$$

where

q_S = soil load,

q_L = traffic load and other live load,

q_V = vacuum pressure.

Assuming a k safety factor, the pressure q must be satisfied by eqn (8)

$$Kq \leq q_{cr} \qquad (8)$$

In pipe design, the shear strength of the pipe wall core must be sufficient. The long-term strength and modulus must be considered and their degradation should be further investigated.

4 ANALYSIS OF THE JOINT

In considering many operational factors, a certain relative angle of rotation in the joint must be allowed; so that it is easy to install for the pipeline, and it is well suited to hydrotesting without filling the line, etc., so we adopt the spigot-and-socket joint with double watertight packing, shown in Fig. 4.

In the structural design of the joint, in addition, the loading conditions which are carried by the pipe should be considered, we are also concerned about the additional action which results in the bending deformation of the

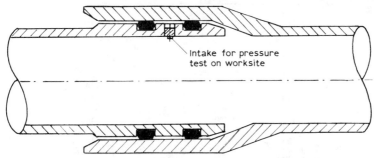

FIG. 4. Spigot-and-socket joint with double watertight packing.

joint, which is caused by non-normal loading conditions. The additional action is very important for the strength of joint. In fact, the failure of many underground pipes (e.g. cast iron pipes) results from a longitudinal crack on the socket, which is produced by the additional excessive hoop stress when the relative rotational angle in the joint is excessive.

We analyse the distribution of stresses in the joint by FEM. Consideration must be given to severe conditions which may result in giving a relative unit displacement of the ends between the spigot and socket joint.

Figure 5 shows the unit cell, on which the symmetry conditions of the system are superimposed. At the beginning of the calculation, two pairs of nodes which respectively lie in spigot and socket are contacted. This means that there are only equal normal displacements for the pair of contact points. For this purpose, a pseudo element is applied between the two contact nodes. First as the external load increases, the contact points of spigot and socket are increased. However, after the external load reaches a

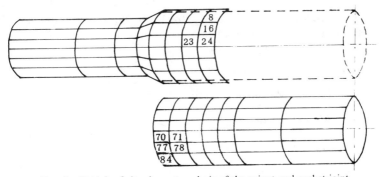

FIG. 5. Grid for finite element analysis of the spigot and socket joint.

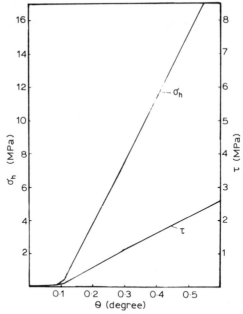

FIG. 6.   Relations between maximum hoop stresses and rotational angle of the section.

certain value, the contacted nodes are not increased. The relationships between maximum hoop stresses and the rotational angle of the section at the socket end are shown in Fig. 6.

For example, for a joint for a normal diameter 500 mm pipe, when the rotational angle of the section is 0·5 degree, the calculated results are as listed in Table 1.

Figure 7 shows the expanding process of the contacted area of spigot and socket joint. Figure 8 and Table 2 show the acting forces between spigot and socket on contact points, when the rotational angle is 0·256 degree.

TABLE 1

	Maximum hoop stress		Maximum shear stress	
	MPa	Element	MPa	Element
Socket	15·2	8	2·1	23
Spigot	−14	84	2	71

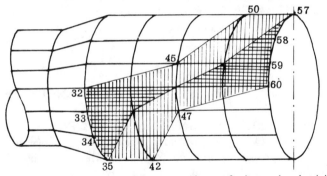

FIG. 7. Expanding process of the contacted area of spigot and socket joint.

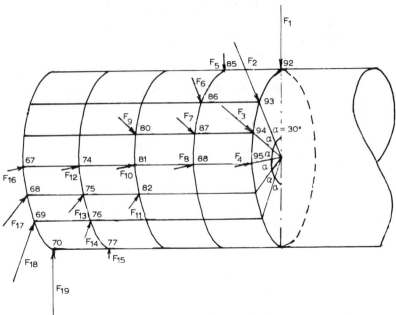

FIG. 8. Acting forces between spigot and socket on contact points.

TABLE 2

Acting forces between spigot and socket (N)

F_1	2 422	F_5	458·4	F_9	694·8	F_{13}	726·9	F_{17}	1 261
F_2	2 065	F_6	800·0	F_{10}	575·9	F_{14}	658·9	F_{18}	2 015
F_3	1 318	F_7	741·2	F_{11}	673·9	F_{15}	276·2	F_{19}	2 392
F_4	530·5	F_8	608·0	F_{12}	582·2	F_{16}	516·1		

With a suitable safety factor against additional stresses, we can lay the additional fibre on the spigot and socket joint to resist corresponding severe conditions.

5 TESTING AND STRUCTURAL PERFORMANCE OF PIPES

In order to examine the structural performance of GRP pipes with mixed wall construction, two types of test should be carried out. One is to measure the mechanical properties of all composed materials. And the other is to examine the structural performance of the pipes.

5.1 Tests for Measuring the Mechanical Properties of Composed Materials

5.1.1 To determine the elasticity constants (E_L, E_T, μ_{LT}) and strength (σ_{LB}, σ_{TB}) of GRP laminate

These tests are carried out according to GB 1447–83. The stress–strain curves for GRP laminate are shown in Fig. 9.

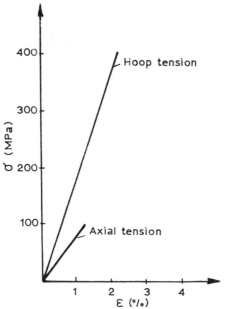

FIG. 9.   Stress–strain curves for GRP laminate.

5.1.2 To determine the mechanical properties of polymer concrete

There is no testing standard for polymer concrete. Considering many factors, such as the size of sand, the ratio of height to width and instruments, etc., we selected a specimen size to determine its tensile, compressive, bending modulus and strength. With a cast bar of the polymer concrete with diameter of the cross-section 12 mm, the shear strength is measured. Figure 10 shows the shape of specimens of polymer concrete. The effect of sand content has been studied. As sand content by weight is increased, the modulus and strength in tension change slightly, while both modulus and strength in compression increase.

For example, the test results of polymer concrete with 75% sand content are shown in Table 3. The stress–strain curves for polymer concrete are shown in Fig. 11.

The test results for polymer concrete show that the rupture takes place along the cross-section in tension and along the longitudinal profile in compression and their failure is brittle. Therefore, when there is a lack of twisted testing data we can make use of the maximum tensile stress

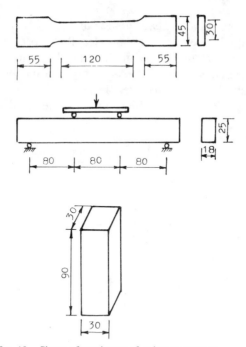

FIG. 10. Shape of specimens of polymer concrete.

TABLE 3

	Strength (MPa)	Elasticity modulus (GPa)	Poisson's ratio
Tension	10·4	8·75	0·27
Compression	64	11·8	0·4
Bending	18·5		
Torsion	10		

criterion, the maximum tensile strain criterion and Mohr's criterion to predict shear strength. That is,

$$\tau_B = \left(1 \sim \frac{1}{1 + \dfrac{\mu_c E_t}{E_c}} \right) \delta_t$$

and

$$\tau_B = \frac{\sigma_t \sigma_c}{\sigma_t + \sigma_c}$$

5.1.3 To determine the welding strength of PVC plate

Two groups of welded and unwelded PVC specimens are tested in tension. The tensile strength and modulus are determined, as shown in Fig. 12. If the minimum strength of the welded specimen is greater than 70% of the PVC itself, the weld is accepted. In fact, under service conditions, the strain of the GRP laminate is about 0·5% and the shrinkage of GRP curve

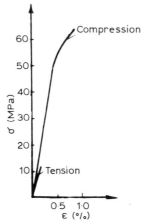

FIG. 11.   Stress–strain curves for polymer concrete.

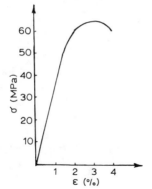

FIG. 12.   Stress–strain curve of PVC.

reaches $0.5 \sim 1.0\%$, therefore, the stress of PVC is low under service conditions.

5.2 Tests for Determining Structural Performance

5.2.1 Bending tests of long span and short span beam with sandwich structure

The sandwich beams are made of GRP laminates in top layer and bottom layer with a polymer concrete core and its build-up is similar to the pipe wall construction. This kind of specimen is used to determine the strength and modulus of GRP, shear strength of core and to inspect phenomenon of damage.

Figure 13(a) shows that shear failure occurs in the short span beam, where the oblique crack is about an angle of $45°$ to the axis of the beam and the shear stress is 5 MPa. Figure 13(b) shows that the compressive failure of the GRP top layer occurs in the long span beam. The compressive strength of GRP is about 200 MPa and the modulus is obtained from the test data by the method mentioned in Section 5.1.1.

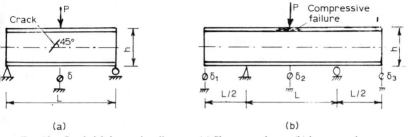

(a)   (b)

FIG. 13.   Sandwich beams bending test. (a) Short span beam; (b) long span beam.

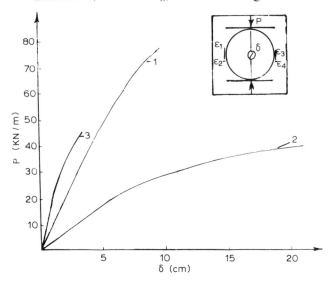

FIG. 14.   Parallel-plate external loading test.

Observing the failed specimens, we can find out if the bond between the GRP layer and the core layer is perfect.

5.2.2 Parallel-plate external loading test

Figure 14 shows the parallel-plate test. A pair of compression forces along diameter loaded a pipe ring, the results verified the relation between external load P and the deflection δ in the direction of the load. The specific stiffness of the pipe is calculated by formula

$$S = \left(\frac{EI}{D^3}\right) = 0.0186\,\frac{P}{\delta}$$

The measured value of the pipe stiffness is slightly higher than the predicted one, which indicates that the polymer concrete may play a certain role in pipe bending deformation. In addition, strain gauges were placed on the outer surface of the pipe wall at hoop and axial directions. Then, the load–strain curve is obtained. It is analogous to the load–deflection curve. Two types of failure occur in the ring samples with various stiffnesses. As the specific stiffness increases, the shear failure of the core layer occurs, where the oblique crack is about an angle of 45° to the direction of load and the shear stress is 5 MPa. Another is the compressive failure of the GRP layer with a compressive strength of 220 MPa. Yet, before failure takes

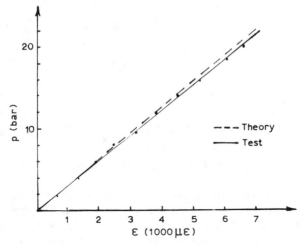

FIG. 15.   Relation between internal pressure and hoop strain.

place, the integral bond between the GRP layer and the core layer is present without cracking, delamination, etc.

Figure 14 shows three load–deflection curves of mixed wall pipes, 1 and 2 are the compressive failure of the GRP layer, 3 is the shear failure of the core of the polymer concrete. The deflection of curve 3 is the smallest. In the design, we should avoid the occurrence or core shear failure.

5.2.3 Internal pressure and burst test

This is a short-term destructive test carried out on a full-scale pipe segment, while the strain gauges are placed on the outer surface of the pipe along the hoop direction at four positions.

First, according to the requirement, we retain the testing pressure for at least 5 min against leak tightness. And we increase the pressure gradually until the pipe fails.

Figure 15 shows the relation between hoop strain and internal pressure. The result shows that when the error between the test value and that theoretically predicted is less than 4%, the maximum pressure reaches the design level. It is full proof that the manufacturing process mentioned above can ensure the structural performance of the pipe.

5.2.4 Long-term flexural test

The corresponding long-term test for specific stiffness is carried out by parallel-plate loading. A ring sample is initially deflected at a level of

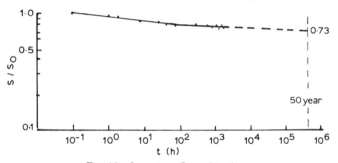

FIG. 16.   Long-term flexural testing.

$2.5\%D$. Then, we keep the deflection constant and record the variation of load as a function of time. Because the ring sample behaves both elastically and linearly in the loading region, the equation $S = 0.0186P/\delta$ is still correct for the relation of load P and specific stiffness S. Using a log–log plot of (S/S_o) versus t, we can fit two segments of straight lines and the experimental points are close to them. By extrapolating to 50 years, we can get $(S/S_0) = 0.73$. This result is shown in Fig. 16.

FIG. 17.   GRPPC pipes with 1000 mm diameter, 6 m long.

5.2.5 Joint watertightness test

To examine the watertightness of the joints, the hydrotesting is carried out on the region with two watertight packings. The test result shows that the internal pressure is greater than 2·5 WP.

6 CONCLUSION

Based on the analysis and tests, we can obtain the results:

(1) A new type of GRP pipe, that is named GRPPC pipe, has a reasonable wall build-up and its material is fully utilized.

(2) According the working conditions presented by the user, we can design and manufacture GRPPC pipes that are reliable.

(3) In the pipe design, Spangler's model fits analysis of the flexible pipes and the simple model fits analysis of the rigid pipes.

(4) The cost of GRPPC is acceptable and is lower than other GRP pipes.

BIBLIOGRAPHY

1. ZHU YILING, Discussion on large size GRP pipelines. *GRP/Composite* (1987) 3.
2. KANONA, M. A. and CURRIE, R. D., Structural Performance of Centrifugally Cast Glass Reinforced Plastic Pipes (1987) 7.
3. ROLSTON, J. ALBERT and GILBU, AGNAR, *Design, Production and Testing of Large Diameter Reinforced Plastic Underground Effluent Pipe*. SPI, 28th, Section 6-A, 1973, 10 pp.

35

Optimum Design of Circular Cylindrical Pipes Combined with GFRP/Steel

ANLIN YAO

Department of Mechanical Engineering,
Southwestern Petroleum Institute, Nanchong,
Sichuan, People's Republic of China

ABSTRACT

This paper puts the emphasis on studying the optimum design of a thick-walled circular cylindrical pipe combined with GFRP/steel (called 'combined pipe' below) and subjected to internal high pressure. Firstly, considering a few of the boundary conditions, the formulas of three principal stresses in the wall of the combined pipe in the state for plane strain are derived by elastic theories. Secondly, a mathematical model for the optimization of the combined pipe is set up, in which the total wall thickness of the combined pipe, the wall thickness of GFRP out-sleeve and the wound angle of the glass fibre are chosen as design variables, the material cost of the combined pipe per unit length is taken as an objective function, and the strength requirement of steel pipe and the deformation limitation of GFRP are considered as constraints as well. Finally, the model for optimization is solved by using the Complex Method. Also, the author discusses the influences, caused by the changes of inner diameter of the combined pipe, the internal pressure as well as the strength class of steel pipe, on the optimum values of design variables. A numerical example shows that, under the same pressure, the material cost of the combined pipe is less than that of a pure steel pipe, and the weight evidently lighter.

INTRODUCTION

Recently, because of the need for anticorrosion properties of oil and gas pipelines buried underground, a method, in which glass fibre is wound on a steel pipe, has been utilized to prevent further corrosion of locally corroded pipelines for collecting and transporting high pressure oil and gas, and reinforce them. This practice may not only save a lot of high-quality steels, but also prevent outside corrosion of the buried pipes. This paper makes a principal study of the optimum design of a thick-walled circular cylindrical pipe that is combined with GFRP/steel and subjected to internal high pressure.

STRUCTURAL MODEL AND ANALYSIS

Figure 1 shows the cross-sectional shape of a circular cylindrical pipe combined with GFRP/steel. The inner layer in the combined pipe is an ordinary steel pipe, and the outer layer is a GFRP pipe, which is wound with glass fibre cladding with the winding method as shown in Fig. 2. In an established cylindrical coordinate system, let x, r and θ be measured in the axial, normal and circumferential directions of the combined pipe respectively and the origin of coordinates be located at the central point of the cross-section of the combined pipe. Assuming that there are not sealed caps at the two ends of the combined pipe and its axial deformation is restricted, then the mechanical model for the combined pipe could be undertaken as an axisymmetrical stress question in the state for plane strain. From Ref. 1, the expressions for the normal stresses in the walls of the lining steel pipe and GFRP out-sleeve may be assumed to be

$$\sigma r = \frac{A}{r^2} + 2C \qquad \sigma_\theta = -\frac{A}{r^2} + 2C \qquad (1)$$

and

$$\sigma r' = \frac{A'}{r^2} + 2C' \qquad \sigma_\theta' = -\frac{A'}{r^2} + 2C' \qquad (2)$$

Utilizing the boundary conditions

$$(\sigma r)_{r=a} = -p \qquad (\sigma r')_{r=b} = 0 \qquad (\sigma r)_{r=d} = (\sigma r')_{r=d} \qquad (3)$$

three equations for the undetermined parameters A, C, A', C' in eqns (1) and (2) are obtained as follows

$$\frac{A}{a^2} + 2C = -p \qquad \frac{A'}{b^2} + 2C' = 0 \qquad \frac{A - A'}{d^2} + 2(C - C') = 0 \qquad (4)$$

In order to solely determine the four undetermined parameters above, it is

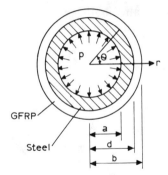

FIG. 1. The cross-sectional shape of a combined pipe.

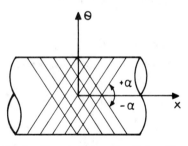

FIG. 2. The winding mode of glass fibre cladding.

necessary that a deformation conformability equation is added. Because it was assumed that lining steel pipe and GFRP out-sleeve are in a state of 'perfect contact' on their contact layer, they should meet the following relation of deformation conformability on the contact layer:

$$(\varepsilon_\theta)_{r=d} = (\varepsilon'_\theta)_{r=d} \tag{5}$$

where $(\varepsilon_\theta)_{r=d}$ denotes the circumferential strain of lining steel pipe on the contact layer. From Ref. 1, it can be expressed as

$$(\varepsilon_\theta)_{r=d} = \frac{1+\mu}{E}\left(2(1-2\mu)C\frac{A}{d^2}\right) \tag{6}$$

and $(\varepsilon'_\theta)_{r=d}$ shows that for the GFRP out-sleeve, using the stress–strain relation[2] of an orthotropic body, its expression could be written as

$$(\varepsilon'_\theta)_{r=d} = \lambda\left(2C' - \frac{A'}{d^2}\right) + \eta\left(2C' + \frac{A'}{d^2}\right) \tag{7}$$

in which

$$\lambda = \frac{S_{\theta\theta} - CxrS_{\theta\theta}Sxr + CxrS_\theta xSr_\theta}{1 - CxrSxr - Cx_\theta S_\theta x} \tag{8}$$

$$\eta = \frac{S_\theta r - CxrSxrS_\theta r + CxrS_\theta xSrr}{1 - CxrSxr - Cx_\theta S_\theta x}$$

Here those characters S and C having subscripts indicate softness coefficients and stiffness coefficients of the orthotropic body respectively, and the subscripts denote the coordinate directions. These coefficients are functions of elastic moduli and Poisson's ratio in the principal axes directions of the GFRP out-sleeve and their expressions are[2]

$$S_{\theta\theta} = 1/E_\theta$$
$$S_\theta r = -v_\theta r/E_\theta = -vr_\theta/Er$$
$$S_\theta x = -v_\theta x/E_\theta = -vx_\theta/Ex$$
$$Srr = 1/Er \tag{9}$$
$$Sxr = -vxr/Ex$$
$$Sr_\theta = S_\theta r$$

$$Cx_\theta = \frac{E_\theta Ex}{\Delta}(Ervx_\theta + Exvxrvr_\theta)(Er - E_\theta v^2 r_\theta)$$

$$Cxr = \frac{ErEx}{\Delta}(Exvxr + E_\theta vr_\theta vx_\theta)$$

$$\Delta = (Er - E_\theta v^2 r_\theta)(ErEx - E_\theta Exv^2 r_\theta \tag{10}$$
$$- E_\theta Erv^2 x_\theta - Ex^2 v^2 xr - 2E_\theta Exvr_\theta vx_\theta vxr)$$

From eqns (5)–(7), we have

$$\frac{1+\mu}{E}\left(2(1-2\mu)C - \frac{A}{d^2}\right) = (\eta - \lambda)\frac{A'}{d^2} + 2(\eta + \lambda)C' \tag{11}$$

Simultaneously solving eqns (4) and (11), the undetermined parameters could be derived as follows:

$$A = -\frac{\zeta a^2 d^2}{\xi}p \qquad\qquad C = \frac{1}{2}\left(\frac{\zeta d^2}{\xi} - 1\right)p$$

$$A' = -\frac{b^2 d^2[\xi - \zeta(d^2 - a^2)]}{\xi(b^2 - d^2)}p \qquad C' = \frac{d^2[\xi - \zeta(d^2 - a^2)]}{2\xi(b^2 - d^2)}p \tag{12}$$

in which

$$\zeta = \frac{(1+\mu)(1-2\mu)}{E}(b^2 - d^2) - b^2(\eta - \lambda) + d^2(\eta + \lambda)$$

$$\xi = \frac{1+\mu}{E}(b^2 - d^2)[a^2 + d^2(1-2\mu)] + b^2(a^2 - d^2)(\eta - \lambda) \tag{13}$$
$$+ d^2(d^2 - a^2)(\eta + \lambda)$$

E and μ are the elastic modulus and Poisson's ratio of steel respectively. Those elastic coefficients in the principal axes directions of the GFRP out-sleeve, included in eqns (9) and (10), depend on the elastic properties of the unidirectional GFRP layer and the wound angle α of fibre, and they could be calculated on the basis of formulas proposed in Ref. 3.

Substituting eqn (12) into eqns (1) and (2), yields the formulas for normal stresses in the combined pipe, i.e.

$$\sigma r = \left[\frac{\zeta d^2(r^2 - a^2)}{r^2 \xi} - 1\right] p$$

in steel layer (14)

$$\sigma_\theta = \left[\frac{\zeta d^2(r^2 + a^2)}{r^2 \xi} - 1\right] p$$

$$\sigma r' = \frac{d^2(r^2 - b^2)[\xi - \zeta(d^2 - a^2)]}{r^2 \xi(b^2 - d^2)} p$$

in GFRP layer (15)

$$\sigma_\theta' = \frac{d^2(r^2 + b^2)[\xi - \zeta(d^2 - a^2)]}{r^2 \xi(b^2 - d^2)} p$$

where p denotes the uniform pressure acting on the inner wall of the combined pipe.

According to the relevant equations[1,2] on the axisymmetrical question, it would be easy to write the formulas of axial stresses in the walls of the lining steel pipe and GFRP out-sleeve as follows

$$\sigma x = \frac{\mu}{1 - \mu}(\sigma_\theta + \sigma r)$$

in steel layer (16)

$$\sigma x' = \phi \sigma_\theta' + \psi \sigma r'$$

$$\phi = \frac{Cx_\theta S_{\theta\theta} + Cxr Sr_\theta}{1 - Cxr Sxr - Cx_\theta S_\theta x}$$

in GFRP layer (17)

$$\psi = \frac{Cx_\theta S_\theta r + Cxr Srr}{1 - Cxr Sxr - Cx_\theta S_\theta x}$$

OPTIMIZATION

In optimizing the combined pipe, the total wall thickness t of the combined pipe, the wall thickness t_G of the GFRP out-sleeve and the wound angle α of the glass fibre cladding are chosen as design variables, i.e.

$$X = (\alpha, t, t_G)^T \tag{18}$$

Let $\tilde{t} = t/D_i$, $\tilde{t}_G = t_G/t$, we can obtain a non-dimensional form of the design variables as

$$\tilde{X} = (\alpha, \tilde{t}, \tilde{t}_G)^T \tag{18'}$$

The objective function is the material cost of the combined pipe per unit length and its expression is

$$F = \pi\{(t^2 + t_G^2 - 2tt_G + D_i t - D_i t_G)\gamma_s V_s + (D_i t_G + 2tt_G - t_G^2)\gamma_G V_G\} \quad (19)$$

in which D_i denotes an inner diameter of the combined pipe; γ_G, V_G, γ_s and V_s represent the density and price of GFRP and those of steel respectively. If formula (19) is divided by the material cost of a pure steel pipe subjected to the same pressure, the objective function could be also written in a non-dimensional form as follows:

$$\tilde{F} = D_i^2\left\{[\tilde{t}^2 + (\tilde{t}\tilde{t}_G)^2 - 2\tilde{t}\tilde{t}_G + \tilde{t} - \tilde{t}\tilde{t}_G]\right.$$

$$\left. + [\tilde{t}\tilde{t}_G + 2\tilde{t}^2\tilde{t}_G - (\tilde{t}\tilde{t}_G)^2]\frac{\gamma_G V_G}{\gamma_s V_s}\right\}\bigg/(t_p^2 + D_i t_p) \quad (19')$$

here t_p is the wall thickness of a pure steel pipe and it is a positive real root of the following fourth order equation

$$A_0 t_p^4 + B_0 t_p^2 + C_0 = 0 \quad (20)$$

in which

$$A_0 = p^2 + ep + f \qquad e = -\frac{2k^2 - 2k - 1}{k^2 - k + 1}p$$

$$B_0 = \frac{D_i^2}{2}(p^2 - f) \qquad f = p^2 - \frac{(\sigma)^2}{k^2 - k + 1} \quad (21)$$

$$C_0 = \frac{D_i^4}{16}(p^2 - ep + f) \qquad k = \frac{\mu}{1 - \mu}$$

(σ) is the allowable stress of the steel in the design; the other symbols are the same as above.

In this subject, the strength constraint of lining steel pipe and the deformation limitation of GFRP out-sleeve are simultaneously considered, i.e. the strength of lining steel pipe must be submitted to Mises's yield criterion[4] and the maximum deformation of the GFRP out-sleeve adopted to the recommendations of the strain limited (ε) criterion.[5] Consequently, the constraint equations are as follows:

$$(\sigma r - \sigma_\theta)^2 + (\sigma_\theta - \sigma x)^2 + (\sigma x - \sigma r)^2 \leq 2Y^2 \quad (22)$$

$$\left[1 - \frac{\zeta}{\xi}(d^2 - a^2)\right]\left[\lambda\frac{(b^2 + d^2)}{(b^2 - d^2)} - \eta\right]p \leq \bar{\varepsilon}_\theta \quad (23)$$

in which

$$\sigma r = -p \qquad a = D_i/2$$

$$\sigma_\theta = \left(\frac{2d^2\zeta}{\xi} - 1\right)p \qquad b = a + D_i\tilde{i}$$

$$\sigma x = \frac{\mu}{1-\mu}(\sigma_\theta + \sigma r) \qquad d = b - D_i\tilde{i}\tilde{i}_G$$

(24)

Y denotes the yield stress of steel and $\bar{\varepsilon}_\theta$ the allowable strain of the GFRP out-sleeve in the direction of the circumference.

In order to solve the model of optimization of the combined pipe proposed here, a direct method in the optimization approach—Complex Method developed by Box (see Ref. 6)—is adopted.

NUMERICAL EXAMPLE AND DISCUSSION

To design a combined pipe that has an inner diameter of $D_i = 0.300$ m and is subjected to an internal pressure of $p = 10.0$ MPa, it is necessary to use the parameters for the physics and mechanics of steel and GFRP shown in Table 1. The upper and lower values for the design variables are as follows:

$$(\alpha, \tilde{i}, \tilde{i}_G)^u = (1.0472, 0.20, 0.80)$$

$$(\alpha, \tilde{i}, \tilde{i}_G)^l = (0.5236, 0.02, 0.20)$$

Utilizing those parameters given in Table 1, the optimum values of the design variables have been calculated as

$$(\alpha^*, \tilde{i}^*, \tilde{i}_G^*) = (0.6574, 0.024\,67, 0.2007)$$

TABLE 1
Parameters for the properties of physics and mechanics of GFRP and steel

		Young's moduli (GPa)		Poisson's ratio		Shear moduli (GPa)	Density (kg/m³)	Allowable strain of GFRP and yield limit of steel
GFRP	E_L	25.5	v_{LT}	0.20				
	E_T	11.77	v_{Lr}	0.10	$G_{LT} = 2.845$	2000.0	$\bar{\varepsilon}_\theta = 0.0015$
	E_r	3.826	v_{Tr}	0.10			
Steel		206.0		0.30		78.48	7850.0	$Y = 240.0$ MPa

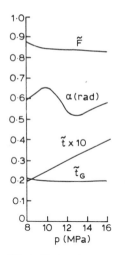

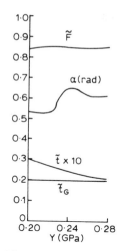

FIG. 3.   Effect of internal pressure on the
optimum design variables.

FIG. 4.   Effect of the glass strength of a
lined steel pipe on the optimum design
variables.

by a computer when the price ratio between GFRP and steel was taken as
5·0. The cost of a combined pipe based on the optimal design parameters is
0·844 17 times of that of a pure steel pipe, and the relative weight of a
combined pipe per unit length is only 0·677 54.

In addition, the optimum values of the design variables and the relative
cost of the combined pipe, which are related to some of the inner diameters
of the pipe, internal pressure as well as glass strength of a lined steel pipe,
have been calculated by means of the optimizing program developed in this
paper. The results show, that with other known parameters unchanged, the
variation of the inner diameter of the pipe would slightly affect the optimum values of the design variables, and the variation of internal pressure
and strength class of lining steel pipe could influence the optimum values
of design variables. The curves showing the relations between the design
variables or objective function and internal pressure as well as glass
strength of a lined steel pipe have been drafted in Figs 3 and 4. The varying
tendencies of curves $\tilde{t}$ in Figs 3 and 4 show that the total wall thickness of
the combined pipe decreases as the strength class of lining steel pipe
increases, and itself, approximately, increases linearly as the internal
pressure increases. The $\tilde{t}_G$ curves in the two figures indicate that the relative
wall thickness of the GFRP out-sleeve varies slightly, and gradually reaches
the lower value of $\tilde{t}_G$, even though both the internal pressure and strength

class of the lining steel pipe are varied. This is because the cost of GFRP per unit weight is much more than that of steel at present.

CONCLUSION

The formulas derived in this paper could be used for the exact analysis of a thick-walled circular cylindrical pipe combined with GFRP/steel. In practical design, the reasonable structural parameters can be provided by using the program on optimum design of the combined pipe. It is fast in calculation, and flexible and convenient to use.

ACKNOWLEDGEMENTS

The author is grateful to Professor Xiao F. C. and Associate Professor Cui Lan of SWPI for their sincere encouragement, and to Mr Huang Kun for tracing all the diagrams.

REFERENCES

1. Xu Zhilun, *Elastic Mechanics*. 2nd edn, Press of People's Education of China, Beijing, 1982 (in Chinese).
2. Cai Siwei, *Structural Mechanics of Composite Materials*. Press of People's Communications of China, Beijing, 1987, pp. 8–13 (in Chinese).
3. Liu Xili and Wang Bingquan, *Basis of Composite Materials Mechanics*. Press of Arch. Industry of China, Beijing, 1984, pp. 244–5 (in Chinese).
4. Archer, Robert R. *et al.*, *An Introduction to the Mechanics of Solids*. 2 edn, 1972, Chinese Translation, Press of People's Education of China, Beijing, 1980, pp. 198–203.
5. Wang Jianxin, Design of corrosion resistant GFRP, *Journal of Engineering Plastics Applications* (1) 1987, 24–8 (in Chinese).
6. Liu Xiashi, *Optimum Design of Engineering Structures—Theorems, Methods and Applications*. Press of Science, Beijing, 1984, pp. 331–3 (in Chinese).

36

A CAD/CAM Program of Filament Winding on a Cylindrical Pipe

Li Xian-Li

Wuhan University of Technology,
Wuhan, Hubei, People's Republic of China

ABSTRACT

A CAD/CAM general program of filament winding on a cylindrical pipe, in which length is limited, is given in this paper. The parameters of the cylindrical pipe, i.e. radius R, length L, width of the fibre wrap b and the filament winding pattern, can be achieved on account of the winding angle, α_0 satisfying mechanical properties and the friction coefficient, μ, friction between the resin impregnated filament and the mandrel or other fibre layers. If necessary, the integral circulatory filament track can be displayed or drawn on the computer, and the whole winding process can also be shown in order to see how the filament is covering the mandrel. These could be the basic data for processing design.

INTRODUCTION

The computer is widely used for many scientific areas with the vigorous development of science and technology. It assists people in drafting, designing, manufacture and analysis. The filament winding regularity on a cylindrical pipe can be analysed with a computer, and a CAD/CAM program is given in this paper.

There are two main types of filament winding on a cylindrical pipe: continuous winding limitless pipe and length-limited pipe. The advantage of the first method is high productivity and suitability for mass production, but its winding regularity should be designed specifically, otherwise the coupling effect may often occur as well as torsional deformation under normal pressure. The resin base can be broken, thereby reducing the ability of leakage resistance for the pipe. The production efficiency of the second method is lower than the first one, but it is flexible and the mechanical properties of the wound pipe are symmetrical. There are no torsional deformations under normal pressure, thus the leakage decreases. The second method can be widely employed in industrial departments. The CAD/CAM program of the limited-length pipe only is discussed in this paper.

FILAMENT PATH EQUATIONS FOR NON-GEODESIC STABLE WINDING

1 Filament Path Equations

According to Ref. 1, it is known that filament path equations for non-geodesic stable windings on the limited-length pipe are:

$$X = \frac{R}{\mu}\left[\frac{1}{\sin \alpha_0} - \text{Ch}\left(\mu\theta + \ln \text{tg}\frac{\alpha_0}{2}\right)\right] \qquad \theta = \frac{1}{\mu}\ln\left(\text{tg}\frac{\alpha}{2}\bigg/\text{tg}\frac{\alpha_0}{2}\right) \quad (1)$$

where R is the radius of the cylindrical pipe; X is the axial length for non-geodesic stable winding; θ is the fibre wrap angle; α_0, α are initial and current winding angle, respectively; μ is the friction coefficient between the filament and mandrel or other fibre layers.

Because $\alpha = 90°$ at the filament back winding, thus the length L_N and the fibre wrap angle θ_N on the non-geodesic winding section are:

$$L_N = \frac{R}{\mu}\left(\frac{1}{\sin \alpha_0} - 1\right) \qquad \theta_N = -\frac{2}{\mu}\ln\left(\text{tg}\frac{\alpha_0}{2}\right) \quad (2)$$

2 Analysis of Parameter Selection

According to eqn (2), we have

$$\alpha = 2\,\text{tg}^{-1}(e^{-\mu\theta_N/2}) \quad (3)$$

Substitute the first formula of eqn (2) into formula (3), we obtain:

$$L_N = \frac{R}{\mu}\left[\text{Ch}\left(\frac{\mu\theta_N}{2}\right) - 1\right] \quad (4)$$

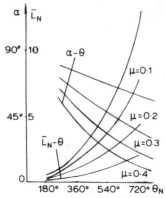

FIG. 1. The relationship between initial winding angle, α_0, length L_N, friction coefficient, μ and the θ_N, the fibre wrap angle.

Let dimensionless parameter $\bar{L}_N = L_N/R$, then

$$\bar{L}_N = \frac{1}{\mu}\left[\text{Ch}\left(\frac{\mu\theta_N}{2}\right) - 1\right] \tag{4a}$$

The initial winding angle α_0 in the non-geodesic winding portion is equal to the winding angle α_0' in the geodesic portion on account of the continuous winding condition, but the α_0' might influence the mechanical properties of the pipe, for example α_0' is equal to 54·75 under internal pressure. Thus selecting adaptable μ and θ_N, the α_0 might be located within predetermined limits.

The curves relating α_0, $\bar{L}_N$ and μ, θ_N are presented in Fig. 1. It is known from Fig. 1 that α_0 decreases and $\bar{L}_N$ increases, i.e. the length of the backstroke portion increases, with μ increasing when the fibre wrap angle θ_N is given. But if μ stays constant, $\bar{L}_N$ increases and α_0 decreases following θ_N increasing.

FILAMENT PATH EQUATION FOR GEODESIC WINDING

1 Filament Path Equation

The helical curve on the cylindrical pipe is geodesic. Its path equation is

$$X = h\theta_G \tag{5}$$

where θ_G is the fibre wrap angle; h is the axial displacement in the unit wrap angle, and designating s as thread pitch, then h is equal to $s/2$.

Li Xian-Li

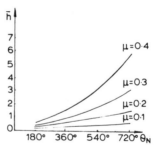

FIG. 2. The relationship between axial displacement in the unit wrap angle, $\bar{h}$, the friction coefficient, μ and θ_N the fibre wrap angle.

2 Analysis of Parameter Selection

Because the winding angle at the junction of the geodesic winding with the non-geodesic winding is equal, thus

$$\text{tg}\,\alpha_0 = R/h \tag{6}$$

as a result,

$$h = Rc\,\text{tg}\,\alpha_0 = RSh(\mu\theta_N/2) \tag{7}$$

Assume dimensionless quantity $\bar{h} = h/R$, then

$$\bar{h} = c\,\text{tg}\,\alpha_0 = Sh(\mu\theta_N/2) \tag{7a}$$

The relationship between $\bar{h}$ and μ, θ_N is illustrated in Fig. 2. It is observed from Fig. 2 that h increases along with increasing μ and θ_N, i.e. the axial displacement on the unit wrap angle increases for the geodesic winding portion.

The length of the geodesic portion is

$$L_G = R\theta_G Sh(\mu\theta_N/2) \tag{8}$$

The corresponding dimensionless quantity $\bar{L}_G$ is

$$\bar{L}_G = L_G/R = \theta_G Sh(\mu\theta_N/2) \tag{8a}$$

ANALYSIS OF THE WINDING REGULARITY

The total length of the pipe consisted of one geodesic portion and two non-geodesic portions and is

$$L = 2L_N + L_G = 2R[\text{Ch}(\mu\theta_N/2) - 1]/\mu + R\theta_G Sh(\mu\theta_N/2) \tag{9}$$

The dimensionless quantity $\bar{L}$ is

$$\bar{L} = L/R = 2[\text{Ch}(\mu\theta_N/2) - 1]/\mu + \theta_G Sh(\mu\theta_N/2) \tag{9a}$$

$\bar{L}$ is only a function of θ_N and θ_G when μ is given.

The turning angle of the mandrel during the fibre pay-out eye traversing forth and back once is

$$\theta'_N = 2(\theta_N + \theta_G) \tag{10}$$

Designating $\theta_N = 2p\pi$, $\theta_G = 2m\pi$, then the number of winding turns for non-geodesic and geodesic are described by p and m, respectively. Usually p and m are fractional expressions.

Let

$$S_0 = 2(p + m) = M/n \tag{11}$$

where n is an equidistributed number of cylinders; M is a positive integer.

The winding pattern on the mandrel and the number of crosspoints between the fibres can be described by S_0.

Transforming formula (9), we get

$$\bar{L} = 2[\text{Ch}\,(\mu p\pi) - 1]/\mu + 2m\pi Sh(\mu p\pi) \tag{12}$$

$\bar{L}$ is a function of p and m, their relationships are shown in Fig. 3. It is observed from Fig. 3 that the effect of p on $\bar{L}$ is greater than that of m. A few patterns are shown in Fig. 4 for same S_0 but different combinations of p and m. The number of filament winding turns on the mandrel and the number of cross-points between the fibres are the same for the same given S_0, but L is different and the distance between the winding turns is also unequal in each case. The winding patterns are different. However, the value of α_0 is dictated by the mechanical properties of the pipe, and the value of μ is concerned with the fibre and resin employed and the material of the mandrel for the practically wound pipe. Finally α_0 and μ are known. Because

$$p = -\ln(\text{tg}\,\alpha_0/2)/(\mu\pi) \tag{13}$$

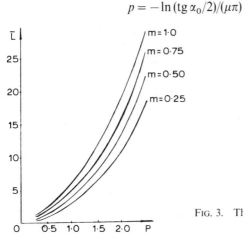

FIG. 3.   The relationship between $\bar{L}$, p and m.

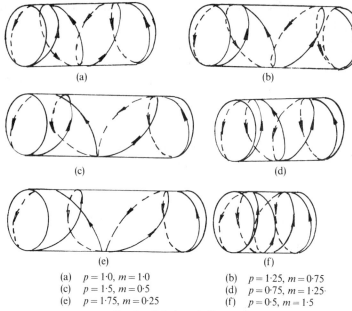

(a) $p = 1\cdot0,\ m = 1\cdot0$	(b) $p = 1\cdot25,\ m = 0\cdot75$
(c) $p = 1\cdot5,\ m = 0\cdot5$	(d) $p = 0\cdot75,\ m = 1\cdot25$
(e) $p = 1\cdot75,\ m = 0\cdot25$	(f) $p = 0\cdot5,\ m = 1\cdot5$

FIG. 4.   Relative winding angles.

thus the value of p is also known for the practical pipe. S_0 and $\bar{L}$ only depend on m. As a consequence, we can describe the fibre winding pattern by S_0.

When given the parameters L and R of the cylindrical pipe, we first calculated L, followed by determination of p and m on the basis of mechanical properties and experiment, respectively, moreover p is obtained from formula (13) and m on account of the transitive formula of eqn (12)

$$m' = \frac{\bar{L} - 2[\text{Ch}\,(\mu p\pi) - 1]/\mu}{2\pi Sh(\mu p\pi)} \tag{14}$$

and obtain S_0' as follows from $S_0' = 2(p + m')$ and make $S_0 = M/n$ approximately equal to S_0'.

The value of m can be obtained from

$$m = M/(2n) - p \tag{15}$$

The length L of the pipe might change because of the difference between m and m'. Substituting formula (15) into formula (12) and calculating $\bar{L}'$, we get the difference $\overline{\Delta L}$ between $\bar{L}'$ and the given $\bar{L}:\Delta L = \bar{L}' - \bar{L}$.

Assuming the fibre bandwidth as bm and its dimensionless $\overline{bm}$ is

$$\overline{bm} = bm/R \qquad (16)$$

In order to cover the whole surface of the mandrel by the fibres, the following formula must be satisfied:

$$\theta_n = \theta'_n \pm \sec \alpha_0/n \qquad (17)$$

where θ_n is the turning angle of the mandrel for a back and forth traverse of

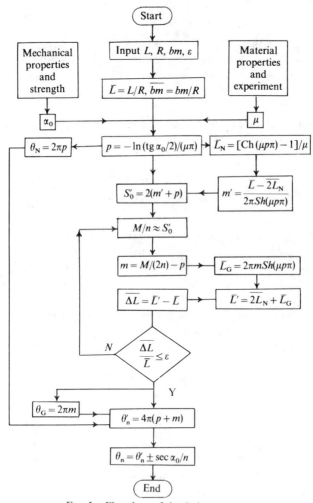

FIG. 5.   Flowchart of the design process.

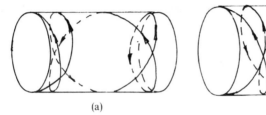

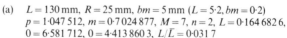

(a)   (b)

(a) $L = 130$ mm, $R = 25$ mm, $bm = 5$ mm ($L = 5·2, bm = 0·2$)
      $p = 1·047\,512$, $m = 0·7\,024\,877$, $M = 7$, $n = 2$, $L = 0·164\,682\,6$,
      $0 = 6·581\,712$, $0 = 4·413\,860\,3$, $L/\bar{L} = 0·031\,7$
(b) $L = 160$ mm, $R = 28$ mm, $bm = 10$ mm ($L = 5·714\,285\,7$, $bm = 0·357\,143$)
      $p = 1·047\,512$, $m = 0·785\,821\,1$, $M = 11$, $n = 3$, $L = 0·020\,439\,6$,
      $0 = 6·581\,712$, $0 = 4·937\,459\,5$, $L/\bar{L} = 0·003\,58$

FIG. 6.   Winding examples showing the pattern for back and forth traverse of the pay-out eye.

the pay-out eye; the plus and minus indicate advance and hysteresis of the fibres, respectively, the minus is usually used; implication of other sign is mentioned above.

A CAD/CAM PROGRAM OF WINDING PIPE

A CAD/CAM program of filament winding pipe has been achieved. The filament winding regularity on the limited-length cylindrical pipe can be displayed on the screen of the computer to facilitate the designer to select, revise and design. A flowchart of the design system is currently implemented as shown in Fig. 5.

Two winding examples are presented in Fig. 6, which show the pattern for a back and forth traverse of the pay-out eye.

The whole winding process is illustrated in Fig. 7.

CONCLUSION

1   This program is a general one. Given the parameters of the cylindrical pipe L and R, the fibre bandwidth bm, the winding angle α_0 determined from mechanical properties, the experimental friction coefficient μ, then the winding pattern can be rapidly determined. With the turning angles θ_N, θ_G, the parameters M, n and

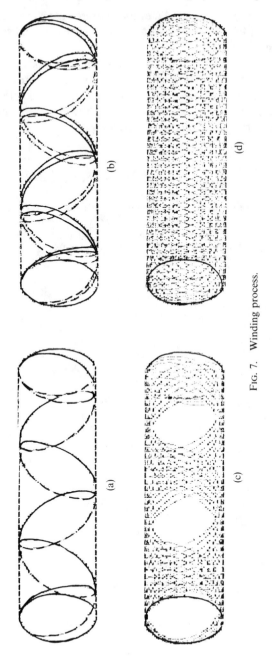

FIG. 7. Winding process.

the revisionary length ΔL of the mandrel, the winding pattern for a back and forth traverse of the pay-out eye can be displayed for the designer's selection.

2   This program can show the integral winding process and the picture of covering the surface of the mandrel with the fibres. Because of use of the axonometric picture, the designer can see the number of crosspoints and its distribution and the decorative pattern of the mandrel fully covered with the fibres.

REFERENCES

1. HUANG YU-SHENG, Non-geodesic winding calculation on the limited-length cylindrical pipe. *Glassfiber Reinforced Plastics*, (4) (1984) 36–42.
2. LI XIAN-LI and LIN DAO-HAI, Non-geodesic winding equations on a general surface of revolution. *ICCM6, ECCM2*, (1) (July 1987) 1.152–1.160.

37

Steady State Creep and Vibrocreep in Orthotropic Composite Materials

R. Pyrz

*Institute of Mechanical Engineering, University of Aalborg,
Pontoppidanstraede 101, 9220 Aalborg Ø, Denmark*

ABSTRACT

A general constitutive equation for steady-state creep deformation is presented based on the concept of dissipation potential. The material parameters required to characterize the stationary static creep behaviour of an orthotropic composite are obtained from the unidirectional tension creep tests and biaxial tension–shear creep tests performed on a glass–xylok woven fabric composite. If an oscillatory load is superimposed on a sustained loading the steady creep rate changes depending on the oscillatory frequency. This effect is shown by conducting biaxial tests with the shear force possessing both static and dynamic components. In order to describe the alteration of the creep rate in the presence of dynamic fluctuations the static dissipation potential is modified to take account of the frequency effects.

INTRODUCTION

The expanding use of composite materials in recent years has resulted in an increase in the number and kinds of characterization tests being performed. Well-defined procedures are used for some types of properties determination of these materials, such as tensile, compressive and shear testing for stiffness and strength determination, fatigue or impact. It is well known[1,2] that polymer-based composites exhibit time-dependent phenomena which are reflected in the form of a continuous increase in deformation as time progresses. Creep in composites is a complicated phenomenon which is manifested by the apparent anisotropy and the local

657

heterogeneity of the material. In order to describe the time-dependent behaviour Brinson and Dillard[3,4] have developed the time–temperature–stress superposition principle to evaluate the long-term compliance data. Another formulation has been proposed by Shapery[5,6] where a single-integral, nonlinear constitutive equation is derived from the thermodynamic theory of irreversible processes. Schaffer and Adams[7] have proposed a micromechanical analysis of composites incorporating time-dependent effects. Only the individual viscoelastic constituent properties need to be determined experimentally and a finite-element model is capable of simulating a unidirectional composite subjected to any combination of longitudinal and transverse loadings. An alternative approach has been proposed by Pyrz[8,9] in order to describe the secondary creep period. The creep rate is derived from a dissipation potential in terms of stress deviator components taking into account a material symmetry constraint.

In practice many composite structures operate not only under the static loading conditions but at the same time are subjected to dynamic loads during the life-time. A pressure vessel is a typical example where the material is subjected to a large mean load and superimposed dynamic pressure fluctuations caused by a supplying compressor and a receiver. It is therefore important to know the influence of fluctuations on different properties such as strain rate, life-time, damage nucleation and fracture. Creep deformation is manifested by two physically different micromechanisms, i.e. matrix flow and internal flaw formation and growth, which in effect leads to an alteration of short-term strength characteristics.[10] It is particularly important to know if dynamic fluctuations are capable of causing considerable change in the creep strain rate. A positive answer to that question would necessitate further investigations aimed toward the strength alteration.

An extensive literature survey was conducted, the conclusion being that very little has actually been done in creep–fatigue testing of composites.[11] The expected difficulties in micromechanical description of the creep–fatigue interaction and lack of an experimental data suggest a phenomenological approach as a first attempt to more detailed analysis. Such an approach is undertaken in the present paper.

STATIC STEADY-STATE CREEP

A customary and often useful approach to the creep of bodies is to assume that secondary or steady creep is of overwhelming importance. Primary creep is ignored, as is the adjustment of strains from an initial elastic or elastic–plastic response to the steady creep state. Guided by the results broadly accepted for isotropic materials the constitutive equation for steady creep rate has been obtained in[8,9] making use of a concept of dissipation potential. A proper use of a representation theorem[12,13] furnished the dissipation potential as a scalar-valued orthotropic function of stress deviator invariants. In the present formulation the steady creep rate is derivable from the dissipation potential Φ, being a function of the stress tensor σ rather than its deviator, i.e.

$$\dot{\varepsilon}^c = \partial\Phi(\sigma)/\partial\sigma \tag{1}$$

The anisotropy of a composite is specified by three tensors:

$$\mathbf{M}_{11} = \bar{v}_1 \otimes \bar{v}_1 \qquad \mathbf{M}_{22} = \bar{v}_2 \otimes \bar{v}_2 \qquad \mathbf{M}_{33} = \bar{v}_3 \otimes \bar{v}_3 \tag{2}$$

where unit vectors $\bar{v}_1, \bar{v}_2, \bar{v}_3$ are aligned with the material privileged directions and determine three mutually perpendicular planes of orthotropic symmetry. The form of the constitutive equation (1) must be invariant with respect to transformations corresponding to the group of material symmetry. If we regard the dissipation potential Φ as an isotropic, scalar-valued function of stress and anisotropy tensors (2) then invariance requirements are automatically satisfied. Referring to the representation theorem[8,12,13] the dissipation potential may be represented in general form by a set of independent invariants I_r of tensors involved:

$$\dot{\varepsilon}^c = \frac{\partial\Phi(\sigma, \mathbf{M}_{ii})}{\partial\sigma} = \frac{\partial\Phi(I_r)}{\partial\sigma} \qquad (i = 1, 2, 3) \tag{3}$$

where the irreducible set of independent invariants can be chosen as follows:

$$I_r: \operatorname{tr}\mathbf{M}_{ii}\sigma, \ \operatorname{tr}\mathbf{M}_{ii}\sigma^2, \ \operatorname{tr}\sigma^3 \qquad (i = 1, 2, 3) \tag{4}$$

Further on we restrict our considerations to the plane stress condition and for that case the number of independent invariants is reduced to three. We postulate the following form of the dissipation potential:

$$\Phi = (1/2n)V^n \tag{5}$$

where the function V is second order in terms of stress components and will be called creep potential

$$V = \beta_1 \operatorname{tr}\mathbf{M}_{11}\sigma \cdot \operatorname{tr}\mathbf{M}_{22}\sigma + \beta_2 \operatorname{tr}\mathbf{M}_{11}\sigma^2 + \beta_3 \operatorname{tr}\mathbf{M}_{22}\sigma^2 \tag{6}$$

with $n, \beta_{1,2,3}$ being material constants, and using the contracted notation

$$\begin{array}{ll} \operatorname{tr} \mathbf{M}_{11}\boldsymbol{\sigma} = \sigma_1 & \operatorname{tr} \mathbf{M}_{11}\boldsymbol{\sigma}^2 = \sigma_1^2 + \sigma_6^2 \\ \operatorname{tr} \mathbf{M}_{22}\boldsymbol{\sigma} = \sigma_2 & \operatorname{tr} \mathbf{M}_{22}\boldsymbol{\sigma}^2 = \sigma_2^2 + \sigma_6^2 \end{array} \qquad (7)$$

Performing differentiation indicated in eqn (3) and bearing in mind (5) and (6) the steady creep rate may be written as

$$\begin{aligned} \dot{\boldsymbol{\varepsilon}}^c = \tfrac{1}{2}V^{n-1}[(\beta_1 \operatorname{tr} \mathbf{M}_{22}\boldsymbol{\sigma})\mathbf{M}_{11} + (\beta_1 \operatorname{tr} \mathbf{M}_{11}\boldsymbol{\sigma})\mathbf{M}_{22} \\ + \tfrac{1}{2}\beta_2(\mathbf{M}_{11}\boldsymbol{\sigma} + \boldsymbol{\sigma}\mathbf{M}_{11}) + \tfrac{1}{2}\beta_3(\mathbf{M}_{22}\boldsymbol{\sigma} + \boldsymbol{\sigma}\mathbf{M}_{22})] \end{aligned} \qquad (8)$$

In order to evaluate material constants we consider the composite subjected to the uniaxial loading σ_x in the direction x inclined at an angle α to the material privilege direction 1. Thus the stress components are

$$\sigma_1 = \sigma_x \cos^2 \alpha \qquad \sigma_2 = \sigma_x \sin^2 \alpha \qquad \sigma_6 = -\sigma_x \sin\alpha\cos\alpha \qquad (9)$$

Substituting (9) in (8) and transforming obtained strain rate components from the material to the loading coordinate system we obtain the strain rate along loading direction as

$$\dot{\varepsilon}_x^c = (\beta_1 \sin^2 \alpha \cos^2 \alpha + \beta_2 \cos^2 \alpha + \beta_3 \sin^2 \alpha)^n \sigma_x^{2n-1} \qquad (10)$$

It is seen from eqn (10) that the creep rate in off-axis uniaxial loading is in a form analogous to the Bailey–Norton creep law for isotropic materials under uniaxial loading:[14]

$$\dot{\varepsilon}_x^c = A\sigma_x^p \qquad (11)$$

where A and p are material constants. On setting $p = 2n - 1$ and A equal to the expression in brackets (10) we reduce (10) to the form (11) with the parameter A now being given as a function of off-axis angle α and material constants $\beta_{1,2,3}$. It follows from (10) that tensile creep experiments at any three distinct values of α will enable the material constants to be evaluated.

DYNAMIC STEADY-STATE CREEP

Two cases must be distinguished while investigating the creep properties under superimposed dynamic loads. If an amplitude of the dynamic load is comparable with a mean value of a static load then the resulting deformation process is a generalized sum of microrearrangements of the material structure caused by fatigue and static loadings. This might be called a creep–fatigue loading configuration. In the case when the amplitude of dynamic fluctuations is very small in relation to the static mean stress, of the order of a few percent say, then the dynamic component will act as a 'catalyst' to the creep process not causing the fatigue accumulation by itself. We call this situation a vibrocreep. Obviously a sharp boundary between creep–fatigue and vibrocreep cannot be traced. However, it is expected that the nature of deformation will be different in these two cases and the life-time and deformation at fracture will be distinct as well. On calculating the rate of dissipation energy K per unit reference volume we see from (6) and (8) that

$$K = \dot{\varepsilon}^c . \sigma = V^n \tag{12}$$

Thus surfaces of constant creep potential V represent constant rate of dissipation in the stress space and the creep strain rate as calculated from (8) is perpendicular to the surface at the corresponding stress point. If a specimen is subjected to dynamic fluctuations the creep resistance of a material as reflected by the creep rate may be altered, i.e. either improved or weakened. It would mean that in order to get the same creep rate as for static creep the dissipation surface in vibrocreep should be distinct from that in static creep. In other words, it generally corresponds to the rotation, translation and eventual distortion of the dissipation surface. This statement is in full analogy with the hardening phenomenon in elasto-plastic materials where an initial yield surface is deformed due to prior cold work and fatigue or a subsequent yield surface (loading surface) is subjected to complicated changes in its shape in the course of plastic deforma-tion.[15−22] Hence an additional factor characterizing at least the frequency and the amplitude of an oscillatory agency must enter the constitutive equation for vibrocreep in order to predict the changes in the dissipation surface induced by the dynamic fluctuations. Thus the creep potential V may be modified in the following way:

$$V^* = \gamma(\beta_1 \operatorname{tr} \mathbf{M}_{11}\sigma^* . \operatorname{tr} \mathbf{M}_{22}\sigma^* + \beta_2 \operatorname{tr} \mathbf{M}_{11}\sigma^{*2} + \beta_3 \operatorname{tr} \mathbf{M}_{22}\sigma^{*2}) \tag{13}$$

where

$$\gamma = 1 + F(\sigma^d) \qquad \sigma^* = \sigma + \mathbf{G}(\sigma^d) \tag{14}$$

F and $\mathbf{G}$ are scalar-valued and tensor-valued functions of the dynamic agency σ^d, respectively. An expansion or contraction of the dissipation surfaces occurs due to the scalar parameter γ, whereas their translation is caused by the term σ^*. The general forms for functions F and $\mathbf{G}$ are not known and can be obtained only from a large amount of experimental data. Therefore in order to get any workable framework for vibrocreep description we suppose that alteration of the dissipation surface depends on loading rate, since negative and positive impulses over one cycle of oscillation will lead to activation of micromechanisms influencing the creep deformation on a macroscale, and these rearrangements are sensitive to the loading rate.[23-25]

RESULTS AND DISCUSSION

The material used in this investigation was glass–xylok woven fabric composite. Two loading configurations were used: off-axis uniaxial loading for the material constants evaluation in the static creep and biaxial tension–shear loading for further check of the static creep theory and for vibrocreep. Figure 1 shows off-axis creep rate in the loading direction for four distinct off-axis angles, where solid lines represent the creep rate calculated according to eqn (10). The material constants appearing in the constitutive equation are listed in Table 1.

The specimen used for biaxial loading is shown in Fig. 2. The privileged material directions 1 and 2 were placed at the angle 45° to the loading directions x and y. An experimental rig allowed for the independent application of the normal force N and the shear force Q.[26] Both static and vibrocreep experiments followed the proportional loading path with the shear force Q always double the normal force N. The relative amplitude of dynamic shear force was also constant and equal to 10% of mean shear force. In Fig. 3 the experimental points from biaxial static creep measured in the material coordinate system are compared with the components of static creep rate calculated from eqn (8), showing fairly good agreement. The experimental creep curves from both static creep and vibrocreep, resolved into material privileged direction 1, are shown in Fig. 4. A remarkable influence of oscillatory frequency is easily seen for both stress levels. The frequency influences not only the total strain level but also significantly changes the creep rate. This is demonstrated in Fig. 5, where again results are resolved for creep strain rate along the material direction 1. Small arrows indicate the loading level corresponding to creep curves in Fig. 4. Thus we may conclude that the frequency of dynamic load induces rotation and translation of static creep rate line into positions determined experimentally for frequencies 7 and 10 Hz. This effect is incorporated in

TABLE 1

β_1	β_2	β_3	n	B	m	k
101.35×10^{-4}	10.99×10^{-4}	7.11×10^{-4}	2.686	-0.519	0.121	2.44

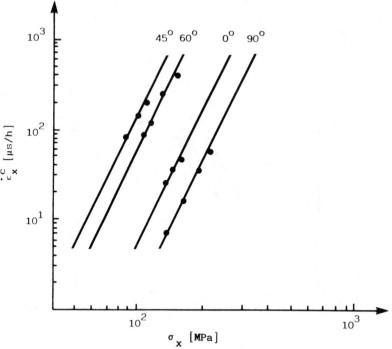

FIG. 1.   Uniaxial off-axis creep rate.

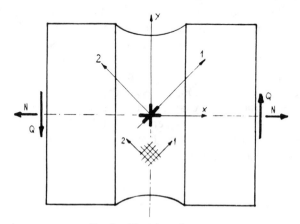

FIG. 2.   Biaxial specimen.

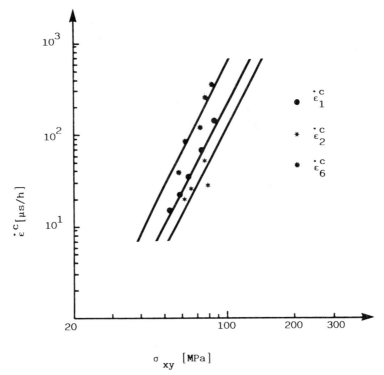

Fig. 3. Biaxial static creep rate.

the constitutive equation (1), developed with the use of modified creep potential V^*, in the following way. It is well documented that vibrocreep deformations of most polymers, and among them an epoxy family,[27] are strongly dependent on the frequency of oscillations. The creep rate dependence on the frequency exhibits a characteristic 'bell'-like form, i.e. for small and large frequencies the vibrocreep rate is very close to static creep rate, while the maximum increase in the vibrocreep rate is observed for moderate frequencies. By large frequencies we understand values 10–15 Hz which do not raise the temperature significantly. It is interesting to notice that a similar form of dependence with respect to fatigue life was observed in pure fatigue studies for $[\pm 45°]$ graphite/epoxy laminated composite.[28]

In the present investigation the stress component corresponding to the shear force Q possesses the dynamic term of the form

$$\sigma_{xy}^* = \sigma_{xy} + f(\omega) = \sigma_{xy} + \bar{f}e^{-[(\omega - \bar{\omega})^2/k]} \qquad (15)$$

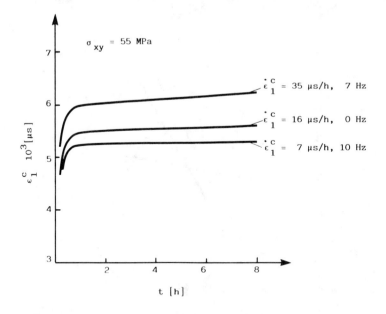

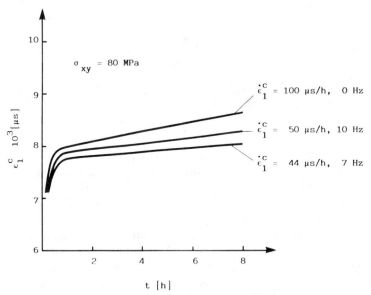

FIG. 4.   Creep curves for the strain in direction 1.

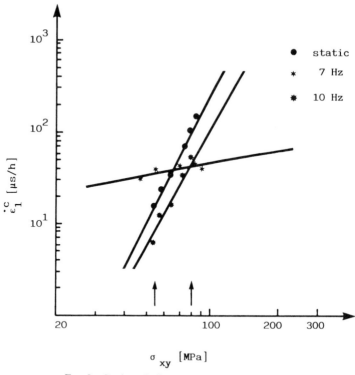

FIG. 5. Static and vibrocreep rate along direction 1.

where σ_{xy} is the mean value of the shear stress and $k, \bar{f}, \bar{\omega}$ are constants. The shape of the function f is shown in Fig. 6 together with the function

$$\gamma = 1 + Bf^m \tag{16}$$

Performing the same calculations as for the static creep, with the creep potential (13) and taking into account (15) and (16), the steady vibrocreep rates along the direction 1 are obtained as shown by solid lines in Fig. 5. It is worth mentioning that constants used for vibrocreep calculations and listed in Table 1 were obtained by best curve fitting and therefore are not material constants. Nevertheless, the modification of the creep potential V by functions f and γ gives the possibility for rotation and translation of the creep rate line in a double logarithmic $\dot{\varepsilon}^c$–σ scale. Moreover, using different values for constant k in small $(\omega \leq \bar{\omega})$ and large $(\omega \geq \bar{\omega})$ ranges of the frequency allows independent control of the slope of the ascending and descending parts of the f curve. It is clearly seen from Fig. 5 that for larger

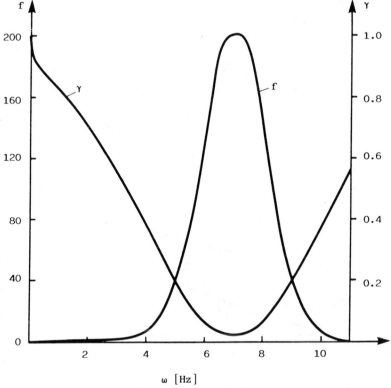

FIG. 6.   Functions f and γ.

stress levels the composite exhibits cyclic hardening. Such effect is probably due to the fact that cyclic impulses block the rheological mechanisms by not giving them time for full development. This effect diminishes with larger frequencies due to increased activation energy by a slightly growing temperature. For lower stress levels, on the other hand, the static creep is very small as the stresses are not large enough to develop a viscoelastic flow. In that case dynamic impulses may help to create relative slips in polymer chains and loosen its side constraints, softening in effect the polymer's resistance to creep. Particular attention must be also paid to the fact that the present investigation did not consider the influence of oscillatory amplitude. Only the relative amplitude was kept constant in the course of the experimental programme, which means that the absolute value of the amplitude changes for each stress level.

The rate of dissipation energy (12) is a very useful parameter in

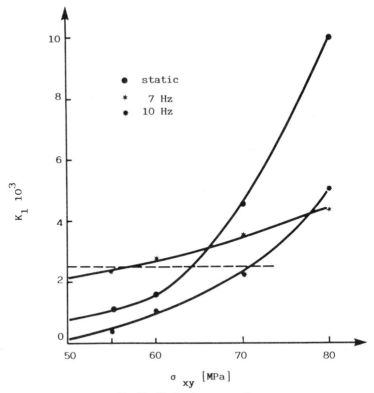

FIG. 7.   Dissipation energy K_1.

characterizing the material creep resistance. Figure 7 shows the relation between the rate of dissipation energy K_1 caused by the stress and strain rate along direction 1 versus the mean value of the shear stress. If we would like to select the loading configuration for maximal load carrying capacity with fixed level of dissipation energy, denoted by the dashed line in Fig. 7, then the best choice will be the vibrocreep with frequency 10 Hz.

The main conclusion drawn from the theoretical/experimental study presented here is that small oscillations are able to change significantly the creep characteristics of composite materials. It has also been shown that properly modified static creep theory is capable of predicting the changes in the vibrocreep rate.

Further developments must be aimed at investigations concerning the larger frequency spectrum, determination of the amplitude influence on the vibrocreep, and vibrocreep tests for other loading configurations.

REFERENCES

1. STURGEON, J. B., Creep of fibre reinforced thermosetting resins. In *Creep of Engineering Materials*, ed. C. D. Pomeroy. Mechanical Engineering Publications Ltd, London, 1978, pp. 175–95.

2. DARLINGTON, M. W. and TURNER, S., Creep of thermoplastics. In *Creep of Engineering Materials*, ed. C. D. Pomeroy. Mechanical Engineering Publications Ltd, London, 1978, pp. 197–214.

3. BRINSON, H. F. and DILLARD, D. A., The prediction of long-term viscoelastic properties of fiber reinforced plastics. In *Progress in Science and Engineering of Composites, Proc. ICCM4*, ed. T. Hayashi, K. Kawata and S. Umekawa. Japan Society for Composite Materials, 1982, Vol. 1, pp. 195–202.

4. DILLARD, D. A. and BRINSON, H. F., A numerical procedure for predicting creep and delayed failures in laminated composites. *ASTM STP 813*, ed. T. K. O'Brien. Philadelphia, 1983, pp. 23–37.

5. SCHAPERY, R. A., Application of thermodynamics to thermomechanical fracture and birefringent phenomenon in viscoelastic media. *J. Appl. Phys.*, **35**(5) (1964) 1451–65.

6. SCHAPERY, R. A. and LOU, Y. C., Viscoelastic characterization of nonlinear fiber reinforced plastics. *J. Comp. Mater.*, **5** (1971) 208–34.

7. SCHAFFER, B. G. and ADAMS, D. F., Nonlinear viscoelastic analysis of a unidirectional composite material. *J. Appl. Mech.*, **48** (1981) 859–65.

8. PYRZ, R., Governing equations for the steady-state creep in orthotropic materials. *Acta Mechanica*, **68** (1987) 251–63.

9. PYRZ, R., Description of steady-state creep in orthotropic materials. In *Mechanical Characterisation of Fibre Composite Materials*, ed. R. Pyrz. University of Aalborg, 1986, pp. 197–212.

10. PYRZ, R., Strength reduction of woven glass fabric composite after creep deformation. In *Proc. 6th ICCM & 2nd ECCM*, Vol. 4, ed. F. L. Matthews, N. C. R. Buskell, J. M. Hodgkinson and J. Morton. Elsevier Applied Science, London, 1987, pp. 272–7.

11. STURGEON, J. B., Creep, repeated loading, fatigue and crack growth in ±45° oriented carbon fibre reinforced plastics. *J. Mater. Sci.*, **13** (1978) 1490–8.

12. BOEHLER, J. P., A simple derivation of representations for non-polynomial constitutive equations in some cases of anisotropy. *ZAMM*, **59** (1979) 157–67.

13. SPENCER, A. J. M., Theory of invariants. In *Continuum Physics*, ed. A. C. Eringen. Academic Press, New York–London, 1971, pp. 239–353.

14. BOYLE, J. T. and SPENCE, J., *Stress Analysis for Creep*. Butterworths, London, 1983.

15. SHRIVASTAVE, H. P., MRóZ, Z. and DUBEY, R. N., Yield criterion and the hardening rule for plastic solids. *ZAMM*, **53** (1973) 625–33.

16. ELLYIN, F. and NEALE, K. W., Effect of cyclic loading on the yield surface. *J. Pres. Ves. Technol.*, **101** (1979) 59–63.

17. DAFALIAS, Y. F., Anisotropic hardening of initially orthotropic materials. *ZAMM*, **59** (1979) 437–46.

18. IKAGEMI, K., Experimental plasticity on the anisotropy of metals. In *Mechanical Behaviour of Anisotropic Solids*, ed. J. P. Boehler. Martinus Nijhoff Publishers, 1982, pp. 201–41.

19. REES, D. W. A., Yield functions that account for the effects of initial and subsequent plastic anisotropy. *Acta Mechanica*, **43** (1982) 223–41.

20. REES, D. W. A., An examination of yield surface distortion and translation. *Acta Mechanica*, **52** (1984) 15–40.

21. BOEHLER, J. P., On a rational formulation of isotropic and anisotropic hardening. In *Plasticity Today*, ed. A. Sawczuk and G. Bianchi. Elsevier Applied Science, London, 1985, pp. 483–502.

22. GUPTA, N. K. and MEYERS, A., Description of initial and subsequent yield surface. *ZAMM*, **66** (1986) 435–9.

23. POTTER, R. T., Repeated loading and creep effects in shear property measurements on unidirectional CFRP. *Composites*, **4** (1974) 261–5.

24. SCHULTZ, J. M. and FRIEDRICH, K., Effect of temperature and strain rate on the strength of a PET/glass fibre composite. *J. Mater. Sci.*, **19** (1984) 2246–58.

25. HIGHSMITH, A. L., STINCHCOMB, W. W. and REIFSNIDER, K. L., Effect of fatigue-induced defects on the residual response of composite laminates. In *Effects of Defects in Composite Materials. ASTM STP 836*, 1984, pp. 194–216.

26. CHRISTENSEN, B., JOHANNESSON, H. and KOEFOED, B., *Creep in Woven-Fabric Composite Materials*. MS Thesis, University of Aalborg, June 1987 (in Danish).

27. ZAWADZKI, J. (ed.), *Fatigue and Strength of Polymers*. Scientific Publishers, Warsaw, 1978 (in Polish).

28. SUN, C. T. and CHAN, W. S., Frequency effect on the fatigue life of a laminated composite. In *Composite Materials: Testing and Design. ASTM STP 674*, ed. S. W. Tsai. ASTM, 1979, pp. 418–30.

38

Influence of the Stacking Sequence and Load Direction on the Fatigue Behaviour of Quasi-isotropic Graphite/Peek Laminates

B. PICASSO and P. PRIOLO

*Department of Mechanical Engineering, University of Cagliari,
Piazza d'Armi, 08100 Cagliari, Italy*

ABSTRACT

Fatigue behaviour of graphite/PEEK quasi-isotropic symmetric laminates, subjected to tension–tension loading with fatigue ratio R = 0·1, has been extensively analysed. Two stacking sequences were considered, i.e. [45/90/ − 45/0] and [0/45/ − 45/90] and two load, and specimen, directions 0° and 22·5°. Results reveal no significant differences, in the S–log N plane, between the two sequences. A relevant different behaviour however is observed varying the load direction. For specimens at 22·5°, data show a decrease in fatigue strength with respect to the 0° direction, that can reach 19% at 100 000 cycles. During the tests the hysteresis cycle was continuously recorded and processed to follow damage progression. The final loss of stiffness seems to have no clear relation with the load direction. For the final maximum strain the hypotheses of independence on the number of cycles formulated by some authors was checked.

1 INTRODUCTION

A great deal of attention has been recently dedicated to thermoplastic resins as matrices for advanced composites. The main reason is that potentially they allow high production rate manufacturing with the same technology used for metals.[1] Graphite/PEEK is a relatively new composite, made by ICI, having a semicrystalline polyetheretherketone (PEEK) matrix, showing a high melting temperature and good mechanical characteristics. As shown in recent papers,[2-5] PEEK is characterized, under static loading and in long term fatigue experiments, by a somewhat brittle behaviour and a small amount of damage, if compared with epoxy based composites.[6] With the aim of a better understanding of the matrix influence on the long term fatigue behaviour of composites, a series of fatigue tests has been performed on quasi-isotropic laminates. In this paper new results are relative to [0/45/−45/90], [90/−45/45/0] and [−22·5/ 22·5/−67·5/67·5] symmetrical laminates, while [45/90/−45/0] and [22·5/ 67·5/−67·5/−22·5] had been analysed in preceding works.[3,4] So we have at this moment a quite complete set of data for the quasi-isotropic laminate. Sequence [−22·5/+22·5/−67·5/67·5] simply indicates a new load direction, taken between 0° and 45°. Comparison between the two load directions can give an idea of the actual anisotropy of the laminate.

2 EXPERIMENTAL

The material used was APC2 with 68% of fibre volume. Fatigue tests were performed on dog-bone and coupon shaped specimens, on a servo-hydraulic machine under load control. Sinusoidal loading was applied with a fatigue ratio $R = 0·1$ at a frequency of 6 Hz.

Several tests at lower frequency, 2 Hz mainly, did not reveal any appreciable influence of the test frequency. Similarly, no difference was observed in results obtained on 2 and 1 mm thick specimens. As previously described,[4] the load and strain signals were continuously monitored during the test and processed to furnish a mean evaluation of the secant modulus and the hysteresis loop area. The gauge length was 50 mm, but in some cases 100 mm was used to include the fracture region.

If a linear viscoelastic model is adopted the storage and loss components of the elastic modulus are easily determined if the hysteresis loop is measured. Generally only two points are required to determine the dynamic modulus components if the material exhibits a linear elastic behaviour. In practice however, composite laminates show a non-linear behaviour and the hysteresis loop measured through continuous scanning of the load and strain signals is irregular and affected by the noise produced by the testing system response to dynamical loading, Fig. 3.

Two linearization formulas were adopted using all the experimental data points.[7] Because of the integral definition of these formulas, scattering in

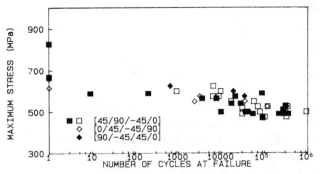

FIG. 1. Fatigue data for loading at 0°. Solid squares refer to a thickness of 1 mm.

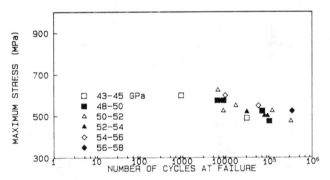

FIG. 2. Some of the preceding data evidencing the modulus value.

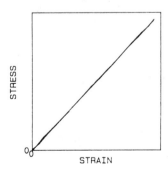

FIG. 3. A typical hysteresis loop.

the modulus components is considerably reduced if compared with the traditional two points method.

3 RESULTS

Figures 1 and 4 show *S*–log *N* diagrams with different stacking sequences and load directions. In Fig. 1 scattering is evident, but no considerable differences between the two basic sequences [45/90/−45/0] and [0/45/−45/90] are present. The third sequence [90/−45/45/0] only corresponds to a 90° rotation of the second one.

Scattering was correlated with the variations of the elastic initial modulus, for the [45/90/−45/0] laminate, as can be seen in Fig. 2, where results for this laminate are displayed according to the elastic modulus

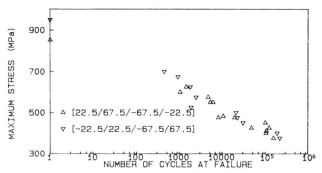

FIG. 4.   Fatigue data for specimens at 22·5°. Thickness is 1 mm.

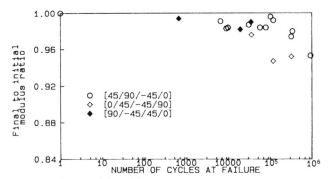

FIG. 5. Normalized modulus with respect to the initial value. Specimen at 0°.

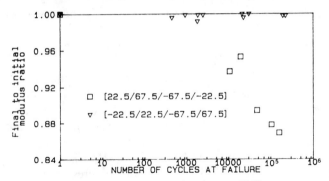

FIG. 6. Normalized modulus. Specimen at 22·5°.

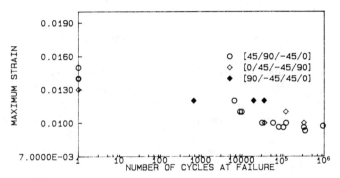

FIG. 7. Strain at failure. Specimen at 0°.

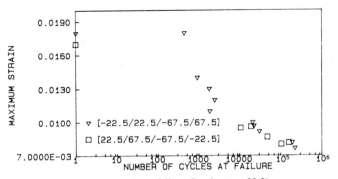

FIG. 8. Strain at failure. Specimen at 22·5°.

range. Very limited scattering and similar results are shown in Fig. 4 relative to specimens and load at 22·5°.

Two indexes of the damage state just before failure are plotted in Figs 5–8 against the number of cycles at failure. Respectively in Figs 5 and 6 the ratio between the axial elastic modulus before failure and at the beginning of the test for both the basic sequences are reported in the ordinates. As can be observed, the final modulus deterioration is always very limited and does not exceed 16%, which is far from the typical values of epoxy based laminates (40%). Again, the increased sensitivity to damage in the 22·5° load direction evidenced on the basic [45/90/−45/0],[3] is not confirmed by the tests on the [0/45/−45/90] laminate. Figures 7 and 8 show the maximum final strain calculated as σ_{max}/E_{fin}. Scattering of data is very limited and results tend to reach a horizontal asymptote with a long term fatigue value lower than the static one.

4 DAMAGE DETECTION

Continuous acquisition of the hysteresis cycle allowed recording of the axial modulus and the loop area. The first parameter seems at present a more reliable damage indicator and we refer to it in this work. Typical curves relating to the four cases treated are shown in Fig. 9, chosen so that the final number of cycles, i.e. the failure number, is of the same order. As can be seen, the shapes are very different. For specimens at 22·5° we have respectively the higher and the lower damage progression. We can provide no clear explanation for this fact. We can suppose that in any case the amount of damage in this material is very limited and it seems to depend on

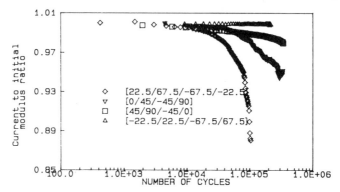

Fig. 9.   Stiffness deterioration for the cases considered.

the state of the interlaminar region which can be influenced by the manufacturing process.

5 CONCLUSIONS

Data from fatigue testing of quasi-isotropic graphite/PEEK laminates, at room temperature, can lead to two different kinds of considerations. Representation in the S–log N plane shows that:

(a) There are no relevant differences between the two sequences analysed, i.e. $[45/90/-45/0]$ and $[0/45/-45/90]$. Dispersion is greater in the first, but this fact can be correlated with scattering in the elastic properties.

(b) The specimens cut in the $22 \cdot 5°$ direction, i.e. the laminates with stacking sequence $[22 \cdot 5/67 \cdot 5/ -67 \cdot 5/ -22 \cdot 5]$ and $[-22 \cdot 5/22 \cdot 5/ -67 \cdot 5/67 \cdot 5]$ present nearly the same S–log N behaviour, but the slope of the possible Woehler curve is greater than in the preceding case. Although starting from higher static values, they tend to have a lower performance in the fatigue range from 10 000 cycles on. Furthermore, the final loss of stiffness seems to have no or very poor correlation with fatigue strength.

REFERENCES

1. WIRSEN, A., Resin development in advanced composites. FOA Report C 20509-F9, September 1983.
2. BISHOP, S. M., The mechanical performance and impact behaviour of carbon-fibre reinforced PEEK. *Comp. Struct.*, **3** (1985) 295–318.
3. PICASSO, B. and PRIOLO, P., Fatigue behaviour of graphite-PEEK laminates. *Advancing with Composites, Int. Conf. on Composite Materials*, Milan, May 10–12, 1988.
4. PICASSO, B. and PRIOLO, P., Damage assessment and life prediction for graphite-PEEK quasi-isotropic composites. *ASME Pressure Vessels and Pipings Conference*, Pittsburgh, June 19–23, 1988.
5. SCHULTE, K., Damage development in composite materials under static loading. *Advancing with Composites, Int. Conf. on Composite Materials*, Milan, May 10–12, 1988.
6. POURSATIP, A., ASHBY, M. F. and BEAUMONT, P. W. R., The fatigue damage mechanics of a carbon fibre composite laminate I—Development of the model. *Comp. Sci. Technol.*, **25** (1986) 193–218.
7. PRISS, L. S., The behaviour of tire vulcanizates under repeated deformation. *Kautschuk ans Gummi-Kunststoffe*, **19**(10) (1966).

39

The Reliability of Composite Structures

Z. K. MA and L. YANG

Aircraft Engineering Department, Northwestern Polytechnical University, Xian, Shaanxi Province, People's Republic of China

ABSTRACT

In this paper the characteristics of reliability analysis of composite materials are discussed. Based on them the method of reliability analysis of laminate and structural systems and the method of enumerating and identifying significant failure modes are presented. Some illustrative examples are given for explaining the application of these methods and they confirm their effectiveness.

1 INTRODUCTION

In aircraft structures fibre-reinforced composite is adopted to obtain a considerable decrease in structural weight. The standard deviations of mechanical properties of composite materials are relatively large and the strength of composite products depends on the manufacturing processes and standards of the workers. Consequently the research on reliability of composite structures possesses special significance.

For the reliability analysis of composite structural systems the study is at the beginning of development and only a few articles have been published in this area.

Composite material is an anisotropic, brittle, laminated construction and there are many characteristics compared with conventional engineering materials. This paper is a comprehensive report of our study. Based on these characteristics the method of reliability analysis of composite laminate and structural systems, and the method of enumerating and identifying significant failure modes, are mainly discussed. Some illustrative examples are given for explaining the application of these methods.

2 THE CHARACTERISTICS OF RELIABILITY ANALYSIS OF COMPOSITE STRUCTURES

A laminate is the basic member of composite structures. In aircraft structures, in general, it consists of four kinds of ply group and a ply group includes several plies of the same orientation. The orientations of the above ply groups are $0°$, $90°$ and $\pm 45°$ respectively. Because the laminate is of inhomogeneous construction along its thickness it is also considered as a structural system. The stiffness and strength of the laminate depend on material properties of the ply, laminated parameters, manufacturing processes and standards of workers, etc. In the reliability analysis of composite materials there are many characteristics compared with metal materials. They are as follows.[1]

2.1 Density Distribution Function of Material Strength

A composite is a kind of constituent material, so its strength distributions have many levels. For the strength of matrix normal distribution is used. For the strength of a single filament or a bundle of fibres, in general, the Weibull distribution simulates an actual one well. For the strength of a unidirectional composite or laminate normal and Weibull are both assumed. For the strength of a composite structure it is necessary to further study what kind of distribution is more reasonable. Now normal distribution is temporarily adopted in this paper.

As mentioned above, the mechanical properties of a laminate depend on many factors, so its coefficient of variation of strength is larger than that of general materials. And its value is about 0·1–0·2.

2.2 The Available Means of Estimating the Mechanical Properties of a Laminate

For a unidirectional composite its properties can be obtained by experiment. But there are so many kinds of laminates that their property data cannot be obtained by experiments because of the cost. There are two ways to approach them. One is only to obtain property data of some typical laminates based on the experiments. Another is based on prediction. The latter is simpler but there is a difference between predictive and experimental results. For the preliminary design step of structure, in general, it is necessary to predict theoretical values and then to carry out experiments with some typical laminates.

2.3 Correlation between the Properties of Unidirectional Composites

The longitudinal strength X depends mainly on the property of the reinforced fibre. The transverse and shear strengths Y and S are determined by matrix material, so the relation between X and Y or X and S may be considered as uncorrelative but Y and S are considered as correlative. If the matrix property is good and manufacturing processes are efficient, the correlation coefficient should be positive.

2.4 Effect of Geometric Size

In general, the capability of loading of the component is influenced by its size. However, when the number of plies is determined, the volume of fibre is invariable. And in this case the different thicknesses of laminate are caused by the different volumes of matrix. For a unidirectional composite plate its thickness difference has no effect on longitudinal capability of loading but only has an effect on transverse and shear capability. For a multidirectional plate its thickness difference is considered to have no effect on the capability of loading in each direction.

2.5 Model of Reliability Analysis

For reliability analysis of composite structural systems there are two typical models. One is the laminated construction model. In this model inhomogeneous characteristics along the laminate's thickness is considered. As mentioned above, under loading in the plane the laminate itself may be considered as a structural system. It consists of several components with different stiffnesses and is a parallel system. This kind of model is adopted in the present paper. Another is the equivalent plate model. The laminate is simplified as anisotropical plate with homogeneity. Its equivalent properties can be obtained by prediction using laminated plate theory or by experiment. Because this kind of equivalent plate is homogeneous in all directions it is considered as a component of a structural system.

3 RELIABILITY ANALYSIS METHOD FOR A LAMINATE

On the basis of these characteristics the full quantity loading method is presented by this paper's author[2] for the reliability analysis of a laminate.

3.1 Theory of Full Quantity Loading Method

The strength of each ply group under the unit load can be obtained by the Tsai–Hill criterion (other criteria can also be chosen):

$$T = \left[\left(\frac{\sigma_1}{X} \right)^2 + \left(\frac{\sigma_2}{Y} \right)^2 - \frac{\sigma_1 \sigma_2}{X^2} + \left(\frac{\tau_{12}}{S} \right)^2 \right]^{1/2} \tag{1}$$

where σ_1, σ_2 and τ_{12} are on-axis stress components. T is called Tsai's number.

Let T_i be the maximum value of T_{ij} among all the remaining ply groups. The subscript j corresponds to the j ply group, and i means that $(i-1)$ kinds of ply groups have failed. It can be formulated as follows:

$$T_1 = \max_{j=1\sim4} (T_{1j}) \qquad T_2 = \max_{j=1\sim3} (T_{2j}) \qquad T_3 = \max_{j=1\sim2} (T_{3j}) \tag{2}$$

The ply group which corresponds to the maximum value of $T_{\max}$ can be taken as a critical failure ply group. Then let

$$S_i = 1/T_i \tag{3}$$

where S_i is the loading of the laminate corresponding to the ply group which has reached the critical failure case; it can be called full quantity loading. The matrix equation of the full quantity loading method is as follows:

$$\begin{Bmatrix} 1 \\ 1 \\ 1 \\ 1 \end{Bmatrix} = \begin{Bmatrix} T_1 & & & 0 \\ & T_2 & & \\ & & T_3 & \\ 0 & & & T_4 \end{Bmatrix} \begin{Bmatrix} S_1 \\ S_2 \\ S_3 \\ S_4 \end{Bmatrix} \tag{4}$$

where subscripts 1–4 represent the ply groups in the sequence of failure. From eqn (4) the value of S_i can be found. And the system strength S_s can be obtained as follows:

$$S_s = \max_{i=1\sim4} (S_i) \tag{5}$$

Since the load and strength are random variables, usually not only the most serious failure sequence of the ply group is considered but more serious failure sequences are also considered as significant failure modes. The ratio of Tsai's number of each ply group to $T_{\max}$ each time can be adopted to enumerate and identify significant modes. Then these modes can be shown in a form of a fault tree.

3.2 Reliability Computation of Laminate

As mentioned above, the laminate is a parallel system composed of four brittle components with different stiffness. The steps of reliability computation of laminate are as follows. First, it is necessary to obtain Tsai's numbers for each ply group. Second, T_{ij} are compared with each other to get maximum value. The ply group corresponding to the maximum value of Tsai's number fails first. After degrading the stiffness of this failure ply group all the remaining ply groups are computed, time and again, until the last ply group fails. Then the fault tree of the laminate under load is obtained. Third, corresponding to these failure modes the safety margin equations are linear:

$$g_{Mi} = S_{Si} - L \tag{6}$$

where L is the load per unit area and g_{Mi} is the safety margin of the ith failure mode.

The safety factor f is chosen as 1·5. The density distribution functions of variables S_{Si} and L are assumed to be normal and then the probability distributions of safety margin g_{Mi} are normal. Let the coefficient of variation of load and strength be 0·2 and 0·1 respectively.

Let the mean value and standard deviation of variable g_{Mi} be μ_{Mi} and σ_{Mi}, then the reliability index is

$$\beta_{Mi} = \mu_{Mi}/\sigma_{Mi} \tag{7}$$

and the failure probability P_i of the ith failure mode can be looked up.

According to Ditlevsen's narrow bound theory of reliability, the system failure probability P_{sf} can be solved. Then the system reliability R_s is

$$R_s = 1 - P_{sf} \tag{8}$$

Example

The reliability of a laminate under axial tensile load N_x is computed. The ratios of thickness of each ply group are: $0°$, 27·8%; $90°$, 16·7%; and $\pm 45°$, 55·5%, respectively. The material properties are as follows:

$$E_1 = 128·83 \times 10^3 \, \text{MPa} \qquad E_2 = 7·9478 \times 10^3 \, \text{MPa}$$
$$G_{12} = 3·5868 \times 10^3 \, \text{MPa} \qquad \nu_{12} = 0·34$$
$$X_t = 1148·364 \, \text{MPa} \qquad X_c = 842·11 \, \text{MPa}$$
$$Y_t = 23·128 \, \text{MPa} \qquad Y_c = 135·926 \, \text{MPa}$$
$$S = 60·956 \, \text{MPa}$$

The mean value of the load μ_L is 250·88 MPa. The computation results of T_{ij} are given in Table 1.

In the third column of Table 1 the superscript ** indicates the first critical ply group and * indicates the second critical ply group ($2·6531/2·7098 = 0·9791 > 0·8$).

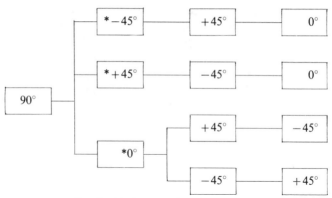

FIG. 1.   Fault tree under axial tensile load.

The fault tree is shown in Fig. 1.

From the matrix equation of full quantity loading it is shown that the strengths of the 1st and 2nd modes are the same:

$$S_1 = S_2 = 369 \cdot 03\,\text{MPa}$$

Similarly, the strengths of the 3rd and 4th modes are the same:

$$S_3 = S_4 = 376 \cdot 92\,\text{MPa}$$

And they are caused by failure of the same ply group. Therefore the original M_4 can be omitted.

The safety margin equations of these three modes are

$$g_{\text{M}i} = S_{\text{S}i} - L \qquad (i = 1, 2, 3)$$

Then

$$P_{\text{sf}} = 0 \cdot 061\,42 \qquad R_{\text{s}} = 0 \cdot 938\,58$$

TABLE 1
Date of T_{ij} under axial tensile load ($\times\,10^{-3}$)

	T_{1j}	T_{2j}	T_{3j}	T_{4j}
0°	2·189 3	2·653 1*	2·981 9	3·132 3
90°	5·943 3			
45°	3·167 3	2·709 8**	5·234 1	
			(16·003 2)	(23·127 7)
−45°	3·167 3	2·709 8**	5·234 1	
			(16·003 2)	(23·122 7)

4 RELIABILITY ANALYSIS METHOD OF A COMPOSITE STRUCTURAL SYSTEM

The method of reliability analysis of a laminate is developed to be used with a composite structural system. It is called the full quantity loading method with a change in stiffness.[3] The principal points and steps of this method are illustrated as follows.

4.1 Computation Method of Ultimate Strength of a Composite Structural System

The finite element method is adapted to carry out stress analysis of a structure. The ply group is taken as the basic member. Under the general loading the structural global stiffness is changed with the occurrence of failure members. When the general loading increases up to a certain value, the stipulated loss efficacy index of structure is reached. Corresponding to this condition, the loading is taken as the structural ultimate strength.

In general, the loss efficacy index is provided by standard or experiment. They can be divided into three kinds of index: (a) strength index, for example the first peak of the load–displacement curve; (b) stiffness index, for example the maximum relative flexibility or torsion angle of structure exceeds a certain value; (c) other index, for example a certain bar fails in the loading direction. When any index is reached the structure is considered as a loss efficacy.

In order to simplify computation the failure members are treated group by group. When the number of failure members reaches M, their member stiffnesses are degraded and the global stiffness is modified each time.

Under a certain general loading the on-axis stress components of each ply group can be solved. Then they are substituted in the Tsai–Hill criterion. It can be expressed as

$$\left(\frac{\sigma_1}{X}\right)^2 + \left(\frac{\sigma_2}{Y}\right)^2 - \frac{\sigma_1\sigma_2}{X^2} + \left(\frac{\tau_{12}}{S}\right)^2 = R \qquad (9)$$

It is noted that where σ_1, σ_2 and τ_{12} are the stresses of each ply group under general loading, R is called the utilization ratio of the member. If R is larger than or equal to 1·0, it means that the ply group fails.

The utilization ratios of all members are arranged in size sequence from large to small and their maximum value is R_{max}. If R_{max} equals 1·0, it means that the most serious critical member just fails.

Let S_i be the ratio of utilization ratio R_i of the ith member to R_{max} and constant c denote the selected range of more critical members. It can be shown that

$$S_i = R_i/R_{max} \qquad 1 \geq S_i \geq C \qquad (10)$$

The value of the constant C may be taken as about 0·8–0·9. Therefore formula (10) is a criterion for selecting critical members. Then the number n_1 of critical members are selected which include the most and more serious members.

These critical members are composed to obtain various significant failure modes, so they can be shown in a form of a fault tree of the structural system.

ND members are chosen and their member stiffnesses are certainly degraded. According to the sequence of S_i the $(M - \text{ND})$ members are taken among other critical members. They are combined with ND members into some group. There are still M members in each group. The stiffness of members in each group is degraded respectively. So the first class branches of the fault tree are constituted. After the load increases the method of selecting critical members n_2 of the second class is similar to the above. Therefore the second class branches of the fault tree are constituted. In accordance with this method every class of branch of the fault tree is obtained.

Some principles to simplify computation of the failure probability of the system are presented. They are as follows.

(i) If L_1 modes are completely relative, it is necessary to retain one significant failure mode and $(L_1 - 1)$ modes can be deleted.

(ii) If the failure probabilities $P_i (i = 1 \sim L_2)$ of L_2 modes are the same, the probability can be multiplied by some coefficient to simplify computation of the system probability. It is expressed as

$$\sum P_i = L_2 \times P_i \qquad \sum P_{ij} = \left[L_2(L_2 - 1) - \sum_{K-1}^{L_2-1} (L_2 - K) \right] P_{ij} \qquad (11)$$

where P_i and P_{ij} are the failure probability of a single mode and the joint probability between i and j modes respectively.

(iii) If L_3 modes approximate to each other, they are considered as completely relative and then can be merged by the principle (i).

Example

A multi-walled structure made of composite material of skin and metallic skeleton is shown in Fig. 2. For the property data see Table 1.

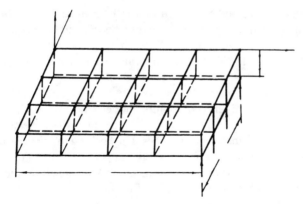

FIG. 2.   Multi-walled structure made of composite material of skin and metallic skeleton.

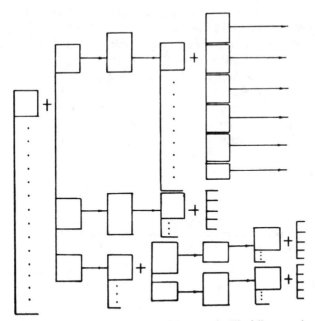

FIG. 3.   Part of the branches of the fault tree of the example. (The failure member is shown as $m_1 m_2$, where m_2 is the number of failure members and m_1 is the number of laminate plates which include this failure member.)

Parameters C, M and ND are chosen as 0·9, 5 and 2. Let the coefficients of variation of load and ultimate strength be 0·2 and 0·1 respectively. The mean value of the load equals 2920 lb. There are metallic bar and shearing panel elements and composite quadrangle panel elements in the finite element model. Part branches of the fault tree are shown in Fig. 3.

Deletion and mergence of failure modes: the first to sixth branches belong to principle (ii) and L_2 equals 6; in Fig. 3 the lower branches are compared with the upper six branches. They belong to principle (iii) and can be merged; the middle branches and the upper sixth are all the same. They belong to principle (i) and can be deleted. After deletion and mergence, there are only six effective modes in Fig. 3.

5 CONCLUSION

1. For composite materials there are many characteristics compared with metal materials and they must be considered in the reliability analysis.
2. For the composite laminate in the structure it is obvious that if the value of required reliability is 0·999 the safety factor f has to be larger than 1·5. This conclusion can be drawn only after doing a reliability computation.
3. The full quantity loading method is used for the reliability analysis of a laminate, and its computation is simpler and the concept is clearer. This method is developed for use in the reliability analysis of a composite structural system. When parameters C, M and ND are properly chosen, one may take less computer running time and obtain better precision. For large-scale structures its effect is especially obvious.

REFERENCES

1. YANG, L. and MA, Z. K., The reliability analysis and design of composite structures. *Journal of Chinese Society of Astronautics*, 1989.
2. YANG, L., Reliability of composite laminate. *Mechanics of Structures and Machines*, **16**(4) (1988).
3. YANG, L., MA, Z. K. and DUAN, Q. M., A method of reliability analysis for composite structural system. *Int. Conf. of Composite Materials—7*, China, 1989.

40

Stress Concentration in a Finite Width Laminate Containing Several Elliptical Holes

LIN JIAKENG and WANG LINJIANG

Nanjing Aeronautical Institute, Nanjing, People's Republic of China

ABSTRACT

In this paper an analytical method for determining the stress distribution in an infinite composite laminate containing multiple different elliptical holes is developed by using the Faber series. By utilizing two narrow elliptical holes to model two straight boundaries of a plate the stress concentration problem in a finite width laminate containing several elliptical holes is also solved. Some experimental and numerical results are presented.

INTRODUCTION

It is well known that stress concentration widely exists in all engineering structures and machinery, and it exists more severely in the astronautical and aeronautical vehicles in which weight saving is very important and composite materials have been successfully used. For isotropic material the stress concentration problem has been solved thoroughly. There are lots of references for engineers to use. But for composite material, although some results have been published, it has not been studied sufficiently. As applications of composites continue to grow at a rapid rate in engineering, the determination of stress distribution has become one of the most important topics which need to be solved urgently in composite material.

The stress distribution in a thin and symmetric composite laminate subjected to in-plane loads can be modeled as a plane stress problem of an anisotropic plate. The anisotropic degree can be characterized by two complex parameters:

$$U_1 = \alpha_1 + \beta_1 i \quad \text{and} \quad U_2 = \alpha_2 + \beta_2 i$$

The stress distribution in an infinite orthotropic plate containing one hole, two holes or many of the same holes in a line have been examined. But as we know the results from an analytical study in a finite anisotropic plate containing multiple holes are published rarely. Based on the method of Ref. 3, a new method for calculating the stress field in an anisotropic plate that contains any number of different elliptical holes is established by using the Faber series in this paper. Then by using a simple but effective method we apply two narrow elliptical holes whose major axes keep a certain length and their minor axes approach zero to simulate two straight boundaries of a finite width laminate; the stress field in a finite width composite laminate containing multiple different elliptical holes is determined. The cases of two holes, for example, are examined in more detail. Numerical calculations and experimental results indicate that the method developed in this paper is exact enough for engineering purposes.

ANALYTICAL METHOD

An infinite orthotropic plate containing multiple different elliptical holes not overlapping each other and subjected to uniform load P_x, P_y and P_{xy} is shown in Fig. 1. The center of the holes can be expressed as

$$Z_{ok} = X_{ok} + i y_{ok}$$

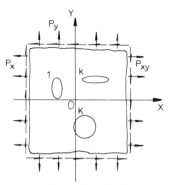

FIG. 1.   An infinite orthotropic plate with multiple elliptical holes subjected to uniform load, P_x, P_y and P_{xy}.

where the subscript o refers to the center of a hole and k is the hole number. The directions of the coordinate axes are assumed to be parallel to the directions of the major and minor axes of the holes but may not coincide with the principal directions of the elasticity of the orthotropic plate. Based on the results for a single hole case,[2] by considering the mutual effects among the holes and utilizing the superposition principle, a pair of new potential functions can be obtained as[3,4]

$$\phi_l(Z_l) = Q_l Z_l + \sum_{k=1}^{K} \sum_{m=1}^{\infty} \phi_{lkm} \zeta_{lk}^{-m} \qquad (l = 1, 2) \tag{1}$$

where

$$Q_1 = [P_x + (\alpha_2^2 + \beta_2^2)P_y + 2\alpha_2 P_{xy}]/D$$
$$Q_2 = [(\alpha_1^2 - \beta_1^2)P_y - 2\alpha_1\alpha_2 P_y - P_x - 2\alpha_2 P_{xy}]/D$$
$$\quad + i\{(\alpha_1 - \alpha_2)P_x + [\alpha_2(\alpha_1^2 - \beta_1^2) - \alpha_1(\alpha_2^2 - \beta_2^2)]P_y$$
$$\quad + [(\alpha_1^2 - \beta_1^2) - (\alpha_2^2 - \beta_2^2)]P_{xy}\}/(\beta_2 D)$$
$$D = 2[(\alpha_1 - \alpha_2)^2 + \beta_2^2 - \beta_1^2]$$
$$Z_l = X + U_l y \tag{2}$$
$$\zeta_{lk} = [Z_l - Z_{olk} + \sqrt{(Z_l - Z_{olk})^2 - a_k^2 - U_l^2 b_k^2}]/(a_k - i U_l b_k) \tag{3}$$

ϕ_{lmn} are undetermined coefficients. a_k and b_k are the semi-major and semi-minor axes of the elliptical hole k. Furthermore, the subscript m is a corrective term number. l is the plane number.

Expression (2) changes all the elliptical holes in plane Z to elliptical holes in planes Z_1 and Z_2, and expression (3) transforms the hole k in plane Z_1 or Z_2 into a unit circle of which the center is located at the origin of the coordinate axes in plane ζ_{lk}. Figure 2 gives an example of those transform processes for two holes, where $U_l = 1.2472 + 0.9405i$, Fig. 2(b) is plane Z_1 and Fig. 2(c) is plane ζ_{lk}.

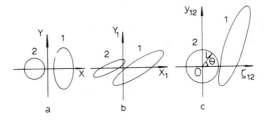

FIG. 2.   Transformation axes.

From eqn (3) the following expression can be obtained:

$$\zeta_{lk}^{-m} = \{(a_k = i U_l b_k)/[Z_l - Z_{olk} + \sqrt{(Z_l - Z_{olk})^2 - a_k^2 - U_l^2 b_k^2}]\}^m \quad (4)$$

It is obvious that function ζ_{lk}^{-m} is univalent and analytical on planes Z_l, so it is sure that ζ_{lk}^{-m} is analytical inside the area of the hole j ($j \neq k$) and continues to the boundary of the hole. That means ζ_{lk}^{-m} satisfy the requirements for expanding them into Faber series,[5] so it can be expressed inside the hole j as follows:

$$\zeta_{lk}^{-m} = \sum_{n=1}^{\infty} C_{jlkmn}(\zeta_{lj}^n + t_{lj}^n \zeta_{lj}^{-n}) \quad (5)$$

where $t_{lj} = (a_j + i U_l b_j)/(a_j - i U_l b_j)$. C_{jlkmn} are the coefficients of Faber series. They can be determined by the coefficients of Fourier series of functions ζ_{lk}^{-m}. n is the term number of the Faber series.

Substituting expression (5) into eqn (1), the following relationship can be obtained:

$$\phi_l(Z_l) = Q_l Z_l + \sum_{\substack{k=1 \\ k \neq j}}^{\infty} \sum_{m=1}^{\infty} \phi_{lkm} \sum_{n=1}^{\infty} C_{jlkmn}(\zeta_{lj}^n + t_{lj}^n \zeta_{lj}^{-n}) + \sum_{m=1}^{\infty} \phi_{ljm} \zeta_{lj}^{-m}$$

$$(l = 1, 2) \quad (6)$$

On the other hand, by solving eqn (3) and using subscript j instead of k, Z_l can be expressed as a function of ζ_{lj}:

$$Z_l = S_{jl1} \zeta_{lj} + S_{jl2}/\zeta_{lj} \quad (7)$$

where $S_{jl1} = (a_j - i U_l b_j)/2$ and $S_{jl2} = (a_j + i U_l b_j)/2$.

As mentioned above, hole j has become a unit circle with center at the origin in plane ζ_{lj}, so there is the following relationship on the contour of hole j:

$$\zeta_{lj} = \cos \theta + i \sin \theta = \sigma \quad (8)$$

Based on the fact that the contour of hole j is free from loads, its boundary conditions can be expressed as follows:

$$\sum_{l=1}^{2} [\phi_l(Z_l) + \bar{\phi}_l(Z_l)] = 0 \qquad \sum_{l=1}^{2} [U_l \phi_l(Z_l) + \bar{U}_l \bar{\phi}_l(Z_l)] = 0 \quad (9)$$

where bar refers to conjugate.

Substituting relationships (6), (7) and (8) into the system of eqns (9), we have

$$
\sum_{l=1}^{2} \sum_{m=1}^{\infty} \left\{ \sum_{\substack{k=1 \\ k \neq j}}^{K} \phi_{lkm} \sum_{n=1}^{\infty} C_{jlkmn}(\sigma^n + t_{lj}^n \sigma^{-n}) + \phi_{ljm}\sigma^{-m} \right.
$$

$$
\left. + \sum_{\substack{k=1 \\ k \neq j}}^{K} \bar{\phi}_{lkm} \sum_{n=1}^{\infty} \bar{C}_{jlkmn}(\sigma^{-n} + \bar{t}_{lj}^n \sigma^n) + \bar{\phi}_{ljm}\sigma^m \right\}
$$

$$
= -\sum_{l=1}^{2} [Q_l(S_{jl1}\sigma + S_{jl2}/\sigma) + \bar{Q}_l(\bar{S}_{jl1}/\sigma + \bar{S}_{jl2}\sigma)] \tag{10}
$$

$$
\sum_{l=1}^{2} \sum_{m=1}^{\infty} \left\{ \sum_{\substack{k=1 \\ k \neq j}}^{K} \phi_{lkm} \sum_{n=1}^{\infty} U_l C_{jlkmn}(\sigma^n + t_{lj}^n \sigma^{-n}) + \phi_{ljm} U_l \sigma^{-m} \right.
$$

$$
\left. + \sum_{\substack{k=1 \\ k \neq j}}^{\infty} \bar{\phi}_{lkm} \sum_{n=1}^{\infty} \bar{U}_l \bar{C}_{jlkmn}(\sigma^{-n} + \bar{t}_{lj}^n \sigma^n) + \bar{\phi}_{ljm} \bar{U}_l \sigma^m \right\}
$$

$$
= -\sum_{l=1}^{2} [Q_l U_l(S_{jl1}\sigma + S_{jl2}/\sigma) + \bar{Q}_l \bar{U}_l(\bar{S}_{jl1}/\sigma + \bar{S}_{jl2}\sigma)]
$$

$$
(j = 1, 2, \ldots, K)
$$

Let $j = 1, 2, \ldots, K$ in turn in the above equations; this means the stress potential functions satisfy the boundary conditions of all holes, then $2k$ equalities in σ can be obtained. The coefficients of σ with the same power on both sides of every equality must be equal, so we have a system of infinite equations with undetermined complex coefficients ϕ_{lkm} and $\bar{\phi}_{lkm}$. In practical calculations a finite number N is assumed for the top limit of m and n. Solving the system of equations, the coefficients ϕ_{lkm} can be determined. Once the stress potential functions are available, then the stress field in the plate can be obtained by

$$\sigma_x = 2R_e \left[\sum_{l=1}^{2} U_l^2 \phi_l'(Z_l) \right]$$

$$\sigma_y = 2R_e \left[\sum_{l=1}^{2} \phi_l'(Z_l) \right] \qquad (11)$$

$$\tau_{xy} = -2R_e \left[\sum_{l=1}^{2} U_l \phi_l'(Z_l) \right]$$

where the prime refers to differentiation.

The stress along the contour of the holes can be calculated by

$$\sigma_t = \tfrac{1}{2}(\sigma_x + \sigma_y) + \tfrac{1}{2}(\sigma_x - \sigma_y)\cos 2\theta + \tau_{xy}\sin 2\theta$$
$$\sigma_n = \tfrac{1}{2}(\sigma_x + \sigma_y) - \tfrac{1}{2}(\sigma_x - \sigma_y)\cos 2\theta - \tau_{xy}\sin 2\theta$$
$$\tau_{tn} = -\tfrac{1}{2}(\sigma_x + \sigma_y)\sin 2\theta + \tau_{xy}\cos 2\theta$$

where σ_t is tangential stress, σ_n is normal stress and τ_{tn} is shearing stress. According to the boundary conditions, τ_{tn} and σ_n must equal zero. The maximal stress of the plate is the maximum of σ_t.

By utilizing the hypothesis that the strains at the same point in every layer of the plate are the same, it is easy to determine the stress distribution in every layer of the composite laminate.

For determining the stress field of a finite width laminated composite two narrow elliptical holes, A and A', are used (Fig. 3(a)). Supposing that their

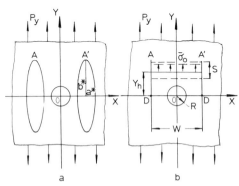

FIG. 3. The stress field of a finite width laminated composite.

TABLE 1
Cross-section stress σ_y/P_y

$\dfrac{x}{y_h}$	0·0	0·4R	0·8R	1·2R	1·6R	2·0R	$\delta\%$
3·0R	0·686	0·719	0·806	0·928	1·039	1·120	27·7
4·5R	0·848	0·854	0·869	0·887	0·901	0·899	3·3
5·5R	0·880	0·880	0·879	0·877	0·873	0·868	1·0
6·0R	0·891	0·890	0·885	0·875	0·861	0·857	2·3
6·9R	0·908	0·907	0·903	0·893	0·865	0·657	25·1

semi-minor axis b^* equals a certain value and semi-minor axis a^* approaches zero, then the elliptical holes become two straight slits to simulate two straight edges of a finite width laminate (Fig. 3(b)). If there were not a circular hole between the two narrow elliptical holes, the stress field would be the same as the uniform stress field at infinity. The stress concentration due to the circular hole only exists in its nearby area. If b^* is chosen large enough, there would be a uniform stress field 'S,' then a model for the finite width plate with a hole (or multiple holes) subjected to a uniform load is established. The larger b^* is, the larger 'S' would be, but more terms of the series are needed in order that the results reach a certain precision. As an example, in Fig. 3(b), $b^* = 7$ and $W = 4$, radius of the circular hole, and $a^* = 10^{-4}R$ are used. In geometry it is reasonable to consider its straightness as good enough to model a straight edge of a part in engineering. In Table 1 some data of the stress distribution are given; relative difference $\bar{\delta} = |(\sigma_{y\max} - \bar{\sigma}_0)/\bar{\sigma}_0|$. It is obvious that the stress field in regions $4·5 \leq y_h \leq 6·0$ is uniform enough. Here $\bar{\sigma}_0$ is the average of the stress σ_y at a cross-section. It is interesting to notice that the average of stress σ_y is $0·877P_y$, less than P_y. $\bar{\sigma}_0$ is used for our following calculation of stress concentration.

RESULTS AND DISCUSSION

Based on the information from the available literature, no result on stress distribution of a finite width orthotropic laminate containing multiple holes has been reported. In order to examine the analytical method presented here, the stress concentration factor K_σ ($K_\sigma = \sigma_t/\bar{\sigma}_0$) along the contour of a circular hole in a finite width quasi-isotropic plate ($U_1 = U_2 = i$) has been calculated (Fig. 4). The results are given in Table 2, where

696   *Lin Jiakeng and Wang Linjiang*

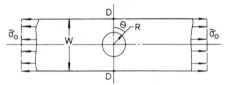

FIG. 4. Calculation of the stress concentration factor, K_σ, along the contour of a circular hole
in a finite width quasi-isotropic plate.

the corresponding analytical solution for isotropic material by Howland[6]
is also included for comparison purposes. The difference δ of K_σ at the
maximum stress point ($\theta = 0°$) is 0·7%, and the point D on the straight
boundary is 1·4%.

An experiment for a carbon fiber reinforced laminate containing two
circular holes (Fig. 5) has been completed. The stacking sequence of the
sample is $(0_2^\circ \pm 45°)_{2s}$ and the fiber volume fraction $\bar{V}$ is 59·6%. The elastic
constants are $E_x = 19·83\,\text{GPa}$, $E_y = 70·19\,\text{GPa}$, $G = 17·98\,\text{GPa}$ and $v = 0·205$. The complex parameters are $U_1 = 0·3043 + 0·6625\text{i}$ and $U_2 = -0·3043 + 0·6625\text{i}$. Strain gauges (0·3 × 0·88 mm) are arranged at points A,
B and C. The results are shown in Table 3.

The corresponding analytical results are completed and the results of

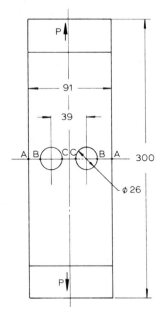

FIG. 5. A carbon fiber reinforced laminate containing two circular holes.

TABLE 2
Stress concentration factor on contour K_σ

θ	0°	15°	30°	45°	60°	75°	90°	Point D
Howland	4·32	3·72	2·32	0·77	−0·51	−1·32	−1·58	0·73
Ours	4·35	3·75	2·32	0·74	−0·53	−1·34	−1·61	0·72
$\delta\%$	0·7	0·8	0·0	3·9	3·9	1·5	1·9	1·4

two methods are compared in Table 4, where $\bar{K}_\sigma^*$ is the average of $\bar{K}_\sigma$ from Table 3, K_σ is the analytical result and $\delta = (K_\sigma - \bar{K}_\sigma^*)/\bar{K}_\sigma^*$.

After having completed the examination, let us see an example (Fig. 6(a)) that for a finite width composite laminate containing two elliptical holes located in series, the stacking sequence is $(0°/90°)_s$, and the parameters are $E_1/E_2 = 1·00$, $E_1/G = 11·43$, $v = 0·058$, $U_1 = 0·2976i$ and $U_2 = 3·3600i$.

In order to observe the variation of stress concentration due to a change of the geometry of the plate, different ratios of L_1/b, a/b and L_2/a are

TABLE 3
Results of the experiment

P (KN)	4·9	9·8	14·7	19·6
$\bar{\sigma}_0$ (MPa)	27·05	54·10	81·14	108·2
$\bar{\varepsilon}_A$ ($\mu\varepsilon$)	307	628	917	1 237
$\bar{\sigma}_A$ (MPa)	21·56	44·10	63·39	86·83
$\bar{K}_A$	0·796	0·814	0·793	0·802
$\bar{\varepsilon}_B$ ($\mu\varepsilon$)	1 176	3 570	5 271	7 303
$\bar{\sigma}_B$ (MPa)	124·7	250·6	370·0	512·5
$\bar{K}_B$	4·067	4·630	4·557	4·736
$\bar{\varepsilon}_C$ ($\mu\varepsilon$)	1 650	3 255	4 872	6 616
$\bar{\sigma}_C$ (MPa)	115·8	228·4	341·9	464·3
$\bar{K}_C$	4·280	4·222	4·213	4·290

TABLE 4
Comparison of results

Point	A	B	C
$\bar{K}_\sigma^*$	0·801	4·633	4·251
K_σ	0·738	4·804	4·524
$\delta\%$	7·87	3·69	6·24

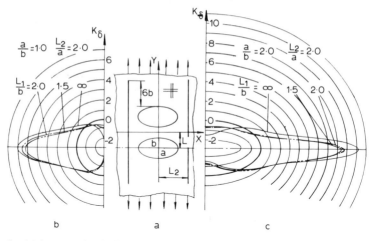

FIG. 6.   (a) An example of a finite width composite laminate containing two elliptical holes located in series; (b),(c) with the stress concentration factors along the contour of the lower hole illustrated.

assumed. The stress concentration factors along the contour of the lower hole are illustrated in Fig. 6(b,c), and the maxima of K_σ are listed in Table 5.

It is clear that the mutual effects of the two holes in series causes a reduction of K_σ. The closer the distance between the two holes is, the larger the reduction would be.

Figure 7 shows another laminate stacking in sequence $(0_2^\circ \pm 45^\circ)_s$ and containing two holes located in parallel. Some analytical results are given in Tables 6 and 7, where $s = (K_{\sigma\max} - K_{\sigma\mathrm{mo}})/K_{\sigma\mathrm{mo}}$ and $K_{\sigma\mathrm{mo}}$ is the $K_{\sigma\max}$ for an infinite plate made of the same material with a single circular hole. The existence of the boundaries and the mutual effects of the holes in parallel

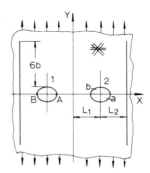

FIG. 7.   A laminate stacking in sequence $(0_2^\circ + 45^\circ)_s$ and containing two holes in parallel.

TABLE 5
Maximum stress concentration factors $K_{\sigma max}$

L_1/b	1·5	2·0	∞
$a/b = 1$	5·290	5·516	6·134
$a/b = 2$	8·307	8·857	10·195

TABLE 6
Maximal stress concentration factor $K_{\sigma max}$, $(L_1 + L_2)/a = 3.5$

L_1/a	1·2		1·5		2·0	
Point	A	B	A	B	A	B
$a/b = 1$	5·814	4·906	4·490	4·775	4·348	5·342
$a/b = 2$	8·708	7·768	7·607	7·573	7·519	8·084

intensify the stress concentration. There is a ratio $(L_1/a)_{opt}$ at which $K_{\sigma max,A} = K_{\sigma max,B}$ for a given width. This ratio is dependent upon the geometry of the holes. From Table 7 it appears that the mutual effects of the holes and the boundaries may be ignored if the ratios $L_1/a \geq 3.0$ and $L_2/a \geq 5.0$.

In this paper an analytical method based on the Faber series has been carried out for determining the stress distribution in an infinite and finite width composite laminate containing multiple different elliptical holes and subjected to in-plane loads. The comparison between calculated and experimental results indicate this method and the corresponding computer program are exact enough for engineering usage. Some conclusions extracted from the parameter study are useful for design engineers. In fact the method can be used for solving many complex and more practical engineering problems such as mechanically fastened joints, fracture, etc.

TABLE 7
Comparison of $K_{\sigma max}$, $L_1/a = 3.0$, $K_{\sigma mo} = 3.459$

L_2/a	3·0	4·0	4·5	5·0	∞
$K_{\sigma max}$	3·827	3·690	3·649	3·618	3·848
$s\%$	10·64	6·68	5·49	4·60	0·72

REFERENCES

1. SAVIN, G. N., *Stress Concentration Around Holes* (English translation by W. Johnson). Pergamon Press, Oxford, 1961.
2. LEKHNITSKII, S. G., *Anisotropic Plate* (English translation by S. W. Tsai and T. Cheron). Gordon and Breach, New York, 1968.
3. LIN, J. K. and UENG, C. E. S., Stress in a laminated composite with two elliptical holes. *J. Comp. Struct.*, **7** (1987) 1–20.
4. UENG, C. E. S. and JIA-KENG LIN, Stress compositions in composite laminates. *J. Engng Mech.*, **113** (1987) 1181–93.
5. CURTISS, J. H., Faber polynomials and the Faber series. *Am. Mathemat. Month.*, **17**(6) (1971) 557–96.
6. HOWLAND, R. C. J., On the stresses in the neighbourhood of a circular hole in a strip under tension. *Phil. Trans. R. Soc.*, *A*, **229** (1930) 48–86.

41

Lamination Theory of Composite Material with Complex Fiber Orientation Distribution

ZEN-ICHIRO MAEKAWA, HIROYUKI HAMADA and ATSUSHI YOKOYAMA

Faculty of Textile Science, Kyoto Institute of Technology, Matsugasaki, Sakyo-ku, Kyoto 606, Japan

ABSTRACT

A generalized laminated theory is developed to estimate the mechanical behaviors of polymeric composites with complicated fiber orientation states. This theory is intended for application to molded products in which the complicated flow of the fibers occurs during the molding process. We propose a new method which indicates various fiber oriented states by use of an incomplete beta function. Next, a generalized laminated plate theory is presented to analyze the stress–strain relations for fiber composites with complex orientation distribution. The effects of fiber orientation distribution on elastic moduli of these composites are discussed. Finally, the deformation behaviors of molded plates can be estimated by connecting this lamination theory to a finite element method. The numerical results on the deformation under uniaxial loading are shown for composite plates with typical orientation states.

INTRODUCTION

It is now clearly evident that polymeric composites have become viable engineering materials. Various molding methods have been developed according to numerous uses. The fiber reinforced plastics with high productivity are usually fabricated by closed mold processes, such as injection molding and matched-die molding. It is well known that the complicated flow of the fibers occurs during the molding process with these mold methods, and undesirable deformation behaviors such as surface sink and warp often occur in molded products. Many researchers have made clear that the mechanical properties of polymeric composites depend on the fiber orientation and the fiber content in addition to the strengths and rigidities of the fiber and matrix.[1,2] These results suggest that a knowledge of the fiber dispersed condition is absolutely necessary to estimate the mechanical behaviors of polymeric composites with complicated fiber flows. The studies on the orientation behavior of fibers have been done actively. The main works are shown as follows: studies on expression method and prediction of the fiber orientation,[3,4] studies on the characteristics of the weld lines which are the weakest part in molded products,[5,6] and studies on the deformation behaviors such as the surface sink and warp.[7,8] However, the procedures on estimation of the complex fiber orientation and deformation behavior in polymeric composites have not been completely established.

This work, carried out to clarify these problems, consists of two parts. First, we propose a new method which expresses complicated fiber orientation distribution in terms of an incomplete beta function. Secondly, a generalized classical laminated plate theory is presented in order to analyze the stress–strain relations for fiber reinforced composite plate and apply to deformation behavior of molded products.

FIBER ORIENTATION DISTRIBUTION

Though fiber reinforced plastics generally may be solid composites with three-dimensional fiber orientation, most molded parts have a sheet-like geometry, with the thickness being much smaller than the other dimensions

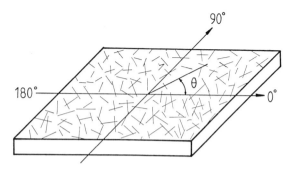

FIG. 1.   Composites with two-dimensional oriented fibers.

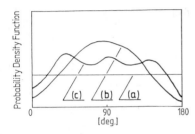

FIG. 2. Typical fiber orientation distribution functions.

of the part. In this paper we deal with the fiber orientation state in the two-dimensional field, where fibers are stratified in the thickness direction, as shown in Fig. 1. It is often observed that the initial orientation of the fiber changes by flow during mold processing and then various fiber orientation states occur in molded parts. These distributions are shown as the probability distribution function for that orientation angle. The fibers have distribution with random orientation in Fig. 2(a), one peak in Fig. 2(b) and several peaks in Fig. 2(c). However, the fiber orientation distribution is characterized by having low and high limits regardless of the complicated shape. The general distributions such as normal and Weibull distributions are not suitable for fiber orientation since the random variable can tend to infinity.

In this study the probability density function with an incomplete beta function, $g(\theta)$, is proposed as the fiber orientation distribution which satisfies the conditions mentioned above. $g(\theta)$ is described as

$$g(\theta) = \frac{\{\sin(\theta/2)\}^{2P-1}\{\cos(\theta/2)\}^{2Q-1}}{\int_{\theta_a}^{\theta_b}\{\sin(\theta/2)\}^{2P-1}\{\cos(\theta/2)\}^{2Q-1}\,d\theta} \tag{1}$$

where $0° \leq \theta_a < \theta_b \leq 180°$, $P \geq 1/2$ and $Q \geq 1/2$.

θ is the orientation angle. θ_a and θ_b are low and high limits, respectively. Using the relation given by

$$X = \sin^2(\theta/2) \tag{2}$$

$$f(X) = \frac{X^{P-1}(1-X)^{Q-1}}{\int_a^b X^{P-1}(1-X)^{Q-1}\,dX} = \frac{X^{P-1}(1-X)^{Q-1}}{B_b(P,Q) - B_a(P,Q)} \tag{3}$$

where

$$a = \sin^2(\theta_a/2) \qquad b = \sin^2(\theta_b/2) \tag{4}$$

Hence the range of x is $0 \leq a \leq X < b \leq 1$. $B_b(P,Q)$ and $B_a(P,Q)$ are incomplete beta functions, shown as follows:

$$B_b(P,Q) = \int_0^b X^{P-1}(1-X)^{Q-1}\,dX$$

$$B_a(P,Q) = \int_0^a X^{P-1}(1-X)^{Q-1}\,dX \tag{5}$$

The distribution function of x, $F(x)$, is given by

$$F(X) = \int_a^x f(X)\,dX = \frac{BX(P,Q) - B_a(P,Q)}{B_b(P,Q) - B_a(P,Q)} \tag{6}$$

Also the mean value, μ, and the variance, σ^2, are represented as

$$\mu = \frac{B_b(P+1,Q) - B_a(P+1,Q)}{B_b(P,Q) - B_a(P,Q)}$$

$$\sigma^2 = \frac{B_b(P+2,Q) - B_a(P+2,Q)}{B_b(P,Q) - B_a(P,Q)} - \left\{ \frac{B_b(P+1,Q) - B_a(P+1,Q)}{B_b(P,Q) - B_a(P,Q)} \right\}^2 \tag{7}$$

We can have various fiber orientation distributions by changing the values of P and Q. Figures 3 and 4 show numerical results of the fiber orientation distribution of eqn (7). These distributions have a low limit of $0°$, a high limit of $180°$ and one peak. In Fig. 3 the curves of $g(\theta)$ in the case of $P=Q$

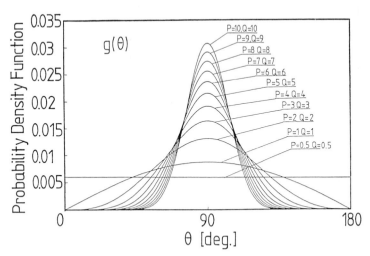

FIG. 3.   Fiber orientation distribution functions of $g(\theta)$ in the case of $P=Q$.

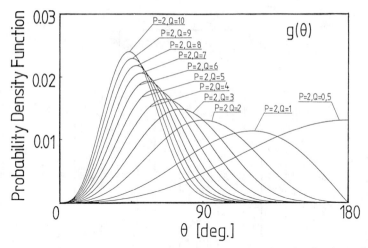

Fig. 4. Fiber orientation distribution functions of $g(\theta)$ in the case of $P = 2$ and various values of Q.

are symmetric about $\theta = 90°$ and have maximum at $\theta = 90°$. The degree of peakedness of these curves are found to increase as the values of two parameters increase.

An example in the case of $P \neq Q$ is shown in Fig. 4. The degree of skewness in these curves has a tendency to increase with an increase in the difference between the two parameters.

Figure 5 shows the relationship between the parameters of $f(x)$ and the various statistic measures, such as mean, mode, standard deviation and maximum value of $f(x)$. Parameter P is held at a constant value of 2. It can be seen from this figure that as the value of Q increases the values of mean and standard deviation decrease, and the maximum value of $f(x)$ and the difference between mode and mean value increase.

In order to apply a proposed theory to fiber composites, two parameter values in $g(\theta)$ have to be determined by using the procedure shown as follows. First, the relative frequency distribution of the fiber orientation is obtained from the observation of the fiber dispersed state. Secondly, two parameter values representing a smooth curve approximating the frequency polygon are determined by the use of statistical inference theory. The flow chart obtaining the values of P and Q are illustrated in Fig. 6. P and Q can be determined by minimizing the sum of the squares of deviations between estimated curve and frequency polygon. Finally, a test of the hypothesis is carried out at the 0·05 level of significance in order to

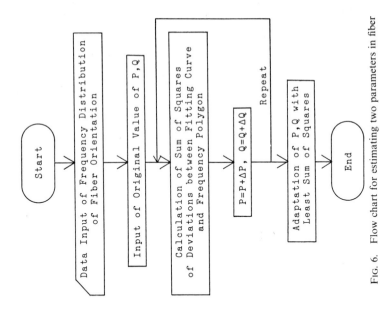

FIG. 6. Flow chart for estimating two parameters in fiber orientation distribution, $g(\theta)$.

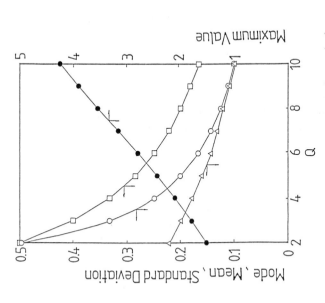

FIG. 5. Influence of Q on the mean, mode, standard deviation and maximum value of $f(x)$. (●, Maximum value; ○, mode; □, mean; △, standard deviation.)

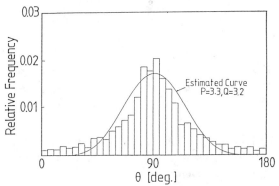

FIG. 7. Comparison of frequency polygon and estimated curve.

examine the significance of parameters. A comparison of the frequency polygon with the estimated curve is shown in Fig. 7.

ESTIMATION METHOD FOR DEFORMATION BEHAVIOR

The composite laminate generally is a stack of laminae with various orientations of principal material directions. A major purpose of this lamination is to tailor the directional dependence of strength and stiffness of a material to match the loading environment of the structural element. In a fiber composite in a molded product, however, it is difficult to tailor mechanical properties because of complicated fiber orientation.

In this study we propose the procedure estimating the deformation behavior of a fiber composite with complex fiber orientation. This estimation method can be performed by connecting a generalized lamination theory to a finite element method.

Generalized Lamination Plate Theory

The fiber composite with complex fiber orientation is assumed to be a stack of unidirectional laminae with various principal material directions. First, consider the stress–strain relation of unidirectional lamina in a plane stress state under the off-axis loading shown in Fig. 8.[9] A lamina is treated as a homogeneous and orthotropic material. In this figure 1 and 2 are the axes of material symmetry, and x and y are the reference coordinate axes of the externally applied stresses. The stress–strain relations in the on-axis system can be given by

$$\{\sigma\} = [Q]\{\varepsilon\} \tag{8}$$

where

$$\{\sigma\} = [\sigma_x, \sigma_y, \sigma_s]^T \qquad \{\varepsilon\} = [\varepsilon_x, \varepsilon_y, \varepsilon_s]^T$$

$$\{Q\} = \begin{vmatrix} Q_{xx} & Q_{xy} & Q_{xs} \\ & Q_{yy} & Q_{ys} \\ \text{Sym.} & & Q_{ss} \end{vmatrix} = \begin{vmatrix} \dfrac{E_x}{1 - v_{xy}v_{yx}} & \dfrac{E_x v_{yx}}{1 - v_{xy}v_{yx}} & 0 \\ & \dfrac{E_y}{1 - v_{xy}v_{yx}} & 0 \\ \text{Sym.} & & E_s \end{vmatrix} \tag{9}$$

where E_x and E_y are Young's moduli in the μ_{xy} and μ_{yx} directions. x and y are Poisson's ratios in the x and y directions. The stress–strain relation in the off-axis system can be given by

$$\{\sigma'\} = [Q']\{\varepsilon'\} \tag{10}$$

where

$$\{\sigma'\} = [\sigma_1, \sigma_2, \sigma_6]^T \qquad \{\varepsilon'\} = [\varepsilon_1, \varepsilon_2, \varepsilon_6]^T$$

$$[Q'] = \begin{vmatrix} Q_{11} & Q_{12} & Q_{16} \\ & Q_{22} & Q_{26} \\ \text{Sym.} & & Q_{66} \end{vmatrix} \tag{11}$$

The transformation equation between the elements of the rigidity matrix in the on-axis system and that in the off-axis system is

$$\{Q'\} = [B_{kl}]\{Q\} \qquad (k = 1, \ldots, 6; \, l = 1, \ldots, 4) \tag{12}$$

where

$$\{Q'\} = \begin{vmatrix} Q_{11} \\ Q_{22} \\ Q_{12} \\ Q_{66} \\ Q_{16} \\ Q_{26} \end{vmatrix} \quad [B_{kl}] = \begin{vmatrix} m^4 & n^4 & 2m^2n^2 & 4m^2n^2 \\ n^4 & m^4 & 2m^2n^2 & 4m^2n^2 \\ m^2n^2 & m^2n^2 & m^4+n^4 & -4m^2n^2 \\ m^2n^2 & m^2n^2 & -2m^2n^2 & (m^2-n^2)^2 \\ m^3n & -mn^3 & mn^3-m^3n & 2(mn^3-m^3n) \\ mn^3 & -m^3n & m^3n-mn^3 & 2(m^3n-mn^3) \end{vmatrix} \{Q\} = \begin{vmatrix} Q_{xx} \\ Q_{yy} \\ Q_{xy} \\ Q_{ss} \end{vmatrix}$$

$$\tag{13}$$

where m is $\cos\theta$ and n is $\sin\theta$.

Consider the laminate consisting of a set of unidirectional laminae with the fiber orientation obeying $g(\theta)$. The transformation equation in the case of this laminate may also be written as

$$\{\bar{Q}\} = [A_{kl}]\{Q\} \tag{14}$$

where

$$A_{kl} = \int_0^{180} g(\theta) H_{kl}\, d\theta \qquad (k = 1, 2, \ldots, 6; \, l = 1, 2, \ldots, 4) \tag{15}$$

Substitution of eqn (1) into eqn (15) leads to an equation in matrix form:

$$
\begin{vmatrix} A_{11} \\ A_{12} \\ A_{13} \\ A_{14} \\ A_{21} \\ A_{22} \\ A_{23} \\ A_{24} \\ A_{31} \\ A_{32} \\ A_{33} \\ A_{34} \\ A_{41} \\ A_{42} \\ A_{43} \\ A_{44} \\ A_{51} \\ A_{52} \\ A_{53} \\ A_{54} \\ A_{61} \\ A_{62} \\ A_{63} \\ A_{64} \end{vmatrix}
=
\begin{vmatrix}
1 & 0 & 0 & 0 & -8 & 0 & 0 & 16 \\
0 & 0 & 0 & 0 & 0 & 0 & 0 & 16 \\
0 & 0 & 0 & 0 & 8 & 0 & 0 & -32 \\
0 & 0 & 0 & 0 & 16 & 0 & 0 & -64 \\
0 & 0 & 0 & 0 & 0 & 0 & 0 & 16 \\
1 & 0 & 0 & 0 & -8 & 0 & 0 & 16 \\
0 & 0 & 0 & 0 & 8 & 0 & 0 & -32 \\
0 & 0 & 0 & 0 & 16 & 0 & 0 & -64 \\
0 & 0 & 0 & 0 & 4 & 0 & 0 & -16 \\
0 & 0 & 0 & 0 & 4 & 0 & 0 & -16 \\
1 & 0 & 0 & 0 & -8 & 0 & 0 & 32 \\
0 & 0 & 0 & 0 & -16 & 0 & 0 & 64 \\
0 & 0 & 0 & 0 & 4 & 0 & 0 & -16 \\
0 & 0 & 0 & 0 & 4 & 0 & 0 & -16 \\
0 & 0 & 0 & 0 & -8 & 0 & 0 & 32 \\
1 & 0 & 0 & 0 & -16 & 0 & 0 & 64 \\
0 & 2 & -2 & 0 & 0 & -8 & 8 & 0 \\
0 & 0 & 0 & 0 & 0 & -8 & 8 & 0 \\
0 & -2 & 2 & 0 & 0 & 16 & -16 & 0 \\
0 & -4 & 4 & 0 & 0 & 32 & -32 & 0 \\
0 & 0 & 0 & 0 & 0 & 8 & -8 & 0 \\
0 & -2 & 2 & 0 & 0 & 8 & -8 & 0 \\
0 & 2 & -2 & 0 & 0 & -16 & 16 & 0 \\
0 & 4 & -4 & 0 & 0 & -32 & 32 & 0
\end{vmatrix}
\begin{vmatrix}
\dfrac{B_b(P,Q)-B_a(P,Q)}{B_b(P,Q)-B_a(P,Q)} \\
\dfrac{B_b(P+1/2,Q+3/2)-B_a(P+1/2,Q+3/2)}{B_b(P,Q)-B_a(P,Q)} \\
\dfrac{B_b(P+3/2,Q+1/2)-B_a(P+3/2,Q+1/2)}{B_b(P,Q)-B_a(P,Q)} \\
\dfrac{B_b(P+1,Q+1)-B_a(P+1,Q+1)}{B_b(P,Q)-B_a(P,Q)} \\
\dfrac{B_b(P+3/2,Q+5/2)-B_a(P+3/2,Q+5/2)}{B_b(P,Q)-B_a(P,Q)} \\
\dfrac{B_b(P+5/2,Q+3/2)-B_a(P+5/2,Q+3/2)}{B_b(P,Q)-B_a(P,Q)} \\
\dfrac{B_b(P+2,Q+2)-B_a(P+2,Q+2)}{B_b(P,Q)-B_a(P,Q)}
\end{vmatrix}
\quad (16)
$$

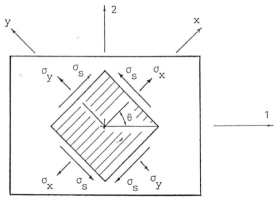

FIG. 8.   Relationship between off-axis and on-axis configurations.

where B_a and B_b are incomplete beta functions defined in eqn (5). Accordingly, the stress–strain relations of a laminate in the off-axis system can be written as

$$\{\sigma'\} = [\bar{Q}]\{\varepsilon'\} \qquad (17)$$

Application of Finite Element Analysis

Next, consider the mechanical properties of molded plates. It is often observed in these plates that there are various fiber orientation states in each part. In order to estimate exactly the deformation behavior of molded plates, therefore, it is necessary to calculate the deformation of the plate composed of each part with the rigidity corresponding to the fiber orientation state. Employing a finite element analysis is considered to be strongly useful in this study;[10] because in this analysis a composite plate is divided into a number of elements and each element can have certain mechanical properties. Therefore we estimated the deformation of the composite plate with complex fiber orientation by the procedure given as follows: (1) division of composite plate into a number of elements, (2) application of the elastic moduli obtained by eqn (14) to each element, and (3) calculation of the deformation behavior of composite plates by two-dimensional finite element analysis.

NUMERICAL RESULTS AND DISCUSSION

Three kinds of elastic moduli of fibrous composites with various fiber orientation distributions were calculated by use of the proposed lamination

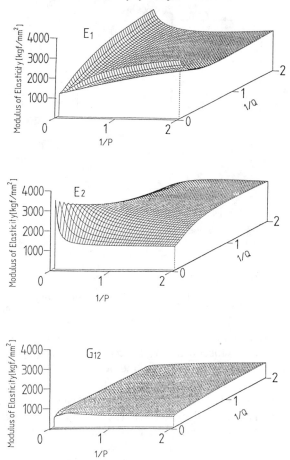

FIG. 9. Relationship between three kinds of elastic moduli and two parameters in $g(\theta)$.

theory. Figure 9 shows the relationship between three kinds of elastic moduli, E_1, E_2, G_{12}, and two parameters of the orientation distribution, where E_1 and E_2 are the elastic moduli at the 1 and 2 axes, and G_{12} is the rigidity modulus at the 1–2 plane. The horizontal axes are the reciprocal of P and Q, respectively. A fiber orientation distribution function can be expressed by a point in a horizontal plane in this figure. The point $(0,0)$ in the horizontal plane, for example, shows the orientation distribution of the unidirectional laminate and the point $(2,2)$ shows that of random oriented laminates. It can be found from this figure that the elastic moduli of E_1, E_2 and G_{12} depend strongly on the fiber orientation function of $g(\theta)$.

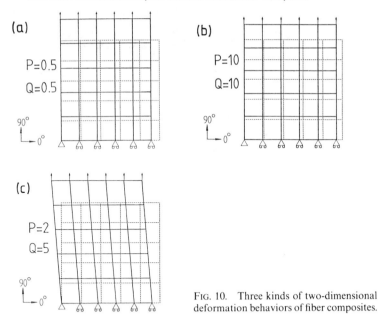

Fig. 10.   Three kinds of two-dimensional deformation behaviors of fiber composites.

Next, the deformation of complex fiber oriented composites was computed in accordance with the procedure connecting the proposed lamination theory to the finite element method. Figure 10 shows three kinds of the numerical results on deformation of a fiber composite plate subjected to uniform tensile load in the 90° direction. In this calculation a rectangular plate is divided into 16 elements and each element has the same fiber orientation distribution and material constants. The fiber orientation distribution in Fig. 10(a) is a uniform distribution curve with $P = Q = 0.5$, that in Fig. 10(b) is the curve with $P = Q = 10$ which has a peak at $\theta = 90°$, and that in Fig. 10(c) is the curve with $P = 2$ and $Q = 5$ which is skewed to the right. The values of the material constants used are $E_x = 181$ GPa, $E_y = 10.3$ GPa, $E_s = 7.17$ GPa and $\nu_{xy} = 0.28$. These values correspond to those of the unidirectional carbon fiber reinforced composite with a fiber content of about 60%. The material in Fig. 10(a) can be observed to have an isotropic property which elongates in the 90° direction and compresses in the 0° direction. The orthotropic material under an on-axis state in Fig. 10(b) shows the same deformation behavior as the isotropic material in Fig. 10(a). But the elongation in the 90° direction is recognized to be smaller than that in Fig. 10(a). It should be noticed in Fig. 10(c) that the orthotropic material under an off-axis state has shear deformation in addition to normal deformation. Though this numerical example may well be simple, the proposed procedure using a finite element method can be applied to a difficult problem such as analysis of deformation behavior of a molded plate which has various fiber oriented states in each part.

CONCLUSIONS

This study was carried out to estimate the deformation behaviors of polymeric composites with complicated fiber orientation states. The main conclusions obtained in this paper may be summarized as follows.

1. A new method which expressed fiber orientation distribution in terms of an incomplete beta function was presented. Various orientation distributions can be represented by selecting the values of two parameters in the beta function.
2. A generalized laminated plate theory was developed by including an orientation distribution function in classical laminated theory. The stress–strain relations for fibrous composites with complicated fiber flow could be formulated.
3. A procedure connecting the proposed laminated theory to a finite element method enabled estimation of the deformation behavior of complex fiber oriented composites.

REFERENCES

1. BERTHELOT, J. M., Moulding influence on the mechanical properties of sheet moulding compounds: Part I—Elastic properties. *Fib. Sci. Technol.*, **17** (1982) 235–44.
2. MAEKAWA, Z. and FUJII, T., Probabilistic design of strength of fiber reinforced composite laminates. *Proc. ICCM—V* (1982) 537–40.
3. LEE, C. C., FOLGAR, F. and TUCKER, C. L. III, Simulation of compression molding for fiber reinforced thermosetting polymers. *ASME Trans., J. Engng Ind.*, **106** (1984) 106–25.
4. SURESH, G. A. and TUCKER, C. L. III, The use of tensors to describe and predict fiber orientation in short fiber composites. *J. Rheol.*, **31** (1987) 751–84.
5. HAMADA, H., MAEKAWA, Z., HORINO, T. and LEE, K., Improvement of weld line strength in injection molded FRTP articles. *J. Int. Polym. Proc.*, **II** (1988) 131–6.
6. JACQUES, M. S., An analysis of thermal warpage in injection molded flat parts due to unbalanced cooling. *Polym. Engng Sci.*, **22** (1982) 241–7.
7. HIRAI, T., Design concept of SMC compression molding to prevent a fault caused by flow state. *Proc. ICCM—V & ECCM—II* (1987) 1.121–1.130.

8. KIM, S. G. and SUH, N. P., Performance prediction of weld line structure in amorphous polymers. *Polym. Engng Sci.*, **26** (1986) 1200–7.
9. AGARWAL, B. D. and BROUTMAN, L. J., *Analysis and Performance of Fiber Composites.* John Wiley, New York, 1979.
10. PANDYA, B. N. and KANT, T., Finite element analysis of laminated composite plates using a higher-order displacement model. *Comp. Sci. Technol.*, **32** (1988) 137–55.

42

Constitutive Modelling of Laminated Fibre-Reinforced Composites

R. VAZIRI, M. D. OLSON and D. L. ANDERSON

Department of Civil Engineering, University of British Columbia, Vancouver, BC, Canada

ABSTRACT

A new macroscopic approach to modelling the nonlinear and failure behaviour of laminated fibre-reinforced material based on incremental plasticity theory is presented. The material model proposed is essentially representative of the mechanical behaviour of a single layer of the laminate. Both unidirectional and bidirectional FRM layers are modelled. The constitutive equations developed are combined using classical lamination theory to form the governing response relations for multilayer laminates. Numerical tests of the model are compared with available experimental data with good success.

1 INTRODUCTION

Analytical modelling of fibre-reinforced material (FRM) has been approached from either the micromechanics level or the macromechanics level. So far the micromechanics approach has proved to be impractical for stress analysis, whereas the macromechanics approach which describes the behaviour of the FRM by continuum models without direct reference to the properties of the individual constituents (fibre and matrix) is appealing to the engineer. Therefore the latter approach is the one followed in this paper.

Existing macroscopic approaches have been based on, for example, nonlinear elasticity,[1-4] incremental plasticity,[5,6] as well as endochronic plasticity theory.[7] The mechanical response of FRMs is very complicated and it seems unlikely that any phenomenological approach will ever be able to embrace all possible variations in material characteristics. The objective of this paper is to propose a relatively simple elastic–plastic failure theory which adequately reflects certain typical trends in FRM behaviour and lends itself to straightforward implementation in finite element codes.

2 ANALYTICAL FORMULATION

We describe the behaviour of FRMs in terms of the classical incremental theory of plasticity for small displacements, ignoring time and temperature effects. The proposed constitutive model, which represents the mechanical behaviour of a single layer, forms the basic building block in the analysis of laminated FRMs. This model is then combined with classical lamination theory to arrive at the governing response equations for symmetric laminates under in-plane loading.

The nonlinear and/or irreversible deformation of FRMs is a complex phenomenon which can be caused by inherent material nonlinearities of the individual constituents, damage accumulation resulting from fibre or matrix cracking, fibre–matrix debonding or any combination of the above. The resulting loss of local structural integrity causes load transferral which is similar to plasticity in pure metals. In light of this, the current work utilizes an elastic–plastic failure model to capture some of the basic response characteristics of laminated FRMs.

The model encompasses three regimes: the elastic, plastic and post-failure regimes, as shown schematically in Fig. 1. Between the initial yielding and the failure states, the constitutive relations are expressed in incremental form based on the associated flow rule of plasticity theory. The initial yield criterion and the failure criterion are assumed to have similar functional forms in stress space. When failure is reached it is ascribed to either matrix or fibre failure depending on the relative magnitude of the various stress ratio terms appearing in the criterion. To simulate post-failure behaviour, two types of failure modes are defined, namely brittle and ductile. For the brittle mode, the layer is assumed to lose its entire rigidity and strength in the dominant stress direction. For the ductile mode, the layer retains its strength but loses all of its stiffness in the failure direction.

2.1 Elastic Regime

The FRM layer is treated as a homogeneous and orthotropic continuum, with x_1 and x_2 being the principal axes of orthotropy in the mid-plane of

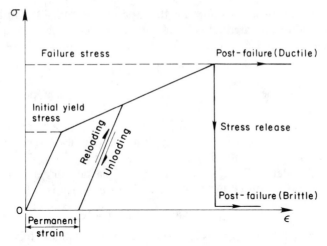

FIG. 1.   Idealized uniaxial stress–strain curve.

the layer. In the linear elastic range, the plane stress incremental stress–strain relationship can be explicitly expressed, in the usual contracted notation, as

$$d\sigma_i = Q_{ij}^e \, d\varepsilon_j \qquad (i, j = 1, 2, 6) \tag{1}$$

where Q_{ij}^e is the standard elastic stiffness tensor given by

$$[Q^e] = \frac{1}{1 - v_{12}v_{21}} \begin{bmatrix} E_1 & v_{12}E_2 & 0 \\ & E_2 & 0 \\ \text{Sym.} & & G_{12}(1 - v_{12}v_{21}) \end{bmatrix} \tag{2}$$

2.2 Plastic Regime

The simplest yield condition for orthotropic plastic materials is the quadratic function given by

$$f(\sigma_i, A_{ij}, k) \equiv A_{ij}\sigma_i\sigma_j - k^2 = 0 \qquad (i, j = 1, 2, 6) \tag{3}$$

where the A_{ij} and k are the orthotropic strength parameters describing the shape and effective size of the yield surface, respectively. Because of orthotropy and from symmetry arguments the only nonzero A_{ij} terms are $A_{11}, A_{22}, A_{12} = A_{21}$ and A_{66}.

To ensure boundedness of the yield surface in the $\sigma_1, \sigma_2, \sigma_6$ stress space, the components of A_{ij} must satisfy the following inequalities:

$$A_{11}A_{22} - A_{12}^2 > 0 \qquad A_{11}A_{66} > 0 \qquad A_{22}A_{66} > 0 \tag{4}$$

Making the usual assumption that the increment of the total strain tensor can be separated into elastic and plastic components, and that the plastic strain increment obeys the associated flow rule, we have

$$d\varepsilon_i^p = d\lambda(\partial f/\partial\sigma_i) \tag{5}$$

where $d\varepsilon_i^p$ is the plastic strain increment and $d\lambda$ is a positive scalar parameter.

An orthotropic hardening rule[8,9] which allows for a nonproportional change of the yield values during plastic flow is used. The underlying assumption in this model is that during plastic loading in any of the principal material directions the amount of plastic work produced is the same as that produced by the effective stress $\bar{\sigma}$, which is defined by

$$\bar{\sigma}^2 = A_{ij}\sigma_i\sigma_j \qquad (i,j = 1,2,6) \tag{6}$$

If the stress–strain diagrams can be represented in bilinear form, then the plastic work W^p can explicitly be written as

$$W^p = \int dW^p = \frac{1}{2E_{p_i}}(\Gamma_i^2 - \Gamma_{0_i}^2) \qquad (i = 1,2,6) \tag{7}$$

where Γ_{0_i} and Γ_i are the initial and subsequent yield values and E_{p_i} ($= d\sigma_i/d\varepsilon_i^p$) is the plastic modulus, all referred to a generic σ_i–ε_i stress–strain diagram.

Equating the plastic work given by eqn (7) to the plastic work done by the effective stress leads to

$$\Gamma_i^2 = (E_{p_i}/H')(\bar{\sigma}^2 - \bar{\sigma}_0^2) + \Gamma_{0_i}^2 \qquad (i = 1,2,6) \tag{8}$$

where H', termed the hardening modulus, is defined as the slope of the effective stress $(\bar{\sigma})$–effective plastic strain $(\bar{\varepsilon}^p)$ curve and $\bar{\sigma}_0$ is the initial effective yield stress. It should be noted that one of the σ_i–ε_i curves can be arbitrarily prescribed as the $\bar{\sigma}$–$\bar{\varepsilon}$ diagram, while the remaining stress–strain curves are then normalized with respect to the prescribed curve. By imposing uniaxial tensile tests along the x_1 and x_2 axes as well as a pure shear test in the plane of the ply, the principal anisotropic parameters at any state of plastic deformation described by $\bar{\sigma}$ can be written as

$$A_{ii} = \bar{\sigma}^2/\Gamma_i^2 \qquad (i = 1,2,6; \text{ summation not implied}) \tag{9}$$

The off-diagonal term A_{12} can usually be obtained from biaxial loading conditions. However, for the particular yield criteria to be defined later, the A_{12} parameter can be expressed in terms of the A_{ii}. From eqns (8) and (9) it can be inferred that the yield values and hence the orthotropic parameters are in general functions of the effective stress reached and are therefore history dependent, i.e. $A_{ij} = A_{ij}(\bar{\sigma})$.

Following the standard formulation of plasticity theory the incremental stress–strain relationship in the plastic region can be obtained as

$$\{d\sigma\} = [Q^{ep}]\{d\varepsilon\} = \left([Q^e] - \frac{[Q^e]\{a\}\{a\}^T[Q^e]}{\mu H' + \{a\}^T[Q^e]\{a\}}\right)\{d\varepsilon\} \tag{10}$$

where $[Q^{ep}]$ is the 3×3 elastoplastic constitutive matrix and

$$\{a\} = 1/\bar{\sigma}[A]\{\sigma\} \tag{11}$$

$$\mu = 1 - 1/2\bar{\sigma}\{\sigma\}^T[\partial A/\partial\bar{\sigma}]\{\sigma\} \tag{12}$$

The above formulation applies to both unidirectional and bidirectional layers. The difference in the two types of material is in the way their yield (and failure) surfaces are defined, which is given below.

2.2.1 Bidirectional layers

The Puppo–Evensen criterion[10] (which was originally suggested for the prediction of failure in multilayer laminates) requires only three material constants for its evaluation and also possesses features that are ideal for modelling a wide range of B/D FRMs. This criterion, which we propose to use as both yield and failure criterion, can be written as

$$\left(\frac{\sigma_1}{\Gamma_1}\right)^2 - \Lambda\left(\frac{\Gamma_1}{\Gamma_2}\right)\frac{\sigma_1}{\Gamma_1}\frac{\sigma_2}{\Gamma_2} + \Lambda\left(\frac{\sigma_2}{\Gamma_2}\right)^2 + \left(\frac{\sigma_6}{\Gamma_6}\right)^2 = 1 \tag{13a}$$

or

$$\Lambda\left(\frac{\sigma_1}{\Gamma_1}\right)^2 - \Lambda\left(\frac{\Gamma_2}{\Gamma_1}\right)\frac{\sigma_1}{\Gamma_1}\frac{\sigma_2}{\Gamma_2} + \left(\frac{\sigma_2}{\Gamma_2}\right)^2 + \left(\frac{\sigma_6}{\Gamma_6}\right)^2 = 1 \tag{13b}$$

where

$$\Lambda = 3\Gamma_6^2/\Gamma_1\Gamma_2 \tag{14}$$

Equations (13a,b) describe a pair of ellipsoids in the stress space $\sigma_1, \sigma_2, \sigma_6$. The yield surface is taken to be the inner surface resulting from the intersection of these ellipsoids. The condition for the boundedness of the yield surface (eqn (4)) now becomes

$$\Gamma_6^2 < \tfrac{4}{3}\Gamma_1^3/\Gamma_2 \quad \text{and} \quad \Gamma_6^2 < \tfrac{4}{3}\Gamma_2^3/\Gamma_1 \tag{15}$$

For strongly anisotropic materials (e.g. unidirectional composites) at least one of the inequalities in eqn (15) may be violated. This makes the Puppo–Evensen criterion unsuitable for prediction of yielding in U/D materials.

2.2.2 Unidirectional layers

Here we propose Azzi and Tsai's[11] extension of Hill's[12] criterion for a transversely isotropic medium. Assuming that the x_1 direction coincides with the fibre direction this criterion reads

$$\left(\frac{\sigma_1}{\Gamma_1}\right)^2 + \left(\frac{\sigma_2}{\Gamma_2}\right)^2 - \frac{\sigma_1\sigma_2}{\Gamma_1^2} + \left(\frac{\sigma_6}{\Gamma_6}\right)^2 = 1 \qquad (16)$$

2.3 Failure

When eqns (13) and (16) are used as failure criteria, the Γ_i $(i = 1, 2, 6)$ quantities must be replaced by their ultimate values Γ_{u_i} $(i = 1, 2, 6)$. Once the state of stress satisfies the failure criterion, fibre or matrix cracking ensues depending on the relative magnitude of various stress ratio terms, e.g. if σ_1/Γ_{u_1} is greatest fibre failure is assumed, and so on.

2.4 Post-failure Regime

During post-failure loading, the elastoplastic constitutive matrix $[Q^{ep}]$ is modified such that the material cannot carry any additional normal or shear stress. This is accomplished by setting to zero the rows and columns of $[Q^{ep}]$ corresponding to the dominant stresses. These modifications apply whether the failure process is assumed to be brittle or ductile. However, in the former case the relevant stress components on the cracked planes occurring just before failure must be removed abruptly and redistributed to adjacent uncracked material of the laminate.

2.5 Multilayer Laminates

The foregoing elastic–plastic failure model for each layer is combined to form the following incremental laminate force–strain relations:

$$\{dN\} = \sum_{k=1}^{n} [Q']_k t_k \{d\varepsilon'\} \qquad (17)$$

where $\{N\} = \{N_x N_y N_{xy}\}^T$ is the vector of the resultant membrane forces in the overall laminate coordinate system (x, y), t_k is the thickness of the kth layer, n is the number of layers and primes denote that quantities are transformed from ply coordinates (x_1, x_2) to laminate coordinates (x, y). In eqn (17), $[Q]$ stands for elastic, elastoplastic and post-failure constitutive matrices, whichever is applicable.

3 RESULTS

The model has been successfully coded in a nonlinear finite element program and used to analyse a variety of composite laminates and loading combinations. The results of comparison with experiments are highly encouraging.

A representative number of laminates of boron/epoxy U/D composites for which uniaxial tensile test data had been obtained by Petit and Waddoups[1] were examined numerically. Best-fit bilinear stress–strain representations for σ_1–ε_1, σ_2–ε_2 and σ_6–ε_6 curves were used. The resulting layer properties used as input to the model are given in Table 1. Note that E_{T_1}, E_{T_2} and G_T are the tangent moduli.

In Fig. 2, the experimental stress–strain curve for a $[0°/90°]_s$ cross-ply boron/epoxy laminate is compared to the results of the present analysis. It appears that the two failure models have provided a good bound to the actual behaviour after matrix cracking in the 90° layer. The crosses in this and subsequent diagrams indicate the point at which ultimate failure occurs.

The results for the case of a $[\pm 45°]_s$ laminate of boron/epoxy are shown in Fig. 3. Clearly, the present analysis predicts very closely the ultimate strength and general shape of the experimental curve. The highly nonlinear

TABLE 1
Single layer properties for various FRMs

Input parameters		U/D boron/epoxy[1] (ksi)	U/D graphite/epoxy[13] (ksi)	U/D boron/aluminium[14] (MPa)	B/D Glass fabric/ polyester resin[15] (ksi)
Elastic	E_1	30 000	20 500	209 700	2 742
	E_2	3 080	1 400	107 000	2 520
	v_{12}[a]	0·3	0·26	0·2	0
	G_{12}	1 000	600	32 000	634
Plastic	E_{T_1}	26 100	20 500	202 700	2 111
	E_{T_2}	2 202	1 400	24 340	2 010
	G_T	180	381	1 486	193
	Γ_{0_1}	132·5	185·6	1 216	34
	Γ_{0_2}	9	7·5	89·2	25·2
	Γ_{0_6}	10	6·75	45	4·88
Failure	Γ_{u_1}	200	185·6	1 662	49·2
	Γ_{u_2}	12·5	7·5	117·6	45·3
	Γ_{u_6}	18·6	11·8	110	13·44

[a] v_{12} is unitless.

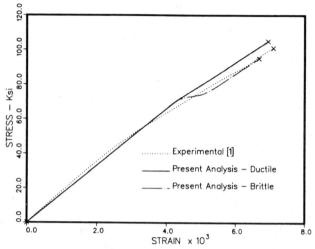

FIG. 2.   Tensile stress–strain curve for a $[0°/90°]_s$ boron/epoxy laminate.

nature of the response in this case is caused by the presence of a considerable amount of shear strain in the plies of the laminate. It should be noted that for this particular configuration of the laminate (i.e. angle-ply) ultimate failure coincides with the failure of one layer, and thus the ductile and brittle post-failure models do not enter the calculations.

Two further test cases were investigated in which the overall loadings

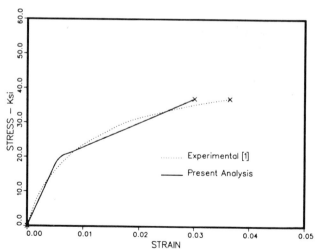

FIG. 3.   Tensile stress–strain curve for a $[\pm45°]_s$ boron/epoxy laminate.

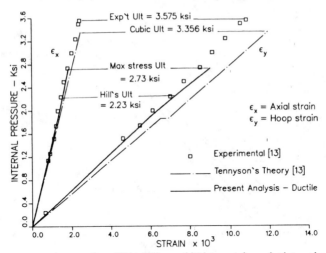

FIG. 4. Pressure–strain curve for a $[0°/\pm 60°]_s$ graphite/epoxy tube under internal pressure.

were biaxial. A graphite/epoxy laminated tube of $[0°/\pm 60°]_s$ construction subjected to internal pressure and torque tested by Tennyson[13] was analysed. The single ply properties are given in Table 1, where it can be seen that nonlinear effects are insignificant. A plot of the pressure–strain curve is shown in Fig. 4 for the case of internal pressure only (proportional stress path). The effects of a constant pre-torque and internal pressure

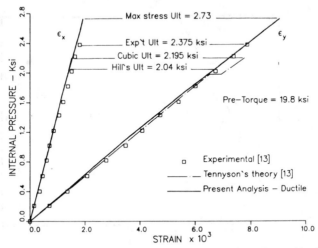

FIG. 5. Pressure–strain curve for a $[0°/\pm 60°]_s$ graphite/epoxy tube under combined torsion and internal pressure.

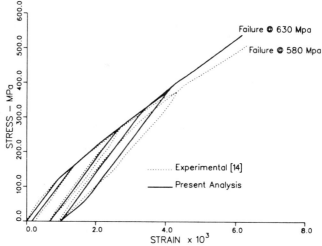

FIG. 6.   Stress–strain curve for a $[0°/\pm 45°]_s$ boron/aluminium laminate subjected to three
loading cycles.

(nonproportional stress path) on the results were also studied and are shown in Fig. 5. The ductile failure model was employed in both analyses.

The effect of cyclic loading is clearly illustrated by the test case documented in Ref. 14. Here a $[0°/\pm 45°]_s$ laminate of boron/aluminium is subjected to three loading cycles. The input parameters are given in Table 1,

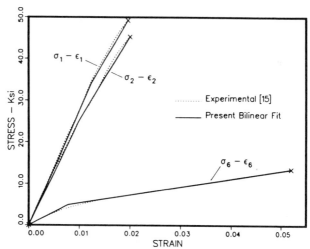

FIG. 7.   Basic input stress–strain curves for a bidirectional layer made with 181 glass fabric
and polyester resin.

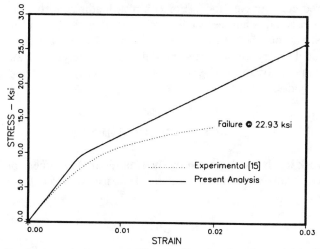

FIG. 8. Tensile stress–strain curve at 45° to the fibre directions for material of Fig. 7.

and the results of the present computations are compared to the experimental results in Fig. 6. The prediction of residual strains and hysteresis loops are seen to be in good agreement with the experimental results. The load cycles did not affect the ultimate stress and strain levels of the laminate, thus supporting the experimental findings of Ref. 14.

Figures 7 and 8 present results for the case of a polyester resin matrix reinforced by bidirectional woven glass fabric. The three basic experimental stress–strain curves[15] and their bilinear representations are shown in Fig. 7. The resulting material constants used as input data are tabulated in Table 1. The uniaxial tensile stress–strain response at 45° to the fibres is shown in Fig. 8, where it can be seen that the experimental curve falls significantly below the predicted one but the quoted failure stress of 22·9 ksi is in reasonable agreement with the predicted value of 26 ksi.

4 SUMMARY AND CONCLUSIONS

A relatively simple elastic–plastic failure model has been developed to represent the nonlinear stress–strain behaviour of laminated FRMs. The comparison between the present predictions and a variety of experimental data illustrates the accuracy and versatility of the model. Even though the experimental results are somewhat limited, the excellent agreement between the numerical and experimental results provides validity to the proposed constitutive model. For complex loadings such as nonproportional stress paths and especially cyclic loadings the present incremental plasticity model should be superior to existing nonlinear elastic models. The model is very simple, requiring only a few material constants as input for its application to various composites and is easy to implement in finite element codes.

ACKNOWLEDGEMENT

This work has been supported by the Canadian Department of National Defence through a contract with the Defence Research Establishment, Suffield.

REFERENCES

1. PETIT, P. H. and WADDOUPS, M. E., A method of predicting the nonlinear behaviour of laminated composites. *J. Comp. Mater.*, **3** (1969) 2–19.
2. HASHIN, Z., BAGCHI, D. and ROSEN, B. W., Nonlinear behaviour of fibre composite laminates. NASA Report No. CR-2313, 1974.
3. HAHN, H. T. and TSAI, S. W., Nonlinear elastic behaviour of unidirectional composite laminates. *J. Comp. Mater.*, **7** (1973) 102–18.
4. JONES, R. M. and MORGAN, H. S., Analysis of nonlinear stress–strain behaviour of fiber-reinforced materials. *AIAA J.*, **15** (1977) 1669–76.
5. GRIFFIN, O. H., KAMAT, M. P. and HERAKOVICH, C. T., Three-dimensional inelastic finite element analysis of laminated composites. *J. Comp. Mater.*, **15** (1981) 543–60.
6. LEEWOOD, A. R., DOYLE, J. F. and SUN, C. T., Finite element program for analysis of laminated anisotropic elastoplastic materials. *Comput. Struct.*, **25** (1987) 749–58.
7. PINDERA, M. J. and HERAKOVICH, C. T., An endochronic model for the response of unidirectional composites under off-axis tensile load. In *Mechanics of Composite Materials, Recent Advances*. Proceedings IUTAM Symposium on Mechanics of Composite Materials, ed. Z. Hashin and C. T. Herakovich. Pergamon Press, Oxford, 1983, pp. 367–81.
8. JENSEN, W. R., FALBY, W. E. and PRINCE, N., Matrix analysis methods for anisotropic inelastic structures. AFFDL-TR-65-220, Air Force Flight Dynamics Laboratory, Wright–Patterson Air Force Base, Ohio, April 1966.
9. WHANG, B., Elasto-plastic orthotropic plates and shells. *Proc. Symp. on Application of Finite Element Method in Civil Engineering*, Vanderbilt University, Tennessee, 1969, pp. 481–515.
10. PUPPO, A. H. and EVENSEN, H. A., Strength of anisotropic materials under combined stresses. *AIAA J.*, **10** (1972) 468–74.
11. AZZI, V. D. and TSAI, S. W., Anisotropic strength of composites. *Exptl Mech.*, **5** (1965) 283–8.

12. HILL, R., *The Mathematical Theory of Plasticity*. Oxford University Press, Oxford, 1950.
13. TENNYSON, R. C., Application of the cubic strength criterion to the failure analysis of composite structures. NASA-CR-165712, 1981.
14. SOVA, J. A. and POE, C. C. Jr, Tensile stress–strain behaviour of boron/ aluminum laminates. NASA-TP-1117, 1978.
15. MIL-HDBK-17, *Plastics for Flight Vehicles, Part I, Reinforced Plastics*. US Government Printing Office, Washington, DC, Nov. 5, 1959.

43

Toughness Testing of Fibre Reinforced Concrete

P. J. Robins, S. A. Austin and C. H. Peaston

*Structural Materials Research Group, Department of Civil Engineering,
University of Technology, Loughborough, Leicestershire LE11 3TU, UK*

ABSTRACT

*An experimental investigation into the effect of beam geometry and test
control method on toughness index is described. The results generally support
the view that the I_5, I_{10} and I_{30} ASTM C1018 toughness indices are
unaffected by variation in test spans between 300 and 500 mm. Different post-
crack load–deflection behaviour was observed using crosshead and strain
(beam deflection) control, resulting in significantly different values of
toughness index. Whilst it is common to test under simple crosshead control
(which is not in accordance with ASTM C1018), this practice should be
reviewed in the light of the different beam behaviour and measured values that
result.*

1 INTRODUCTION

Toughness is generally recognised as the property most enhanced by the
addition of fibre reinforcement to concrete. Historically, prior to the use of
fibres to reinforce concrete, there was no requirement for any standard
concrete tests to assess toughness or its associated properties of post-
cracking strength and ductility, and it has therefore taken time to develop
acceptable methods for testing for these new properties. Toughness index
methods have evolved as the most popular practical technique for
quantitative measurement of toughness. A number of toughness index
definitions have been proposed, all defined as ratios of portions of the area
under load–deflection graphs produced when testing fibre reinforced
concrete specimens. Of these, the ASTM C1018[1] procedure, originally

729

proposed by Johnston,[2] has emerged as the most widely applicable and reproducible test method.

At first sight the ASTM C1018 procedure appears to be a simple adaptation of a traditional modulus of rupture test on a plain concrete beam, the only additional requirement being the production of a load–deflection curve. However, testing in strict accordance with the standard is quite complicated, requiring specimens to be tested under a controlled rate of strain, as well as the use of a minimum of four, and possibly up to eight, deflection measuring devices. This sophistication is beyond most testing laboratories and therefore in most cases a compromise on the rigorous approach of ASTM C1018 is adopted. For example, it has become accepted practice to use a loading apparatus in which the rate of platen movement is controlled rather than the rate of deflection (or strain) of the specimen itself, and to measure deflections in such tests by the use of a single measuring device placed under the centre of the flexural specimen at its mid-span. These compromised approaches to ASTM C1018 clearly have implications for the repeatability of the test and also for the absolute values of the indices obtained.

This paper presents the results of an investigation into the techniques used in flexural toughness testing of fibre reinforced beams and the implications for standardised test methods. The principal objectives of the study were:

(i)   to assess the effect of variation in span/depth ratio on toughness index values;

(ii) to assess the effects of testing under crosshead control as a compromise to the rigorous strain control testing procedure of ASTM C1018;

(iii) to examine the independence of ASTM C1018 toughness index definitions from test geometry; and

(iv) since this work formed part of a larger project on sprayed fibre concrete, to determine the most suitable practical technique for measuring flexural strength and toughness parameters in subsequent tests on sprayed concrete materials.

2 TOUGHNESS MEASUREMENT

2.1 Requirements

The main requirements of any proposed toughness testing technique may be summarised by the following provisions:

(i) specimen adaptability—the method should be independent of size and geometry, enabling testing of specimens relevant to the end use of the material;

(ii) precision—to clearly distinguish levels of performance of different fibre combinations the variability should be low enough to achieve satisfactory repeatability with the same operator–machine combination and reproducibility between different operators in different laboratories using essentially the same equipment; and

(iii) material behaviour—the range of toughness values should be large and be able to quantitatively define performance relative to an easily understandable reference level of material behaviour.

2.2 Development of Toughness Indices

The first formal definition of a toughness index was provided by Henager,[3] and later adopted by ACI Committee 544.[4] Using load–deflection curves produced from modulus of rupture tests on 102×102 mm beams over a 305 mm span in third point loading, toughness index was defined as the area up to a central deflection of 1·9 mm divided by that to first crack (first crack being defined as the limit of proportionality of the initial linear response). The main disadvantage of this index is that the results are only reproducible if the same sized specimens and testing machines are used.

In a paper published in 1982 Johnston[2] proposed a toughness index with the denominator based on the area to first crack as with the ACI index, but with the numerator based on the area up to certain multiples of the first crack deflection. The precise multiple of first crack deflection can be chosen to suit service conditions. This new index also introduced the idea of perfectly elastic–plastic behaviour where the load would hold constant after first crack. The index value corresponding to such behaviour can be calculated at any deflection and used as a reference when analysing toughness indices. Johnston suggested using cut-off deflections of 3, 5·5 and 15·5 times the first crack deflection; corresponding to these cut-offs a perfectly plastic beam would have toughness index values of 5, 10 and 30 respectively. Johnston[5] showed these new indices to be considerably less variable than the ACI definition, and his system of indices was eventually to be incorporated into ASTM Standard C1018,[1] though this was not published until 1984.

In 1982 Barr et al.[6,7] also published a new definition of toughness index. Their definition was also based on a multiple of first crack deflection, in that the numerator is the area up to a deflection of twice that at first crack. The denominator he proposed is essentially the same as in previous definitions except that the area to first crack is multiplied by a factor of 4. Using this multiplying factor means that the toughness index lies in the range 0·25–1·00 and that a value of 0·75 corresponds to perfectly elastic–plastic behaviour.

Barr's definition can in fact be applied to cover any multiple of first crack deflection, the numerator in each case being the area under the load–deflection graph whilst the denominator is taken as the area under the idealised initial linear response (which would have occurred in the absence of cracking) up to the same deflection point. These definitions were illustrated in a paper published in 1985,[8] when a slightly modified definition of the index was also given, the numerator this time being taken as the area under the load–deflection curve *after* first crack.

An alternative toughness definition has been proposed by the Japanese Concrete Institute.[9] In their test method beams are tested under third-point loading and flexural toughness is taken simply as the total area under the load–deflection curve up to a deflection of $\frac{1}{150}$th of the span, expressed in absolute units of energy.

2.3 Dependence on Specimen and Test Geometry

One of the major restrictions on the use of both the ACI[4] and JCI[9] test methods is their inability to be adapted to suit varying test conditions. Both indices specify an arbitrarily prescribed end-point deflection which relates to a particular specimen geometry; the results are therefore only reproducible if exactly the same sized specimen and testing machine are used. Zollo[10] used analysis of variance techniques to show how the ACI index was critically dependent on span/depth ratio.

The ASTM and Barr's proposed indices do not suffer from the same restriction. As they are defined in terms of multiples of first crack deflection, they are potentially applicable to any test from which a load–deflection plot can be produced. Indeed when Johnston first published the proposals that led to ASTM C1018 he also published data[2,5] that inferred that his index was independent of span/depth ratio. To date no further evidence of the independence of the ASTM index from specimen and test geometries has been published.

2.4 Precision

If toughness indices are to enable meaningful comparison of post-crack performance between different fibre reinforced concrete composites, the precision of the proposed index must be high. When Henager[3] first published his proposed index the supporting data detailed results of just one beam each from a number of different fibre reinforced materials, and it was not therefore possible to comment on the reproducibility of the proposed index.

Zollo[10] was the first to investigate the variability of the ACI index and his results were not encouraging. Within batch coefficients of variation for nominally identical beams tested under the same loading conditions varied between 9 and 64%, with an average of 30%. He also found high variation in the area to first crack which ranged between 9 and 47%, averaging 24%.

Johnston[5] provided details of the precision achieved in his testing programme when his index was first published. His work showed that, as might be expected, the variability of the index increased with the area of the load–deflection plot considered. For the I_5, I_{10} and I_{30} indices Johnston obtained within batch coefficients of variation of 12, 14 and 16% respectively. He also evaluated the ACI index, which roughly corresponds to the I_{30} index, for each of his tests and found its coefficient of variation to be 18%.

In 1985 Barr and Hasso studied the precision of toughness indices in a series of modulus of rupture and notched beam tests on steel fibre reinforced concrete[8] and polypropylene fibre reinforced concrete[11] specimens. Their findings were in agreement with Johnston's work in showing increasing variability as larger portions of the load–deflection curves were used. In particular, the coefficients of variation for five different steel fibre concentrations were all below 5% using the index to twice first crack deflection. Coefficients of variation were also shown to decrease with increasing fibre content.

2.5 Load–Deflection Curves

Johnston is the only author to have addressed the problems and practicalities of accurately producing load–deformation curves. All the current index definitions use the individual beam elastic area and as such depend on the accurate determination of the limit of proportionality. ASTM C1018[1] suggests minimum scale sizes for load–deflection plots which help in the accurate determination both of deviation from linearity and of the relevant areas underneath the plots.

In beam flexural tests the use of a single deflection device located at the mid-span is normal practice. However, in order to obtain true beam deflections a minimum of four deflection measuring devices is required. ASTM C1018 suggests a correction procedure in which the straight line portion of the load–deflection curve is projected back to the horizontal axis.

This intercept is taken as the corrected origin and the procedure is said to account reasonably accurately for deflections at the load points. This procedure does not fully account for load point deflection, evidence of this being the low modulus values obtained in flexural tests.

2.6 Effects of Matrix Composition and Fibre Reinforcing System

Johnston and Gray[12] have illustrated that one further disadvantage of definitions based on fixed cut-off criteria is that of extreme sensitivity to changes in matrix stiffness. Because cut-off deflections are inflexible, a small change in stiffness can have a great effect on first crack area while the total area to the specified deflection may remain almost unchanged. This disadvantage is particularly relevant to sprayed fibre concretes produced using the dry-mix process which, because of their method of production, are subject to greater local variability in mix proportions and water cement ratios than conventionally cast mixes. Such variations can all have a significant bearing on matrix stiffness.

Johnston and Gray[12] have also shown that the ASTM index is clearly able to distinguish between the strengthening effects of changes in matrix composition and the toughening effects achieved by the use of a particular fibre system. In a variety of fibre concretes, first crack strengths were shown to be highly dependent on matrix composition and were relatively independent of the fibre parameters of volume fraction and aspect ratio. However, toughness index values determined using the same fibre system in a variety of matrices of widely differing strengths were shown to be independent of the matrix composition.

3 EXPERIMENTAL PROGRAMME

3.1 Mix Details

The mix design chosen for the investigation was a deliberate attempt to approximate the in-situ proportions of a typical fibre reinforced sprayed concrete. A zone 2 sand and OPC were used in the proportions 2·5 parts by weight of sand to 1 part of cement. Dry sand was used throughout and the water/cement ratio was maintained at 0·45 in all cases. 25 mm long melt extract fibres of AISI grade 430 stainless steel were added at 2, 4 and 6% by weight of the dry materials.

A total of twelve beams was made from each mix, four beams being tested on each of the three test spans of 300, 400 and 500 mm. Two mixes were required for the three fibre reinforced mix designs to facilitate testing under both crosshead and strain control. Specimens were compacted using a vibrating table, the higher fibre content mixes requiring a longer period of vibration. No additional water or admixture was used to improve workability because this would have altered the matrix properties. Beams were cured under water and tested at 28 days.

3.2 Testing

All the flexural tests were carried out using an Instron 6025 screw-driven testing machine. The beams were tested with their trowelled face uppermost, the loading rollers being placed directly on the beam surface. The first series of beams were tested under crosshead control, where the crosshead of the testing machine was driven at a constant rate towards the fixed support. In the second series of tests beams were loaded in strain control mode where the rate of deflection of the beam was controlled to increase at a constant rate throughout the test. In each mode of test control a single transducer placed under the centre of the beam at its mid-span was used to measure deflection.

In both series the testing rate (crosshead or deflection movement) was 0·1 mm/min for beams tested on a 300 mm span, in accordance with ASTM C1018. Equivalent rates were used for the 400 and 500 mm span tests, calculated to give the same rate of increase of stress assuming elastic behaviour.

The load–deflection plot was scaled to allow accurate determination of the limit of proportionality, the initial portion of the curve being plotted at a central deflection scale of 250:1 in accordance with ASTM C1018.

4 BEAM LOAD–DEFLECTION BEHAVIOUR

Typical load–deflection curves from tests conducted over a 400 mm span are illustrated in Figs 1 and 2. Figure 1 shows curves from tests performed under crosshead control while those from the strain-controlled tests are shown in Fig. 2. Load–deflection curves from tests over the other spans were similar in shape.

Figures 1 and 2 serve to illustrate the difference in load–deflection response when testing under the two methods of test control. The fundamental difference in behaviour occurs immediately after peak load. In strain-controlled tests, as peak load is reached, the specimen stiffness reduces and the rate of crosshead movement is slowed in order to maintain a constant rate of central deflection. Consequently only a very small

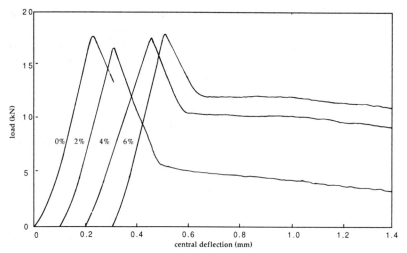

FIG. 1.   Typical load–deflection curves (crosshead control).

increase in deflection occurs as the load drops away after reaching its peak value. Such tests are characterised by a near vertical portion of the load–deflection graph immediately following peak load (Fig. 2).

In crosshead-controlled tests the machine is driven at a rate insensitive to the change in stiffness of the specimen as failure occurs. Beam deflection is uncontrolled and at low fibre contents the central deflection increases

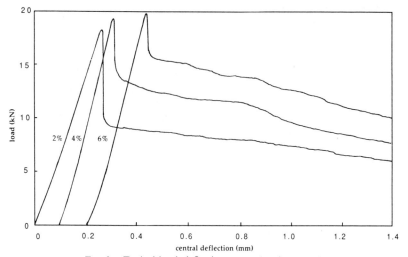

FIG. 2.   Typical load–deflection curves (strain control).

rapidly to two or three times the value at peak load. Consequently the area under the load–deflection curve (and hence toughness index) is greater than would be obtained under strain control.

Testing under strain control (i.e. a controlled rate of beam deflection) is preferable to crosshead control from the viewpoint of standardisation. This is because under crosshead control the machine is able to dissipate stored strain energy into the specimen as it freely cracks and deflects, and so the shape of the load–deflection curve is dependent on machine stiffness. However, testing under strain control (assuming such an option is available) is more difficult, sometimes requiring the control system of the machine to be tuned to take account of the specimen's stiffness which decreases during the test.

5 TOUGHNESS INDEX

5.1 Specimen Geometry

The mean toughness indices obtained under crosshead and strain control, calculated in accordance with ASTM C1018, are given in Tables 1 and 2, together with coefficients of variation. Examination of the data suggests that there is little significant change in toughness index with test span for the range of spans investigated.

The mean value and range of toughness indices for the 4% fibre concrete are shown in Fig. 3 for both methods of test control. The figure emphasises the relatively small changes of mean toughness index with span,

TABLE 1
Toughness index values obtained under crosshead control

Fibre content (%)	Index number	Toughness index (and coefficient of variation, %)		
		300 mm span	400 mm span	500 mm span
2	I_5	3·1 (5·7)	3·1 (3·0)	3·0 (5·0)
	I_{10}	4·8 (10·4)	4·7 (4·4)	4·6 (10·8)
	I_{30}	9·0 (15·6)	9·8 (11·5)	9·1 (13·1)
4	I_5	4·0 (3·0)	4·0 (12·7)	3·8 (4·0)
	I_{10}	7·1 (6·1)	7·2 (17·9)	6·8 (6·2)
	I_{30}	15·3 (10·4)	15·4 (18·7)	14·7 (14·9)
6	I_5	4·2 (10·3)	4·3 (5·1)	4·3 (3·3)
	I_{10}	7·3 (17·3)	7·8 (6·3)	7·8 (5·8)
	I_{30}	13·9 (26·7)	16·1 (10·2)	15·9 (18·8)

TABLE 2
Toughness index values obtained under strain control

Fibre content (%)	Index number	Toughness index (and coefficient of variation, %)		
		300 mm span	400 mm span	500 mm span
2	I_5	2·2 ⎫	2·7 (12·9)[b]	2·5 ⎫
	I_{10}	3·2 ⎬ a	4·3 (19·7)	3·8 ⎬ a
	I_{30}	5·8 ⎭	7·7 (16·2)	7·0 ⎭
4	I_5	3·5 (8·2)	3·6 (8·2)	3·4 (7·5)
	I_{10}	6·2 (8·2)	6·3 (10·2)	5·8 (6·9)
	I_{30}	11·9 (15·5)	12·3 (11·0)	11·1 (11·6)
6	I_5	4·3 (4·0)	4·4 (12·0)	4·0 (9·1)[b]
	I_{10}	7·5 (5·0)	7·9 (15·2)	7·2 (10·5)
	I_{30}	13·5 (9·2)	15·5 (18·5)	15·1 (10·1)

[a] Only one satisfactory test.
[b] Only three beams tested.

particularly when considered in relation to the ranges of the data. If any trend is evident from Fig. 3 (and from similar plots for the other two fibre contents, not presented here), it is of the 400 mm span mean toughness values being slightly greater than the 300 and 500 mm span values in nearly all cases.

A statistical analysis was performed on the toughness data (crosshead control) to objectively test whether the variation between spans is significant in relation to the variation in data for each individual span. An analysis of variance of each set of twelve data points produces a ratio of components of variance between and within samples (F ratio). This ratio is then compared with a critical value of F ratio, which depends upon the

TABLE 3
F-ratio values calculated from analysis of variance tests

Fibre content (%)	Toughness index		
	I_5	I_{10}	I_{30}
2	0·51	0·51	0·55
4	0·73	0·35	0·12
6	0·28	0·43	1·25

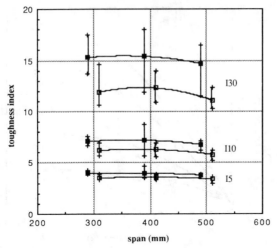

Fɪɢ. 3. Mean toughness index and range, 4% fibres (■, crosshead control; □, strain control).

chosen confidence limit (taken here as 95%) and the number of degrees of freedom of the data set.

The F ratios obtained are given in Table 3 and are all very much less than the critical F ratio of 4·26. Thus we may conclude, with less than a 5% risk of wrong judgement, that there is no difference in the toughness indices obtained on different spans.

5.2 Method of Test Control

It is clear from the foregoing discussion and from Fig. 3 that crosshead and strain control produce different toughness indices. The effect is emphasised in a plot of toughness index against fibre content, shown in Fig. 4. Each data point in this figure is the average of the 12 values of toughness index for each mix on all three spans.

For the range of practical fibre contents of 2–4% by weight, testing under crosshead control overestimates the I_5, I_{10} and I_{30} indices by around 16, 18 and 30% respectively. The overestimate arises because of the different shapes of the load–deflection curves following peak load. Under crosshead control there is a rapid fall off in load with increasing deflection, followed by a relatively flat portion of the curve where load holds with increasing deflection; the tail of the curve is thus offset (Figs 1 and 2), and this is reflected in the large difference between the I_{30} values.

The coefficients of variation of the toughness indices are given in Tables 1

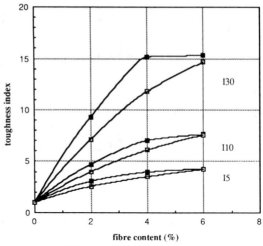

FIG. 4.   Toughness index and fibre content (■, crosshead control; □, strain control).

and 2, and are similar for the two modes of test control. Comparison with the within-laboratory precision values given in ASTM C1018 shows that for each index between one and three coefficients out of the nine exceed the specified precision values for I_5, I_{10} and I_{30} of 12, 14 and 16% respectively.

5.3 Fibre Content

For the range of fibre contents investigated toughness index increased non-linearly with fibre content (Fig. 4). The curved relationships are typical for concrete containing low volumes of steel fibre reinforcement. However, at fibre contents above the critical fibre volume (when multiple cracking can occur) it would be expected that the slope of the curve would increase.

At a fibre content of 6%, the difference between crosshead and strain control toughness values is marginal, due to the smaller drop off in load after peak load.

6 CONCLUSIONS

1. Tests performed over different spans verify that the values of toughness indices calculated according to the definitions of ASTM C1018 are largely independent of test spans over the range of spans investigated.
2. There is a fundamental difference in load–deflection behaviour when testing under the two different methods of control. For the range of fibre

contents investigated there was a significant reduction in load after peak load (coinciding with formation of the first flexural crack). Whilst this occurred without a change in deflection under strain control, specimens tested under crosshead control underwent a sudden large deflection associated with the release of energy stored in the test machine as the specimens cracked.

3. As a consequence of the above, testing under crosshead control (which is not in accordance with ASTM C1018) significantly overestimates toughness index, particularly in the range of practical fibre contents for steel fibre concrete.

4. Testers for fibre concrete consequently face a dilemma as to whether they should test in strict accordance with the standard (which is both more difficult and requires a high level of sophistication in testing machine) or whether to continue the present common practice of testing under crosshead control, which is easier and less restrictive, but produces toughness values that may possibly be machine-dependent. If the industry favours the latter approach then Sub-committee C090304 of the American Society for Testing Materials should reconsider the testing method specified in its standard.

REFERENCES

1. AMERICAN SOCIETY FOR TESTING AND MATERIALS, Standard test method for flexural toughness and first-crack strength of fibre reinforced concrete. *ASTM Standards for Concrete and Mineral Aggregates*, Vol. 0402, No. C1018, August 1984, pp. 637–44, revised June 1985.

2. JOHNSTON, C. D., Definition and measurement of flexural toughness parameters for fibre reinforced concrete. *Cement, Concrete and Aggregates*, 4(2) (1982) 53–60.

3. HENAGER, C. H., A Toughness Index for Fibre Concrete. Testing and Test Methods of Fibre Cement Composites, RILEM Symposium, Construction Press, Lancaster, 1978, pp. 79–86.

4. AMERICAN CONCRETE INSTITUTE, Measurement of properties of fibre reinforced concrete. *ACI Journal*, 75(7) (1978) 283–9.

5. JOHNSTON, C. D., Precision of flexural strength and toughness parameters for steel fibre reinforced concrete. *Cement, Concrete and Aggregates*, 4(2) (1982) 61–7.

6. BARR, B. I. G. and LIU, K., Fracture of GRC material. *International Journal of Cement Composites and Lightweight Concrete*, 4(3) (1982) 163–71.

7. BARR, B. I. G., LIU, K. and DOWERS, R. C., A toughness index to measure the energy absorption of FRC. *International Journal of Cement Composites and Lightweight Concrete*, 4(4) (1982) 221–7.

8. BARR, B. I. G. and HASSO, E. B. D., A study of toughness indices. *Magazine of Concrete Research*, **37**(132) (1985) 162–74.

9. JAPAN CONCRETE INSTITUTE, Method of test for flexural strength and flexural toughness of fibre reinforced concrete. Standard SF4. *JCI Standards for Test Methods of Fibre Reinforced Concrete*, 1983, pp. 45–51.

10. ZOLLO, R. F., Fibrous concrete flexural testing—developing standardised techniques. *ACI Journal*, **77**(5) (1980) 363–8.

11. BARR, B. I. G. and HASSO, E. B. D., The precision of toughness indices based on multiples of first-crack deflection. *ACI Journal*, **82**(5) (1985) 622–9.

12. JOHNSTON, C. D. and GRAY, R. J., Flexural toughness and first-crack strength of fibre-reinforced concrete using ASTM Standard C1018. RILEM Symposium, Developments in fibre-reinforced cement and concrete, Sheffield, 1986, paper 5.1.

44

Creep in SFRC Elements under Long-Term Excentric Compressive Loading

A. M. BRANDT

Institute of Fundamental Technological Research,
00-049 Warsaw, Swietokrzysk, Poland

and

L. HEBDA

Technical University, 25-314 Kielce, Tysiaclecia PP7, Poland

ABSTRACT

The long-term measurements of strain in steel fibre reinforced concrete blocks subjected to excentrical compression are presented and discussed. The blocks were cast with two types of concretes: with basalt and limestone aggregate, and with two fibre volumes: 0·7% and 1·3%. The measurements were executed during 500 and 1000 days on two series of specimens. After unloading the recovery strain development was also recorded. Conclusions concern the influence of fibres and of aggregate types on the creep and other components of concrete strain.

1 INTRODUCTION

Steel fibre reinforced concretes (SFRC) are used for several structural and non-structural applications where their durability and appropriate long-term behaviour are required. These materials have been tested over many years in various laboratories and it is supposed that their properties are described to a sufficient degree to design conventional SFRC elements. The influence of long-term loading on SFRC is however not completely known until now and some doubts still exist as it concerns the influence of the fibre reinforcement on the long-term deformations of SFRC.

743

The aim of the paper is to analyse a few different opinions on the above mentioned problem and to present the results of the tests on the creep strain in SFRC elements subjected to tensile stresses in a part of its cross-section. The influence of the aggregate type and the fibre content on the long-term behaviour of tested elements is discussed.

2 LONG-TERM OBSERVATIONS PUBLISHED BY VARIOUS AUTHORS

The long-term properties of SFRC elements may be considered in two different aspects:

—long-term stability under exposure on various environmental factors like moisture and corrosive agents;
—behaviour under long-term stress and resulting creep strain.

The first one, i.e. the durability of SFRC, is not directly studied here and only the second aspect mentioned above is considered, however both are closely interrelated.

The tests on the influence of the fibre reinforcement on the deformations under long-term load were carried out by several authors but their conclusions are not entirely convergent. Edgington et al.[1] did not observe any decrease of creep and shrinkage due to reinforcement of 2–4% per volume with steel fibres. Malmberg and Skarendahl[2] found that fibres decrease the shrinkage strain by 10–20% with respect to that of plain mortar. The specimen's resistance against cracking was also increased. Swamy and Theodorakopoulos[3] compared mortar and concrete reinforced with glass and polypropylene fibres and observed a decrease of shrinkage strain by about 10% and creep by 40–80%. Fukuchi et al.[4] tested mortars in 1980 and found that the considerable reduction of drying shrinkage due to steel fibres added up to 2% by volume.

Also recently Balaguru and Ramakrishnan[5] carried out tests on concretes reinforced with 0·5% of Bekaert fibres per volume. They have found that:

—shrinkage strain was smaller but the difference was not substantial;
—shrinkage stopped in SFRC after 500 days and only after 600 days in plain concrete;
—creep strain was considerably higher for SFRC;
—creep recovery was substantially the same for reinforced and plain concretes.

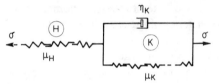

FIG. 1. An elementary rheological model for concrete representing deformations due to microcracking (after Glücklich[7]).

The problem why authors obtained essentially different results may be partially explained using a simple rheological model proposed already in 1959 by Glücklich[6] and discussed also later[7] in 1965.

The model is composed of imperfect Kelvin and Hookean elements shown in Fig. 1. The imperfections of the model are due to elements which transmit forces by friction. With a constant external load the part transmitted to the elastic skeleton is increasing with time. At a certain value of that load slippages take place, representing microcracks which begin to propagate. Some cracks are stabilised after having reached a certain length, but new ones are opening at various places.

Glücklich has shown that the stress σ_{HK} in the Hookean elements which represent the solid phase of the material increases with the time t:

$$\sigma_{HK} = \sigma[1 - e^{-(\mu_K/\eta_K)t}]$$

tending to the full load σ even though the external load is unchanged. Here μ_K and η_K represent the elastic and viscous material characteristics.

The microcracks represented by imperfections were present since the beginning of loading but the intensity of microcracking is increasing with load due to uneven repartition of stress in the elastic skeleton and consequently to local high stress concentrations.

The differences in results obtained by various authors may be therefore attributed to different compositions of their respective materials in the meaning of Glücklich's model. Not only was the distribution of the elements which follow Kelvin or Hookean behaviour in the material structures of tested elements certainly different, but also the particular values of μ_K, η_K and μ_H which characterise these elements. In most cases the published data concerning the materials used are not sufficient for any thorough analysis of their deformability even in the sense of the approximate model.

It is obvious that all reasoning based on phenomenological models have only a speculative character and cannot replace the test results. Such a model with its simplified interpretation may help, however, to understand better the mechanisms of the studied phenomenon and in that case to explain the non-consistency of the experimental data published by different authors.

On the basis of all the divergent results obtained by distinguished experimenters the following conclusion may be proposed: it is difficult, if not impossible, to predict correctly what quantitative result in long-term behaviour would give the fibres introduced to a certain concrete mix without experimental verification. Furthermore, the specimens with and without fibre reinforcement cannot be directly compared. The presence of thin dispersed fibres influence not only the crack propagation but also other mix properties such as porosity and workability.

The aim of the test reported below was to observe how the creep strain in SFRC elements subjected to excentrical compression was developing with time.

3 SPECIMENS AND TEST PROCEDURE

Four similar but slightly different mix compositions were prepared in which two fibre volumes and two kinds of aggregates were used. The tests on two series of specimens were carried out during 1080 and 540 days, respectively.

The ordinary portland cement '35' was used with two type of coarse aggregate: crushed basalt and crushed limestone, both with maximum grains of 8 mm. The limestone aggregate was made of compact rock of pelite or microcrystalline structure. As fine aggregate natural river sand was used throughout.

Mix proportions of cement:sand:aggregate equal to $1:1\cdot07:3\cdot56$ for basalt concrete and to $1:0\cdot97:3\cdot24$ for limestone concrete were used. Both proportions were designed for equal mix density with 400 kg of cement per $1\,\mathrm{m}^3$ and with water/cement ratio equal to $0\cdot6$. As reinforcement the plain round straight steel fibres $\phi0\cdot4 \times 40\,\mathrm{mm}$ were used with two volume contents: $0\cdot7\%$ and $1\cdot3\%$.

The concretes were prepared in a forced horizontal mixer with a vibration grid for introduction of the fibres to avoid their balling. The specimens were cast as blocks of dimensions shown in Fig. 2. The horizontal grooves were cast along both lateral sides to guide the crack. The variable width of the internal core was designed to assure an approximately constant nominal stress from the notch tip along about 50 mm in the tensile zone.

After 48 h the specimens were demoulded and put into a foggy room at

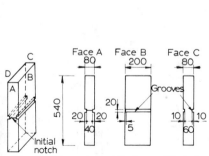

FIG. 2. Dimensions and shape of tested specimens.

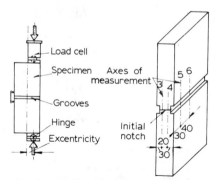

FIG. 3. Loading of specimens and strain measurement, points 1 and 2 are on face D symmetrically to points 6 and 5.

constant temperature of $+18°C$. At an age of 28 days in the mid-height of each specimen tensile zone an initial notch of 5 mm was sawn.

The short-term tests were executed at an age of 60 days. The specimen position similar to that used by Entov and Yagust[8] is shown in Fig. 3 and the load was increased step-wise till the rupture. The load and the deformation at the notch mouth were recorded on an $x-y$ plotter.

The specimens for long-term tests were loaded at an age of 90 days in spring creep-testing machines stored at constant conditions of 60% of RH and $+18°C$. Deformations were measured with the LVDT gauges at the notch mouth and across the grooves along the lateral faces. The specimens with different aggregates and reinforcement volumes were loaded at 0·5 and 0·8 of their cracking loads P_{cr}, respectively.

Because of the storing conditions and the concrete age at loading no influence of drying shrinkage has been taken into account. The following properties of concrete were determined at an age of 28 days:

Static	Young modulus for basalt concrete	18 600 Mpa
	for limestone concrete	26 700 MPa
	Compressive strength for basalt concrete	27·1 MPa
	for limestone concrete	31·8 MPa

4 SHORT-TERM TEST RESULTS

The rupture of the specimens was produced by cracks which started at the notch tip and propagated along the grooves. The crack length was only approximately determined.

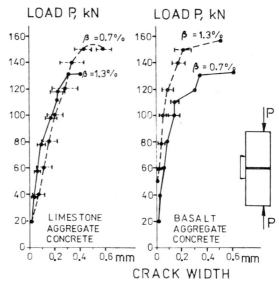

FIG. 4.   Load–strain curves in short-term tests; arrows indicate the crack initiation.

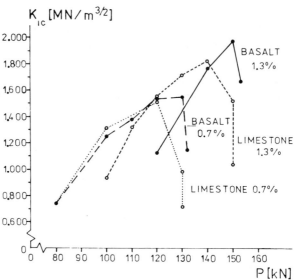

FIG. 5.   Values of K_{Ic} plotted against the load in short-term tests.

The results of measurements of deformation versus load are shown in Fig. 4 for both types of aggregate and for both fibre contents. On the diagrams the average values for 3 or 4 specimens are given together with their dispersions. More information on the tests may be found out in Refs 9 and 10.

It may be observed from these tests that the higher fibre content was more efficient for basalt concrete. On the other hand, the lower fibre volume assured higher strength of the limestone aggregate concrete.

The estimated crack length and corresponding load which initiated crack propagation at each consecutive loading step were used to calculate the critical values of the stress intensity factor K_{Ic} for Mode I. Here the axial compression and pure bending of a specimen were superposed and the formulae published by Tada *et al.*[11] were used. The results are given in Fig. 5.

It was observed during long-term measurements that in basalt concrete nearly total crack length was reached during the first day of loading. The crack propagation was slightly slower in the limestone concrete specimens. The variation of the K_{Ic} with the crack length and load in short-term tests was significant.

It is also interesting to note that approximately the same values of K_{Ic}, meaning similar resistance against crack propagation, may be obtained using an appropriate combination of the aggregate quality and the fibre content. That observation may be useful for material design and optimisation.

5 LONG-TERM TEST RESULTS

The creep strain measured at the tensile face is plotted against time for both series of tests in Figs 6 and 7. Using the proposal by Rüsch and Jungwirth[12] the long-term deformation during loading and unloading may be decomposed into several components. In the considered case the total strain may be considered as a sum:

$$\varepsilon = \varepsilon_i + \varepsilon_d$$

where ε_i is instant strain and ε_d is its delayed value.

After unloading a recovery instant strain ε_{ri} appeared and next the recovery creep ε_{rd}. The permanent strain ε_p is expressed below

$$\varepsilon_p = \varepsilon - \varepsilon_{ri} - \varepsilon_{rd}$$

A. M. Brandt and L. Hebda

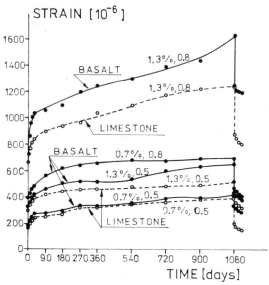

FIG. 6.   Strain versus time in the first series of specimens of basalt and limestone aggregate concretes.

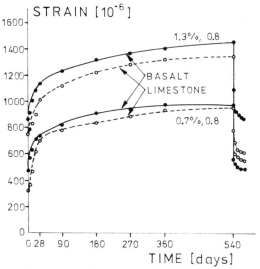

FIG. 7.   Strain versus time in the second series of specimens of basalt and limestone aggregate concretes.

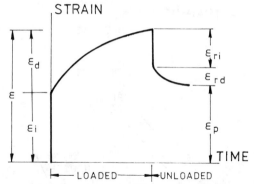

FIG. 8.   Schematic diagram of concrete strain versus time.

The symbols used above are shown on a schematic diagram in Fig. 8. The numerical values of all the strain components are given in Table 1.

The data from Table 1 were used for Figs 9 and 10 where strain measured at various ages and loading situations are plotted at appropriate points of the specimen cross-section. It means that mean values from measurements obtained at symmetrical points 1 and 6, 2 and 5, 3 and 4 (Fig. 3) are shown.

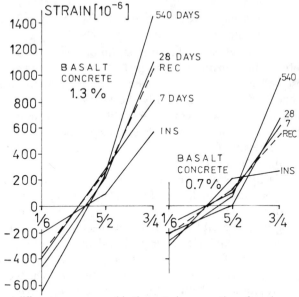

FIG. 9.   Strain at different ages measured in the central cross-section of specimens of basalt aggregate concrete. Points of measurement are indicated in Fig. 3; INS means instant strain ε_i, REC means strain after unloading, $\varepsilon - \varepsilon_{ri}$.

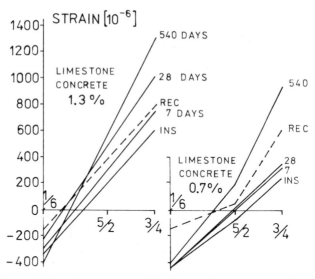

FIG. 10. Strain at different ages measured in the central cross-section of specimens of
linestone aggregate concrete. Symbols as in Fig. 9.

The above results may serve to estimate the quantitative influence of fibres
in different concrete mixes.

As expected, in all cases the creep strain ε_d was consistently lower for
higher fibre volume in concrete mixes (1·3%) but a larger part of the strain
was observed as ε_i immediately after loading than for lower volume
reinforcement (0·7%). Consequently, the corresponding creep ε_d was more
important for the lower reinforcement volume. The difference is more
pronounced for concretes with basalt aggregate than for those with
limestone aggregate. The permanent strain ε_p was slightly higher for basalt
concretes than for limestone ones.

No regularity was observed for recovery strain. In absolute values all
strain components for higher reinforcement were higher, but this was due
certainly also to the higher values of related P_{cr}. The above conclusions
confirm an opinion established after previous tests that steel fibres extend
the concrete deformability. The increased deformability may be used for
better distribution of local stress due to loadings of long-term origin, such
as temperature variation or foundation settlement. On the other hand,
larger deformations and displacements of elements subjected to flexion
should be taken into account.

The general behaviour of fibre reinforced concrete in time is essentially

TABLE 1
Strain components at various stages of loading ($\times 10^{-6}$)

Strain components		First series							Second series			
		Type of aggregate										
		Basalt				Limestone			Basalt		Limestone	
		Load level										
		$0{\cdot}5P_{cr}$		$0{\cdot}8P_{cr}$		$0{\cdot}5P_{cr}$		$0{\cdot}8P_{cr}$				
		Fibre volume										
		0·7	1·3	0·7	1·3	0·7	1·3	1·3	0·7	1·3	0·7	1·3
Total	ε	420	660	700	1 640	400	520	1 240	980	1 460	960	1 340
strain	%	100	100	100	100	100	100	100	100	100	100	100
Instant	ε_i	160	320	280	720	180	320	660	460	860	320	780
strain	%	38	48	40	44	45	62	53	47	59	33	58
Delayed	ε_d	260	240	420	960	220	200	580	520	600	640	560
strain	%	62	52	60	56	55	38	47	53	41	67	42
Instant recovery	ε_{ri}	80	180	260	400	200	180	360	420	380	280	580
strain	%	19	27	37	24	50	35	29	43	26	29	42
Recovery delayed	ε_{rd}	40	60	40	40	40	60	80	60	200	120	60
strain	%	10	9	6	2	10	12	6	6	14	13	4
Total recovery	ε_r	120	240	300	440	240	240	440	480	580	400	620
strain	%	29	36	43	26	60	46	35	49	40	42	46
Permanent	ε_p	300	420	400	1 200	160	280	800	500	880	560	720
strain	%	71	64	57	73	40	54	65	51	60	58	54

similar to that of plain concrete, however the proportions between particular strain components are different and related to fibre volume, type of aggregate and probably to other mechanical characteristics.

It should be also noted that fibre concrete with limestone aggregate may be considered as a valuable structural material.

REFERENCES

1. EDGINGTON, J., HANNANT, D. J. and WILLIAMS, R. I. T., Steel fibre reinforced concrete. Bui. Res. Estab. CP 69/74, July 1974, p. 17.
2. MALMBERG, B. and SKARENDAHL, Å., Method of studying the cracking of fibre

754 *A. M. Brandt and L. Hebda*

concrete under restrained shrinkage. In *Testing and Test Methods of Fibre Cement Composites*. RILEM Symp., Construction Press, London, 1978, pp. 173–9.

3. SWAMY, R. N. and THEODORAKOPOULOS, D. D., Flexural creep behaviour of fibre reinforced cement composites. *Int. J. Cem. Compos.*, 1(1) (May 1979) 37–48.
4. FUKUCHI, T., OHAMA, Y., NISHIMURA, T. and SUGAHARA, T., Effects of steel fiber reinforcement on drying shrinkage of mortar. *Trans. Jap. Concr. Inst.*, 2 (1980) 195–202.
5. BALAGURU, P. and RAMAKRISHNAN, V., Properties of fiber reinforced concrete: workability, behavior under long-term loading, and air-void characteristics. *ACI Mater. J.*, TP 85–M23 (May–June 1980) 189–96.
6. GLÜCKLICH, J., The influence of sustained loads on the strength of concrete. *RILEM Bull.*, No. 5 (December 1959) 14–17.
7. GLÜCKLICH, J., The effect of microcracking on time-dependent deformations and the long-term strength of concrete. In *The Structure of Concrete and its Behaviour under Load. Proc. Int. Conf.*, London, Sept. 1965, Cement and Concrete Association 1968, pp. 176–89.
8. ENTOV, V. M. and YAGUST, V. I., Experimental study on the rules of the microcrack propagation in concrete. *Mekhanika Tverdovo Tiela*, (in Russian) 10(4) (1975) 93–103.
9. BRANDT, A. M. and HEBDA, L., Example of the experimental design method in the long-term testing of SFRC. In *Developments in Fibre Reinforced Cement and Concrete*. RILEM Symp. 2, Sheffield, July 1986.
10. BRANDT, A. M., BURAKIEWICZ, A. and HEBDA, L., On the crack propagation in the fibre concrete element subjected to long-term loading. In *Mechanics and Technology of Composite Materials. Proc. Vth Conf.*, Varna 1988, pp. 679–83.
11. TADA, H., PARIS, P. C. and IRWIN, G. R., *The stress analysis of cracks, Handbook*, Del Res. Corp., Hellertown, PA, USA, 1973.
12. RÜSCH, H. and JUNGWIRTH, D., *Stahlbeton—Spannbeton, Band 2 Berucksichtigung der Einflusse von Kriechen und Schwinden auf das Verhalten der Tragwerke.* Werner-Verlag, Dusseldorf, 1976.

45

Behaviour of SFRC Slabs under Dynamic Loads

Wojciech Radomski

Warsaw University of Technology, Warsaw, Poland

ABSTRACT

The paper is based on the test results obtained by the author and other investigators during the last few years. The differences between the behaviour of RC and SFRC slabs under the impact loads are presented. The problem concerning the perforation of slabs by projectiles is particularly discussed. An author's original formula for prediction of so-called perforation thickness of SFRC slabs is proposed. Moreover, some general research problems concerning the dynamic behaviour of RC and SFRC structural elements are discussed, especially the modelling laws.

1 INTRODUCTION

There is no doubt that steel fibre reinforced concrete (SFRC) is a particularly appropriate material for applications in the various structures subjected to impact or impulsive loads as well as in the structures in which these types of loads may occur accidentally. It has been confirmed by the development of such applications during the last few years. However, the knowledge on dynamic behaviour of SFRC is rather incomplete compared with the knowledge of static properties of this material.

The majority of the dynamic investigations of SFRC carried out has been limited to the determination of only one dynamic characteristic of the material, its impact resistance. Some problems concerning the determination of this resistance have been discussed elsewhere.[1,2]

The impact resistance is a very important material characteristic but it is not sufficient for full estimation of the behaviour of SFRC structural elements under dynamic loads. It is necessary to know other dynamic effects produced in the structural elements by blows, for example load and wave phenomena, fracture mechanisms, dissipation of the impact energy and so on. Investigations of these effects with reference to SFRC have not developed much so far.[3]

The dynamic behaviour of SFRC depends on many factors characterising the material, i.e. matrix and fibres, the impact loads, and the local and overall deformability of the tested specimens or structural elements.[2] Many mechanical characteristics of SFRC cannot be considered as fully defined physical properties.[1]

The results of dynamic tests of SFRC are generally well scattered due to the heterogeneity of the material. Moreover, the phenomena produced by the impact loads in the tested members are very complex. The interpretation of the test results may therefore be difficult and the theory concerning the dynamic behaviour of SFRC has not been well developed. It should be emphasised that the results of dynamic tests and theoretical analysis concerning plain concrete cannot be in general extended to SFRC because of the different properties of these materials.

Hence the research problems concerning the dynamic behaviour of SFRC are very complex. Moreover, the experiments require special and expensive equipment.

In spite of this a development in dynamic research on SFRC may have been noticed during the last few years in consequence of its applications in the various structures subjected to impact loads. The experiments are performed not only on the specimens of this material but also on the structural elements, mostly on beams and slabs. The tests are carried out on the members with different dimensions, matrices and types of fibres. Various measuring techniques are used. For that reason the test results are in general not comparable with each other. Moreover, the so-called scale effect is a very important problem, especially when the dynamic tests are performed for practical purposes. This problem is touched on in the next section of the paper.

The state of dynamic research on SFRC was well summarised by the general reporter Dr Paramasivam during the International Symposium on Fibre Reinforced Concrete in Madras, India, in December 1987. He said: 'In conclusion, the behaviour of fibre reinforced composites under impact and fatigue loading is quite complex and difficult to quantify. The authors have tried to present the results in a simplified manner.'[4]

In this paper some chosen problems concerning the behaviour of SFRC slabs under impact loads are presented. The essential differences between the dynamic behaviour of SFRC and plain (C) or conventionally reinforced concrete (RC) slabs are discussed on the basis of the tests carried out by the author and other investigators. The problem of perforation of the slabs is especially emphasised as very important in practice. The presentation of these topics is followed by some general remarks on testing of the SFRC structural elements under impact loads.

2 GENERAL REMARKS ON TESTING OF SFRC STRUCTURAL ELEMENTS UNDER IMPACT LOADS

Because of the strong heterogeneity of the material, very complex physical phenomena produced in the tested members by blows and the lack of corresponding theoretical solutions, the impact tests of the SFRC structural elements with so-called 'full-scale' are the most appropriate and reliable for practical purposes. However, this type of test is very expensive and requires special full-scale facilities and therefore cannot be widely used. The dynamic tests are mainly performed in laboratories on specimens of material or on the models of structural members with different geometric scales of basic prototypes. The following two conditions should be fulfilled.

First, the tested specimens or models should be representative of the material applied with respect to its heterogeneity. It requires that the minimum dimension of the tested pieces should not be less than twice the fibre length and five times the maximum aggregate size used for a matrix,[5] unless the specimens are cut out from larger blocks.

Second, the applications of the results of model tests to the prototype need to be done with scaling laws which should be derived on a solid physical basis.[6] The determination of these laws with reference to the impact tests of concrete or RC elements is very difficult and complex. In this case a reduction in size is restricted by the grain size of the aggregate and the diameter of the reinforcing bars. Moreover, the strain rate sensitivity of the material can distort the results and many other dynamic effects, for example wave and local phenomena, cannot be modelled according to the scaling laws. The modelling problems concerning the experimental investigations of SFRC members are more complicated due to the random distribution of fibres in a concrete matrix.

In general, in dynamic problems, in order to have a true model, three

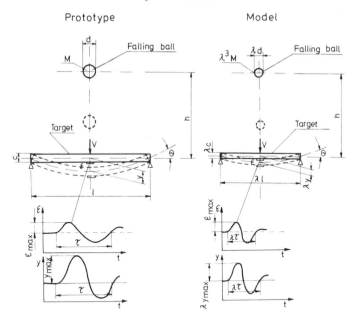

Fig. 1.   Replica scaling of the response of a structural member to impact produced by a falling ball.

independent scale factors can be chosen: length (λ_l), mass or force (λ_f) and time (λ_t). All other scale factors result from dimensional analysis or from differential equations if available for the problem concerned.[6] The replica scaling of response of a structural member to impact loading is schematically shown in Fig. 1.

Unfortunately, the validity of the scaling laws for dynamically loaded SFRC structural elements has not yet been studied in detail and observed in the dynamic tests carried out so far.

The most important problems of concrete and SFRC subjected to impact loads concern the local damage, such as penetration, perforation, spalling, scabbing, and other inelastic phenomena, such as cracking, bond failure and so on. These phenomena are in general very non-linear and depend on geometric, mass (or force) and time factors.

In conclusion, the results of the impact tests carried out on the models of SFRC structural elements should be considered as qualitative only. They allow us to study various complex dynamic effects very important for material science and practical applications of SFRC. However, the prediction of the behaviour of SFRC prototypes under impact loads on the basis of model tests seems to be very risky.

In spite of this, the results of the impact tests carried out so far, for example on SFRC slabs, indicate some interesting dynamic properties of this material which confirm its suitability for applications in the various structures subjected to impacts. These results are discussed in the next two sections of this paper.

3 THE INFLUENCE OF STEEL FIBRE ADDITION ON THE BEHAVIOUR OF CONCRETE AND REINFORCED CONCRETE SLABS SUBJECTED TO IMPACT AND IMPULSIVE LOADS

This problem has been experimentally studied and presented in several publications.[7-13] Slabs of different dimensions, concrete matrices and with different types and contents of steel fibres have been tested. During the tests various forms of impact or impulsive loading have been applied, such as falling hammers of different masses and shapes, a blast-simulator and striking projectiles of different shapes, materials and masses. As a consequence, different impact velocities from 7 m/s up to 700 m/s have been used. For that reason it is very difficult and risky to compare and summarise the results of these tests. In spite of this, the tests carried out so far indicate that the addition of steel fibres considerably improves the dynamic behaviour of C or RC slabs. The most important information on this problem is given below.

Uchida *et al.*[10] found out that the addition of steel fibres to the concrete slabs with conventional reinforcement in bending is very effective in decreasing the formation of cracks under impact loading. They also found that the ratio of energies absorbed in the slabs under impact (E_d) and static (E_s) loading reaches almost 2 in the case of RC slabs and is more than 4 when steel fibres are used. The test results concerning this problem are shown in Fig. 2.

Hülsewig *et al.*[9] observed that the fragmentation zone produced in SFRC slabs by impacts with velocities from about 100 to 220 m/s decreased with increasing fibre content, as shown in Fig. 3. This relation is roughly linear. However, the observations from the tests performed by Anderson *et al.*[7] indicate that the increase of fibre content effects the reduction of front crater volume whilst the change in normal penetration depth may not be significant.

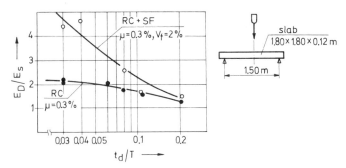

FIG. 2.   Comparison of absorbed energy under static and impact loading of RC slabs and RC slabs with steel fibres (SF). μ, reinforcement ratio; V_f, volume content of fibres; t_d, duration of impact; T, first natural period of slabs.

The differences in the behaviour of RC and SFRC slabs were investigated by Arockiasamy *et al.*[13] They found that in the case of RC slabs the failure mode is predominantly produced by the punching shear effect whereas the failure modes of SFRC slabs are combinations of punching shear, bending and wave propagations effects.

The failure mode of SFRC slabs produced by the impact loads was also investigated by the author and Glinicki. The main test results were

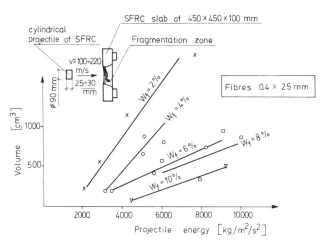

FIG. 3.   Volume of fragmentation zone versus energy of impact for different weight contents of fibres.

discussed in detail in the previous publications.[11,12] In this paper some examples of the failure patterns not yet published are presented in Figs 4 and 5. Structural and geometrical parameters of the tested slabs as well as the impact parameters provided a variety of failure modes of the slabs, i.e. from almost 'pure' perforation to the perforation accompanied by extensive cracking over the whole surface of the slab. Square slabs of 0·50 × 0·50 m with variable thicknesses from 5 to 25 mm were centrally loaded by steel cylinders with a hemispherical nose of radius 25 mm and masses of 1·495, 2·610 and 4·175 kg. The impact velocity was the same throughout the tests and equal to 7 m/s. Straight, smooth, steel fibres 0·4 × 40 mm were used in volume fractions $V_f = 0·5\%$ (15 slabs) and $V_f = 1·0\%$ (15 slabs). The slabs of series I were subjected to single impact whereas the slabs of the other five series were impacted repeatedly until complete perforation occurred. In the case of slab thicknesses of 20 and 25 mm and $V_f = 1·0\%$ it was observed that the formation of a cone-shaped plug of material in the vicinity of the impact point was followed by its movement and crushing as a consequence of repeated blows. As far as the thinner slabs of the same V_f are concerned it was found that the formation and movement of similar cone-shaped plugs was accompanied by circular cracking around the point of impact in a fan pattern. This suggests that a complex failure mechanism, being a result of punching shear conditions as well as bending conditions, was involved. A similar failure pattern with circular cracks in a fan form was observed for slabs of $V_f = 0·5\%$ irrespective of their thickness. An overall structural response of slabs resulted in cracks extending towards the edges. A significant contribution of fibres in impeding the progress of the cracks was noticed. The cracking was visibly smaller in the case of slabs with $V_f = 1·0\%$ in comparison to slabs of $V_f = 0·5\%$. An increase of slab thickness also resulted in smaller width and length of cracks.

The experimental observations from impact tests carried out so far by the author and other investigators indicate that steel fibres evidently reduce spalling and scabbing effects in the slabs, and also the fibres reduce the velocity of any ejected matrix particles.

An attempt to predict the ultimate failure load for SFRC slabs subjected to impacts, using various available empirical formulae concerning RC slabs, indicated that these formulae underpredict this load.[13] An analytical approach for determining the impact toughness of SFRC slabs is very difficult because of very complex dynamic phenomena and heterogeneity of the material. No theoretical solutions are yet available for practical purposes.

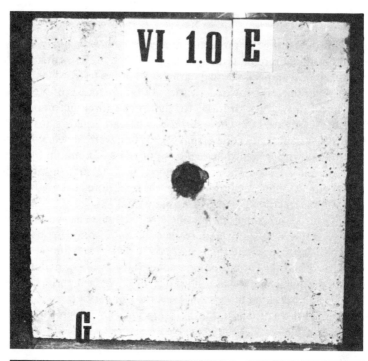

FIG. 4. Failure modes of slabs produced by a projectile with mass 2·610 kg. (a) Slab of $h = 25$ mm and $V_f = 1·0\%$ ($n = 9$).

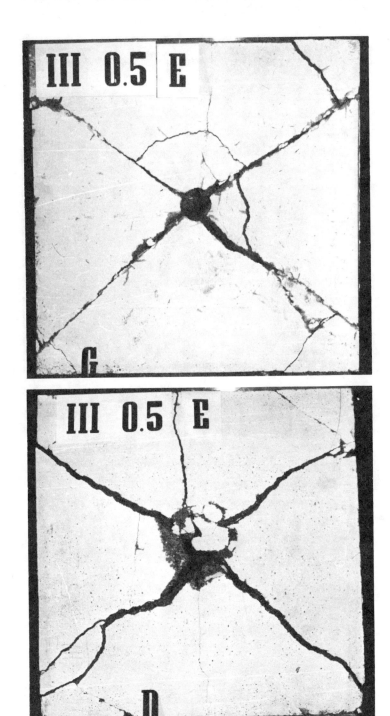

FIG. 4—*contd.* (b) Slab of $h = 25$ mm and $V_f = 0.5\%$ ($n = 5$); n, number of blows to perforation; G, front face; D, rear face.

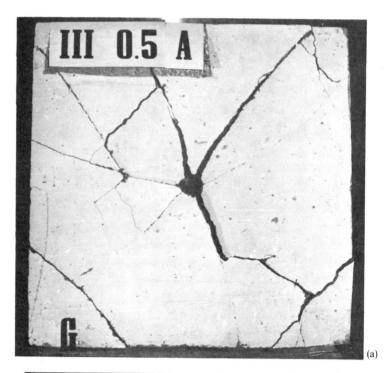

(a)

(b)

FIG. 5. Failure modes of slabs produced by a projectile with mass 2·610 kg. (a) Slab of $h = 5$ mm and $V_f = 0.5\%$ ($n = 1$); (b) slab of $h = 5$ mm and $V_f = 1.0\%$ ($n = 1$).

4 PERFORATION OF SFRC SLABS

One of the most important problems concerning the behaviour of the structural elements subjected to impact loads is their perforation. Unfortunately, this problem has not been studied very much with reference to SFRC slabs. It has been studied up to now on concrete or RC slabs and various empirical formulae for prediction of so-called perforation thickness are available.[14-17]

The perforation thickness is defined as the minimum thickness of a slab which a missile with a given velocity will completely penetrate.

The perforation problem concerning SFRC slabs has been studied by the author and Glinicki during the tests briefly described in the previous section of this paper. The theoretical analysis has been reported elsewhere.[11,12] The analytical consideration based on the test results and on the conservation of energy for penetrating projectiles have led to the following original formula for prediction of the perforation thickness (h_p) of SFRC slabs:

$$h_p^{SFRC} = \frac{U}{D\{(\pi/\sqrt{3})\lambda f_{fl} u_0 + \eta V_f l[k_1(l/d)\tau_f + k_2 f_y]\}} \tag{1}$$

where $U = 0 \cdot 5mv^2$ denotes the kinetic energy of a projectile with mass m and impact velocity v, while the dimensionless parameters are as follows:

$$k_1 = 1 + (2/\pi)p \qquad k_2 = \tfrac{1}{4}(\pi + p) \qquad p = S_r/lD$$

The other symbols are: λ, strain rate factor;[18] D, diameter of projectile; f_{fl}, static flexural strength of concrete matrix; u_0, depth of crater formed in the impact zone; V_f, volume content of fibres as a percentage; l, d, length and diameter of fibres respectively; η, coefficient of fibre effectiveness, equal to 0·637 for random plane distribution of fibres;[19] τ_f, fibre pull-out strength; f_y, yield stress of steel fibres; S_r, surface of structural cracks, i.e. product of crack length summed for all the cracks in slab.[19]

The appropriate calculations of the penetration thickness have been also performed according to the CEA–EDF (Commissariat à l'Energie Atomique–Electricité de France) empirical formula[14,17] with reference to RC slabs. This formula is expressed as follows:

$$h_p^{RC} = 0 \cdot 765(f_c')^{-0 \cdot 375}(W/d)^{0 \cdot 5}v^{0 \cdot 750} \tag{2}$$

where the symbols denote: f_c', static compressive strength in psi; W, mass of projectile in lb; d, diameter of projectile in in.; v, velocity of impact in ft/s.

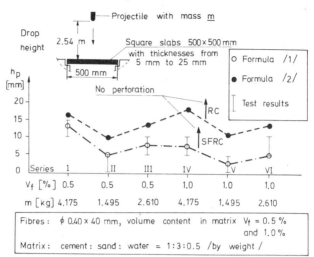

Fig. 6.   Comparison of theoretical and experimental values of perforation thicknesses of slabs.

The results of the calculations and experiments are summarised in Fig. 6. They allow us to formulate two following remarks. First, a good correlation between the test observations and the calculations performed according to formula (1) is obtained. Second, the perforation thickness of an RC slab is more than the thickness of an SFRC slab when they both are subjected to impacts with the same energy. However, this observation should be confirmed by further investigations.

5 CONCLUSIONS

The research problems concerning the behaviour of SFRC slabs under impact loads are not yet very developed. Because of the incomparability of the tests performed it is very difficult to summarise their results. In spite of this, the following conclusions may be obtained.

1. SFRC has very good properties with respect to impact loads due to its high energy absorption and a suppression of the scabbing and spalling phenomena.
2. While the failure mode seems to remain the same for RC slabs it varies considerably for SFRC slabs depending mostly on the fibre content and the slab thickness as well as on the impact characteristics.

3. The empirical formulae available for the prediction of ultimate failure load and the perforation thickness of RC slabs subjected to impacts are not suitable for the prediction of these quantities in the case of SFRC slabs.

4. The author's original formula for the prediction of perforation thickness of SFRC slabs seems to be very promising but requires further experimental verification.

5. The scaling laws concerning the dynamic experiments carried out on the models of SFRC structural elements should be studied because they are as yet almost unknown.

REFERENCES

1. RADOMSKI, W., Application of the rotating impact machine for testing fibre-reinforced concrete. *The International Journal of Cement Composites and Lightweight Concrete*, 3 (1981) 3–12.
2. RADOMSKI, W., Some problems concerning the determination of the impact resistance of steel reinforced concrete. *Brittle Matrix |Composites 2*, ed. A. M. Brandt & I. H. Marshall. Elsevier Science Publishers, London, 1989, p. 412.
3. RADOMSKI, W., Some properties of steel fibre reinforced concrete subjected to impact loads. *Trans. of the 9th Int. Conf. on Structural Mechanics in Reactor Technology*, Lausanne, Aug. 1987, Vol. H, pp. 285–91.
4. *Proc. of the Int. Symp. on Fibre Reinforced Concrete*, Madras, India, Dec. 1987, Vol. III, pp. 2.82–2.83.
5. HIBBERT, A. P. and HANNANT, D. J., *The design of an instrumented impact test machine for fibre concretes*. Testing and Test Methods of Fibre Cement Composites, RILEM Symposium 1978. The Construction Press Ltd, Hornby, 1978, pp. 107–20.
6. REINHARDT, H. W., *Testing and monitoring techniques for impact and impulsive loading of concrete structures*. RILEM–CEB–IABSE–IASS Interassociation Symposium on Concrete Structures under Impact and Impulsive Loading, Berlin (West), June 1982, Introductory Report, pp. 63–87.
7. ANDERSON, W. F., WATSON, A. J. and ARMSTRONG, P. J., *High velocity projectile impact on fibre reinforced concrete*. RILEM–CEB–IABSE–IASS Interassociation Symposium on Concrete Structures under Impact and Impulsive Loading, Berlin (West), June 1982, pp. 368–78.
8. MAYERHOFER, C. and THORN, H. J., *Dynamic response of fibre and steel reinforced concrete plates under simulated blast-load*. RILEM–CEB–IABSE–IASS Interassociation Symposium on Concrete Structures under Impact and Impulsive Loading, Berlin (West), June 1982, pp. 279–88.
9. HÜLSEWIG, M., STILP, A. and PAHL, H., *Behaviour of fibre reinforced concrete slabs under impact loading*. RILEM–CEB–IABSE–IASS Interassociation Symposium on Concrete Structures under Impact and Impulsive Loading, Berlin (West), June 1982, pp. 322–8.

10. UCHIDA, T., TSUBATA, H. and YAMADA, T., Experimental investigations on reinforced concrete slabs subjected to impact loading. *Trans. of the 8th Int. Conf. on Structural Mechanics in Reactor Technology*, Brussels, Aug. 1985, Vol. J, pp. 173–8.

11. GLINICKI, M. A. and RADOMSKI, W., *Fracture of steel fibre reinforced concrete slabs produced by impact loads*. Proc. of the RILEM Symp. on Developments in Fibre Reinforced Cement and Concrete, Sheffield, 1986, Vol. 2, paper 6.6.

12. GLINICKI, M. A. and RADOMSKI, W., Investigation of failure mode of fibre reinforced concrete slabs loaded by impacts. *Archiwum Inżynierii Lądowej*, **XXXII**(3) (1986) 483–502 (in Polish).

13. AROCKIASAMY, M., SWAMIDAS, A. S. J. and MUNASWAMY, K., *Behaviour of fibre reinforced concrete panels subjected to impact loads*. Proc. of the Int. Symp. on Fibre Reinforced Concrete, Madras, India, Dec. 1987, Vol. I, pp. 2.95–2.107.

14. BERRIAUD, C. *et al.*, Comportement local des enceintes en beton sous l'impact d'un projectile rigide. *Nuclear Engng and Design*, **45** (1978) 457–69.

15. BRANDES, K., *Behaviour of critical regions under soft missile impact and impulsive loading*. RILEM–CEB–IABSE–IASS Interassociation Symposium on Concrete Structures under Impact and Impulsive Loading, Berlin (West), June 1982, Introductory Report, pp. 91–111.

16. KENNEDY, R. P., A review of procedures for analysis and design of concrete structures to resist missile impact effects. *Nuclear Engng and Design*, **37** (1976) 183–203.

17. SLITER, G. E., Assessment of empirical concrete impact formulas. *Jnl of the Struct. Div., Proc. of the American Soc. of Civil Engineers*, **106** (1980) 1023–45.

18. HUGHES, G., Hard missile impact on reinforced concrete. *Nuclear Engng and Design*, **77** (1984) 23–35.

19. BRANDT, A. M., *On the optimization of the fiber orientation in cement-based composite materials*. Fiber Reinforced Concrete—International Symposium, ed. G. C. Hoff. ACI SP-81, Detroit, 1984, pp. 23–8.

46

Existence of Interphase Cracks in Polymer Concrete

Mieczysław Jaroniek

Institute of Applied Mechanics,
Technical University of Łódź, Zwirki Poland

ABSTRACT

Polymer concrete should be considered as material composed of two media; the polymer mortar and the grains. The existence of interphase cracks between the polymer mortar and the grains (the components of the material) is assumed. Experimental tests were carried out using DCB with specimens of the polymer concrete, composed of two media; polymer mortar and the grains. The critical value of the J-integral (the strain energy release rate G) was determined from load–displacement curve obtained using an X–Y plotter. The stress intensity factors K_{Jc} were determined by applying the Irwin method and photoelastic measurement results. The displacements and the stresses were also determined by applying strain gauges. Since an analytical solution for this problem is not available the distribution of stresses and displacements has been calculated using the finite element method.

The crack initiation always occurs in the interface region between the components of the specimens. Using the critical values of the strain energy release rate G_{cl} and the ultimate strength of the polymer concrete (obtained experimentally) and after calculation of Young's modulus (applying a double phase compound material law), the lengths of interphase cracks 'a' can be determined. After examination of the fracture of the specimens failed under tension the existence of interphase cracks has been confirmed with their lengths range agreeing with values calculated above.

PROPERTIES OF THE MATERIAL

Polymer concrete should be considered as macroscopically homogeneous characterised by Young's modulus E_{pc} and Poisson's ratio v_{pc}. These values can be determined experimentally or by applying a double phase compound material law drom the following expressions:

(a) Parallel model

$$E'_c = E_1 v_1 + E_2 v_2 \tag{1}$$

(b) Series model

$$E''_c = \left(\frac{v_1}{E_1} + \frac{v_2}{E_2}\right)^{-1} \tag{2}$$

(c) Hirsch's model[15]

$$E^H_{pc} = \left[\kappa \frac{1}{E_1 v_1 + E_2 v_2} + (1 - \kappa)\left(\frac{v_1}{E_1} + \frac{v_2}{E_2}\right)\right]^{-1} \tag{3}$$

where

v_1, v_2 = volume contents of the grains and the mortar, respectively,
E_1, E_2 = Young's moduli of the grains and the mortar, respectively,
$\kappa = 0.5$, found experimentally.

For properties of the grains '1' and the mortar '2':

v_1 (%)	v_2 (%)	E_1 (MPa)	v_1 (1)	E_2 (MPa)	v_2 (1)
0·8	0·2	32 000	0·16	5 850	0·3

From (3) we obtain

$$E^H_{pc} \simeq 20\,700\,\text{MPa}$$

and experimentally:

$$E_{pc} = 21\,800\,\text{MPa} \qquad R_m = 16\,\text{MPa} \qquad v_{pc} = 0.21$$

where $R_m = f_c$, ultimate strength, v_1, v_2, Poisson's ratios of the grains and the mortar, respectively.

THEORETICAL MODEL

In the case where the specimens failed under tension it is assumed that the initiation of the crack propagation occurs along the interphase cracks between the mortar and the grains (Fig. 1). For an infinite bi-material plate with a central crack along the interface subjected to biaxial loading (Fig. 2) the analytical solution was given in Refs 1–6.

The solution to the problem can be obtained using the complex variable approach. For $k = 0$

$$\sigma_{\theta\theta}^{(1)} = \frac{\sigma\sqrt{\pi a}}{(1+\alpha)\sqrt{2\pi r}} (e^{\varepsilon\theta} F_1 + \alpha e^{-\varepsilon\theta} f_2) + \sigma \frac{1-\alpha}{2(1+\alpha)} (\cos 2\theta - 1) \tag{4}$$

$$\sigma_{\theta\theta}^{(2)} = \frac{\sigma\sqrt{\pi a}}{(1+\alpha)\sqrt{2\pi r}} r(\alpha e^{\varepsilon\theta} F_1 + e^{-\varepsilon\theta} f_2) + \mu\sigma \frac{1-\alpha}{2(1+\alpha)} (\cos 2\theta - 1)$$

where

$$\mu = \frac{\mu_2(1+\chi_1)}{\mu_1(1+\chi_2)} \qquad \alpha = \frac{\mu_1 + \mu_2\chi_1}{\mu_2 + \mu_1\chi_2}$$

$$\varepsilon = \frac{\ln \alpha}{2\pi} \qquad E_i = 2\mu_i(1+\nu_i)$$

and

$$F_1 = 2f_1 - \tfrac{1}{2}(f_3 + f_4)$$

where E_i, ν_i = Young's moduli and Poisson's ratios.

$$\chi_i = \frac{3-\nu_i}{1+\nu_i} \text{ for plane stress} \qquad \chi_i = 3 - 4\nu_i \text{ for plane strain}$$

$$f_1 = \cos\left(\varepsilon \ln \frac{r}{2a} + \frac{\theta}{2}\right) - 2\varepsilon \sin\left(\varepsilon \ln \frac{r}{2a} + \frac{\theta}{2}\right)$$

$$f_2 = \cos\left(\varepsilon \ln \frac{r}{2a} + \frac{3\theta}{2}\right) - 2\varepsilon \sin\left(\varepsilon \ln \frac{r}{2a} + \frac{3\theta}{2}\right)$$

$$f_3 = (1 - 4\varepsilon^2)\cos\left(\varepsilon \ln \frac{r}{2a} - \frac{3\theta}{2}\right) - 4\varepsilon \sin\left(\varepsilon \ln \frac{r}{2a} - \frac{3\theta}{2}\right)$$

$$f_4 = (1 + 4\varepsilon^2)\cos\left(\varepsilon \ln \frac{r}{2a} + \frac{\theta}{2}\right)$$

The critical stress intensity factors $K_{1c}^{(i)}$ can be determined from the following conditions:

$$\begin{aligned} K_{1c}^{(1)} &= \sqrt{2\pi r_0}\, \sigma_{\theta\theta}^{(1)}(\theta_0^{(1)}, r_0) & 0 \le \theta_0^{(1)} \le \pi \\ K_{1c}^{(2)} &= \sqrt{2\pi r_0}\, \sigma_{\theta\theta}^{(2)}(\theta_0^{(2)}, r_0) & -\pi \le \theta_0^{(2)} \le 0 \\ K_{bc} &= \sqrt{2\pi r_0}\, \sigma_{\theta\theta}(0, r_0) & \theta_0^{(2)} = \theta_0^{(2)} = 0 \end{aligned} \tag{5}$$

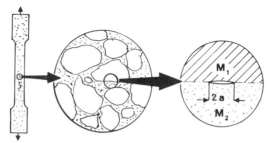

FIG. 1.   Following augmentation of the crack initiation.

where $\theta_0^{(1)}$, $\theta_0^{(2)}$ are the angles corresponding to the maximum values of $\sigma_{\theta\theta}$ and $K_{Ic}^{(1)}$, $K_{Ic}^{(2)}$ are the critical stress intensity factors of these two media and K_{bc} denotes the critical bonding factor.

The strain energy release rate G is calculated using the crack closure method[5] by employing the following relation:

$$G = \frac{(\sigma\sqrt{\pi a})^2(1 + 4\varepsilon^2)\alpha}{4(1 + \alpha)^2}\left(\frac{\chi_1 + 1}{\mu_1} - \frac{\chi_2 + 1}{\mu_2}\right) \tag{6}$$

For the homogeneous material $\alpha = 1$, $\varepsilon = 0$ and the strain energy release rate G is reduced to the expression:

$$G = \frac{(\sigma\sqrt{\pi a})^2}{E} = \frac{K_I^2}{E} \tag{7}$$

Using the critical values of the strain energy release rate $G_c = J_c$ and the ultimate strength of the polymer concrete (obtained experimentally) and after calculation of Young's modulus (applying a double phase compound material law), the lengths of interphase cracks a can be determined from:

$$a = \frac{G_c E^*}{\pi f_c^2} \tag{8}$$

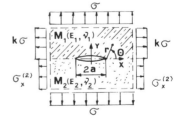

FIG. 2.   General geometry and central crack between two isotropic media: the grains (M_1) and the mortar (M_2).

where

$$E^* = \frac{4(1+\alpha)^2}{(1+4\varepsilon^2)\alpha}\left(\frac{\chi_1+1}{\mu_1} + \frac{\chi_2+1}{\mu_2}\right)^{-1}$$

$f_c = \sigma_{crit}$, critical value of the ultimate strength.

NUMERICAL AND EXPERIMENTAL ANALYSIS OF STRESS AND STRAIN DISTRIBUTION

The distribution of stresses and displacements has been calculated for the double cantilever beam (DCB) specimen. The configuration of this specimen is given in Fig. 3. The geometry and materials of the models were chosen to correspond to the actual specimens used in the experiments. Since an analytical solution for this problem is not available the distribution of stresses and displacements has been calculated using the finite element method. The strain energy release rate G equals in this case the Rice J-integral which can be found from the following expression:

$$J = \int_s \left(\frac{1}{2}\sigma_{ij}\varepsilon_{ij}\,dx_2 - T_i^{(n)}\frac{\partial u_i}{\partial x_1}\,ds\right) \tag{9}$$

or from numerical calculations using the relation

$$J = \sum \left\{\frac{1}{2}\left[\frac{1}{E}(\sigma_{y_i}^2 - \sigma_{x_i}^2) - \tau x_{y_i}^2\right]n_{1i} - \left[\frac{\tau x_{y_i}}{E}(\sigma_{x_i} - v\sigma_{y_i}) - \sigma_{y_i}\frac{\Delta v_i}{\Delta x_i}\right]n_{2i}\right\}\Delta s_i \tag{10}$$

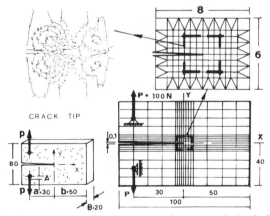

FIG. 3. General view of the DCB specimen, mesh for a numerical calculation and the isochromatic fringe pattern at the crack tip.

TABLE 1
Results

No. Spec.	Numerical			Photoelastic
	P_{crit} (N)	K_{Ic} (MN/m$^{3/2}$)	J_c (N/m)	K_{Ic}^{PH} (MN/m$^{3/2}$)
1	1 620	1·33	90·5	1·42
2	1 700	1·39	99·7	1·3
3	1 690	1·38	98·5	1·25
4	1 540	1·26	81	1·1
5	1 580	1·3	86	0·94
6	1 650	1·36	93·9	0·85

The J-integral can be also found from the results of the calculation on the basis of the force-displacement curve P:

$$J = \frac{2A}{b - B} \tag{11}$$

where
    A = area under load–displacement curve,
    B = thickness of the specimen,
    b = ligament depth.

A finite element mesh of a DCB specimen used for numerical simulation of the stress distribution and the isochromatic fringes was as shown in Fig. 3 and the numerical results are given in Table 1. Using FEM the isochromatic fringe pattern has been also calculated. The pattern obtained from the numerical calculations for a DCB specimen and the distribution of the isochromatic fringes obtained experimentally in a photoelastic coating is shown in Fig. 4.

The strain energy release rate G and the stress intensity factors K_I were evaluated from (10), (11) and (7) for the tearing forces $P = 100$ N. The critical values of the stress intensity factors K_I and G_c were evaluated using critical values of tearing forces P_c obtained experimentally

$$K_{Ic} = \beta K_I \qquad G_c = \beta^2 G \tag{12}$$

where

$$\beta = P_c/P$$

The K_{Ic} values can be determined by Irwin's method[10] using isochromatic

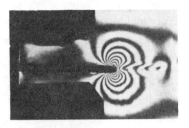

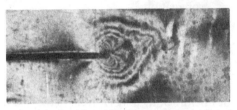

FIG. 4. Isochromatic fringes obtained from FEM and experimentally using a photoelastic coating in the region of a crack tip.

fringe loops which occur in the region adjacent to the crack tip by employing the following relation:

$$K_{Ic} = 2\tau_m v 2\pi r_i \phi(\theta_i) \qquad (13)$$

where

$$\tau_m = 0.5 f_i n_i = \tfrac{1}{2}(\sigma_1 - \sigma_2)$$

θ_i = angle corresponding to point in the isochromatic fringe loop which occurs in the region adjacent to the crack tip where $\partial\tau_m/\partial\theta_i = 0$

$$\phi(\theta_i) = \frac{1}{\sin\theta_i}\left[1 + \left(\frac{2}{3\,\mathrm{tg}\,\theta_i}\right)^2\right]^{-1/2}\left(1 + \frac{2\,\mathrm{tg}\,3\theta_i/2}{3\,\mathrm{tg}\,\theta_i}\right)$$

$f_i = f\dfrac{E_{pc}(1 + v)}{E(1 + v_{pc})}$ = photoelastic material fringe value on the PC surface,

f = material fringe value of the photoelastic coating,

E_{pc}, E, v_{pc}, v = the Young's moduli and Poisson's ratios of the PC and photoelastic coating, respectively,

n_i = isochromatic fringe order obtained experimentally or from numerical calculations.

EXPERIMENTAL RESULTS

Experimental tests have been carried out using the following specimens:

(1) photoelastic model of a DCB specimen (Fig. 5) and type 2 DCB specimens (Fig. 3),

(2) photoelastic model.

In order to visualise the stress distribution and to analyse the influence of the grain distribution on the crack development and its propagation in the neighbourhood of the grains the tests have been carried out at critical load and during crack propagation.

Components of the photoelastic model are:

the grains: 'plexiglass' ($E_{px} = 6000$ MPa, $v_{px} = 0.33$)
the mortar: epoxy resin ($E_r = 3540$ MPa, $v_r = 0.34$)

In Fig. 5 the isochromatic fringe pattern representing the concentration at the tip of the crack before branching is given. In the case of the photoelastic model the grains are a satisfactory barrier for changing the direction of the propagating crack. In this case: $G_c = 191$ N/m, $K_c = 0.82$ MN/m$^{3/2}$.

DCB Specimens of Polymer Concrete

The critical value of the J-integral (the strain energy release rate G) was determined according to expression (11) from the load–displacement curve obtained using an X–Y plotter. The stress intensity factors K_{Ic} were determined by applying the Irwin method from (13) and photoelastic measurement results. The displacements and the stresses were determined by applying the strain gauges.

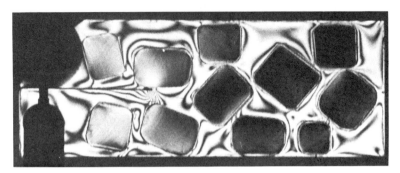

FIG. 5.   Photoelastic model of DCB specimen. Isochromatic fringes in the mortar epoxy resin and the grains.

TABLE 2
Experimental results

No. Spec.	P_{cr} (N)	b (mm)	B (mm)	G_c (N/m)	K_{Ic} (MN/m$^{3/2}$)	a (mm)
1	1 620	50	20·5	105	1·51	1·33
2	1 700	49	21	124	1·64	1·58
3	1 690	49	23	114	1·57	1·45
4	1 540	50	20	87	1·37	1·11
5	1 580	50	19·5	92	1·24	1·16
6	1 650	51	18	61	1·06	0·77

The experimental and calculated values of the K_{Ic} factors and critical values of the J_c-integrals are given in Table 2. Using the critical values of the strain energy release rate $(G_c = J_c)$ and the ultimate strength $(R_m = f_c)$ of the polymer concrete obtained experimentally, the lengths of interphase cracks a can be determined substituting the values of G_c and ultimate strength $(f_c = \sigma_{crit.})$ into (8). After examination of the fracture of the specimens failed under tension the existence of interphase cracks had been confirmed with their length range agreeing with the values calculated above.

CONCLUSIONS

Generally a good agreement between numerical, photoelastic and experimental results shown in Tables 1 and 2 has been observed. In the case of polymer concrete the crack initiation occurs always in the interface region between the components of the specimens. Using the critical values of the strain energy release rate G_c and the ultimate strength f_c of the polymer concrete obtained experimentally the lengths of interphase cracks a were determined. After examination of fracture of the specimens failed under tension the dimension of the interphase cracks have been confirmed with their lengths range agreeing with values calculated above.

ACKNOWLEDGEMENTS

This research was supported by CPBP PROBLEM 02.21 in cooperation with the Institute of Fundamental Technological Research, Polish Academy of Sciences. The author is most grateful to Professor A. M. Brandt for many valuable remarks with regard to the content of this paper.

Preparation of the polymer concrete specimens by Professor L. Czarnecki, Warsaw Technical University, is gratefully acknowledged.

REFERENCES

1. SNEDDON, I. N., *The Use of Integral Transforms*. McGraw-Hill, New York, 1972.
2. WILLIAMS, M. L., The stresses around a fault or crack in dissimilar media. *Bull. Seismol. Soc. Amer.*, **49** (1959) 199–204.
3. RICE, J. R. and SIH, G. C., Plane problem of crack in dissimilar media. *J. Appl. Mech.*, **E32** (1965) 418–23.
4. CLEMENTS, D. L., A crack between dissimilar anisotropic media. *Int. J. Engng Sci.*, **9** (1971) 257–65.
5. WILLIS, J. R., Fracture mechanics of interfacial cracks. *J. Mech. Phys. Solids*, **19** (1971) 353–68.
6. PIVA, A. and VIOLA, E., Biaxial load effects on a crack between dissimilar media. Nota Tecnica n.35.1979, Istituto di Scienza delle Costruzioni, University of Bolonia, Italy.
7. SZMELTER, J., *The Finite Element Method Programs* (in Polish). Arkady, Warszawa, 1973.
8. BRANDT, A. M., Influence of the fibre orientation on the energy absorption at fracture of SFRC specimens. In *Brittle Matrix Composites 1*, ed. A. M. Brandt and I. H. Marshall. Elsevier Applied Science Publishers, London, 1986, pp. 403–20.
9. IRWIN, G. R., DALLY, J. W., KOBAYASHI, T., FOURNEY, W. L., ETHERIDGE, M. J. and ROSSMANITH, H. P., On the determination of the a–K relationship for birefringent polymers. *Exp. Mech.*, **19**(4) (1979) 121–8.
10. IRWIN, G. R., The dynamic stress distribution surrounding a running crack—a photoelastic analysis. *Proc. SESA*, **16**(1) (1958) 92–6.
11. ZIENKIEWICZ, O. C., *The Finite Element Method in Engineering Science*. McGraw-Hill, London, 1971.
12. SIH, G. C. and LIEBOWITZ, H., *Mathematical Theories of Brittle Fracture in Fracture*, Vol. II, Academic Press, New York and London, 1968, pp. 67–190.
13. CZARNECKI, L. and LACH, V., Structure and fracture in polymer concretes: some phenomenological approaches. In *Brittle Matrix Composites 1*. ed. A. M. Brandt and I. M. Marshall, Elsevier Applied Science Publishers, London, pp. 241–61.
14. JARONIEK, M. and NIEZGODZIŃSKI, T., Studies of fracture and the crack propagation in the concrete and the polymer concrete. In *Brittle Matrix Composites 1*. ed. A. M. Brandt and I. M. Marshall, Elsevier Applied Science Publishers, London, pp. 355–70.
15. HOLLIDAY, L., *Composite Materials*, Elsevier, Amsterdam, 1966, p. 540.

47

Micro-to-Macro Analysis of Viscoelastic Laminated Plates

Gabriel Cederbaum and Jacob Aboudi

Department of Solid Mechanics, Materials and Structures,
Faculty of Engineering, Tel-Aviv University, Ramat-Aviv,
69978, Israel

ABSTRACT

This paper deals with the analysis of viscoelastic laminated plates subjected to static, dynamic or random loads. The methodology is based on a micromechanics approach by which the time-dependent properties of the composite are established, followed by a macromechanics analysis of the laminated plate. The following problems are analyzed: quasi-static deflection, quasi-static buckling, dynamic response and random vibration.

1 INTRODUCTION

Various theories appear in the literature for the analysis of laminated composites beams, plates and shells, differing mainly in whether the shear deformation and rotary inertia effects are included. In applying these theories, the composite is effectively represented by anisotropic elastic material. However, in the case of polymeric matrices, in particular, strong viscoelastic behavior is exhibited, implying time-dependent properties which should be incorporated. A comprehensive review of viscoelastic behavior and analysis of composite materials was given by Schapery.[1] It appears that the determination of the five time-dependent functions which characterize the effective transversely isotropic behavior of viscoelastic unidirectional fiber composites is very difficult. However, these functions can be readily obtained by using a micromechanical analysis of the composite which takes into account the interaction between fiber and matrix.

Recently, the second author developed a micromechanical analytical model for the prediction of the effective elastic moduli,[2] thermoelastic properties,[3] strength[4] and fatigue failure[5] of fiber-reinforced materials. Furthermore, this theory was extended for the prediction of the yield surfaces[6] and the elasto-plastic response of metal matrix composites by establishing constitutive equations which related the average stresses, strains and plastic strains.[7,8] The micromechanics theory is based on the analysis of a periodic array of square fibers imbedded regularly in the matrix. It should be emphasized that this micromechanics theory provides in a closed form manner all the five independent effective elastic properties of the unidirectional composite from the knowledge of the fiber and matrix constants and the reinforcement volume ratio. In applying this micro-mechanics analysis on viscoelastic composites, the time-dependent properties of the phases should be provided. The prediction of this theory for a unidirectional viscoelastic composite was shown recently to correlate well with experimental results.[9,10]

By adopting the above methodology, we first analyze the behavior of viscoelastic laminated plates subjected to a quasi-static loading. The single unidirectional lamina is composed of a viscoelastic matrix (epoxy), reinforced by perfectly elastic fibers (graphite). The five time-dependent functions of the composite are determined from the above micromechanics analysis and the correspondence principle. The inversion of the Laplace transform (using the procedure of Bellman *et al.*[11]) yields the required response function in the time domain. The deflections of a square orthotropic plate and of a semi-infinite cross-ply (6 layers) plate are analyzed by using the classical lamination theory (CPT),[12] as well as the first-order shear deformation theory (FSDT).[13,14] These two theories, in conjunction with the above procedure, are also used for the prediction of the quasi-static buckling load of a square orthotropic plate. It is shown that the viscoelasticity affects significantly the plate behavior, and when employing the shear deformation theory these effects are dominant even at short times, and for high length-to-thickness ratio.

Next, the dynamic response of viscoelastic laminated plates under impulsive loading is investigated. The Boltzmann representation of the viscoelastic phases is given in the frequency domain by the application of the Fourier transform. By utilizing the above micromechanics analysis, the Fourier transform of the five time-dependent functions are obtained. These functions enable us to express the governing equations of the plate in the frequency domain as well. The inversion of the response functions to the time domain is performed by using the Fast Fourier Transform algorithm

(FFT). The analysis of the plates is carried out by using the FSDT and a high-order shear deformation theory (HSDT).[15] The case of a square, cross-ply viscoelastic plate is considered. A point load with a timewise function in the form of a half sine pulse is applied, and the dynamic transverse response is derived. It is shown that due to the viscoelastic effects the dynamic response is rapidly decaying, as compared with the nondecaying behavior of the elastic counterpart. It is also shown that the behavior of the plate when predicted by the two shear deformation theories is very much the same. In addition, the complex eigenvalues are also investigated. The real part of the first (fundamental) eigenvalue is found to be asymptotic to the elastic natural frequency. The ratio between the imaginary and the real parts, η, is found to be related to the level of the material damping, and to vary for different values of length-to-thickness ratio and fiber orientation.

The last part of this study deals with the response of angle-ply laminates to random excitation. The results are obtained by considering the material damping mechanism only. Two cases of stationary random fields are considered: (i) band-limited white-noise, and (ii) ideal white-noise. In both cases a closed form solution is derived for the mean-square response. Since the eigenvalues were found to be well-separated, and since η is small, Laplace's asymptotic method can be applied for the calculation of the mean-square response due to other excitations.

2 COMPOSITE EFFECTIVE ELASTIC CONSTANTS

Let (x_1, x_2, x_3) denote a Cartesian system of coordinates with x_1 oriented in the fiber direction of a unidirectional fiber-reinforced elastic composite. The constitutive law for the effective transversely isotropic behavior of such a composite is determined from micromechanics analysis in the form[8]

$$\bar{\sigma} = \mathbf{E}\bar{\varepsilon} \tag{1}$$

where $\bar{\sigma} = (\bar{\sigma}_{11}, \bar{\sigma}_{22}, \bar{\sigma}_{33}, \bar{\sigma}_{12}, \bar{\sigma}_{13}, \bar{\sigma}_{23})$ is the average stress, $\bar{\varepsilon} = (\bar{\varepsilon}_{11}, \bar{\varepsilon}_{22}, \bar{\varepsilon}_{33}, 2\bar{\varepsilon}_{12}, 2\bar{\varepsilon}_{13}, 2\bar{\varepsilon}_{23})$ is the average strain, and

$$\mathbf{E} = \begin{bmatrix} e_{11} & e_{12} & e_{12} & 0 & 0 & 0 \\ & e_{22} & e_{23} & 0 & 0 & 0 \\ & & e_{22} & 0 & 0 & 0 \\ & & & e_{44} & 0 & 0 \\ & \text{symmetric} & & & e_{44} & 0 \\ & & & & & \frac{1}{2}(e_{22}-e_{23}) \end{bmatrix} \tag{2}$$

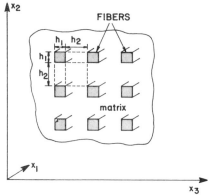

FIG. 1. A composite with a doubly periodic array of square fibers extending in the x_1 direction.

where e_{ij} are the effective elastic constants (their explicit expressions in terms of the fibers and matrix properties and the reinforcement volume ratio can be found in Aboudi[8]). The above equations were derived from a micromechanical analysis in which the fibers of a square cross-section h_1^2 are arranged in the matrix phase in a doubly periodic array at distances h_2 apart, see Fig. 1. A representative cell of this arrangement contains four subcells, the first one of which is occupied by the fiber region, while the remaining subcells are occupied by the matrix region.

3 COMPOSITE VISCOELASTIC REPRESENTATION

With $\mathbf{E}$ in (1) involving the five independent elastic constants of the equivalent transversely isotropic material which represent the unidirectional composite, it is possible to obtain the five time-dependent functions which characterize the viscoelastic composite whose phases are viscoelastic materials. Each phase (fiber and matrix) is represented by the Boltzmann superposition principle[16] in tensorial notation as follows

$$\sigma_{ij}^{(p)}(t) = \int_{-\infty}^{t} c_{ijkl}^{(p)}(t - \tau)\dot{\varepsilon}_{kl}^{(p)}(\tau)\,\mathrm{d}\tau \tag{3}$$

where $c_{ijkl}^{(p)}(t)$ are the relaxation functions of the phase—p. By applying the Laplace transform on (3) we obtain

$$L[\sigma_{ij}^{(p)}] = sL[c_{ijkl}^{(p)}]L[\varepsilon_{kl}^{(p)}] \tag{4}$$

where $L[\cdot]$ denotes the Laplace transform and s is the transform parameter. Thus, the use of $sL[c_{ijkl}^{(p)}]$ in the micromechanics analysis (instead of $c_{ijkl}^{(p)}$ in the elastic case), provides $sL[\mathbf{E}(t)]$ (instead of $\mathbf{E}$ in the elastic case). It follows that in the transform domain, the stress–strain constitutive law of the composite is given according to (1) and (4) by

$$L[\bar{\sigma}(t)] = sL[\mathbf{E}(t)]L[\bar{\varepsilon}(t)] \tag{5}$$

The inversion of $L[\mathbf{E}(t)]$ back to the time domain provides the relaxation functions $\mathbf{E}(t)$, while the solution of any laminated composite structure problem yields the corresponding solution in the transform domain by replacing $\mathbf{E}$ with $sL[\mathbf{E}]$. The inversion to the time domain can be performed by adopting the numerical method of Bellman *et al.*[11]

The above representation can be also obtained in the Fourier transform domain. In this case, the functions $c_{ijkl}^{(p)}(t)$ can be represented in the following form[17]

$$c_{ijkl}^{(p)}(t) = c_{ijkl}^{(p)}(0)[1 - \psi_{ijkl}^{(p)}(t)] \tag{6}$$

in which $c_{ijkl}^{(p)}(0) \equiv c_{ijkl}^{(p)}(t=0)$, and $\psi_{ijkl}^{(p)}(t)$ are zero for $t \leq 0$. It should be noted that in the case of an elastic phase $\psi_{ijkl}^{(p)}(t) \equiv 0$. Substituting eqn (6) into (3) yields

$$\sigma_{ij}^{(p)}(t) = c_{ijkl}^{(p)}(0)\varepsilon_{kl}^{(p)}(t) - c_{ijkl}^{(p)}(0)\int_{-\infty}^{t} \psi_{ijkl}^{(p)}(t-\tau)\varepsilon_{kl}^{(p)}(\tau)\,\mathrm{d}\tau \tag{7}$$

Integrating by parts the convolution integral provides

$$\sigma_{ij}^{(p)}(t) = c_{ijkl}^{(p)}(0)\varepsilon_{kl}^{(p)}(t) - c_{ijkl}^{(p)}(0)\int_{-\infty}^{t} \dot{\psi}_{ijkl}^{(p)}(t-\tau)\varepsilon_{kl}^{(p)}(\tau)\,\mathrm{d}\tau \tag{8}$$

and by defining $\zeta = t - \tau$ we get

$$\sigma_{ij}^{(p)}(t) = c_{ijkl}^{(p)}(0)\varepsilon_{kl}^{(p)}(t) - c_{ijkl}^{(p)}(0)\int_{0}^{\infty} \dot{\psi}_{ijkl}^{(p)}(\zeta)\varepsilon_{kl}^{(p)}(\tau-\zeta)\,\mathrm{d}\xi \tag{9}$$

The lower limit of integration can be extended to $-\infty$. Applying the Fourier transform to eqn (9) and using the convolution theorem, we get

$$F[\sigma_{ij}^{(p)}(t)] = c_{ijkl}^{(p)}(0)\{1 - F[\dot{\psi}_{ijkl}^{(p)}(t)]\}F[\varepsilon_{kl}^{(p)}(t)] \tag{10}$$

Thus, the use of eqn (10) in the micromechanics analysis (instead of $c_{ijkl}^{(p)}$ in the elastic case), provides the Fourier transform of the effective time dependent viscoelastic properties, $F[\mathbf{E}(t)]$, of the composite (instead of $\mathbf{E}$ in the elastic case). These properties enable us to express the governing equations of the laminated viscoelastic plates in the frequency domain. To this end, the imposed dynamic load is incorporated in these equations in its Fourier transform form. The inversion to the time domain is achieved by utilizing the FFT.

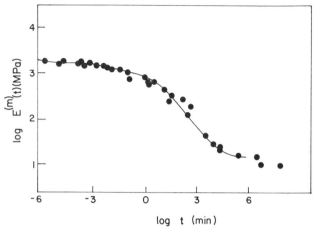

F_IG. 2.   Measured data[18] of an epoxy matrix at 50°C. The solid line is its representation by cubic splines.

In this paper we used a viscoelastic epoxy matrix (Epon 815 mixed with Versamid 140),[18] whose measured time-dependent Young's modulus $E^{(m)}(t)$ at 50°C is shown in Fig. 2. This is a relatively low performance resin, whose viscoelastic behavior is significant. For high performance materials, appreciable viscoelastic effects can be obtained at higher temperatures. Cubic splines were fitted to these measured data in the region $0 \le t \le 10^6$ min. The correspondence between the measured values and the cubic splines representation, as shown in Fig. 2, is excellent. The Poisson ratio of the matrix was assumed to be constant, $v_m = 0.38$, in accordance with Ref. 1, in which it is stated that the Poisson ratio of epoxy is very nearly constant with values in the range of 0.38 ± 0.03. It should be noted, however, that the present micromechanics analysis allows the use of time-dependent Poisson's ratios whenever they are available. T-300 graphite elastic fibers were chosen to reinforce the epoxy matrix with 60% volume ratio. The elastic constants of the transversely isotropic graphite fibers are $E_A = 220$ GPa, $v_A = 0.3$, $E_T = 22$ GPa, $v_T = 0.35$ and $G_A = 22$ GPa. Here E_A and v_A denote the axial Young's modulus and Poisson's ratio, E_T and v_T are the transverse Young's modulus and Poisson's ratio, and G_A is the axial shear modulus.

Results at a temperature other than the reference one (50°C) can be obtained by determining first the value of the shift factor a_T at the desired temperature[18]

$$\log a_T = -0.1049 - 0.2071(T - 50)$$
$$- 1.243 \times 10^{-3}(T - 50)^2 + 4.013 \times 10^{-5}(T - 50)^3 \qquad (11)$$

and then the abscissa, log t, of the master curve of Fig. 2 is shifted by log a_T, resulting in a new abscissa, $\log t^* = \log t - \log a_T$.

4 MACROMECHANICAL ANALYSIS OF VISCOELASTIC LAMINATED PLATES

The analysis of laminated plates can be carried out by several theories which differ mainly in whether the effects of the shear deformation and rotary inertia are incorporated. The CPT[12] is based on the Kirchhoff assumption that the inplane transverse normal remains normal while loading the plate, and thus the effect of the shear deformation is not included. The FSDT was developed by Yang et al.[13] and leads to a linear distribution of the inplane normal and shear stresses through the thickness. This yields nonzero transverse shear stresses on the plate bounding planes, and therefore the need of a shear correction factor. HSDT, on the other hand, leads to a nonlinear distribution of the shear stresses through the thickness. Reddy's theory[15] provides a parabolic distribution, by which the conditions on the boundary planes are fulfilled and the need of a shear correction factor is removed. In the following, the FSDT will be mainly used, while in some cases results obtained from the CPT and HSDT will be given for a matter of comparison.

Within FSDT, the following displacement field across the plate thickness is considered:

$$U_1 = U + z\psi_x \qquad U_2 = V + z\psi_y \qquad U_3 = W \qquad (12)$$

where U_1, U_2 and U_3 are the components of the three-dimensional displacement vector in the x, y and z directions, respectively, U, V and W denote the displacements of a point (x, y) on the mid-plane $(z = 0)$, while ψ_x and ψ_y are the rotations of the normals to the mid-plane about the y and x axes, respectively. The equations of motion are[14]

$$N_{1,x} + N_{6,y} = I_1 \ddot{U} + I_2 \ddot{\psi}_x$$
$$N_{6,x} + N_{2,y} = I_1 \ddot{V} + I_2 \ddot{\psi}_y$$
$$Q_{x,x} + Q_{y,y} + P_1 W_{,xx} + P_2 W_{,yy} + P_3 = I_1 \ddot{W} \qquad (13)$$
$$M_{1,x} + M_{6,y} - Q_1 = I_2 \ddot{U} + I_3 \ddot{\psi}_x$$
$$M_{6,x} + M_{2,y} - Q_2 = I_2 \ddot{V} + I_3 \ddot{\psi}_y$$

where N_i, M_i, Q_i, I_i and P_i are the in-plane stress resultants, stress couples, shear stress resultants, inertia terms and the applied loads, respectively.

Exact solution of (13) can be obtained for two types of rectangular $(a \times b)$, simply supported laminated plates: cross-ply (CP) and anti-symmetric angle-ply (AP).

Using the procedure given in Refs 19 and 20, one obtains the following general relation for the corresponding viscoelastic problem

$$\{[K(\xi)]_{mn} - \lambda^2_{mn}(\xi)[M]\}\{\Delta\}_{mn} = \{P\} \tag{14}$$

where $[K(\xi)]$, $[M]$, $\{\Delta\}$ and $\{P\}$ are the stiffness matrix in the transform domain (ξ is the transform parameter), the inertia matrix, the generalized modal displacements vector and the force vector, respectively, while $\lambda^2_{mn}(\xi)$ depend on the problem to be solved (quasi-static, dynamic, etc.).

5 APPLICATIONS

5.1 Quasi-Static Deflection

Let the only static force, applied on the laminated plate, be in the form

$$P_3 = q_0 \sin\frac{\pi x}{a} \sin\frac{\pi y}{b} \tag{15}$$

where q_0 is an amplitude factor. In this case $\lambda^2_{mm}(\xi) \equiv 0$. For an orthotropic square elastic plate the maximum deflection is given in Refs 12 and 21 as obtained by using the CPT and the FSDT, respectively. The maximum deflection of an antisymmetric cross-ply laminate in cylindrical bending (CB) is given in Ref. 14. The corresponding viscoelastic results for the above two cases, obtained within the CPT and the FSDT, normalized to $\bar{W}$ where

$$\bar{W} = W_{\max}E^{(m)}(t=0)h^3 10^3/q_0 a^4 \tag{16}$$

(h is the total thickness of the plate), are shown in Fig. 3 and in Fig. 4 (CPT only), in which the shear correction factor is 5/6 and the length-to-thickness ratio, a/h, is 20. It is clearly seen that the viscoelastic effects predicted by the FSDT are significant as compared with those of the CPT. This significant difference occurs already at a short time of about 10 min. Moreover, this difference is relatively high even for high length-to-thickness ratio. For example, for $a/h = 50$ it was found that $\bar{W}(t = 10^6)/\bar{W}(t = 0) = 1.8$ in the case of cylindrical bending within the FSDT, while within the CPT it is 1.2 only. Thus, in the case of viscoelastic laminated plates, the use of FSDT is necessary even for thin plates.

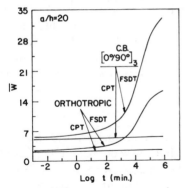

FIG. 3. Normalized maximum deflections versus $\log t$ of viscoelastic orthotropic and cross-ply laminates, $a/h = 20$.

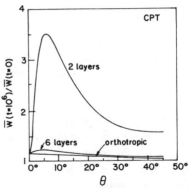

FIG. 4. The ratio between the viscoelastic deflection at $t = 10^6$ min and the elastic one versus θ for angle-ply laminates.

5.2 Quasi-Static Buckling

In this case $\lambda_{mn}^2(\xi) \equiv 0$, and for a uniaxial loading $P_1 \neq 0$, $P_2 = P_3 = 0$. The determination of the buckling loads of elastic laminates can be found in Ref. 12. The critical buckling load, P_1 is obtained by solving eqn (14). Figure 5 shows the variation of P_1 versus $\log t$ for a $45°$ angle-ply square viscoelastic laminate, in the orthotropic case (large number of layers). The results are normalized to $\bar{N}$, where

$$\bar{N} = P_1 a^2 / E^{(m)}(t = 0)h^3 \tag{17}$$

The importance of the FSDT even for $a/h = 50$ can be easily noticed.

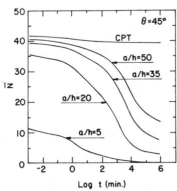

FIG. 5. Normalized buckling loads versus $\log t$ of a viscoelastic $45°$ angle-ply laminate (orthotropic case).

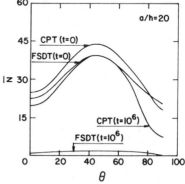

FIG. 6. Normalized buckling loads at $t = 0$ min and at $t = 10^6$ min versus θ for four layers angle-ply viscoelastic laminates, $a/h = 20$.

The values of $P_1(t = 0)$ and $P_1(t = 106)$ for $0° \leq \theta \leq 90°$ and $a/h = 20$ are shown in Fig. 6. In all cases the maximum critical load is obtained at $\theta = 45°$, while the minimum at $\theta = 90°$ (due to the change of the buckling mode). For both the CPT and the FSDT, qualitative as well as quantitative differences of the viscoelastic behavior with respect to the elastic one (at $t = 0$) are seen.

5.3 Dynamic Response

Setting $\{P\} \equiv 0$, eqn (14) represents the free vibrations problem, in which $\lambda_{mn}(\omega)$ are the complex eigenvalues in the frequency domain. In the forced vibrations problem, the expression for the generalized coordinates in the frequency domain, $T_{mn}(\omega)$, is obtained by using the modal analysis technique, in conjunction with the orthogonality condition,[20] as follows

$$-\omega^2 T_{mn}(\omega) + \lambda_{mn}^2(\omega) T_{mn}(\omega)$$

$$= \frac{P(\omega) \int_0^a \int_0^a [P_3(x, y) W_{mn}(x, y)] \, dx \, dy}{\int_0^a \int_0^a [I_1 W_{mn}^2(x, y)] \, dx \, dy} \equiv \frac{Q_{mn}(\omega)}{J_{mn}} = F_{mn}(\omega) \quad (18)$$

from which

$$T_{mn}(\omega) = \frac{1}{J_{mn}} \frac{Q_{mn}(\omega)}{-\omega^2 + \lambda_{mn}^2} \equiv \frac{1}{J_{mn}} \frac{Q_{mn}(\omega)}{L_{mn}(\omega)} \quad (19)$$

The dynamic part of the load is chosen as a half sine pulse extending from zero to t_1, and normalized to a unit maximum value. Its Fourier transform is given by

$$P(\omega) = \frac{\pi}{t_1} [\exp(-i\omega t_1) + 1] \Big/ \left[\left(\frac{\pi}{t_1}\right)^2 - \omega^2 \right] \quad (20)$$

Substituting $T_{mn}(\omega)$ into the assumed solution function one obtains the final solution of the frequency-dependent transverse displacement. The inverse of this solution into the time domain is performed by using the FFT algorithm.[22]

The response of a square, viscoelastic cross-ply laminate ($[0/90]_s$), obtained within the FSDT and the HSDT, is shown in Fig. 7 at two periods of time. A unit point load located at the mid-point of the plate with a pulse duration of $t_1 = 10^{-6}$ s was considered. The result are given for a

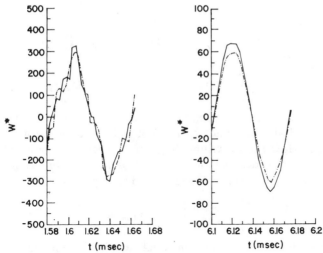

FIG. 7. Normalized mid-plate displacements versus t at two time intervals of a four layers viscoelastic cross-ply laminate, $a/h = 20$. (——, FSDT; –·–·–, HSDT).

temperature of 78°C, at the mid-point of the plate, and in a normalized form

$$W^* = W/h \qquad (21)$$

where h is the thickness of a single ply.

The difference between the results obtained by the two theories are relatively small for both periods of time. It can be seen that the response attenuates rapidly, although only material damping was considered. Moreover, it can be verified that the contribution of the non-fundamental modes $(m, n \neq 1)$ becomes negligible and the response behavior tends to that of a single mode.

5.4 Random Vibrations

The cross-spectral density function of the transverse displacement W can be written in the form

$$
S_w(x_1, y_1, x_2, y_2, \omega) = \sum_{m,n} W_{mn}(x_1, y_1) W_{mn}(x_2, y_2) \frac{1}{J_{mn}^2 |L_{mn}^{(\omega)}|^2} S_{mn}^{(\omega)}
$$
$$
+ \sum_{\substack{m,n \\ m,n \neq p,q}} \sum_{p,q} W_{mn}(x_1, y_1) W_{pq}(x_2, y_2) \frac{1}{J_{mn} J_{pq} L_{mn}^{*(\omega)} L_{pq}^{(\omega)}} S_{mnpq}^{(\omega)}
$$
$$(22)$$

where

$$S_{mn}^{(\omega)} = \int_{A_1} \int_{A_2} S_p(x_1, y_1, x_2, y_2, \omega) W_{mn}(x_1, y_1) W_{mn}(x_2, y_2) \, dA_2 \, dA_1$$

$$S_{mnpq}^{(\omega)} = \int_{A_1} \int_{A_2} S_p(x_1, y_1, x_2, y_2, \omega) W_{mn}(x_1, y_1) W_{pq}(x_2, y_2) \, dA_2 \, dA_1$$

and $S_p(\cdot)$ is the spectral function of the load. The first sum in eqn (22) is associated with the modal autocorrelation and the second with the cross correlations. However, by writing the complex eigenvalue for each m,n as $\lambda = \omega' + i\omega''$ and by defining the damping parameter $\eta = \omega''/\omega'$, the cross-correlation terms can be disregarded if the following condition is fulfilled[23]

$$\left| 1 - \frac{(\omega'_{mn})^2}{(\omega'_{pq})^2} \right| \gg \max(\eta_{mn}, \eta_{pq}) \tag{23}$$

The above condition holds if the material damping is light and if the real parts of the eigenvalues are well separated, which is the case of the laminated plates considered.[23]

When the system is driven by a point load at the mid-plane of the plate, and for the case of a band-limited white noise, where

$$S(\omega) = \begin{cases} S_0 & |\omega| \leq \omega_c \\ 0 & \text{otherwise} \end{cases} \tag{24}$$

and with $\bar{\omega}^2 = (\omega')^2 - (\omega'')^2$, the mean-square transverse displacement is found to be

$$E[W^2] = S_0 \sum_{m,n} \frac{1}{J_{mn}^2} \int_{-\omega_c}^{\omega_c} \frac{d\omega}{[\bar{\omega}_{mn}^2 - \omega^2]^2 + 4\omega'_{mn}\omega''_{mn}} \tag{25}$$

Following Elishakoff,[24] and by writing $\tilde{\omega}^2 = (\omega')^2 + (\omega'')^2$, the final expression of the mean square is

$$E[W^2] = S_0 \sum_{m,n} \frac{1}{J_{mn}^2 \tilde{\omega}_{mn}^2}$$

$$\times \left\{ \frac{1}{2\omega'_{mn}} \ln \frac{(\omega_c + \omega'_{mn})^2 + (\omega''_{mn})^2}{(\omega_c - \omega'_{mn})^2 + (\omega''_{mn})^2} + \frac{1}{\omega''_{mn}} \right.$$

$$\left. \times \left[\tan^{-1} \left(\frac{\omega_c - \omega'_{mn}}{\omega''_{mn}} \right) + \tan^{-1} \left(\frac{\omega_c + \omega'_{mn}}{\omega''_{mn}} \right) \right] \right\} \tag{26}$$

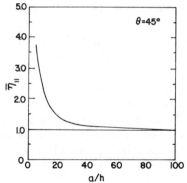

FIG. 8. Relative (to CPT) $\bar{\eta}_{11}$ versus a/h of four layers, viscoelastic angle-ply laminates, $\theta = 45°$.

The case of ideal white noise is obtained from the previous one by setting $\omega_c \to \infty$. The final expression for the mean-square is then

$$E[W^2] = \sum_{m,n} \frac{S_0}{J_{mn}^2} \frac{\pi}{2\omega''_{mn}\tilde{\omega}_{mn}^2} \tag{27}$$

(see the similarity with the elastic solution of the above problem,[25] where proportional viscous damping was considered with the damping ratio ξ. Since $(\omega'')^2 \ll (\omega')^2$, one may infer that $\eta \sim \xi$). The dependence of the material damping parameter, η_{11}, on the length-to-thickness ratio is shown in Fig. 8, for the case of a square orthotropic plate at $T = 78°C$ and shear correction factor $5/6$. The parameter $\bar{\eta}$ is the normalized value of η_{11}

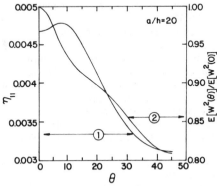

FIG. 9. Line 1, variation of η_{11} versus θ, line 2, relative (to the value at $\theta = 0°$) mean-square displacements versus θ. Both lines are for four layers, viscoelastic angle-ply laminates, $a/h = 20$.

with respect to η_{11} obtained by using the CPT. It can be seen that even at $a/h = 50$ the two theories differ by $7\cdot2\%$. The variation of η_{11} versus the angle orientation is shown in Fig. 9, line 1, for a four layers laminate where $a/h = 20$. The maximum value is obtained at $\theta = 7°$, and the ratio between this value and the minimum one (at $\theta = 45°$) is $1\cdot535$. The mean-square values, normalized to the value obtained at $\theta = 0°$ are shown in Fig. 9, line 2. Despite the reduction of η, the results are continuously decreasing as the value of θ increases. Therefore, it remains for the designer to decide whether a laminate with a low mean square and small damping, high mean square with high damping or something in between is preferred.

6 CONCLUSIONS

It is established that the use of a micromechanical approach for the determination of the effective time-dependent properties of unidirectional composites, enables one to carry out a macromechanical analysis of viscoelastic laminated plates. Generalization of the micro-to-macro mechanics analysis to viscoelastic laminated shells can be performed along the same lines.

REFERENCES

1. SCHAPERY, R. A., Viscoelastic behavior and analysis of composite materials. In *Composite Materials*, Vol. 2, ed. G. P. Sendeckyj. Academic Press, New York, 1974.
2. ABOUDI, J., Effective behavior of inelastic fiber-reinforced composites. *Int. J. Engng Sci.*, **22** (1984) 439–49.
3. ABOUDI, J., Effective thermoelastic constants of short-fiber composite. *Fibre Sci. Technol.*, **20** (1984) 211–25.
4. ABOUDI, J., Micromechanical analysis of the strength of unidirectional fiber composites. *Composite Sci. Technol.*, **33** (1988) 79–96.
5. ABOUDI, J., Micromechanics prediction of fatigue failure of composite materials. *J. Reinf. Plast. Comp.* (in press).
6. PINDERA, M. J. and ABOUDI, J., Micromechanical analysis of yielding of metal matrix composites. *Int. J. Plasticity*, **4** (1988) 195–214.
7. ABOUDI, J., Elastoplasticity theory for composite materials. *Solid Mech. Arch.*, **11** (1986) 141–83.
8. ABOUDI, J., Closed form constitutive equations for metal matrix composites. *Int. J. Engng Sci.*, **25** (1987) 1229–40.
9. YANCEY, R. N. and PINDERA, M. J., Micromechanical analysis of the time-dependent response of unidirectional composites. ASME Winter Annual Meeting, Chicago, Illinois, 1988.

10. YANCEY, R. N. and PINDERA, M. J., Radiation and temperature effects on the time-dependent response of T300/934 graphite/epoxy, CCMS-88-02, VPI-E-88-5. Virginia Polytechnic Institute and State University, 1988.

11. BELLMAN, R., KALABA, K. E. and LOCKETT, J. A., *Numerical Inversion of the Laplace Transform.* Elsevier Science Publishers, Amsterdam, 1966.

12. WHITNEY, J. M., *Structural Analysis of Laminated Anisotropic Plates.* Technomic Publishers, 1987.

13. YANG, P. C., NORRIS, C. H. and STAVSKY, Y., Elastic wave propagation in heterogeneous plates. *Int. J. Solids Struct.*, **2** (1966) 665–84.

14. WHITNEY, J. M. and PAGANO, N. J., Shear deformation in heterogeneous anisotropic plates. *J. Appl. Mech.*, **37** (1970) 1031–6.

15. REDDY, J. N., A simple higher-order theory for laminated composite plates. *J. Appl. Mech.*, **21** (1984) 745–51.

16. CHRISTENSEN, R. M., *Theory of Viscoelasticity.* Academic Press, New York, 1982.

17. ABOUDI, J., Propagation of transient pulses from a spherical cavity in a viscoelastic medium. *Int. J. Num. Meth. Engng*, **4** (1972) 289–99.

18. MOEHLENBAH, A., ISHAI, O. and DiBENEDETTE, A. T., The effect of time and temperature on the mechanical behavior of epoxy composites. *Polymer Engng Sci.*, **11** (1971) 129–38.

19. ABOUDI, J. and CEDERBAUM, G., Analysis of viscoelastic laminated composite plates. *Composite Structures* **12** (1989) (in press).

20. CEDERBAUM, G. and ABOUDI, J., Dynamic response of visoelastic laminated plates. *J. Sound Vib.*, (in press).

21. DOBYNS, A. L., Analysis of simply-supported orthotropic plates subject to static and dynamic loads. *AIAA J.*, **19** (1971) 642–50.

22. IMSL, FFTRC, Tel-Aviv University, 1984.

23. CEDERBAUM, G., Random vibrations of viscoelastic laminated plated. (Submitted.)

24. ELISHAKOFF, I., *Probabilistic Methods in the Theory of Structures.* John Wiley, New York, 1983.

25. CEDERBAUM, G., ELISHAKOFF, I. and LIBRESCU, L., Random vibrations of laminated plated modelled within the first-order shear deformation theory, *Composite Structures* **12** (1989) 97–111.

48

Flexural Behaviour of Multi-directional Glass/Carbon Hybrid Laminates

G. Kretsis, F. L. Matthews, G. A. O. Davies

*Department of Aeronautics, Imperial College of Science,
Technology and Medicine,
Prince Consort Road, London SW7 2BY, UK*

and

J. Morton

*Department of Engineering Mechanics,
Virginia Polytechnic Institute and State University,
Blacksburg, Virginia, USA*

ABSTRACT

The results are presented of a combined experimental and theoretical investigation into the flexural mechanical properties of hybrid laminates. Specimens (16-ply) covering a range of material and orientation stacking sequences were fabricated from carbon fibre and glass fibre/epoxy prepreg and tested in four-point bending. Theoretical modelling accounted for non-linear behaviour, neutral axis shift, thermal stresses and ply mechanical interactions. In general, there was satisfactory agreement between predicted and measured values.

1 INTRODUCTION

Composites containing more than one type of fibre are commonly known as 'hybrid composites'. Extensive research has been carried out in the past[1] on uni-directional hybrid materials, but practical requirements call for additional work on multi-directional configurations. The main reason for using hybrid composites in a particular structure lies in the ability to mix the two or more types of fibre in such a way as to 'tailor' the material to the exact needs of the structural application. Advantages found with one type of fibre can be used to complement the poor performance of the other type for a particular property. For example, the impact performance of carbon fibre-reinforced plastic (CFRP) can be improved by hybridisation with glass fibre-reinforced plastic (GFRP).

The present paper describes experimental and computational research work carried out on hybrid glass/carbon multi-directional laminates loaded in flexure. This work formed part of an extensive study of uni- and multi-directional glass/carbon hybrid laminates tested in tension, compression and flexure.[2-5] The main aim of the multi-directional phase was to establish the extent to which the results of a uni-directional test programme could be used to predict the behaviour of hybrid multi-directional laminates.

2 MATERIALS AND METHODS

Multi-directional hybrid composites were fabricated by laminating pre-impregnated sheets of either XAS carbon or E-glass fibres in Ciba-Geigy 913 epoxy resin. Symmetric laminates (16-ply, nominal ply thickness 0·125 mm) were produced in three orientation stacking sequences which originated from a basic $0°/\pm45°/90°$ lay-up (see Table 1). Most specimens were tested in a dry condition, having been kept over silicon gel at room temperature for four months. A few were tested 'wet', after being kept for about one year at room temperature and 85% relative humidity.

Four-point displacement-controlled loading was used for all tests. For a specimen of unit thickness, the distances between the rollers were 12·5, 15 and 12·5 units, thus yielding a span-to-depth ratio of 25:1 in the outer segments. While this ratio has to be kept relatively high in order to reduce shear stresses, the overall length had to be kept small because of the very large displacements arising when testing to failure; large displacements lead to geometric non-linearity and complicate the manipulation and understanding of the results. Therefore the above dimensions were a compromise between having low shear stresses and avoiding excessive non-linearity for most of the stacking sequences tested. Specimen width-to-thickness ratio was 10 and the overhang, beyond the outside rollers, at least 20 times specimen thickness.

In preliminary flexure tests, it was found that failure was in compression, with damage being initiated at the point of contact with the inner loading

TABLE 1
Laminate stacking sequences

Stacking sequence number	Stacking sequence	Number of specimens	
		D	S
	$[/|\backslash 0/|\backslash 0]_s$		
32	$[cccccccc]_s$	15	15
33	$[gggccgcc]_s$	15	
34	$[gggcgggc]_s$	15	
35	$[gggcgggg]_s$	15	15
36	$[cgcgcgcg]_s$	15	15
37	$[gggggggg]_s$	15	15
	$[/|\backslash 00|00]_s$		
38	$[cccccccc]_s$	15	
39	$[gggccgcc]_s$	15	
40	$[gggccggg]_s$	15	
41	$[cgcgggggg]_s$	15	
42	$[gggggggg]_s$	15	
	$[/0\backslash |/|\backslash 0]_s$		
43	$[cccccccc]_s$	15	
44	$[gcggcgcc]_s$	15	
45	$[gggggggg]_s$	15	

D = dry; S = saturated.
$/ = +45°$; $\backslash = -45°$; $| = 90°$; $0 = 0°$; c = carbon; g = glass.

rollers. In order to avoid this undesirable failure mode, rubber sleeves were placed on the 6-mm diameter inner rollers. Grease was also used to encourage slipping between the specimen and the rubber. As a result the failure mode changed to one of tension, away from the rollers.

During each test five parameters were monitored: the vertical load (P) on the inner rollers, the tensile and compressive surface strains (ε_t and ε_c respectively) at mid-span, the relative displacement (δ) of the inner and outer rollers, and the slope at the free end of the specimen.

3 EXPERIMENTAL RESULTS

3.1 Failure Modes

Macroscopic examination revealed that the failure modes closely resembled those obtained in the tensile tests.[3] There was delamination, splitting parallel to the fibres, and fracture of the 0° fibres. Most of the damage was found to be in the outermost ±45° layers on the tensile side, this being in the form of resin cracking. No damage was found further inwards than the −45° ply closest to the laminate symmetry plane, on the tensile side. It seemed that failure initiated in the ±45° plies adjacent to a 0° ply, propagating as delamination damage.

Problems were encountered with the dry all-glass stacking sequences 37 and 42, because the 0° layer nearest to the surface was the fourth layer into the laminate, thus rendering it very difficult to test to failure due to the extremely high curvature required. Therefore some of these specimens did not fail but merely produced marginal damage in the outermost ±45° and 90° layers, this being mostly resin cracking and edge delamination. However, similar samples that had been wet conditioned failed without any problems, because of the reduced fibre–resin bond strength.

3.2 Test Data

Although the same cure schedule was used for all laminates, there were variations in the moulded thickness of each panel, according to the absolute content of carbon and glass present. The carbon-rich lay-ups were consistently thicker than nominal (i.e. 2 mm), whilst the opposite was true for the glass-rich laminates with, presumably, consequent variations in fibre volume fraction (nominal value 0·60). These thickness changes caused difficulties when interpreting and presenting the results.

All experimental values were based on actual thicknesses. The bending moments allowed for non-linear behaviour, which becomes significant once the surface strain exceeds 1·5%. The calculations included the influence of the horizontal components of the roller reactions, which

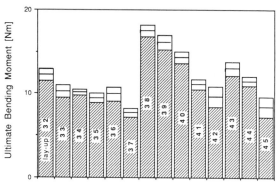

FIG. 1.   Variation of ultimate bending moment with stacking sequence.

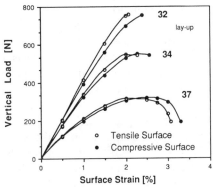

Fig. 2. Typical load–strain curves.

become important at large deflections. Friction at the rollers was taken to be zero.[6] Ultimate bending moments are summarised in Fig. 1.

The following observations can be made on the experimental results.

(a) The trends between the various stacking sequences exhibited in flexure were largely similar to those found in the tensile tests,[3] i.e. failure strain increased but stiffness and strength decreased with decrease in carbon content.

(b) Placing the 0° plies closer to the surface of the laminate resulted in considerably higher stiffness and a small increase in strength.

(c) The maximum tensile strain in the outermost 0° ply at initial failure was lower than that in tensile tests[3] of the same stacking sequence, unlike the uni-directional results.[2,4]

(d) The effect of moisture uptake was to decrease the flexural strength by a notable amount. This is believed to have been caused by early failure of the outer tensile off-axis layers when saturated, this in turn causing early failure of the outermost tensile 0° ply (failure of the coupons always occurred on the tensile side).

Typical load–strain curves are presented in Fig. 2, where the strong non-linearity in the measured vertical load is shown.

4 THEORETICAL ANALYSIS

4.1 Introduction

For uni-directional laminates Weibull statistics and the maximum strain criterion were found to be adequate for the prediction of laminate strength.

Furthermore, for stiffness predictions, the rule of mixtures (RoM) yielded satisfactory results in tension[2] and compression,[5] whereas classical laminate theory was equally good in flexure.[4] In the case of multi-directional stacking sequences, however, the effects of resin cracking, thermal stresses, edge stresses and mechanical interactions between layers were thought to be important and have, therefore, been investigated. In particular, it was found necessary to include non-linear stiffness effects in the analysis, thus complicating the modelling considerably.

In the present paper the effects mentioned above are described. Good agreement between experiment and theory has already been found, both for stiffness and strength, for the tensile results.[3] The theoretical investigation was limited to the dry laminates, which constituted the bulk of the experimental work.

4.2 Non-linear Stress–Strain Formulation

The method adopted for the solution of the non-linear stress–strain behaviour of the multi-directional laminates was that of simple incremental loading, assuming linear stiffness behaviour over each load increment, and using the updated local strain values to calculate the laminate tangent stiffness. This method is subject to cumulative error unless a reasonably small load increment is used; in this case 1–4% of the ultimate value was chosen.

4.3 Stiffness Properties

From other parts of the programme it was found that, for the uni-directional materials, the longitudinal tensile modulus of carbon laminates increased notably with strain, while that of glass remained virtually constant. Similarly, the longitudinal compressive modulus of carbon decreased with increasing loading, while that of glass was roughly equal to that in tension, remaining constant with increasing strain.

In order to allow for this behaviour, classical laminate theory was used to calculate the initial stiffness (i.e. at zero strain) of the multi-directional laminates, in combination with experimental data for the principal stiffnesses of the uni-directional materials. The complete stress–strain curve was then obtained from the incremental analysis using a mixture of micromechanics,[7] experimental data and assumed values.[6]

4.4 Thermal Stresses

The thermal stresses generated in the multi-directional laminates due to the curing procedure were calculated at the start of the incremental stress–strain calculations. The data used for the material expansion coefficients were obtained by averaging values from several sources. The change of temperature imposed was $-100°C$, calculated as the difference between the curing temperature (120°C) and room temperature (20°C). The thermal stresses generated were found to be small (as compared to strength) in the fibre direction, but were often large across the fibres. Therefore resin cracking was expected to occur at different axial strain levels for the various stacking sequences. Moreover, as layers of carbon were replaced by layers of glass at the same orientation the negative thermal fibre stress in the axial layers became larger, thus indicating a positive hybrid effect. The opposite behaviour was found for the stresses in the axial glass plies, i.e. when glass layers were replaced by carbon layers at the same orientation.

4.5 Failure Criterion and Strength Data

Because it was believed to work reasonably well for the stress states investigated here, it was decided to use the Tsai–Hill failure criterion. The data required for this criterion are the tensile and compressive ply strengths in the longitudinal and transverse directions, as well as the shear strength. These values were obtained experimentally, except for the transverse compressive strength which has been shown to be equal to between 1 and (at least) 4 times the equivalent value in tension. In the present calculations this quantity (Y_c) was varied so that sensitivity to its effects on the tensile strength could be studied for a multi-directional, single-phase, stacking sequence. It was found that a value of three times the transverse tensile strength was satisfactory.

The values used were calculated to the nominal lamina thickness. The small amount of scatter present was not included in the theoretical analysis. Volume (statistical) effects on the longitudinal tensile strength (as included for the uni-directional material) were ignored, since they were believed to be of small importance for the multi-directional laminates in which early resin cracking in the off-axis layers affects the strength of the 0° layers.

Finally, the possibility of the 90° layers being stronger when part of a laminate than when tested in a simple transverse tensile test was also not included in the model, since it required a complex statistical analysis. The effect of accounting for such behaviour would be to delay the stiffness reduction due to 90° ply failure, and therefore to decrease slightly the predicted ultimate strain.

4.6 Post-failure Stiffness Reduction

The various plies in any particular stacking sequence did not fail simultaneously. Typically, the 90° layers failed first, then the $\pm 45°$ layers and finally the 0° layers. During a test, once microcracks form in a particular lamina, its stiffness contribution to the laminate is undoubtedly reduced. The most common source of stiffness reduction arises from resin cracking in the 90° and ±45° layers. In the present model the behaviour of these layers was assumed to be bi-linear, i.e. linear until ply failure, then linear again but with the transverse and shear stiffnesses reduced by a certain common factor. This factor was derived by fitting theoretical to experimental tensile stiffness results, for either an all-carbon or all-glass stacking sequence. For CFRP a reduction of 50% to the stiffness was adopted, while for GFRP this value was as high as 90%. It was assumed that identical reduction factors applied in tension and compression.

In similar work by other researchers[8,9] the magnitude of the stiffness reduction factor ranged between 0 and 100%, with the best results obtained near the latter extreme. Furthermore, in Ref. 8, upon lamina failure the stress–strain calculation was re-started from the origin, using the updated stiffness values (note that axial stiffness was also reduced). Thus a drop in the stress was found whenever a particular layer failed. In the present work this technique was only adopted if a 0° ply failed, because it was believed that the behaviour of the 90° and ±45° plies would be pseudo-plastic.

4.7 Effect of Edge Stresses

It is well known that, because of the nature of composite materials, small through-thickness direct and shear stresses develop near the edges of a multi-directional specimen under axial load. These stresses are small compared to the applied stress, but can be important because of the low strength of the composite in these directions. However, it has been reported that, although the edge stresses may produce edge cracking and delamination, their effect on the tensile strength is small. On the other hand, such delamination is important in compression, since it undermines the resistance of the composite against buckling.

A three-dimensional finite element mesh composed of brick elements was used to calculate the stress field near the free edge of some of the stacking sequences tested. Three distinct cases were examined, investigating the effects on the edge stresses of (a) altering the stacking sequence of all-carbon laminates, (b) replacing layers of carbon by layers of glass at the same orientation, and (c) changing the loading for an all-carbon laminate from tensile to flexural (lay-up 32 only).

The results of the latter investigation (which are of interest here) showed that the outermost layers are under considerable edge stresses in flexure. Hence we would expect flexural specimens to be more prone than tensile specimens to delamination and resin cracking at the edges.[6]

5 DISCUSSION

The deflected shape of the complete flexural specimen was not modelled (as it was for the uni-directional laminates[4]), because it required an extremely complex analysis in order to account for both geometric and stiffness non-linearity. Instead only the effects of the latter were investigated, and comparisons between theory and experiment were made in terms of bending moment or strain, and not in terms of load. Thus the effects of changes in geometry were included only in the experimental results.

Classical laminate theory was successful in predicting the gradual shift in the neutral axis as the material properties changed with increasing loading, thus producing an unbalanced laminate. The position of the neutral axis was predicted using exact equations for the variation of the axial modulus through the laminate thickness, instead of the piecewise representation (i.e. constant value within each layer) used with classical laminate theory.[6] The exact method was shown to be much easier to use than laminate theory, and also more accurate, if used with caution, and faster in terms of computing time.

TABLE 2
Theoretical ply failure sequences

Stacking sequence	Failure sequence (until mult.)		Ultimate moment		
	T	C	M (N mm)	ε_t (%)	ε_c (%)
$[/|\backslash 0/|\backslash 0]_s$					
32. $[ccccccccc]_s$	213 854	76	13 500	2·56	−2·65
33. $[gggccgcc]_s$	213 745	6	9 660	2·60	−2·68
34. $[gggcgggc]_s$	213 754	6	9 240	2·60	−2·69
35. $[gggcgggg]_s$	213 764	5	9 200	2·70	−2·84
36. $[cgcgcgcg]_s$	13 285	674	9 600	3·86	−3·78
37. $[gggggggg]_s$	213 854	675	7 040	3·92	−3·70
$[/|\backslash 00|00]_s$					
38. $[ccccccccc]_s$	2 137·4	65	17 150	2·53	−2·65
39. $[gggccgcc]_s$	2 135·4	5	13 160	2·55	−2·66
40. $[gggccggg]_s$	2 136·4	5	13 160	2·67	−2·83
41. $[cgcgggggg]_s$	2 138·5	674	10 780	3·86	−3·80
42. $[gggggggg]_s$	2 138·4	675	8 550	3·92	−3·77
$[/0\backslash |/|\backslash 0]_s$					
43. $[ccccccccc]_s$	2 431		13 340	1·76	−1·87
44. $[gcggcgcc]_s$	1 432		11 040	1·78	−1·91
45. $[gggggggg]_s$	163 254	4	6 900	2·77	−2·68

c = carbon; g = glass.

TABLE 3
Comparison of experiment and theory (dry specimens only)

Stacking sequence	M_{ult} (N mm) Experiment, Theory, Diff. T − E (%)	$\varepsilon_{t.ult}$ (%) Experiment, Theory, Diff. T − E (%)	$\varepsilon_{c.ult}$ (%) Experiment, Theory, Diff. T − E (%)	M/k_x (N m²) $\frac{1}{4}$% secant Experiment, CLT, Non-linear Diff. T − E (%)
$[/|\backslash 0/|\backslash 0]_s$				
32. $[\text{cccccccc}]_s$	12 248	2·08	2·47	0·503
	13 500 +10·2%	2·56 +23·1%	2·65 +7·3%	0·643 +27·8%
				0·629 +25·0%
33. $[\text{gggccgcc}]_s$	10 220	2·10	2·49	0·429
	9 660 −5·5%	2·60 +23·8%	2·68 +7·6%	0·460 +7·2%
				0·429 0%
34. $[\text{gggcgggc}]_s$	10 126	2·27	2·53	0·413
	9 240 −8·7%	2·60 +14·5%	2·69 +6·3%	0·434 +5·1%
				0·402 −2·7%
35. $[\text{gggcgggg}]_s$	9 342	2·10	2·56	0·378
	9 200 −1·5%	2·70 +28·6%	2·84 +10·9%	0·396 +4·8%
				0·374 −1·1%
36. $[\text{cgcgcgcg}]_s$	9 844	2·98	3·57	0·308
	9 600 −2·5%	3·86 +29·5%	3·78 +5·9%	0·376 +22·1%
				0·339 +10·1%
37. $[\text{gggggggg}]_s$	7 634	3·02	3·22	0·252
	7 040 −7·8%	3·92 +29·8%	3·70 +14·9%	0·266 +5·6%
				0·240 −4·8%
$[/|\backslash 0 0|0 0]_s$				
38. $[\text{cccccccc}]_s$	17 417	2·21	2·62	0·696
	17 150 −1·5%	2·53 +14·5%	2·65 +1·1%	0·809 +16·2%
				0·788 +13·2%
39. $[\text{gggccgcc}]_s$	16 058	2·48	2·81	0·582
	13 160 −18·0%	2·55 +2·8%	2·66 −5·3%	0·611 +5·0%
				0·572 −1·7%
40. $[\text{gggccggg}]_s$	14 299	2·37	2·85	0·519
	13 160 −8·0%	2·67 +12·7%	2·83 −0·7%	0·583 +12·3%
				0·547 +5·4%
41. $[\text{cgcggggg}]_s$	11 072	2·70	3·37	0·333
	10 780 −2·6%	3·86 +43·0%	3·80 +12·8%	0·391 +17·4%
				0·357 +7·2%
42. $[\text{gggggggg}]_s$	9 594	3·00	3·23	0·263
	8 550 −10·9%	3·92 +30·7%	3·77 +16·7%	0·298 +13·3%
				0·271 +3·0%
$[/0\backslash|/|\backslash 0]_s$				
43. $[\text{cccccccc}]_s$	12 926	1·53	1·92	0·757
	13 340 +3·2%	1·76 +15·0%	1·87 −2·6%	0·895 +18·2%
				0·868 +14·7%
44. $[\text{gcggcgcc}]_s$	11 424	1·54	1·94	0·622
	11 040 −3·4%	1·78 +15·6%	1·91 −1·5%	0·698 +12·2%
				0·672 +8·0%
45. $[\text{gggggggg}]_s$	8 860	2·46	2·87	0·280
	6 900 −17·5%	2·77 +12·6%	2·68 −6·6%	0·308 +10·0%
				0·292 +4·3%

c = carbon; g = glass.

When using laminate theory for the iterative calculations, it was assumed that the anti-clastic curvature (k_y) is not suppressed during a test. However, it was assumed that the twist (k_{xy}) is suppressed, based on the experimental boundary conditions. Its effects, however, were studied and it was found that suppressing it resulted in slightly stiffer as well as stronger laminates.

The theoretical sequence of ply failures is presented in Table 2. The axial compressive strength data for carbon were doubled here, so as to avoid the ultimate failure being predicted on the compressive side of the laminates. This was imposed because it was believed that the inherent curvature in flexure inhibited fibre micro-buckling on the compressive side. In addition, the entire group of multi-directional flexural specimens actually failed on the tensile side. It can be seen from Table 2 that it is usually the tensile (T) 90° and ±45° layers that fail first, followed by the equivalent layers on the compressive (C) side. Ultimate bending moment was reached upon the failure of the outermost tensile axial layer. It is possible, however, that in the tests no damage whatsoever occurred in the layers under compression. Therefore the theoretical stiffness reduction procedure may well have resulted in the underestimation of the tangent flexural stiffness towards the last stages of loading.

The experimental and theoretical values for M_{ult}, $\varepsilon_{t.ult}$, $\varepsilon_{c.ult}$ and M/k_x (initial) are compared in Table 3. It can be seen that the strains to failure are overestimated, as is the stiffness for the all-carbon laminates. Note, however, that the experimental values of M_{ult} and M/k_x have been calculated assuming zero frictional forces. If the effect of friction is included, these values will in general decrease.

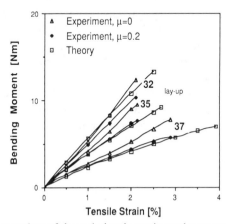

FIG. 3. Comparison of theoretical and experimental moment–strain curves.

The overestimation of the flexural strength is believed to be due to early failure of the 0° plies in the tests, arising from stress-raising resin cracking in the outermost tensile off-axis layers. Typical moment–strain curves are shown in Fig. 3, where it is made clear that frictional forces are almost definitely present in the experimental results and must be accounted for. It is difficult, however, to comment on how successful the iterative theoretical solution has been in predicting the shape of these curves, since uncertainty exists over the experimental data.

6 CONCLUSIONS

Failure in the flexural tests took place away from the rollers, and on the tensile side of the specimen. The trends exhibited in flexure between the various stacking sequences were largely similar to those found in tensile tests, i.e. strain increasing but stiffness and strength decreasing with decreasing carbon content. However, the maximum tensile strain in the outermost 0° ply at ultimate failure was lower than that in tensile tests of the same stacking sequence, in contrast to the uni-directional results.

The effect of moisture uptake was to decrease the flexural strength by a significant amount.

Microscopic examination showed most of the damage occurring in the outermost ±45° layers on the tensile side, this being resin cracking. Delamination damage as well as 0° ply splitting and fibre cracking were also apparent.

Classical laminate theory calculations were found to slight overestimate the flexural stiffness of the all-carbon laminates, but predictions were otherwise good. The non-linear shape of the flexural moment–strain curves was also modelled, but it is uncertain whether good agreement exists between experiment and theory. This is due to the unknown influence of the frictional forces present in the tests.

The flexural strain to failure (using the Tsai–Hill failure criterion) was in general overestimated, possibly due to early failure of the 0° plies in the tests, arising from stress-raising resin cracking in the outermost tensile off-axis layers.

REFERENCES

1. Kretsis, G., A review of the tensile, compressive, flexural and shear properties of hybrid fibre-reinforced plastics. *Composites*, **18** (1987) 13–23.

2. KRETSIS, G., MATTHEWS, F. L., MORTON, J. and DAVIES, G. A. O., Tensile behaviour of unidirectional glass/carbon hybrid laminates. *Proc. Reinforced Composites 1986* (Paper C47/86), University of Liverpool, UK, April 1986. Institution of Mechanical Engineers, London, 1986.
3. KRETSIS, G., MATTHEWS, F. L., MORTON, J. and DAVIES, G. A. O., Tensile behaviour of multi-directional glass/carbon hybrid laminates. *Proc. ICCM— VI/ECCM—2*, Imperial College, London, July 1987. Elsevier Applied Science, London, 1987.
4. KRETSIS, G., MATTHEWS, F. L., MORTON, J. and DAVIES, G. A. O., Flexural behaviour of unidirectional glass/carbon hybrid laminates. In *Engineering Applications of New Composites*, ed. S. A. Paipetis and G. C. Papanicolaou. Omega Scientific, Wallingford, UK, 1988.
5. KRETSIS, G., MATTHEWS, F. L., MORTON, J. and DAVIES, G. A. O., Compressive behaviour of unidirectional glass/carbon hybrid laminates. *Proc. ECCM—3*, Bordeaux, France, March 1989. European Association for Composite Materials, Bordeaux, France, 1989.
6. KRETSIS, G., Mechanical characterisation of hybrid glass/carbon fibre-reinforced plastics. PhD thesis, University of London, 1987.
7. JONES, R. M., *Mechanics of Composite Materials*. McGraw-Hill, New York, 1975.
8. TAKAHASHI, K., BAN, K. and CHOU, T.-W., Non-linear stress–strain behaviour of carbon/glass hybrid composites. *Proc. ICCM—V*, San Diego, CA, USA, 1985. The Metallurgical Society, Warrendale, PA, USA, 1985.
9. TALREJA, R., Transverse cracking and stiffness reduction in composite laminates. *J. Comp. Mater.*, **19** (1985) 355–75.

49

Application of Continuum Damage Mechanics to Post Failure Analysis of Notched Cross Plied Laminates

D. Valentin, G. Cailletaud

*ENSMP—Centre des Matériaux PM Fourt,
B.P. 87-91003 Evry Cedex, France*

and

M. Renault

Aerospatiale—78130 Les Mureaux, France

ABSTRACT

A finite element analysis of the damage accumulation during uniaxial loading of notched CFRP has been performed. Two lay-ups were considered: $(0, 90)_s$ and $(0, 90 \pm 45)_s$. The elastic case has been considered first, by using shell elements with superimposed layers. The results are compared with the experimental ones obtained at low loads. At higher loads, non-linearities are observed due to mainly intralaminar and interlaminar damage accumulation. To modelize this behaviour, a numerical simulation of the tests has been made with the finite element code ABAQUS. The capability of the code to account for a user defined material was used, first to define the elastic behaviour of the laminates, then to introduce damage with the help of the continuum damage mechanics formalism. Only progression of interlaminar damage is considered, but a fixed delamination zone has been examined and its influence on stress profile, stress strain curves are compared with the experimental results.

INTRODUCTION

Carbon fibre reinforced plastic is currently used in the aeronautical and spatial industries for load bearing structures where requirement of safety needs an ultimate load analysis. For this kind of material, stress concentrations and damage mechanisms in the vicinity of holes are very different from those usually found for isotropic materials. Many authors[1-3] have been interested in this problem and have described the behaviour of similar materials in the elastic case, though many tests have shown that it is not realistic to estimate the failure of notched samples without taking into account damage evolution during the loading. The aim of this study is to introduce in the calculation of samples the different kinds of encountered damages and to evaluate their influence on the observed failure load values. Only the case of intralaminar matrix failures and fibre breakage is studied and tests are conducted with different lay-ups $(0, 90)_s$, $(0, 90 \pm 45)_s$ to follow damage evolutions and strain profiles.

EXPERIMENTAL METHOD

Flat specimens were made from Ciba Geigy prepreg containing 914 C epoxy resin and Toray T6K carbon fibres and were cured in an autoclave at 180°C for 2 h. The thicknesses of the samples were equal to 0·5 mm and 1 mm respectively for the four layers $(0, 90)_s$ and the eight layer $(0, 90 \pm 45)_s$ plates. The gauge length of the sample was equal to 100 mm. Tensile tests were conducted in a screw driven Instron tensile testing machine at a speed of 0·2 mm/s. The ratio $W/2R$ was equal to 6, where $2R$ and W are respectively the diameter of the hole and the width of the sample, the last parameter taking values equal to 18, 32 and 60 mm. A ratio of 6 leads (in the elastic case), to a correcting factor CF due to the finite width of the sample equal to 1·03 defined as follows[4]

$$CF = \frac{K_T}{K_T^\infty} = \frac{2 + (1 - 2R/W)^3}{3(1 - 2R/W)} \quad \text{where } \sigma_N^\infty = \left(\frac{K_T}{K_T^\infty}\right)\sigma_N \tag{1}$$

with

$$K_T^\infty = 1 + \left\{2\left[\left(\frac{E_y}{E_x}\right)^{1/2} - v_{yx}\right] + \frac{E_y}{G_{xy}}\right\}^{1/2} \tag{2}$$

σ_N^∞ is the remote failure stress (based on gross cross-sectional area) for a notched plate of infinite width. During the tests, strains are measured using photoelasticimetry or strain gauges. It should be specified that in the vicinity of the hole strain measurements were performed with gauges having very little grid surface (1·5 mm × 1·5 mm). Moreover, observations of replicas made on the edge of the samples and X-radiography enable one to follow damage evolution.

DAMAGE MODELING

1 Failure Criterion

To separate the nature of the damage two failure criteria are used:

$$\text{fibre failure} = \left(\frac{\varepsilon_1}{\varepsilon_{1u}}\right)^2 = 1 \tag{3}$$

$$\text{intralaminar matrix failure:} \left(\frac{\sigma_2}{\sigma_{2u}}\right)^2 + \left(\frac{\tau}{\tau_{2u}}\right)^2 = 1 \tag{4}$$

where ε_1, σ_2 and τ are respectively the strain in the direction of fibres, the stress perpendicular, and parallel to the fibres. The subscript u indicates ultimate values.

2 Damage Accumulation

The consequence of these failures are of two kinds: firstly there appears a fall in modulus in the direction perpendicular to the fibres, secondly there is a reduction of the stress transfer by shear. To represent these phenomena, it is considered that the material properties are modified by a damage state as follows:

$$E_1 = E_1^{\cdot}(1 - D_1) \qquad v_{12} = v_{12}^{\cdot}\frac{E_1}{E_{10}} = v_{12}^{\cdot}(1 - D_1)$$

$$E_2 = E_2^{\cdot}(1 - D_2) \qquad G_{12} = G_{12}^{\cdot}(1 - D_6) \tag{5}$$

and the proposed values of D_1, D_2 and D_6 are:

$$D_1 = 1 - \exp\left[-\left\langle\left(\frac{\sigma_1^e}{\sigma_{1u}}\right)^p - 1\right\rangle\right] \tag{6}$$

$$D_2 = 1 - \exp\left[-\left\langle\left\{\left(\frac{\sigma_2^e}{\sigma_{2u}}\right)^2 + \left(\frac{\tau_{12}^e}{\tau_{12u}}\right)^2\right\}^m - 1\right\rangle\right] \tag{7}$$

$$D_6 = 1 - \exp\left[-\left\langle\left\{\left(\frac{\sigma_2^e}{\sigma_{2u}}\right)^2 + \left(\frac{\tau_{12}^e}{\tau_{12u}}\right)^2\right\}^n - 1\right\rangle\right] \tag{8}$$

where σ_1^e, σ_2^e, τ_{12}^e are fictive elastic stresses defined as follows: $\sigma_i^e = Q_{ij}^e \varepsilon_j$.[9]

The parameters m, n, p have been identified on a $(\theta, 90 + \theta)_s$ laminate tested with θ equal to $15°$, $30°$ and $45°$.

Figure 1 gives a comparison of the theoretical and experimental results obtained. Figure 2 gives a representation of mechanical behaviour of the individual ply for different loadings with $m = 0.5$, $n = 0.29$, $p = 3$.

The authors are aware of the empiricism in the identification of the

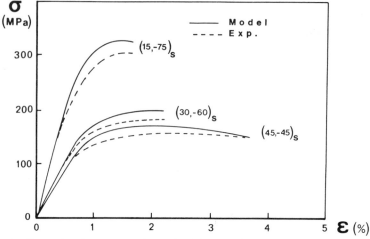

FIG. 1.   Comparison between experimental and theoretical curves according to the presented model.

behaviour. For this reason, the decrease of the transverse modulus E_2 of a $(0, 90)_s$ laminate deduced from our modelling has been compared to the analytical expression proposed by Kamimura.[5] This modelling is based on the shear lag theory

$$\sigma_2 = E_2^{\cdot}(1 - D_2)\varepsilon_c \qquad (9)$$

where σ_2 is the transverse stress in the $90°$ ply, ε_c is the composite deformation with

$$D_2 = \frac{\sinh Y}{Y \cosh Y} \quad \text{and} \quad Y = \frac{\lambda}{2D} \qquad (10)$$

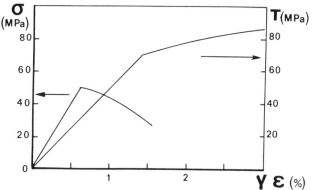

FIG. 2.   Behaviour of the individual layer in shear or in tensile loading.

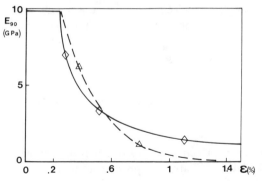

FIG. 3. Comparison of the evolution of the transverse modulus of a $(0, 90)_s$ laminate according to ($\Diamond$) Kamimura model[5] and according to ($\triangle$) our predictions.

where D is the crack density and λ is a function of $E_1^{\cdot}$, $E_2^{\cdot}$, $G_{12}^{\cdot}$, $G_{23}^{\cdot}$ of the constitutive plies and of their respective thicknesses t_1 and t_2.

$$D = \left[\frac{2}{\lambda} \cosh^{-1} \left[1/(1 - \sigma_{cf}/\sigma_c) \right] \right]^{-1} \tag{11}$$

where σ_c is the stress applied to the composite and σ_{cf} is the corresponding critical stress in the composite when the first ply failure occurs: it can be seen from Fig. 3 that the behaviour is similar.

3 Calculation Principles

A numerical simulation of the tests have been performed with the FE code ABAQUS. Shell elements with superimposed layers and reduced integration were used. Each layer has a thickness and is considered as an orthotropic material defined with the values of the constants E_1, E_2, E_3, G_{12}, v_{12}, v_{13} and v_{23} in its orthotropy axis and the value of θ, the angle between these axes and those of general reference.

The capability of the code to account for a user defined material was used to introduce damage. If damage occurs, the values taken by the elastic constants defined in the orthotropy axis are modified and the damage rigidity matrix and the Jacobian defined as follows:

$$[J_{ij}] = \left[\frac{\partial \Delta \sigma_i}{\partial \Delta \varepsilon_j} \right] \tag{12}$$

are calculated in the general reference axis for these new values, with

$$[\sigma_i] = [Q_{ij}][\varepsilon_i] \quad \text{and} \quad Q_{ij} = Q_{ij}(\varepsilon_1, \varepsilon_2, \varepsilon_6) \tag{13}$$

Using superimposed shell elements for each layer, it was possible to force the superimposed corresponding nodes to have the same displacement or not (in case of delamination). A subroutine was developed to manage such conditions as a function of delamination criteria.[6]

RESULTS

1 Elastic Behaviour

The mesh used is presented in Fig. 4. The results obtained with the finite element calculations (Table 1) present a good correlation with the experimental results at low loading and with theoretical results obtained by using expression (2) or the Lekhnitskii[7] approach

$$K_T = 1 + \frac{B_1^2 + B_2^2}{(B_1 - B_2)}$$

where B_1 and B_2 are positive solutions of

$$\frac{1}{E_x} B^4 + \left(\frac{1}{G_{xy}} - \frac{2v_{xy}}{E_x}\right) B^2 + \frac{1}{E_y} = 0 \tag{14}$$

Elastic stress profiles in the weakest section, stress contours obtained from the calculations and strain contours obtained from photo-elasticimetry measurement have been compared.[8] These comparisons show that the calculations provide a good approximation of the sample behaviour at low stress level.

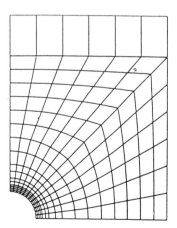

FIG. 4.   Mesh used for the calculation.

TABLE 1
Values of K_T for a sample with a hole

	Expression 2	Expression 12	FE calculation
$(0, 90)_s$	4·98	4·95	5
$(0, 90 \pm 45)_s$	3	3	2·95

2 Damaged Behaviour

When the samples' behaviour is no longer elastic, different non-linearities on the curves (strain–load) may be observed. Practically all the behaviour modifications are the consequences of local loss of material properties and of stress redistributions due to damage evolution:[9] matrix failure parallel to the fibres in the different ply orientations and delamination near the hole and in areas where high fissuration densities exist in two adjacent plies (Fig. 5).

In Fig. 6, the experimental behaviour and the corresponding modeling of the $(0, 90 \pm 45)_s$ sample are presented and a good correspondence can be found. It should be noted that the behaviour is almost linear: the accumulation of damage before catastrophic failure is rather low compared to the amount of accumulated damage in the $(0, 90)_s$ sample. For this latter material, the non-linearities are much more important. It has been found that the above modelization was able to describe qualitatively the behaviour (Fig. 6). It has been thought that the difference could be explained by delamination as shown in Fig. 4.

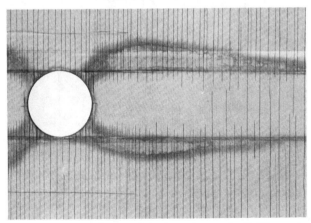

FIG. 5. X-ray investigation of the damage state of a $(0, 90)_s$ sample.

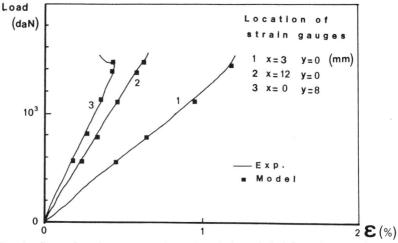

FIG. 6. Comparison between experimental and theoretical deformations obtained at different locations around the hole in the $(0, 90, \pm 45)_s$ sample. No delaminations are taken into account.

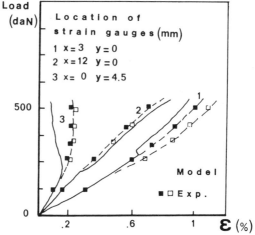

FIG. 7. Comparison between experimental and theoretical deformations obtained at different locations around the hole in the $(0, 90)_s$ with (- -■- -) or without (- -□- -) delamination taken into account.

To account for this phenomenon, a new simulation has been undertaken with different areas of delamination determined by X-ray observations. One can notice a greater tendency (Fig. 7). The persistent difference between the experimental results and the modelization could be explained by the fact that the behaviour of the damaged layer is not intrinsic but depends on the lay-up in which it is located. If the environment of the layer is changed—by delamination for example—its behaviour will change and the corresponding parameters m, n, p for example will be different from those determined in uniaxial tension on samples where no delamination has been found.

3 Application to the Failure Prediction

The calculations can be used to predict the failure of the samples. Two methods can be considered. The first one consists of using the theoretical

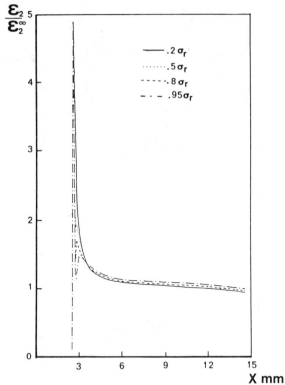

FIG. 8. Evolution of the theoretical stress profile in the case of $(0,90)_s$ laminate when delamination is taken into account.

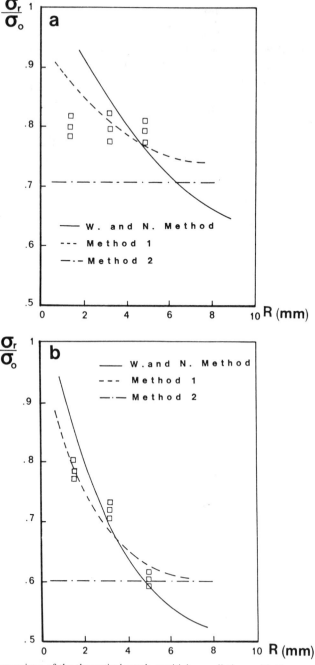

FIG. 9. Comparison of the theoretical notch sensitivity predictions with the experimental data (□). R = hole radius. (a) for $(0, 90)_s$ material and (b) for $(0, 90 \pm 45)_s$ material.

stress obtained (Fig. 8) profile and to apply the criteria proposed by Whitney and Nuismer[10,11] which use the point stress criterion. The point stress criterion assumes that failure occurs when the stress σ_y over some distance d_0 ahead from the discontinuity is equal to or greater than the strength of the unnotched specimen σ_0. The two parameters σ_0 and d_0 have to be determined by experience by varying the ratio $W/2R$. This first method leads to the definition of a parameter d_0 similar in nature to σ_0, the second one is based on the fact similar that when the fibre failure criteria (eqn (3)) is reached over some Gauss point, there is a numerical instability and the failure sample can be assumed. The results obtained with methods 1 and 2 are compared in Fig. 9 with the classical Whitney and Nuismer approach. It can be seen that the second method is not sensitive to the hole radius and the failure stresses obtained are always below the experimental ones.

It is thought that the introduction of a volume effect due to the statistic aspect of the fibre failure would improve this second method.

The first method gives better results especially with the $(0, 90)_s$ sample. However tests on larger hole size should be necessary in the case of the $(0, 90, \pm 45)_s$ material to clearly establish the better accuracy of the method.

CONCLUSION

During these investigations, the complexity of the interaction between the different mechanisms of degradation appeared very clearly both from the numerical and from the experimental point of view. This approach could be used to determine the load at failure of notched structures for example. Two possibilities could be considered: the first one is to improve the average or the point stress criterion proposed by Whitney and Nuismer[10] by using the real stress profile in the weakest section, the second one could be based on an instability criteria.

ACKNOWLEDGEMENT

The authors gratefully acknowledge the CNES for its support during this study.

REFERENCES

1. Pipes, R. B., Wetherold, R. C. and Willespie, J. W., Notched strength of composite materials. *J. Comp. Mater.*, **13** (1979) 148–60.

2. PRABHAKMAN, R., Tensile fracture of composite with circular holes. *J. Mater. Sci. Engng*, **41** (1979).

3. ERICSON, K., PERSON, M., CARLSON, I. and GUSTAVSON, A., On the prediction of the initiation of delamination in a $(0, 90)_s$ laminate with a circular hole. *J. Comp. Mater.*, **18** (1984) 495–506.

4. MORRIS, DON M., Modelling the failure strength of notched graphite/polyimide plates subjected to extreme temperature. In *Composite Structures 3*. ed. I. H. Marshall. Elsevier Applied Science, London, 1985, pp. 356–68.

5. KAMIMURA, K., Modélisation théorique de la croissance d'endommagement appliquée à la théorie des plaques stratifiées. *J. Theoret. Apl. Mech.*, **4**(4) (1985) 537–53.

6. RENAULT, M., Tolérance à l'endommagement de composites carbone-résine stratifié T300/914. PhD Thesis, Ecole des Mines de Paris, 1988.

7. LEKHNITSKII, S. G., *Theory of Elasticity of an Anisotropic Body*. Holdenday, San Francisco, 1963.

8. RENAULT, M., VALENTIN, D. and CAILLETAUD, G., A finite element study of damage behaviour of notched crossplied laminates. *Proc. Int. Conf.: CADCOMP*, Southampton, ed. C. Brebria. 1988, pp. 429–42.

9. RENAULT, M., VALENTIN, D. and PEREZ, F., A damage tolerance study of notched CFRP laminates. *ECF7 Proceedings,* Budapest, ed. E. Czobaly. 1988, pp. 442–50.

10. NUISMER, R. J. and WHITNEY, J. W., Uniaxial failure of composite laminates containing stress concentrations. *ASTM, STP 593*, 1975, pp. 117–42.

11. WHITNEY, J. M. and NUISMER, R. J., Stress fracture criteria for laminated composites containing stress concentration. *J. Comp. Mater.*, **8** (1974) 253–65.

50

Design and Manufacture of Radar Antennae with Composite Materials. Effects of Curing Temperature and Characterization of a Very Low Curing Temperature Resin System

PASQUALE CIRESE

Divisione Ingegneria, SMA, via del Ferrone 5, 50124 Firenze, Italy

and

ANDREA CORVI

Mechanics Department, University of Ancona,
via delle Brecce Bianche, 60131 Ancona, Italy

ABSTRACT

In this paper some problems regarding the design and manufacture of radar antennae made with composite materials are briefly outlined. After a description of the reasons that recommend the use of composites in this kind of production, the paper deals with the general design problems and the types of characterization of the materials that are necessary for the required reliability of the product.

Specific design, technology and calculation problems are then introduced.

Particular attention is paid to the choice of the resin system in order to minimize the problems related to the curing temperature.

INTRODUCTION

The introduction of composite materials considerably modifies the conventional approach to structural design. In fact, the concepts of *material* and *structure*, distinct and independent in the traditional design techniques, become intimately joined. To design a structure to be

821

manufactured with composites, involves design of the most suitable material, at lamina and laminate level, that acquires its own *essence* related to that structure.

This concept is the focal point of the technologies relevant to the composites, making the difference with respect to the past, and requiring a new kind of learning, necessary for a non-trivial use of these materials. Radar antennae are, in many cases, very difficult structures to design and manufacture, due to the extremely high severity of the surface accuracy, stiffness and strength requirements.

The tendency in today's radar technology to increase antenna performance levels by using higher and higher frequencies, increases the structural requirements too. In fact, the closer the actual reflector's geometry approaches the ideal, the greater the optimization in microwave propagation. The displacements from the ideal surface are due both to surface inaccuracies consequent to manufacturing operations and elastic deformations under applied service loads and/or environmental conditions. The allowable entity of such displacements is a very small fraction of the reflector's characteristic length.

In structural applications with metallic materials, the design criteria are sufficiently defined, the behaviour of the materials is outlined with simple models, in the elastic case, and their mechanical characteristics are known.

For composite materials, these approach methodologies are inadequate and can lead to errors of evaluation. With materials which can be designed following the specific requirements, it may be necessary to run through the ways of the design from the origin, both from a point of view of the calculations and specifications, and the geometry of the structures. In fact, for these materials the concept of thickness is usually replaced by the percentage of fiber and matrix, number of layers of the laminate and the orientation of the laminae.

The first step in the design of a structure with composites, and for every type of design for that matter, is the definition or the requisites of the component: this step includes the evaluation of the physical, mechanical and chemical properties that the component to be designed must possess and consider all the environmental and non-environmental factors such as stress, temperature and the agressiveness of the environment.

This step is followed by the initial base of the design during which it is necessary to define first the data on the geometric shape of the component. It is now appropriate to remember the fundamental advantages of composite materials: the possibility of being free of the bases of traditional design with conventional materials, such as steel, and then to implement particular shapes, and shape resistant structures, to incorporate, during construction, stiffenings in such a way as to obtain, for very case, the required performance. In this stage, two methods can be used: stress analysis or empirical design using existing standards.

In the case of the advanced design of components, such as radar antennae, a specific stratified experience does not exist and every case must be discussed starting from stress analysis. Stress analysis avails itself of the relations between stress and strain in the formulation for anisotropic materials and is used to evaluate whether the design choices are verified.

During this step, the choice of the materials is formulated taking into account the basic principles which govern the use of composite materials. Mechanical characteristics, in fact, depend on the combined effect of the content and the orientation of the fibers in the finished product. The chemical, electrical and thermal characteristics of reinforced plastics mainly depend on the type and the chemical formulation of the resin which constitutes the matrix. The final characteristics of the component, from the point of view of the performance and cost, depend on the manufacture and materials used.

Then, the mechanics of composite materials constitute a well defined theory which enables determination of the component's structural response on the basis of knowledge of the matrix and fiber characteristics and the stacking sequence. The structural behaviour of the material can be controlled by the choice of suitable technology: in this way it is possible to obtain desired properties such as local stiffness, strength, toughness and other structural and non-structural properties by controlling the type of fiber, the type of matrix, and the volume percentages between matrix and reinforcing material.

Once the material's structural response is known, the *shape* of the structure may be determined on the basis of performance requirements and available manufacturing technologies. (Note that the term *shape* includes the material's geometry, any insert, etc.) The next iterative design step is to return to the material so that it can be tailored to performance requirements: the result is a circular optimization process from micro to macro and back again. Each of the design steps is accompanied by analytical modeling (Fig. 1).

According to the rule-of-mixtures, a composite's mechanical properties may be determined at micro level when the mechanical properties of the single components are known. At that macro level, the structural behaviour of the material (laminate) may be evaluated according to the classical lamination theory (CLT), while the overall response of the structure can be

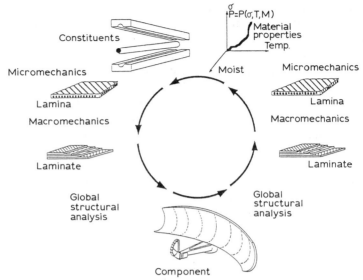

FIG. 1.   The circular design process.

determined through use of continuous orthotropic mechanics implemented on a computer code.

To be successful, a circular design process must possess a flow of reliable information regarding the micro and macro characteristics of the material, validated and easy-to-use computation tools and a specific knowledge that interacts with the design process as a guide for subsequent choices.

It is necessary to note how a design with composites requires interdisciplinary expertise: the designer, or his team, must possess experience in materials mechanics, stress analysis and technology. In the future, it is desirable that the entire design process be automatically managed by an *expert system*. In the meantime, the designer's experience is the only base on which the design with composites can be developed. To deal with the problem of structural analysis, useful tools are the finite element method (FEM) and other calculation procedures (either as a routine in a FEM code or developed separately) by which the laminate's structural behaviour may be examined with the CLT.

Sets of data regarding material characteristics for general-purpose use (such as that for metallic materials) are normally unavailable. This is due to several factors, mainly the variety of existing and potential composite materials, the huge array of composite constituents, and the variability deriving from differences in application. Moreover, the fact that the technologies employed do not possess 100% repeatability can cause variance in the material performance levels. The result is that for composites the behavior of the finished structures is even less predictable than in the case of traditional materials to obtain technical reliability, even when validated codes and experienced analysts are utilized in computation. Consequently, experimental testing allowing characterization of the material constituting the structure and verification of the component, has a major role in the design process.

Once the loads the laminate is subjected to are known in every section, the design can be outlined in two steps: the first includes the choice and characterization of the laminae and the definition of their number and orientation; the second is based on the definition of the stacking sequence. The problem of optimization, in the case of failure design, for example, can be solved by looking for the minimum of the sum

$$\sum_1^t s_i$$

under the conditions

$$\sum_1^t s_i c(\theta_i) > C$$

where C is a function of the load state of the laminate in the considered section and of the design coefficient, $c(\theta_i)$ depends on the mechanical characteristics of the laminae which are oriented at an angle θ_i with respect to the direction x of the laminate, s_i is the number of these laminae and t is the number of the different orientations of the laminae. Moreover, it is necessary to determine the stacking sequence according to appropriate criteria. Generally speaking, a design methodology must define the number and the orientation of the laminae of a laminate which is subjected to a set of known loads. The problem, formulated in this way, is not solvable because without knowing *a priori* the structure of the laminate (and thus its stiffness matrix), it is impossible to solve the stress–strain relations of the laminate and, consequently, to evaluate the strain and stress states of the laminate itself.

It is possible to obtain solutions to the design problem by assuming different stacking sequences, verifying them and then optimizing them with subsequent approximations. In actual practice, all the possible solutions which have a minimum number of layers, can be found without solving the minimum problem, but using an iterative process to define the stratification and to verify that the laminate has the proper characteristics to satisfy the design conditions.

ADVANTAGES AND DISADVANTAGES OF COMPOSITES

Composite materials are particularly well suited to the manufacture of complex geometry items, such as double curvature radar antennae, because they are much easier to form than metallic materials. Moreover, composite materials allow one to achieve especially accurate surface finishes and afford the possibility of obtaining the same stiffness and strength as metallic structures at reduced weight, by virtue of the higher specific stiffness and strength, lower specific weight and the possibility of tailoring the material in order to obtain the desired engineering properties, as highlighted above. Among the advantages of the tailorability of the material, is the possibility of obtaining higher dimensional stability, tuning the coefficients of thermal expansion. Besides the important advantages just described, there are some disadvantages, the most significant of which are the effects of the curing temperature and of the environmental service conditions, such as humidity and temperature. Such effects have repercussions on the precision of the shape, the dimensional stability and the degradation of mechanical properties.

The epoxy resin systems, generally used for the impregnation of composite materials, require curing temperatures of about 120–180°C. The curing process takes place in an autoclave; during the heating stage, the mold (generally a metallic one) expands; the resin, still in a semifluid state, follows it. When the curing temperature has been reached and maintained for a few hours, the composite attains its final mechanical properties. The coefficient of thermal expansion of the composite reflector is generally different from that of the mold, so their contraction during the cooling stage is not the same. This results in possible distortions of the reflector, that must be kept to a minimum.

A radical way to avoid these problems is to use low curing temperatures of the resin systems, possibly only slightly higher than room temperature. One of these resin systems has been experimentally characterized by SMA.

TECHNOLOGY

The most used structural solution for the antenna reflector is the sandwich, manufactured by co-curing or secondary bonding. The core is generally aluminum or Nomex honeycomb, the laminated skins are from UD or fabric layers of carbon, kevlar or glass/epoxy (Fig. 2).

When carbon composites are used together with an Al honeycomb core, isolation to prevent galvanic corrosion must be provided. In general, it is obtained by means of the insertion of one or two layers of glass/epoxy between Al and C.

The reflector surface requires to be conductive. A variety of methods exists to render conductive the materials that are not conductive, such as:

(1) metalization, applied directly on the reflector, or on the mold before the lay-up process;
(2) a thin copper grid inserted among the layers of the front skin of the reflector;
(3) correct thickness of graphite composite in the front skin of the reflector;
(4) coating of the fibers with metals, in order to obtain or enhance their conductivity; unfortunately, these fibers haven't yet sufficient mechanical properties and can only be used in place of the copper grid.

The lay-up operations are completely manual; the fabric is extensively used, because it's cheap, easy and quick. Of course, the stiffness in the fiber direction is much lower than for the UD.

The molds are generally metallic; carbon steel and mehanite cast iron are the most suitable materials. Graphite molds, whose use was not

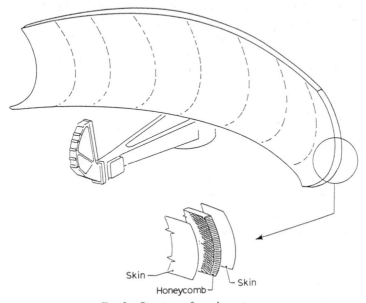

Skin
Honeycomb
Skin

FIG. 2. Structure of a radar antenna.

recommended in the past, seemed to have become popular, after recent improvements. The surface finishing of the mold is related to the desired surface finishing of the reflector; the tolerances in the surface irregularities of the mold must be very small.

The polymerization of the resin is generally carried out in an autoclave, with temperature and pressure in control. Typical parameters of curing cycles are:

—hot cycle: 120–180°C, 400 kPa, 2–4 h;
—cold cycle: 40–60°C, 400 kPa, 6 h.

SPECIFIC PROBLEMS WITH STRUCTURAL CALCULATIONS OF THE ANTENNAE

Special computer codes were developed by the University of Florence, and then the University of Ancona and SMA to meet the specific needs of composite design and structural evaluation. Each of the codes has been validated by experimental testing conducted on specially created test structures. An example is DESCOMP developed by the University of Florence, whose functions include:

—Structural verification of any kind of laminate (including laminae having diverse orientations and different materials, orthotropic and isotropic components, sandwiches).
—Hypotheses of structurally optimum stacking sequences.
—Evaluation of delamination danger indices.

In addition, subroutines have been implemented to determine laminate thermal expansion on the basis of knowledge of the temperature variations and thermal expansion coefficients of the laminate's constituent layers and the noise absorption levels through composite panels.

As mentioned above, the radar antenna is a challenging structure whose analysis and design has to be carried out by means of sophisticated tools. Normally, as already stated, the following methods are used:

—finite element method and computer codes;
—computer codes, based on the Classical Lamination Theory, for the preliminary analysis of the composites.

Special problems in antenna design are mainly related to the requirements relevant to the stiffness, precision of the shape and dimensional stability.

Some of these problems are:

—minimization of the thermal and mechanical distortions; among the former which are not negligible are those due to the honeycomb springback;
—effect on the achievable precision of the materials constituting the reflector, including skin laminates, core, adhesives;
—effect of the curvatures of the faces;
—effect of the metallization and paint.

Their influence on the calculations is mainly:

—necessity for refined meshes to match the precision levels required;
—difficulty in schematizing the loads, for which, generally, deterministic models cannot be defined, while envelopes are too heavy;
—modeling of the honeycomb;
—evaluation of the residual stresses;
—reliability and precision of the material characteristic to be input into the calculations.

EXPERIMENTAL CHARACTERIZATION OF A VERY LOW CURING TEMPERATURE RESIN SYSTEM

An experimental program has been undertaken by SMA, with the cooperation of University Laboratories, in order to characterize the materials (lamina level), validate the analysis methods and tools (laminate and structure level), develop, tune and check the test methodologies.

The range of the tested materials is shown in the following diagram.

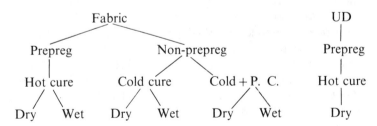

The aforesaid materials have been subjected to:

—tensile tests in order to evaluate the elastic engineering properties;
—measurements of the CTE using two different methodologies, i.e.

electronic differential dilatometer and strain gauges, tests of water absorption, 45 days long, and desorption;
—simultaneous differential calorimetry, consisting of the measurement of the variation of the specific heat and weight of the material sample versus the temperature increase. It allows evaluation of the glass transition temperature (T_g), the degree of complete or incomplete polymerization, the temperature at which the thermal stability degradation begins (T_d).

A sketch of the test typology is shown in Table 1.

The main object of this activity was to check the characteristics of the materials obtained by means of the impregnation of the fibers with a resin system able to polymerize at low temperatures.

The results of the tests are synthetically reported in Table 2.

Mechanical Properties

The hot cured prepregs and the cold cured materials with postcuring have practically the same properties. The cold cured materials without postcuring show slightly lower properties, about 10% with respect to the cold cured with postcuring, about 12% with respect to the hot cured materials. They are in general acceptable, but it's necessary to bear in mind that not only a decrease in mechanical properties occurs, but also the allowable service temperature is bounded by a lower value than with the hot curing prepregs, as will be underlined in the next paragraph.

Simultaneous Differential Calorimetry

- The glass transition temperature is 35% lower in the cold cured materials, with respect to the hot cured and the cold cured with postcuring ones, that show quite the same values. This means that an incomplete degree of polymerization has been reached. The service temperature of the structure is then limited to 70–80°C. The same tests performed on samples subjected to water absorption for 45 days show a general decrease of the values of about 10%.
- The chemical thermal stability begins to fail, for cold cured materials, at temperatures 10% lower than for the other; in other words the degradation takes place at temperatures 10% lower. As regards the *wet* case, the specimens tested after water absorption for 45 days show characteristics 10% lower with respect to the *dry* case.

Thermal Expansion

The measurement of the thermal expansion, performed by means of the electronic differential dilatometer, leads to not only deduction of the CTE

TABLE 1
Typology of the performed tests

Material	Curing process	Mechanical properties	Measurement of CTE		Simultaneous difference calorimetry	Water absorption/desorption
			Electronic	Strain gauge		
Glass fabric + epoxy prepreg Narmco 3203/7781	125°C 0.35 MPa 1.5 h	×	×	×	×	×
Graphite fabric Gividi CC 201 prepreg	125°C 0.35 MPa 1.5 h	×	×	×	×	×
Glass fabric 7781 + XB 3052 hand impr.	50°C 0.4 MPa 7 h	×	×		×	×
Graphite fabric Gividi CC 90 + XB 3052 hand imp.	50°C 0.4 MPa 7 h	×	×		×	×
Glass fabric 7781 + XB 3052 hand impr.	As 3 + postcuring 110°C 0.1 MPa 4 h	×	×	×	×	×
Graphite fabric Gividi CC 90 + XB 3052 hand imp.	As 3 + postcuring 110°C 0.1 MPa 4 h	×	×	×	×	×

Table 2
Results of the tests

Material	Curing process	E_1 (GPa)	E_2 (GPa)	G_{12} (GPa)	ν_{12}	X (MPa)	Y (MPa)	S (MPa)	α_1 (ppm/°C)	α_2 (ppm/°C)	% H_2O	T_g (0°)	T_g (45°)	T_D (0°)	T_D (45°)
Glass fabric + epoxy prepreg Narmco 3203/7781	125°C 0·35 MPa 1·5 h	24·5	20·9	4·5	0·09	290	270	100	9·3[a] 15·5	10·9[a] 16·7	0·45	100	85	300	270
Graphite fabric Gividi CC 201 prepreg	125°C 0·35 MPa 1·5 h	57·2	56·4	3·9	0·05	570	580	120	0·72[a] 4·7	0·62 4·2	0·063	125	115	300	280
Glass fabric 7781 + XB 3052 hand impr.	50°C 0·4 MPa 7 h	22·3	18·9	—	0·14	—	—	—	7·8[a] —	— —	0·068	75	70	280	260
Graphite fabric Gividi CC 90 + XB 3052 hand imp.	50°C 0·4 MPa 7 h	51·4	49·0	—	0·06	—	—	—	16·6[a] —	— —	0·93	75	70	270	240
Glass fabric 7781 + XB 3052 hand impr.	As 3 + postcuring 110°C 0·1 MPa 4 h	24·2	21·0	3·5	0·09	390	500	70	8·4[a] 14·2	9·8[a] 14·6	0·73	105	98	300	290
Graphite fabric Gividi CC 90 + XB 3052 hand imp.	As 3 + postcuring 110°C 0·1 MPa 4 h	56·2	55·3	3·6	0·05	600	520	110	1·52[a] 4·4	0·92[a] 4·6	0·98	115	105	300	280

[a] Electronic differential dilatometer.

but also to discovery of the beginning of the phenomena of chemical degradation or transition. The results confirm those found by means of differential calorimetry. The tests for the measurement of the CTE have been performed also by means of strain gauges. The two methods, applied to samples of different sizes, give results that, even if not conflicting, are quite different. They fit two separate bands around the data generally found in the literature. The data obtained by means of the differential dilatometer, albeit useful for an engineering design approach, do not give sufficient accuracy; for this reason, they are still under investigation, also involving other experimental activities.

Water Absorption/Desorption

The weight increase due to water absorption reaches the maximum value of 1% in the case of cold cured graphite with or without postcuring against the 0·6% for the hot cured graphite; as regards glass, 0·7% is reached for the cold cured, 0·45% for the hot cured. Maximum absorption or desorption is reached in about 30–40 days; half of the maximum value is reached in 1–3 days.

CONCLUSIONS

The introduction of composite materials into the radar antenna technology is not only possible and advantageous, but even essential to obtain the surface accuracy required with the high frequencies used in some applications. On the other hand it is necessary to bear in mind some problems related to their use.

The technologies have to be correctly used and/or improved to avoid or minimize possible collateral effects, like distortions due to the high curing temperatures.

Special care must be taken when dealing with data relevant to material properties. The structural analysis involves more complex tools than for the traditional structures, due to the anisotropic characteristics of the materials and the higher amount of data to be processed.

The design is, in general, more strictly related to the manufacturing technologies, involving a higher degree of interdisciplinarity.

51

A Practical Methodology for Assessing the Validity of Strain Gauge Data Obtained from Advanced Composite Structures

STEPHEN R. HALL

The Structures and Materials Laboratory, The National Aeronautical Establishment, Montreal Rd, Ottawa, Ontario, K1A 0R6, Canada

ABSTRACT

The sensitivity of aircraft structural analysis methods to load history definition is well documented. Inaccurate definitions of operational load histories can produce misleading, non-conservative predictions for component lives. The accuracy of load histories used to evaluate critical component lives must be verified. Electrical resistance strain gauges remain one of the most practical and cost-effective means of verification.

Unfortunately the application of strain gauges to composite structures is more complex than for metal structures. The anisotropic nature of many composite structures greatly increases the sensitivity of the strain gauge data to errors such as gauge/load direction misalignment.

This paper proposes a method of validating strain gauge data which are obtained from symmetric laminates under in-plane loading. Errors are discussed and criteria for defining acceptable gauge alignment tolerances are proposed. It is shown that these criteria must account for variations in both lay-up and loading.

1 INTRODUCTION

The structural substantiation of aircraft components is accomplished by a combination of experimental and analytical stress analysis. An important part of this process is the long-term durability testing of coupon specimens, sub-components and full-scale components. To maximize the benefits of this testing, and to ensure the validity of results, a number of parameters are monitored continuously. One of the most useful parameters is the analogue history of the specimen strain levels as recorded by strain gauges which are bonded to test specimens.

Strain gauges provide a cost-effective method of determining the strain distribution in a specimen under applied loads. The data obtained from strain gauges can also be used to determine the in-service strain spectrum at a specific structural feature.

In contrast to metallic structures, the limitations of applying strain gauges to composite structures are not widely understood. The anisotropic behaviour of many composite components can result in 'unexpected' structural responses when loads are applied to a structure. If metal strain gauge technology is applied directly to composites, incorrect interpretations of the strain gauge data may occur.[1]

To ensure that valid conclusions are reached the accuracy of composite strain gauge data has to be established. Unfortunately erroneous strain data from gauges which are attached to composite structures are not easy to detect. For even a small number of gauges the interdependency of many of the parameters to which the strain gauge is sensitive make it difficult to intuitively validate the output data. This problem is further exacerbated when large quantities of data are involved.

This paper proposes a method of assessing the validity of strain gauge data obtained from composite structures. Methods of identifying and, in some instances, correcting erroneous data are discussed. Finally, a technique of estimating acceptable gauge misalignment angles is presented, and the importance of considering both the composite lay-up and the applied load is emphasized.

2 AN OVERVIEW OF THE PROPOSED METHOD

The method discussed in this paper hypothesizes that any strain gauge errors are attributable to gauge misalignment alone and hence is termed the 'misalignment method'. Based on this hypothesis, strain gauge data are compared with calculated results for known loads cases and predicted angles of gauge misalignment are computed. Data from gauges whose predicted misalignment angles lie within an acceptable tolerance limit are considered valid whereas data from gauges with misalignment angles outside the acceptable tolerance limit are flagged for further investigation. The actual misalignment angle of each gauge with an excessive predicted misalignment angle is then measured. If the measured and predicted angles are in agreement, the initial hypothesis is proven and the experimental data can be corrected. Conversely, if the measured and predicted misalignment angles differ, the initial hypothesis is incorrect and other sources of error have to be systematically examined.

3 STRAIN GAUGE ALIGNMENT

Generally strain gauge alignment is far more critical for composite structures than for metallic structures. The main reason for this is that a composite structure's axes of principal stress and strain are not always coincident. Consequently a 'small' misalignment of the gauge with respect to the direction of the principal load can cause a large error in the measured strain.[2]

Prior to evaluating the misalignment of a gauge relative to a reference axis it is necessary to calculate two parameters, namely:

(i) the location of the laminate principal strain axis relative to the reference axis, and

(ii) the orientation of the strain gauge relative to the laminate principal strain axis.

Although any arbitrary reference axis can be chosen, it is convenient to use the 'laminate reference axis' (LRA) of a composite structure. The LRA is defined as the axis about which the laminate stiffness properties have been calculated and typically corresponds to the zero degree ply direction.

3.1 The Location of the Laminate Principal Strain Axis

Consider a laminate whose reference directions are defined in Fig. 1. Using the plane stress assumptions of classical lamination theory, the

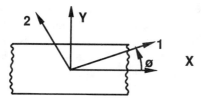

where *X–Y* denotes the global axis coordinates, 1–2 denotes the local axis coordinates and ϕ denotes the strain transformation angle.

FIG. 1. Definition of laminate reference directions and strain transformation angles.

following equation for the orientation of the principal plane relative to the LRA can be derived:[2]

$$\phi = \tfrac{1}{2}\tan^{-1}\left[\varepsilon_{xy}/(\varepsilon_x - \varepsilon_y)\right] \tag{1}$$

where ε_x = laminate strain in the X direction, ε_y = laminate strain in the Y direction and ε_{xy} = laminate shear strain in the X–Y plane.

3.2 The Orientation of a Strain Gauge Relative to the Plane of Principal Strain

The orientation of a strain gauge relative to the plane of principal strain can be determined either by iteration or by calculation. If the gauge under consideration is a uniaxial or biaxial gauge the problem must be solved using an iterative technique whose convergence is extremely sensitive to the actual laminate lay-up and applied loading. However, if the gauge is a three-leg rosette gauge, the problem may be solved using a closed-form solution. This closed-form solution will always locate the position of the laminate principal axis relative to the strain gauge for a given lay-up and load case.

3.2.1 The iterative technique

The iterative technique uses strain transformation relationships[2,3] to calculate the angular variation of strain with respect to a laminate's LRA. Each strain gauge reading is compared with the calculated values until the angle at which the two results are equal is located. This angle is the misalignment angle of the gauge relative to the LRA. Unfortunately there are two limitations associated with this technique:

(i)   in many instances it is not possible to ascertain whether the gauge misalignment angle is positive or negative, and

(ii) unless there is good agreement between the actual and calculated loads the results may never agree and convergence will not be attained.

Although these limitations can be reduced by using rosette gauges, practical constraints on the number of available recording channels often require the use of uniaxial gauges. In such instances the iteration technique provides the only practical method of estimating the location of the plane of principal strain relative to the gauge axis.

3.2.2 The calculation technique

Three-element rosettes measure the strain at a point in three different directions and are specifically designed for applications where the direction of the principal strains are unknown.[4] The tridirectional values of strain

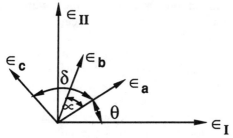

where ε_I and ε_{II} denote the axis of principal strain, ε_a denotes the orientation of the rosette gauge reference leg, ε_b denotes the orientation of the rosette gauge middle leg and ε_c denotes the orientation of the rosette gauge third leg.

FIG. 2.   Strain gauge rosette orientation.

which are obtained from the gauge allow the angle between the principal strain and gauge axes to be determined.[5]

The orientation of a three-leg strain rosette relative to a plane of principal strain can be shown to equal[2]

$$\theta = \tfrac{1}{2} \tan^{-1} \frac{(1 - \cos 2\alpha) - K(1 - \cos 2\delta)}{(K \sin 2\delta - \sin 2\alpha)} \tag{2}$$

where $K = (\varepsilon_a - \varepsilon_b)/(\varepsilon_a - \varepsilon_c)$, and α, δ and θ are defined in Fig. 2.

Calculation of the angle of strain gauge misalignment from rosette data. In accordance with the sign conventions previously adopted (see Figs 1 and 2), the misalignment angle (μ) of a three-leg rosette relative to the LRA is calculated by summing ϕ and θ, i.e.

$$\mu = \theta + \phi \tag{3}$$

4 DATA VALIDATION AND ERROR IDENTIFICATION

In many practical situations (e.g. a full-scale aircraft fatigue test or a flight loads survey) an analyst is required to validate the output from many strain gauges. The misalignment method allows erroneous gauges to be quickly identified.

The strain gauge output from known load cases are used to calculate an angle of misalignment for each gauge. If a calculated misalignment angle lies within an acceptable tolerance (see Section 5), the data obtained from the gauge are considered valid. Conversely, if a calculated misalignment angle exceeds the acceptable tolerance limit, the gauge is considered to require further examination. Once all the gauges with calculated 'out-of-tolerance' misalignment angles have been identified, the reason(s) for the misalignment have to be determined.

For each gauge requiring further examination the first issue that needs to be addressed is the validity of the initial hypothesis, i.e. is the error actually due to gauge misalignment alone? This is determined by measuring the actual gauge misalignment angle and comparing it to the calculated misalignment angle. If a gauge's measured and calculated misalignment angles are within reasonable agreement, the initial hypothesis is shown to be true and the reason for the out-of-tolerance reading is confirmed to be gauge misalignment. All the recorded data from the gauge may then be corrected and used in subsequent analyses.[2] Conversely, if there is little or no correlation between the measured and calculated misalignment angles, the data are most likely erroneous and other potential sources of strain gauge error have to be considered.

Many of the potential sources of error for strain gauges attached to composite materials have previously been identified[1] and can be categorized as follows:

(a)   improper strain gauge selection criteria;
(b)   inadequate gauge bonding procedures;
(c)   transverse sensitivity effects;
(d)   temperature compensation effects; and
(e)   strain gauge misalignment.

Additionally, the use of the misalignment method requires that two other sources of error be considered:

(i)   incorrect lay-up, and
(ii)   incorrect applied loading.

Having previously eliminated strain gauge misalignment as being the source of error, the remaining sources have to be checked systematically until the cause of the error is established. Depending on the nature of the error, the data from the gauge in question may have to be discarded.

As the errors attributable to categories (i) and (ii) are particular to the misalignment method, they will be discussed further in the following paragraphs. A detailed discussion of the errors defined in categories (a)–(d) can be found in Ref. 1 and will not be duplicated here.

4.1 Incorrect Laminate Lay-ups

Incorrectly orientated plies will effect the laminate compliance terms and cause a discrepancy between the measured and calculated misalignment angles. However, as the in-plane compliance terms are independent of the stacking sequence, the misalignment method will be insensitive to incorrect stacking sequences unless the resultant laminate is unsymmetric and significant coupling between in-plane axial, bending and twisting deformations occurs.

4.2 Incorrect Value of Applied Loading

The misalignment method will detect inconsistencies in the applied load and resultant strain readings. A number of factors, such as incorrect data processing or grip slippage, can cause errors of this nature.

5 ACCEPTABLE GAUGE ALIGNMENT TOLERANCES

Acceptable alignment tolerances for strain gauges that are attached to composite structures are dependent on the relative angular location of the principal strain axis with the assumed axis of strain gauge alignment, the applied loading and the laminate lay-up. Appropriate tolerances can only be defined by using polar strain diagrams to assess the variation of strain with angle for all critical load cases. It cannot be automatically assumed that alignment tolerances which were acceptable for metal structures will be acceptable for composite structures.

The scope of this problem can best be illustrated by a series of examples.

5.1 Examples of the Effects of Gauge Misalignment

Consider a strain gauge which is monitoring a structural element that is subjected to combinations of the applied loads N_x, N_y and N_{xy} (Fig. 3).

The sensitivity of the strain gauge to different structural element lay-up configurations can best be illustrated by considering the following four examples:

(a) a balanced symmetric graphite/epoxy laminate with a stacking sequence of $[0, 90, 45, -45, 0]_s$;

(b) an unbalanced symmetric graphite/epoxy laminate with a stacking sequence of $[0, 90, 45_2, 0]_s$;

(c) a unidirectional graphite/epoxy laminate with a stacking sequence $[0]_{10}$; and

(d) an aluminium block.

The mechanical properties of the graphite/epoxy and aluminium are summarized in Table 1.

TABLE 1
Material properties used in the strain gauge analysis

	Graphite/epoxy	Aluminium
E_x (Msi)	20·30	10·30
E_y (Msi)	1·19	10·30
G_{xy} (Msi)	0·90	3·96
v_{xy}	0·30	0·30

In all cases the variation of the strain along the gauge axis ('1' direction) will be considered as the gauge and is rotated either side of the LRA.

To evaluate the angular variations of strain with respect to the LRA it is necessary to determine the in-plane compliance terms of each laminate and to define the applied loading. The in-plane compliance terms were calculated using the laminated plate theory program LAMCAL[6] and the results of these calculations are summarized in Table 2.

Four load cases, as defined in Table 3, were applied to each of the 'example laminates'. Whilst the magnitude of the load cases were arbitrarily chosen, they are typical of the loading that would be applied during the

TABLE 2
In-plane compliance terms for the example lay-ups

Lay-up type	In-plane compliance terms ($\times 10^{-6}$)					
	a_{11}	a_{12}	a_{16}	a_{22}	a_{26}	a_{66}
Balanced	1·926	−0·562	0·000	2·960	0·000	7·609
Unbalanced	2·103	−0·250	−1·354	3·509	−2·382	10·340
Unidirectional	0·985	−0·296	0·000	16·810	0·000	22·220
Aluminium	1·942	−0·583	0·000	1·942	0·000	5·048

TABLE 3
Applied load cases used in the strain gauge analysis

Load case number	Applied load (lbf/in)		
	N_x	N_y	N_{xy}
1	5 000	0	0
2	0	5 000	0
3	0	0	5 000
4	5 000	5 000	5 000

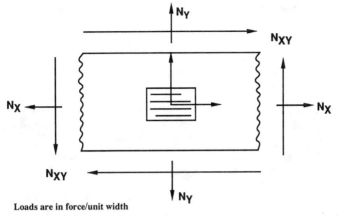

Loads are in force/unit width

FIG. 3. Loads applied to the example strain gauge element.

calibration and actual use of a strain gauge. Cases 1–3 simulate point load cases that would be applied to check gauge calibrations. Case 4 simulates the recording of actual data from a structure subjected to a combination of in-plane axial and shear loading.

The analysis of each lay-up/load combination was undertaken with the computer program CSGAGE.[7] Some of the results obtained from the analysis are contained in Figs 4–8 inclusive.

Figures 4–8 show how the percentage strain gauge error varies with the angular misalignment of the gauge from the LRA.

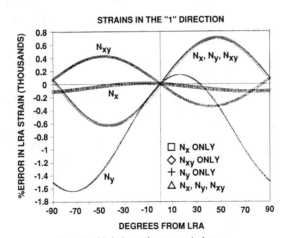

FIG. 4. Unbalanced symmetric lay-up.

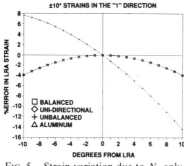

FIG. 5. Strain variation due to N_x only.

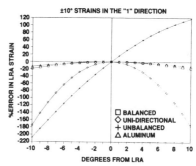

FIG. 6. Strain variation due to N_y only.

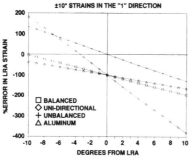

FIG. 7. Strain variation due to N_{xy} only.

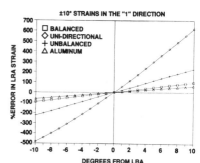

FIG. 8. Strain variation due to combined loading.

The term 'percentage error' merits some explanation. In the context of this analysis the percentage error for a strain gauge is defined as follows:[2]

$$\% \text{ error} = (\varepsilon_G - \varepsilon_0)/\varepsilon_0 \times 100 \cdot 0 \tag{4}$$

where $\varepsilon_G = $ the strain reading recorded by the gauge and $\varepsilon_0 = $ the calculated strain along the LRA.

As the % error can be both positive and negative some care has to be taken when interpreting the results. When ε_0 is tensile ($+$ve), an overestimate of tensile strain gives a positive % error and an underestimate gives a negative % error. Similarly, when ε_0 is compressive ($-$ve), an overestimate of compressive strain gives a positive % error and an underestimate gives a negative % error.

Figure 4 provides a global view of the problem for the unbalanced lay-up. The % error is plotted over an angular range of $\pm 90°$ from the LRA. It can be seen that for a gauge nominally aligned with the X axis significant errors can result from small gauge misalignments when N_y, N_{xy} or a

TABLE 4

Load case number	Percentage strain gauge error							
	Balanced		Unbalanced		Unidirectional		Aluminium	
	$+3°$	$-3°$	$+3°$	$-3°$	$+3°$	$-3°$	$+3°$	$-3°$
1	−0·35	−0·35	−3·67	3·06	−0·36	−0·36	−0·36	−0·36
2	−1·72	−1·72	45·66	−53·89	−15·85	−15·85	−1·19	−1·19
3	−129·37[a]	−70·63[a]	−39·70	40·12	−185·77[a]	−14·23[a]	−119·49[a]	−80·51[a]
4	29·37	−28·95	69·39	−68·98	174·69	−162·12	19·41	−19·41

[a] Denotes a relative error as the actual value is zero.
Values have been compared to the LRA axis strains of the unbalanced lay-up.

combination of N_x, N_y and N_{xy} are applied to the structural element of Fig. 3.

Figures 5–8 illustrate the percentage errors that are incurred for gauge misalignments of $\pm 10°$ from the LRA. For the reasons detailed in Ref. 1 it is not unreasonable to anticipate strain gauge misalignment tolerances of the order of $\pm 3°$. An inspection of Figs 5–8 indicates that the percentage error that can occur at these angles is significant. The actual values of percentage error resulting from gauge misalignment angles of $\pm 3°$ are summarized in Table 4.

It should be noted that while Figs 5, 6 and 8 give 'absolute' percentage errors three of the plots in Fig. 7 are plots of 'relative' percentage error. The need to use relative percentage error is a result of the 'balanced', 'unidirectional' and 'aluminium block' lay-ups being balanced (i.e. $a_{16} = a_{26} = 0·0$). For a balanced lay-up that is subjected to in-plane shear loading only (i.e. $N_x = N_y = 0·0$) ε_0 will equal zero.[2] Therefore the percentage error calculation of eqn (4) will always produce an answer of infinity and it will be impossible to assess the significance of an error. To overcome this problem the program CSGAGE[7] permits the user to define an alternative, non-zero, strain against which the off-axis strains may be compared. The alternative strain used to generate Fig. 11 was $-6770\mu\varepsilon$. This corresponds to the ε_0 value obtained from the unbalanced lay-up when it was subjected to N_{xy} only.

The results of these analyses illustrate that a 'universally' acceptable misalignment tolerance for strain gauges that are attached to composite structure cannot be defined. Instead methods of determining actual gauge misalignment angles have to be developed so that recorded data may be corrected for any misalignment which occurs. As discussed previously, one reliable method of doing this is to use strain gauge rosettes.

5.2 A Design Approximation for Balanced Uniaxial Test Specimens

An examination of Fig. 5 indicates that over the range of $\pm 10°$ the percentage error for the 'balanced', 'unidirectional' and 'aluminium block' lay-ups are almost coincident when they are subjected to a uniaxial load (i.e. $N_y = N_{xy} = 0$). Once again this is a result of the laminates being balanced as, in all three cases, $a_{16} = a_{26} = 0$. This result is of interest as it leads to a useful 'design approximation' for the percentage error in strain that will be obtained from a misaligned strain gauge which is mounted parallel to the loading axis of a balanced uniaxial test specimen.

The % error in the X direction (loading axis) strain for a balanced symmetric laminate subject to a uniaxial load is equal to[2]

$$\% \text{ error} = \frac{a_{11}(\cos^2 \phi - 1) + a_{12} \sin^2 \phi}{a_{11}} \times 100 \tag{5}$$

where a_{ij} denote laminate in-plane compliance terms.

Generally a_{12} will be an order of magnitude smaller than a_{11}. Furthermore, as ϕ tends towards zero $\sin^2 \phi$ also tends towards zero and the second term in the numerator can be considered negligible. Therefore eqn (5) reduces to

$$\% \text{ error} = (\cos^2 \phi - 1) \times 100 \tag{6}$$

Thus, for small angles, the percentage strain gauge error can be considered to be independent of the specimen lay-up and the magnitude of the applied load. Table 5 shows the accuracy of this approximation for the

TABLE 5
Evaluation of approximating the strain gauge error with a $(\cos^2 - 1)$ term

Angle	$\cos \theta$	$\cos^2 \theta$	Approximate % error $[(\cos^2 \theta - 1) \times 100]$	Actual % error Balanced lay-up	Unidirectional lay-up	Aluminium block
10	0·9848	0·9698	−3·02	−3·90	−3·92	−3·92
9	0·9877	0·9755	−2·45	−3·16	−3·18	−3·18
8	0·9903	0·9806	−1·94	−2·50	−2·52	−2·52
7	0·9925	0·9851	−1·49	−1·92	−1·93	−1·93
6	0·9945	0·9891	−1·09	−1·41	−1·42	−1·42
5	0·9962	0·9924	−0·76	−0·98	−0·99	−0·99
4	0·9976	0·9951	−0·49	−0·63	−0·63	−0·63
3	0·9986	0·9973	−0·27	−0·35	−0·36	−0·36
2	0·9994	0·9988	−0·12	−0·16	−0·16	−0·16
1	0·9998	0·9997	−0·03	−0·04	−0·04	−0·04
0	1·0000	1·0000	0·00	0·00	0·00	0·00

'balanced', 'unidirectional' and 'aluminium block' lay-ups over an angular range of $\pm 10°$ from the LRA.

6 CONCLUSIONS

A practical method for validating large quantities of strain gauge data obtained from composite structures has been proposed. The method is termed the 'misalignment method' and uses an initial hypothesis that any strain gauge errors are attributable to gauge misalignment alone. Based on this hypothesis, gauges which are producing readings outside an acceptable tolerance limit are identified and examined in detail. If the hypothesis is found to be true the data are corrected for gauge misalignment. Conversely, if the hypothesis is found to be false, other possible causes of the error have to be examined and if necessary the data rejected.

A discussion of the significance of strain gauge misalignment has also been presented. It is shown that it is not possible to define a 'universal' tolerance for gauges bonded to composite structures. Acceptable tolerances can only be defined by studying a polar strain diagram of all critical load cases, thereby accounting for the effects of both load and lay-up. For uniaxially loaded, balanced, symmetric coupon specimens a simplified version of the misalignment method has been developed. This method provides a good initial estimate of strain gauge error for *small* misalignment angles.

ACKNOWLEDGEMENTS

The author is indebted to many colleagues at the Structures and Materials Laboratory for their comments and guidance. In particular, he would like to thank Mr J. P. Komorowski, Mr P. C. Conor (presently seconded as a guest worker from the New Zealand Defence Scientific Research Establishment) and Mr D. L. Simpson.

REFERENCES

1. TUTTLE, M. E. and BRINSON, H. F., Resistance-foil strain-gage technology as applied to composite materials. *J. of Exp. Mech.*, **24**(1) (March 1984) 54–65.
2. HALL, S. R., An assessment of the validity of strain gauge data that have been obtained from composite structures. *NRCC/NAE Aeronautical Note* (in press).

3. JONES, R. M., *Mechanics of Composite Materials*. Scripta Book Company, USA, 1975.
4. DALLY, J. W. and RILEY, W. F., *Experimental Stress Analysis*. McGraw-Hill, New York, 1978.
5. MEGSON, T. H. G., *Aircraft Structures for Engineering Students*. Edward Arnold, London, 1972.
6. HALL, S. R., LAMCAL: an iterative design package for performing a laminated plate theory analysis of composite materials. *NRCC/NAE LTR-ST-1656*, November 1987.
7. HALL, S. R., CSGAGE: a program to evaluate the misalignment of strain gauges which have been applied to symmetric, unbalanced composite structures. *NRCC/NAE LTR-ST-1692*, December 1988.

52

Optimum Design for Knee Prosthesis Using Laminated Composite Materials—Stress Transmission Pattern and its Estimation

Tsuneo Hirai, Tsutao Katayama

Department of Mechanical Engineering, Doshisha University, Kamigyo-Ku, Kyoto, Japan

Nozomu Inoue

Department of Orthopaedic Surgery, Kyoto Prefectural University of Medicine, Kamigyo-Ku, Kyoto, Japan

and

Yasuo Kida

Doshisha University, Kamigyo-Ku, Kyoto, Japan

ABSTRACT

The purpose of our investigation was to devise a procedure for the optimum design for knee prosthesis (artificial knee joint) using laminated composites. This paper describes the numerical approach to the stress transmission from the tibial component (prosthesis) to the cancellous bone using the finite element method and the usefulness of composite materials for knee prosthesis.

1 INTRODUCTION

Recently, knee prostheses have been widely used in the field of orthopaedic surgery for knee joint diseases such as rheumatoid arthritis and osteoarthritis. But the problems of prosthetic loosening and sinking, that is, the failure of prosthesis/bone interface, have not been solved sufficiently.[1,2]

849

Stress concentration and micro-movement at the prosthesis/bone interface appear to be important factors which cause sinking or loosening of the prosthesis. The earlier tibial component of the total knee prosthesis was made of high density polyethylene (HDP); and as for the bonding substance between the prosthesis and bone, polymethylmethacrylate (PMMA) 'bone cement' is used. But, because of the stress concentration just beneath the loading point of the prosthesis and chemical or thermal problems during polymerization of PMMA, cementless metal-backed HDP tibial component have been developed. To prevent micro-movement at the prosthesis/bone interface, many configurations and surface properties have been developed.[3−5] But it seems difficult to prevent the micro-movement under metal prosthesis because of the shearing stress caused by the large difference of the elastic moduli between the metal prosthesis and bone. The ideal prosthesis must have the ability to distribute stress and at the same time to transmit the compressive stress to the bone without causing shearing stress.

The human body is optimally designed. The bone tissue, as a loaded transmission system, is reasonably arranged in all respects. In particular, the arrangement of the bone at the joint part is the most suitable in response to all kinds of loads. The load goes to the cortical bone through the cancellous bone. Cancellous bone is of heterogeneous nature and anisotropic, due to the orientation of the trabeculae and its three-dimensional structure. It is clear that homogeneous and isotropic materials, as previously mentioned, are unsuitable as the materials for replacing the cancellous bone. So, the purpose of our investigation was to consider the optimum fabrication of the knee prosthesis using composite materials. Composite materials have been designed as tailored materials. In this study, laminated composites consisting of unidirectional lamina with various ply angles were applied to the prosthetic construction.

In this paper, there is an attempt to establish two things: first, numerical analysis for the stress transmission from a T-shaped, laminated composite material prosthesis to cancellous bone, taking into consideration the interface condition; secondly, a frame work for effective design in ply construction of a laminate prosthesis which will allow optimum transmission of stress.

Though the cancellous bone has heterogeneous and anisotropic characteristics, epoxy resin is used in place of the cancellous bone to obtain the stress distribution in the part by the photoelasticity method since the numerical mode for the laminate and the interface condition should be investigated. The material replacing the cortical bone is steel. The uniform or concentrated load is applied to the upper side of the T-shaped prosthesis instead of actual loading the human body. In the analysis, a two-dimensional finite element method is applied to a modeled knee prosthesis.

2 NUMERICAL ANALYSIS

2.1 Finite Element Model

The two-dimensional finite element method (FEM) is applied to interpret the stress condition at the prosthesis/bone interface, and to develop the design of a laminated composite knee prosthesis. A typical finite element model is shown in Fig. 1. The finite element model consists of about 1400 triangular elements. Plane stress conditions are assumed in the finite element analysis. The deformations between A and B, and between C and D are fixed. As an adhesive layer, about 10 μm in thickness, existed at prosthesis/bone in the experimental model, elements of 10 μm in thickness are interposed between the prosthesis and the bone in the numerical model. The material properties of the laminated prosthesis are obtained from laminated plate theory. The Poisson's ratio of laminated composite materials in the thickness direction is assumed in the following section.

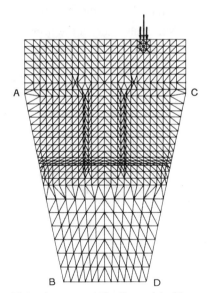

FIG. 1.   Finite element mesh for the model of knee prosthesis.

2.2 Estimation of the Poisson's Ratio in the Orthogonal Thickness Direction of Laminated Composite Materials

The elastic modulus of laminated composite materials in the X–Y plane is given by[6]

$$E_1 = \frac{1}{C_{11}} \qquad \nu_{12} = -\frac{C_{22}}{C_{12}} \qquad G_6 = \frac{1}{C_{66}} \tag{1}$$

where

$$C_{11} = \frac{l^4}{E_X} + \frac{m^4}{E_Y} + l^2 m^2 \frac{1}{G_{XY}} - 2\frac{\nu_{XY}}{E_X}$$

$$C_{12} = -\frac{\nu_{XY}}{E_X} + l^2 m^2 \frac{1}{E_X} + \frac{1}{E_Y} + 2\frac{\nu_{XY}}{E_X} + \frac{1}{G_{XY}}$$

$$C_{66} = -\frac{\nu_{XY}}{E_X} + 4l^2 m^2 \frac{1}{E_X} + \frac{1}{E_Y} + 2\frac{\nu_{XY}}{E_X} - \frac{1}{G_{XY}}$$

and where the orientation of reinforcement is shown as

$$l = \cos\gamma \qquad m = \sin\gamma$$

and the moduli on-axis,

$$E_X = \text{Young's modulus along the fibre direction}$$
$$E_Y = \text{Young's modulus across the fibre direction}$$
$$\nu_{XY} = \text{Poisson's ratio in } X\text{–}Y \text{ plane}$$

Poisson's ratio, in width, ν_{12}, and in thickness, ν_{13}, is given by

$$\nu_{12} = \frac{dY/Y}{dX/X} \qquad \nu_{13} = \frac{dZ/Z}{dX/X} \tag{2}$$

where, dx, dy and dz are small alteration in dimensions X, Y and Z during uniaxial tensile load (Fig. 2).

The next equation (eqn (3)) is obtained on the assumption that the rate of variation in volume, R, is constant, even if the ply angle (ϕ, γ) is changed.

$$R\left(1 + \frac{dX}{X}\right)\left(1 - \nu_{12}\frac{dX}{X}\right)\left(1 - \nu_{13}\frac{dX}{X}\right) = 1 \tag{3}$$

From eqns (2) and (3), ν_{13} is given by

$$\nu_{13} = 1 - \frac{1}{2R(1 - \nu_{12})} \tag{4}$$

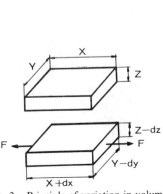

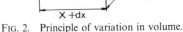

FIG. 2. Principle of variation in volume.

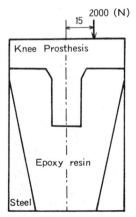

FIG. 3. Illustration   of   photo-elastic
model.

From eqns (1) and (2), v_{13} using C_{11} and C_{12} is obtained by

$$v_{13} = 1 - \frac{1}{2R(1 + C_{12}/C_{11})} \tag{5}$$

From eqn (4) and $v_{12}(\phi = 0°, \gamma = 0°) = v_{13}(\phi = 0°, \gamma = 0°)$, the rate of variation in volume R is determined by

$$R = \frac{1}{2(1 - v_{12})(1 - v_{13})} = \frac{1}{2(1 - v_{12})^2} \tag{6}$$

In the case of GFRP, the Poisson's ratio in thickness, v_{13}, is given by

$$v_{13} = 1 - \frac{1}{2 \times 1\cdot11 \times (1 + C_{12}/C_{11})} \tag{7}$$

where, $R = 1\cdot11$ obtained by substitution from $v_{12} = 0\cdot33$ ($\phi = 0°$, $\gamma = 0°$) into eqn (5).

3 PHOTO-ELASTIC EXPERIMENT

The model of the stem type tibial component of the knee prosthesis is made of three different materials. These materials, laminated composites (glass/epoxy) with various ply configurations, aluminium showing high stiffness and homogeneity, and epoxy resin showing low stiffness and homogeneity are used as the material of the prosthesis. Epoxy resin is used

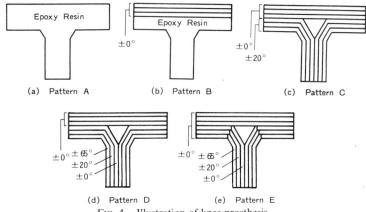

(a) Pattern A   (b) Pattern B   (c) Pattern C

(d) Pattern D   (e) Pattern E

Fig. 4.   Illustration of knee prosthesis.

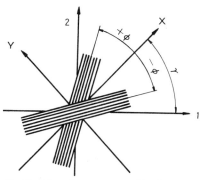

Fig. 5.   Coordinate system of laminate.

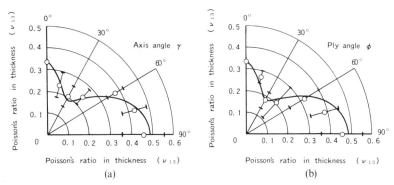

Poisson's ratio in thickness (ν_{13})

(a)

(b)

Fig. 6.   Relation between Poisson's ratio (ν_{13}) and several angles (ϕ, γ) for GFRP, (a) Ply angle $\phi = 0°$; (b) Axis angle $\gamma = 0°$. ——, Calculated result; $\female$, experimental result.

in place of the cancellous bone. Figure 3 shows the shape of the photo-elastic model of the proximal tibia. Materials and laminate configurations of some prostheses are shown in Fig. 4. In this model, the part corresponding to the cortical bone is made of steel (Fig. 3).

The photo-elastic experiments are carried out under two different conditions: (1) a uniform load of 2000 N on the joint surface of the photo-elastic specimen and (2) a load of distributed fraction of 2000 N on the joint surface shown in Fig. 2. Various fringe patterns are observed under several loading conditions (Fig. 3).

4 MEASUREMENT OF THE POISSON'S RATIO OF LAMINATED COMPOSITE MATERIALS

Laminates are made by a hand lay-up method using GFRP prepreg (TR19-200EC) supplied by NITTOBO Co. Ltd. Tensile test specimens are machined from the laminates using a diamond tipped cutting wheel.

As shown in Fig. 5, γ is the axis angle and ϕ is the ply angle of laminated composite materials. The dimensions of the test specimens are 3 mm width, 4 mm thickness and 200 mm length.

Tensile tests are performed to measure the Poisson's ratio in thickness of laminated composite materials using an Instron testing machine, model 1122, at a crosshead speed of 2·0 mm/min. The strains in the direction of length and thickness are measured by an Instron Strain Gage Extensometer (2630-002) and Instron Transverse Strain Sensor (057-11).

The results from eqn (7) are compared with experimental results.

Elastic moduli of GFRP are $E_X = 35·0$ [GPa], $E_Y = 9·0$ [GPa], $G_{XY} = 3·2$ [GPa] and $v_{XY} = 0·33$, and these elastic moduli are used in eqn (1).

As shown in Fig. 6(a) and (b), the results calculated from eqn (7) are found to be in good agreement with the experimental results. It is confirmed that Poisson's ratio of laminated composite materials in thickness can be estimated from eqn (7). Their values calculated from eqn (7) can be used in the numerical analysis to design the laminated knee prosthesis.

5 RESULTS AND DISCUSSION

5.1 Stress Distribution in Natural Proximal Tibia

In optimum design, the first step is to consider what constitutes rational design in nature. Therefore, a two-dimensional finite element method is applied to consider the stress transmission in natural proximal tibia.

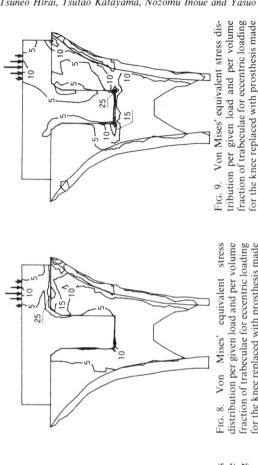

FIG. 7. Von Mises' equivalent stress distribution per given load and per volume fraction of trabeculae for uniform loading for the natural tibia.

FIG. 8. Von Mises' equivalent stress distribution per given load and per volume fraction of trabeculae for eccentric loading for the knee replaced with prosthesis made of chromo-cobalt.

FIG. 9. Von Mises' equivalent stress distribution per given load and per volume fraction of trabeculae for eccentric loading for the knee replaced with prosthesis made of high density polyethylene.

The material properties of the cancellous bone are as follows. Authors reported that the elastic moduli could be estimated from the volume fraction of trabeculae and the distribution function of the trabecular orientation by using image analysis in consideration of heterogeneous and anisotropic properties using ordinary simplified assumption as follows.[7]

$$E_L = \alpha_1 E_f V_f \qquad E_t = \alpha_2 E_f V_f \qquad G_{LT} = 4077 V_f 0.5358^{(\alpha_1/\alpha_2)}$$

$$\alpha_1 = \sum_{\theta=0°}^{180°} \cos^4(\theta - \theta_m) f(\theta) \qquad \alpha_2 = \sum_{\theta=0°}^{180°} \sin^4(\theta - \theta_m) f(\theta)$$

where

E_f = the modulus of the cortical bone
V_f = volume fraction of the trabeculae
$f(\theta)$ = distribution function of the trabecular orientation

Figure 7 shows von Mises' equivalent stress distribution per given load and per one bone tissue in the consideration of the volume fraction of the trabeculae. The stress concentration can't be found in this figure. It is considered that the stress is distributed by variation of volume fraction of the trabeculae and the trabecular orientation in real bone.

Figures 8 and 9 show the same stress distribution as in the above case in which the chromo-cobalt prosthesis replaces the cut off portion from the previous model, and where HDP prosthesis replaces its cut off portion, respectively. The load is applied to the right side in consideration of the eccentric loading.

The stress concentration seems to occur beneath the loading point because of the low stiffness of the prosthesis in Fig. 9. Figure 8 shows the stress concentration around the corner of the stem because of the high stiffness of the prosthesis. These results agree with the observed inherent faults of chromo-cobalt and HDP.

5.2 Stress Distribution along Prosthesis/Bone Interface Depending on Laminate Configuration

It is found that uniform stress distribution is a fundamental factor in the design of knee prosthesis. In particular, attention should be paid to the stress distribution along the interface between the prosthesis and the bone. The laminate prostheses shown in Fig. 4 are considered here. The mean stiffness of the stem in patterns C, D and E is designed to be the same value, 20 GPa.

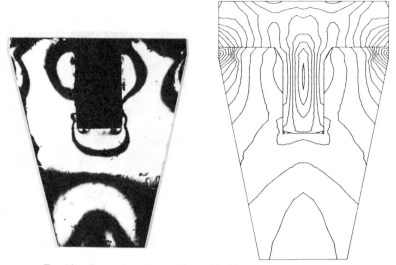

FIG. 10.   Comparison between isopachic fringe and numerical result.

To confirm that this numerical system is valid for the design of the laminate knee prosthesis, the numerical data are compared with the photo-elastic experimental results on Pattern E shown in Fig. 10. The numerical results are shown as the principal stress difference and experimental results as the isopachic fringe respectively. There is good agreement in both results.

In other cases, the load is applied eccentrically. The effect of anisotropy at the horizontal part of the T-shaped prosthesis on the stress distribution along the interface is considered by comparing pattern B with pattern A. Both results are shown in Fig. 11 using the principal stress difference per loading stress. The solid line is for pattern B and the dotted line is for pattern A. In the case of pattern B, the stress concentration beneath the

FIG. 11.   Comparison between result in the case of pattern A and pattern B. Distribution of stress ($\sigma_{12} = \sigma_1 - \sigma_2$) per given stress ($\sigma_G$) at the knee prosthesis/epoxy resin interface.

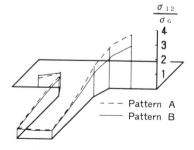

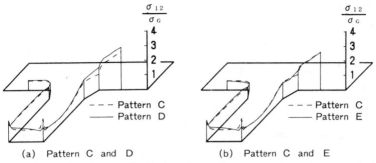

(a) Pattern C and D   (b) Pattern C and E

FIG. 12. Comparison between distribution of stress ($\sigma_{12} = \sigma_1 - \sigma_2$) per given stress ($\sigma_G$) at the knee prosthesis/epoxy resin interface for laminated prosthesis.

loading point is lessened by transmitting stress to the horizontal direction because of the anisotropy at the upper part of the prosthesis.

To consider the effect of the stiffness at the outer layer along the bone on the stress concentration, the result for pattern C is compared with one for pattern D (Fig. 12(a)) and for pattern E (Fig. 12(b)). The dotted lines show the case of pattern C. The lower stiffness in the perpendicular to the thickness direction at the outer layer, pattern D, distributes the stress around the corner of the stem, but concentrates the stress beneath the loading point (Fig. 12(a)).

To improve the stress concentration at the horizontal part, the higher

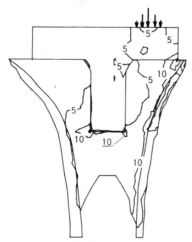

FIG. 13. Von Mises' equivalent stress distribution per given load and per volume fraction of trabeculae for eccentric loading for the knee replaced with prosthesis made of FRP.

stiffness laminates in the perpendicular to the thickness direction are arranged at this part such as in pattern E. It can be seen that the stress is distributed efficiently (Fig. 12(b)).

5.3 Stress Distribution in Using Effective Laminate Prosthesis for Actual Bone

Figure 13 shows the stress distribution in the case when the effective laminated prosthesis on stress transmission to the homogeneous and isotropic body is applied to the actual bone. Also it is confirmed that the effective laminate in the prosthesis in the previous section is suitable for the actual bone, by comparing with HDP prosthesis (Fig. 9) and chromo-cobalt prosthesis (Fig. 8).

6 CONCLUSION

The stress concentration at the corner of the stem is shown in the case of the chromo-cobalt knee prosthesis model. This observed result can explain the failure of this prosthesis. Though a similar stress concentration cannot be found in the HDP prosthesis, the stress concentration beneath the loading point in the HDP prosthesis is shown because of its low stiffness. In consideration of the above, the reinforcement orientation at the loading side is laid in the normal direction to loading and the lower stiffness is chosen along the boundary against the cancellous bone. The developed composite structure for the knee prosthesis is found to be effective on the above weak points. The numerical results are found to be in good agreement with the experimental results for a modelled knee prosthesis. Therefore, this numerical model for laminated composite material and stress transmission while considering interface condition is shown to be useful.

REFERENCES

1. BARTEL, A. and WALKER, P. S., The effect of the interface on the bone stresses beneath tibial component. *J. Biomech.*, **19** (1986) 957–67.
2. DUCHEYNE, P., KAGAN, A. II and LACEY, J. A., Failure of total knee arthroplasty due to loosening and deformation of the tibial component. *J. Bone Jt Surg.*, **60-A** (1978) 1026–33.
3. MERKOW, R. L., SOUNDLY, M. and INSALL, J. N., Patellar dislocation following total knee replacement. *J. Bone Jt Surg.*, **67-A** (1985) 1321–7.

4. BARTEL, D. L., BURNSTEIN, A. H., SANTAVICCA, E. A. and INSALL, J. N., Performance of the tibial component in total knee replacement. *J. Bone Jt Surg.*, **64-A** (1982) 1026–33.
5. LEWIS, J. L., ASKEW, M. J. and JAYCOX, D. P., A comparative evolution of tibial component design of total knee arthroplasty. *J. Bone Jt Surg.*, **67-A** (1985) 1321–7.
6. TSAI, W., *Program of composite material computation workshop.* 1.1–7.12, 1986.
7. HIRAI, T., KATAYAMA, T., INOUE, N. *et al.*, The elastic modulus of cancellous bone: dependence on trabecular orientation. *Biomech. Basic Appl. Res.* (1986) 207–12.

Index of Contributors

863

Subject Index